WITHDRAWN

Control System Design

Control System Design

Second Edition

C. J. Savant, Jr., Ph.D.

*Manager, Systems Department
Hughes Tool Company
Aircraft Division
and
University of California
at Los Angeles*

McGraw-Hill Book Company

New York   San Francisco   Toronto   London

TJ
216
S3
1964

Control System Design

Copyright © 1958, 1964, by McGraw-Hill, Inc. All Rights Reserved. Printed in the United States of America. This book, or parts thereof, may not be reproduced in any form without permission of the publishers.
Library of Congress Catalog Card Number 63-20725

54959

preface

Once I watched an office manager attempt to teach a secretary to use a ditto machine. He told her the characteristics of the fluid and how the fluid chemically activates the material deposited on the ditto master. Explaining the operation, he said, "The drum is rotated at constant speed by the motor. The master copy is first brought into contact with the fluid, then in contact with the blank paper...." "But how do I do it?" was the doubtful secretary's interruption. About this time another secretary came over and said, "Put the stack of paper here, slip in the purple thing with the dirty side up, like this...." In a few minutes the secretary was making copies and getting her job done.

I believe that control-system design can best be taught to engineers through the technique of "learning by doing" just as the new secretary learned to use the ditto machine by watching the other girl, then by running the machine. This book attempts to teach the fundamentals of control-system design by means of practical examples and from the student's own point of view. The book was written by and for students taking the control-system courses. Prepared notes were distributed to the students and were freely criticized by them. Their criticisms were integrated and used in revising the manuscript. Thus, this resulting book, which presents control-system design from a fundamental point of view, is written for undergraduate engineering students and for graduate practicing engineers—by fellow students.

Contact with practicing engineers through courses taught at locations close to industrial plants and on the company's premises, when security permitted, has led me to deviate somewhat from convention in the method of presentation.

The first difference is in the use of the root-locus method as a basic tool for control-system design. From my own experience the trial-and-error method is presented as being basic in control-system design. Each trial is analyzed, and the results of the analysis are then examined to determine the next trial. The engineer varies system quantities until satisfactory performance is obtained. The design technique, which I term "synthesis by analysis," requires that the engineer be able to analyze each subsequent trial rapidly. For this reason, the root-locus technique, based strongly upon sketching methods, is recommended for system synthesis.

Frequency-analysis techniques, especially the Bode construction, i.e., a plot of gain in decibels and of phase vs. logarithmic frequency, are a necessary part of the complete design for two reasons:

1. The governing equations of certain components, such as pneumatic valves, hydraulic actuators, and jet engines, are found experimentally from frequency-analysis methods. The engineer converts the experimental data into an approximate analytic form. The root-locus synthesis procedure is then used to design the control system.

2. For physically constructed equipment, frequency analysis provides a useful experimental method of checking the design.

The second difference is that the design approach presented in this text, although rigorous, does not require an elaborate mathematical background and requires only junior-level circuit theory. Emphasis is placed upon the Laplace-transform method of solution of differential equations, and an appendix treats the mechanics of this method because of its increasing importance in more advanced system thinking.

Problems are included at the end of each chapter. Detailed derivations are presented in the appendixes. Like the secretary mentioned in the example, the student can read directly through the text without studying derivations in detail until he knows how to use the material.

The widespread use of the first edition has proved that this book satisfies the need for a fundamental text on control-system design. With the use of this text during the past five years numerous improvements have been suggested by the students at both the University of Southern California and the University of California at Los Angeles. The following changes have been incorporated into this present text:

1. The Laplace-transform variable s is used rather than p.
2. Many new problems, some with answers, are included in the text.
3. More material on the frequency-response method and on system equalization is presented.
4. More design examples are included in the text.

I would like to express my appreciation to the students of EE482 at the University of Southern California and XL136A at the University of California at Los Angeles for suggesting the improvements in the forerunner of this present text. Sincere thanks are due to my associates at the University of Southern California and University of California at Los Angeles. Particular thanks are due to Dr. Charles H. Wilts under whom I studied at the California Institute of Technology. The ideas for many of the examples and problems came from his excellent class lectures. I would like to express my gratitude to the Hughes Tool Company, Aircraft Division, for providing the typing and reproduction facilities necessary for such a project. I wish to express sincere thanks to the following:

Bowmar Instrument Company, Fort Wayne, Indiana
DeJur-Ansco Corporation, Long Island City, New York
Diehl Manufacturing Company, Somerville, New Jersey
Electro-Mec Laboratory, Inc., New York
Fairchild Controls Corporation, Hicksville, New York
Ford Instrument Company, Long Island City, New York
Genisco, Inc., Los Angeles, California
Helipot Division, Beckman Instruments, Fullerton, California
Kearfoot Division, General Precision Inc., Little Falls, New Jersey
Minneapolis-Honeywell, Minneapolis, Minnesota
Reeves Instrument Corporation, New York
Statham Laboratories, Los Angeles, California
Summers Gyroscope Company, Santa Monica, California
Whittaker Gyro, Division of Telecomputing Corp., Van Nuys, California
Wiancko Engineering Company, Pasadena, California

These companies kindly supplied glossy prints and technical data for many of their products. I would like to express my appreciation to Prof. T. J. Higgins, Robert Mullen, and the other reviewers of the manuscript. Particular thanks are due Garnet Fulmer, who typed the manuscript.

C. J. Savant, Jr.

contents

Preface v
List of Symbols xv

Chapter 1 Introduction to Feedback Control Systems   1

1-1 The Control System   1
1-2 A Mechanical System   7
1-3 Equations for a Position Servo   9
1-4 Damping Ratio and Undamped Natural Frequency   12
1-5 Laplace-transform Solution of a Differential Equation   13
1-6 Transfer Functions and Block Diagrams   17
1-7 Block-diagram Algebra   20
1-8 Signal-flow Graphs   22
1-9 Transfer Functions and Convolution   24
1-10 Control-system Conventions   27
1-11 Stability and the Location of the Roots of the Characteristic Equation   27
1-12 Conclusions   30
Problems   31

Chapter 2 Obtaining the System Differential Equation   35

2-1 Introduction   35
2-2 Loop Analysis of Electrical Networks   35
2-3 Loop Analysis of Electrical Networks with Mutual Inductance   43
2-4 Nodal Analysis of Electrical Networks   44
2-5 Loop Analysis of Active Networks   51

x   Contents

2-6   Node Analysis of Active Networks   54
2-7   Mechanical Networks; Linear Motion   56
2-8   Mechanical Systems; Rotational Motion   61
2-9   Mechanical Coupling; Gear Trains   63
2-10 Electromechanical Networks   66
2-11 Analogies   69
2-12 Separable and Nonseparable Networks   71
Problems   72

Chapter 3   Steady-state Errors   78

3-1   The Steady-state Component   78
3-2   Steady-state Errors Due to Input Disturbances   78
3-3   Classification of Feedback Control Systems According to Type   81
3-4   Static-error Constants   82
3-5   Generalized Steady-state Errors   88
3-6   Steady-state Errors Due to Output Load Disturbances   90
3-7   Control-system Specifications   93
Problems   96

Chapter 4   The Root-locus Method   99

4-1   Introduction   99
4-2   The Second-order System   99
4-3   The Locus of Roots   102
4-4   Rules for Rapid Construction of Root-locus Diagrams   108
4-5   Example of the Application of the Above Rules   117
4-6   Measurement of Gain   119
4-7   Proof of the Root-locus Construction Rules   121
4-8   The Spirule   130
4-9   Use of the Spirule to Sum Angles   131
4-10 The Spirule Used to Find Lengths   133
4-11 The Spirule Used to Find Damping Ratio   134
4-12 Root-locus Plots with Variable Other than Gain   135
4-13 Root Locus Used to Factor a Polynomial   140
4-14 Root Locus Applied to Multiple-loop Systems   143
Problems   144

Chapter 5 Stability; the Frequency-analysis Method   149

5-1   The Impedance Concept   149
5-2   Generalized Impedance Functions   152
5-3   Plotting Impedance and Frequency Transfer Functions   155
5-4   The Asymptotic Approximation   158
5-5   Polar Plots and Amplitude vs. Phase-shift Plots   168
5-6   The s Plane and the $GH(s)$ Plane   169
5-7   The Nyquist Stability Criterion   174
5-8   Stability Criterion with the Decibel Gain and Phase vs. Frequency Diagrams   182
5-9   Examples of Bode Construction   183
5-10 M and N Contours; Nichols Charts   186
5-11 Nichols Charts   191
5-12 Determining the Degree of Stability of a System by Use of the Frequency-analysis Method   194
5-13 Comparison of Various Methods of Finding ζ 198
5-14 Closed-loop Frequency Response   201
5-15 Experimental Data   202
Problems   204

Chapter 6 Design of Feedback Control Systems   210

6-1   Introduction   210
6-2   Equalization by Gain Adjustment   210
6-3   Equalization by Inserting a Network   211
6-4   Method of Control-system Equalization   217
6-5   Passive Lead Network   218
6-6   Passive Lag Network   220
6-7   Summary of Various Equalizer Networks   223
6-8   Bridged-T Network   223
6-9   Adjustable Networks   225
6-10 Design of a Position Servo with Lead Network Equalization   226
6-11 Elementary Lattice and Ladder Synthesis   229
6-12 Active Network Synthesis   238
6-13 Parallel Equalization   240
6-14 Design of a Position Servo with Rate Feedback Equalization   243
6-15 Multiple-loop Control Systems   247
6-16 Comparisons of Various Equalizers   250

Contents

6-17 A-C Control Systems   252
6-18 Suppressed-carrier Modulation   254
6-19 A-C System Equalization   256
6-20 Carrier Networks   258
6-21 Electromechanical Networks   262
6-22 Various Methods of Stabilizing A-C Systems   265
6-23 Design of a Position Servo with an Acceleration Damper   268
6-24 Practical Considerations in A-C Control-system Design   271
Problems   277

Chapter 7 Servomechanism Components   286

7-1 Introduction   286
7-2 Potentiometers   287
7-3 Induction-type Position Indicators   295
7-4 Resolvers   296
7-5 Synchros   302
7-6 Reluctance Pickoffs   305
7-7 Measurement of Velocity   306
7-8 Induction Tachometers   306
7-9 D-C Tachometers   309
7-10 Permanent-magnet Tachometers   310
7-11 Measurement of Acceleration   310
7-12 Mechanical Accelerometer   311
7-13 Force-balance Accelerometers   314
7-14 Pressure Transducers   316
7-15 Control Motors   317
7-16 A-C Control Motors   325
7-17 D-C Control Motors   326
7-18 Forcers   327
7-19 Gear Trains   327
7-20 Choice of Optimum Gear Ratio   328
7-21 Gyroscopes   330
7-22 More Complete Mathematical Treatment of a Gyroscope   332
7-23 The Free Gyro   335
7-24 The Rate Gyro   336
7-25 The Restrained Gyro   338
7-26 The Vertical Gyro   341
7-27 Subtractors   342
7-28 The Differential Gearbox   342
7-29 Transformers   343

7-30 Difference Amplifier 344
7-31 Resistance Subtraction 344
7-32 Demodulators and Modulators 345
7-33 Magnetic Amplifiers 347
Problems 348

Chapter 8 Nonlinearities in Control-system Design 352

8-1 Classification of Nonlinearities 352
8-2 Linearization of Small Nonlinearities 354
8-3 Equivalent Damping 354
8-4 Equivalent Spring Constant 356
8-5 The Describing-function Method 356
8-6 Describing Functions for Common Nonlinearities 358
8-7 Describing Function for Saturation 358
8-8 Describing Function for Threshold 360
8-9 Use of the Describing Function in Control-system Design 363
8-10 Limitations of Describing Functions 365
8-11 Topological Solution of Feedback Control Systems 367
8-12 The Phase Plane 367
8-13 The Method of Isoclines 369
8-14 The Lienard Construction 372
8-15 Determination of the Time Markers 378
8-16 Comparison of Several Methods 380
Problems 381

Appendix I Laplace-transform Method 385

I-1 Introduction 385
I-2 The Laplace Transform of Functions 386
I-3 The Laplace Transform of Operations 386
I-4 Solution of Ordinary Linear Differential Equations Utilizing Laplace Transforms 389
I-5 Partial-fraction Expansion 391
I-6 Additional Properties of the Laplace Transform 395

Appendix II Roots of Equations 399

II-1 Introduction 399
II-2 The Quadratic 399

xiv   Contents

    II-3 The Cubic   399
    II-4 Descartes' Rules   402
    II-5 Higher-degree Algebraic Equations   402

Appendix III Use of Determinants   403

    III-1 Definition of a Determinant   403
    III-2 Expansion of a Determinant   403
    III-3 Theorems Concerning Determinants   405

Appendix IV Routh-Hurwitz Stability Criterion   406

    IV-1 Introduction   406
    IV-2 Routh-Hurwitz Stability Criterion   407

Appendix V Frequency-response Derivations   415

    V-1 Derivation of the Nyquist Criterion   415
    V-2 Derivation of M and N Circles   417
    V-3 Relations between ζ and Other Stability Quantities   419

Appendix VI Design of Bridged-T and Parallel-T Networks   422

    VI-1 Introduction   422
    VI-2 Design Procedure for the Bridged-T Networks   423
    VI-3 Design Procedure for the Infinite-attenuation Parallel-T Network   432
    VI-4 Design of Loaded Bridged-T Networks   432

Appendix VII A Rule for Determining Stability of a Linear System   437

Bibliography   447
Index   449

list of symbols

A	Amplifier gain or amplitude of a complex quantity
a	Resistance ratio
B	Damping coefficient
$B(s)$	Feedback signal, Laplace-transformed
b	A constant
$b(t)$	Feedback signal, function of time
C	Capacitance
$C(s)$	Output signal, Laplace-transformed
$c(t)$	Output signal, function of time
D	Gyro damping coefficient
d	Resistance value used in synthesis
$E(s)$	Voltage, function of s (or $j\omega$)
e	Base of the Napierian logarithm
$e(t)$	Voltage, function of t
$F(s)$	Force, Laplace-transformed
$f(t)$	Force, function of time
G_n	Describing function
$G(s)$	Forward transfer function
g	Translational mechanical-admittance function or acceleration of gravity $= 32.2$ ft/sec^2
$H(s)$	Feedback transfer function
h	Rotational mechanical-admittance function
I	Current, Laplace-transformed, or moment of inertia
$i(t)$	Current, function of time
J	Moment of inertia
j	Imaginary operator $\sqrt{-1}$
K	Loop gain or spring constant
K_a	Acceleration-error coefficient
K_0	Position-error coefficient
K_v	Velocity-error coefficient

xv

xvi List of Symbols

$\mathcal{K}$	A constant representing a product of other constants
k	Integer values used in a summation
L	Inductance
L	Torque, Laplace-transformed
$\mathcal{L}$	Symbolic representation of the Laplace-transform process
$\mathcal{L}^{-1}$	Symbolic representation of the inverse Laplace-transform process
l	Fixed linear length or torque, time-dependent
M	Mass, magnitude of $C(j\omega)/R(j\omega)$, mutual inductance, or momentum
M_p	Peak magnitude of $C(j\omega)/R(j\omega)$
m	Slope of the speed-torque curve
N	Ratio of Im $[C(j\omega)/R(j\omega)]$ to Re $[C(j\omega)/R(j\omega)]$, number of turns, or gear ratio
N_p	Number of poles
N_z	Number of zeros
n	Notch width in network design
$P(s)$	Polynomial in s
Q	Applied torque used in connection with gyro equations
$Q(s)$	Polynomial in s
$q(t)$	Charge, function of time
R	Resistance
$R(s)$	Input signal, Laplace-transformed
r	Notch ratio (depth) in network design or an integer
r_p	Plate resistance
$r(t)$	Input signal, function of time
s	Laplace-transform operator
t	Time variable
U	Undesirable torque used in connection with gyro equations
u	Frequency ratio ω/ω_n
$V(s)$	Voltage, Laplace-transformed; or velocity, Laplace-transformed
$v(t)$	Voltage, function of time; or velocity, function of time
W	Weight or energy
w	Weighting function
$X(s)$	Linear-motion position variable, Laplace-transformed
$x(t)$	Linear-motion position variable, function of time
x, y, z	Linear coordinates
$Y(s)$	Admittance function
$y(t)$	Dependent variable, function of time
$Z(s)$	Impedance function
α	Damping factor or ratio of rate to position feedback signal
β	Real frequency component of the root or ratio of load resistance to potentiometer resistance
γ	Fraction of total potentiometer rotation

List of Symbols

Δ	Symbolic representation of a determinant
$\delta(t)$	Unit impulse function
$E(s)$	Error signal, Laplace-transformed
$\epsilon(t)$	Error signal, function of time
ζ	Damping ratio
θ	Argument of complex quantity, angle of a vector on the s plane, or gyroscope gimbal angle
λ	Slope on the phase trajectory
μ	Vacuum-tube amplification factor
π	3.1416
ρ	Ratio of load resistance to potentiometer resistance
Σ	Symbolic representation of summation
σ	Dummy variable of integration
τ	Time constant
ϕ	Phase angle
ϕ_m	Phase margin
ψ	Flux variable
ω	Frequency variable used with sinusoidal driving functions or angular velocity in mechanical systems
ω_n	Undamped resonant frequency
ω_0	Network resonant frequency
ω_r	Frequency of transient oscillation

chapter 1

INTRODUCTION TO FEEDBACK CONTROL SYSTEMS

1-1 The Control System. Automatic control systems are assuming an ever-increasing importance in all phases of our modern way of life. Automatic washers and dryers, thermostatically controlled furnaces, automatic guidance systems, autopilots, missiles, jet transports, space satellites, and nuclear power plants are just a few examples.

Control systems can be classified as "open-loop" and "closed-loop." An "automatic" washing machine is an open-loop system because the machine is controlled by a timer. The wash time is determined by estimating the time required and setting the timer accordingly. In addition to the timer, satisfactory operation depends upon using the correct amount of soap, hot water, and bleach. If, accidently, a smaller amount of soap is used, the clothes will not be clean even if the timer is set properly. A closed-loop automatic washer would continuously measure the degree of cleanliness of the clothes and would turn itself off when the desired cleanliness was achieved. With a closed-loop automatic washer the desired result would be achieved even if the wrong amount of soap were used.

Suppose a gasoline engine is used to drive a load, say a large pump. If the engine delivers several hundred horsepower, a small change in engine throttle setting causes a large change in power output. Engine shaft speed for a constant load is a function of the throttle position. The throttle, carburetor, and engine comprise a type of control system wherein large power output is controlled with small power input. Figure 1-1 depicts a typical control system. A small input control signal is amplified electrically, pneumatically, or mechanically. This signal controls the power magnitude delivered from the source through the actuator to the output. The throttle setting of the gasoline engine–pump system is the input control signal. Gasoline, which is the external power source needed by the engine, is controlled by the throttle and is used to develop engine power. A typical plot of engine speed vs. throttle

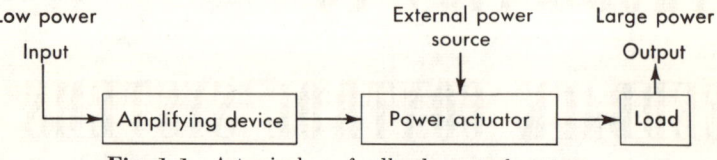

Fig. 1-1 A typical nonfeedback control system.

position is sketched in Fig. 1-2. This "calibration curve" gives the operator the engine speed for a given throttle setting.

In general, the type of control system pictured in Fig. 1-1 is termed an "open-loop," or nonfeedback, system. The input drives the output directly through intermediate components. Many examples of this type of nonfeedback control system can be cited: human muscles, voltmeters, electric generators, pneumatic valves and actuators, magnetic amplifiers, timers, and programmers.

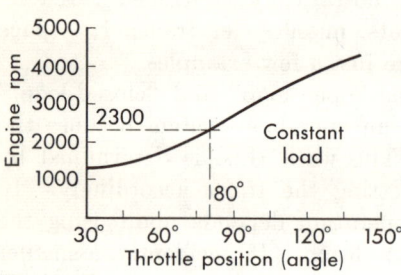

Fig. 1-2 A typical engine speed vs. throttle position curve.

One of the consequences of an open-loop system is the dependence of the output upon component calibration. For example, if the carburetor on the gasoline engine should become dirty and the fuel-air ratio should change, the curve of output rpm versus throttle position (Fig. 1-2) would change. Since the control system depends directly on intermediate components, variations of temperature, humidity, vibration, age, etc., tend to change the carburetor action and hence the calibration curve.

In many applications (e.g., automatic washing machine, automobile engine) maintaining an accurate calibration curve is unimportant. In other applications it suffices to calibrate the system at intervals. To "calibrate" means to establish or reestablish the input-output relation often enough to obtain the desired accuracy. For example, a d-c galvanometer is "zeroed" (balanced) before it is used. Many types of laboratory instrumentation systems are calibrated or balanced before a run. The operator depends upon the balance remaining constant during the period of usage. In many systems (e.g., driving an automobile) a human operator is capable of making the necessary corrections; the human operator closes the loop. When one first drives another's automobile, a new sense of "feel" must be established because, for example, the accelerator setting on no two automobiles produces exactly the same engine performance.

Another consequence of open-loop systems is that the controlled output varies as the load changes. As an example, suppose the pump attached to the engine output began pulling mud. If the mud was not stopped or the throttle angle increased, the engine output would decrease and, perhaps, eventually stop.

Thus, two of the disadvantages of open-loop systems are:

1. Controlled output varies as intermediate components change.
2. Controlled output varies as the load changes.

Open-loop systems have several advantages which should be considered in the planning stages of a control system. Some of them are:

1. Simplicity of operation
2. Fewer components
3. Stable operation

If the system requirements cannot be satisfied with an open-loop control system, as is often the case with accurate control devices, a closed-loop system may be required. In order to obtain the desired accuracy of control, the system output is determined, "fed back," and compared with the system input. The difference between the "desired" output, represented by the system input, and the actual output is termed the "error." A closed-loop system is actuated by the error signal:

$$\text{Error} = \text{desired output} - \text{actual output} \qquad (1\text{-}1)$$

For an open-loop system, the engine throttle is positioned directly. If a certain rpm is desired, the operator refers to the calibration chart of Fig. 1-2 and sets the throttle to the correct angle. For example, if it is desired to operate at 2,300 rpm, a throttle setting of 80° is required.

Suppose that the system comprising a gasoline engine driving a pump is used in a closed-loop manner. In such a system, the quantity that is to be controlled, in this case the engine rpm, is measured. The *desired* rpm, the input signal, is compared with the *actual* rpm, the output signal, and the difference, the error signal, is used to position the throttle. One possible closed-loop engine control system is shown in Fig. 1-3. A tachometer,* which produces a voltage proportional to shaft speed, is used to measure output rpm. The input voltage, which is the desired speed, is varied by means of a potentiometer,* and the two voltages are subtracted electrically. The difference, or error voltage, is amplified and used to position the throttle through an appropriate amplifier.

The important difference between the closed-loop method (Fig. 1-3) and the open-loop method (Fig. 1-1) lies in the feedback link and its effects. The signal that changes the throttle setting is the difference between the desired rpm and the actual rpm. When the output-shaft

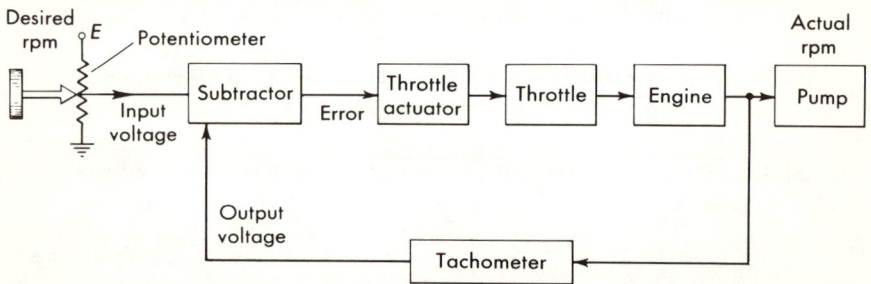

Fig. 1-3 One possible closed-loop engine control system.

speed is equal to the desired speed, i.e., the difference or *error* is zero, the throttle actuator (which could be a reversible electric motor whose output is geared to the throttle arm) remains fixed. If any change, such as a change in load or a change in the engine components, should occur in the system so that the actual rpm is no longer equal to the desired rpm, an error will result. This error will cause the throttle setting to change until the actual rpm corresponds to the desired rpm.

The concept of "feedback" is the essence of control-system design. The precise definition of feedback is not simple, and even the existence of feedback in a system is often difficult to determine. Feedback is associated with the comparison of the actual value of the control variable (output) with the desired value of the control variable (input). More generally, feedback is said to exist when a closed sequence of cause-and-effect relations exists between the system variables.

When feedback is intentionally introduced, as with most control systems, the existence and the purpose are readily identified. The effects

* These elements are discussed in greater detail in Chap. 7.

of feedback, which can also be classed as the advantages of closed-loop systems, can be summarized as follows:

1. *Increased accuracy.* Because the system continues to function until the error is reduced to zero, the accuracy is maintained high in spite of variation of load, intermediate-component variation, temperature variation, etc. Introduction of feedback can reduce or eliminate the errors within the loop. However, components which are outside the loop, such as the transducing elements (e.g., those which measure the actual output and convert it to a voltage), contribute to the closed-loop system error. Although the variations of elements within the loop

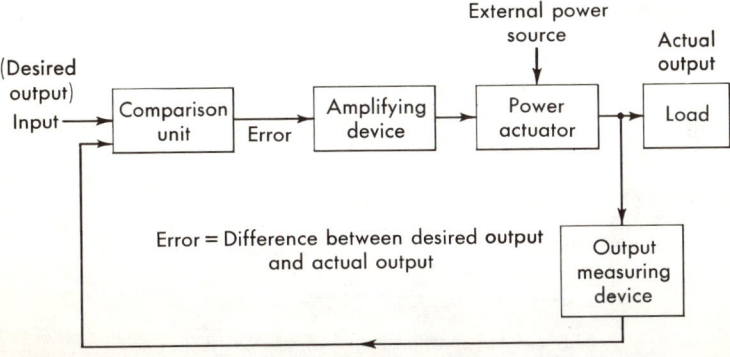

Fig. 1-4 A typical closed-loop system.

vary the system time response, feedback reduces the errors caused by these variations.

2. *Reduced effects of distortion and nonlinearities.* Feedback reduces the effects of distortion and nonlinearities which occur within the loop. The dynamic performance is affected by these nonlinearities; however, the effects of some nonlinearities, such as amplifier saturation, are greatly reduced by feedback.

3. *Increased bandwidth.* Feedback increases the band of frequencies over which the system will respond (the bandwidth). At the same time, however, the gain over this band of frequencies is reduced.

4. *Increased or decreased impedance.* Depending upon the characteristics desired and the type of feedback used, the impedance presented the load can be either increased or decreased.

A typical "feedback control system," shown in Fig. 1-4, comprises one or more feedback loops which combine functions of the desired output signal with functions of the command signal to tend to maintain pre-

6   Control System Design

scribed relationships between the actual output and desired output. The same amplifying device and power actuator are used as in the system of Fig. 1-1. It is only necessary to add a device to measure the output and a device to subtract the actual output from the desired output. The advantages of closed-loop control on this system can be summarized as follows:

1. Variations in load and in major components have little effect upon system accuracy.
2. It is unnecessary to calibrate the system periodically except for the tachometer.

Although closed-loop systems offer many advantages over open-loop controllers, one difficulty that frequently arises is instability (system

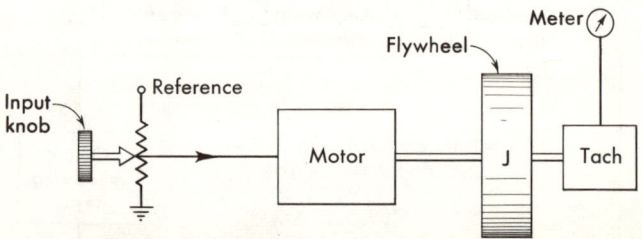

Fig. 1-5 A manual speed regulator system.

oscillation). For example, consider the problem, shown schematically in Fig. 1-5, of setting the speed of an electric motor driving a large inertia. As the operator turns the input knob, the motor starts to rotate. Because of the large load inertia, some time is consumed before the flywheel comes up to speed. In the interest of hurrying the process, the operator sets the input voltage beyond the value necessary to obtain the correct output. As the motor comes up to speed, the energy in the flywheel carries the tachometer beyond the desired speed. The operator now reduces the voltage to lower the speed. By the time the motor speed begins to reduce, the input voltage is set too low and the speed drops below that which is desired. This process could continue—the speed increasing while the operator is decreasing the voltage, and vice versa. A good operator would begin to anticipate the performance and soon bring the machine to the proper speed. If he did not introduce some anticipation control, the system might become unstable, i.e., build up indefinitely. It is usually phase lags (produced by moments of inertia, inductors, human reaction-time lags, etc.) in systems that give rise to instability.

The functioning of an oscillating system (such as a voltage oscillator) depends upon incorporation of intentional limited instability. A portion of the output of a high-gain amplifier with internal energy sources is fed back to the input. If just the right rate of energy feedback is achieved, the input can be removed and the system will continue to have an output with no external input; it will derive the output energy from its internal sources.

Owing to many causes, to be considered later in this book, a feedback control system may be rendered useless because of instability. A formal definition that is applicable for feedback-control-system stability can be written: *A system is absolutely stable if the transients that result when a quiescent system is disturbed approach zero (die out) as time approaches infinity and is limitedly stable if the output, though not zero, remains bounded.* Because of the importance of stability in the operation of feedback control systems, particular emphasis is placed on the stability problem:

1. What is the degree of stability of the system?
2. How can system stability be improved without degrading the error performance?

The basic approach to the design of control systems is through analysis based on the system differential equations. For physical systems the behavior is determined by the differential equation solutions. A control engineer must be capable of obtaining and solving the equations characterizing the system elements. Since differential equations contain the essential information about the system, considerable attention is devoted in this text to formulating and solving differential equations characterizing elements that commonly occur in practice.

1-2 A Mechanical System. As a first example of finding a differential equation, consider the mechanical system shown in Fig. 1-6. A mass M is caused to roll (without bearing friction) on a level track. A spring restrains the motion on the left, and a damper on the right. Small motion is assumed, so that the spring force is linear, as shown in Fig. 1-7a. As the spring is extended $+x$, a force $-F_K$ is produced by the spring to oppose the displacement. The equation for this force is

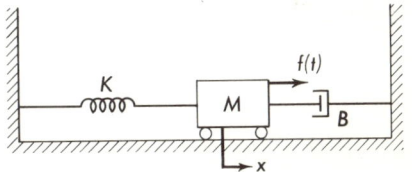

Fig. 1-6 A mechanical system.

$$F_K = -Kx \tag{1-2}$$

8   Control System Design

where K is the proportionality constant for the spring in force units per unit length, e.g., pounds per inch. As the spring is compressed $-x$, the spring force F_K opposes the displacement. The origin, or point at which $x = 0$, is taken as the rest position of the spring.

The mass is restrained on the right with a dashpot (which resembles an automobile shock absorber). This element is termed a "viscous damper," since it applies a restoring force proportional to velocity dx/dt. The linear curve is shown in Fig. 1-7b. As the dashpot is stretched or

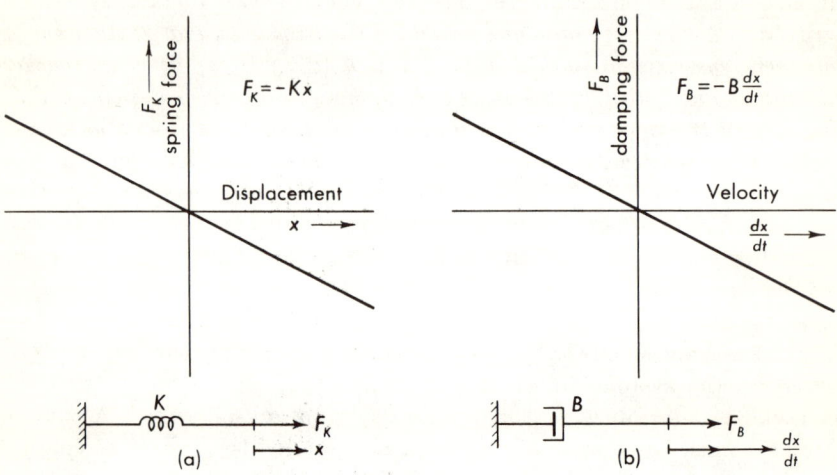

Fig. 1-7   (a) Spring force vs. displacement; (b) damping force vs. velocity.

compressed, the unit opposes this motion according to the approximate relationship

$$F_B = -B \frac{dx}{dt} \tag{1-3}$$

where B is the proportionality constant for the damper in force units per unit length per unit time, e.g., pounds per inch per second.

Application of Newton's second law of motion (i.e., for constant mass, the product of mass and acceleration is equal to the sum of the forces acting on the mass) yields the desired differential equation. Motions to the right of the equilibrium position are taken as positive $(+x)$, and motions to the left as negative $(-x)$. From Fig. 1-6,

$$M \frac{d^2x}{dt^2} = f(t) + F_B + F_K = f(t) - B \frac{dx}{dt} - Kx \tag{1-4}$$

or, rearranged,

$$\frac{d^2x}{dt^2} + \frac{B}{M}\frac{dx}{dt} + \frac{K}{M}x = \frac{f(t)}{M} \tag{1-5}$$

which is the differential equation characterizing the motion of the system.

Equation (1-5) is representative of equations encountered in linear-control design. It is a "linear differential equation with constant coefficients." The equation (and also the system) is "linear" because the dependent variable x and the derivatives of x, namely, dx/dt and d^2x/dt^2, appear in the equation only to the first power. No products of terms, such as $x(dx/dt)$ and $x(d^2x/dt^2)$; or terms raised to higher powers, such as $(dx/dt)^2$, x^2, and $(d^2x/dt^2)^3$; or functions of x, such as sin x, log x, and cos dx/dt, are included. The term "constant coefficients" implies that the quantities multiplying the dependent variable or derivatives of the variable are constants that are independent of the time variable t or displacement variable x. That is, M, B, and K are constants. Before Eq. (1-5) is solved, a position servo, with the same differential equation, is considered.

1-3 Equations for a Position Servo. The closed-loop system of Fig. 1-8 is used to position a large antenna—effectively a mass having a

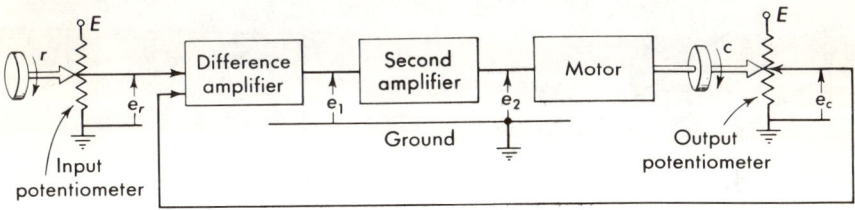

Fig. 1-8 A position servo.

large moment of inertia. An output potentiometer measures the output-shaft position and converts this to voltage according to the equation

$$e_c = K_p c \tag{1-6}$$

where c is the output-shaft angle in radians and e_c is the output voltage. K_p, which has the units volts per radian, is the "transfer function" of the potentiometer and can be derived by dividing the total voltage E by the maximum rotation of the potentiometer:

$$K_p = \frac{E}{c_{\max}} \quad \text{volts/rad} \tag{1-7}$$

The input-knob position r is converted to a voltage by a potentiometer identical with the output potentiometer. A difference amplifier, the theory of which is discussed in Chap. 7, amplifies the difference between the two signals:

$$e_1 = A_1(e_r - e_c) = A_1 K_p(r - c) \tag{1-8}$$

where e_1 = error voltage output from difference amplifier
A_1 = amplification constant of difference amplifier
e_r = input voltage = $K_p r$

This voltage is amplified with A_2 and appears at the motor terminals as

$$e_2 = A_2 e_1 = A_1 A_2 K_p(r - c) \tag{1-9}$$

where e_2 = amplified error voltage
A_2 = amplification constant of the second amplifier

Approximate equations for the motor are determined from its speed-torque curve. Experimental speed-torque curves for servomotors* are linearized, as shown in Fig. 1-9. The equation for each of the straight lines of Fig. 1-9 is

$$L = a\omega + b \tag{1-10}$$

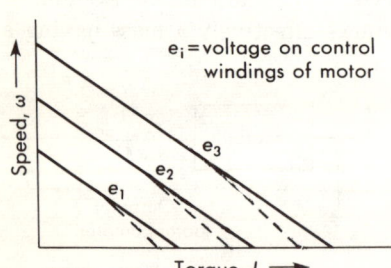

Fig. 1-9 Linear speed-torque curves for a control motor.

where L is the torque delivered by the motor, ω is the angular velocity of the motor shaft, and a and b are constants. Equation (1-10) is the general equation of a straight line. Depending upon the type of control motor (d-c or a-c), different constants a and b are obtained. The slope of all lines is negative and is represented by the same constant

$$a = -m \qquad m > 0 \tag{1-11}$$

The quantity b, Eq. (1-10), depends upon the control voltage. For $\omega = 0$, the intersections of the straight lines with the horizontal axis of Fig. 1-9 yield values from which are plotted the stall torque vs. voltage curve shown in Fig. 1-10. The slope of the linear portion of the curve

---

* Chapter 7 presents more detailed information on servomotors.

(near the origin) is k. The equation of this curve is

$$L_0 = ke_2 \tag{1-12}$$

where e_2 is the voltage applied to the motor and L_0 is the stall torque ($\omega = 0$). Substituting into Eq. (1-10),

$$b = L_0 = ke_2 \quad \text{for } \omega = 0 \tag{1-13}$$

The equation for a motor ($L \neq 0$) is rearranged and written

$$L + m\omega = ke_2 \tag{1-14}$$

where L = torque delivered by motor
e_2 = applied voltage
ω = motor velocity = dc/dt

In this present example the motor drives an inertia load $L = J(d^2c/dt^2)$ and the differential equation is written

$$J\frac{d^2c}{dt^2} + m\frac{dc}{dt} = ke_2 \tag{1-15}$$

where e_2 is the voltage applied at the motor terminals. When e_2 is substituted from Eq. (1-9), the complete system equation results:

$$\frac{d^2c}{dt^2} + \frac{m}{J}\frac{dc}{dt} = \frac{k}{J}[A_1A_2K_p(r-c)] \tag{1-16}$$

which simplifies to

$$\frac{d^2c}{dt^2} + \frac{m}{J}\frac{dc}{dt} + \gamma c = \gamma r \tag{1-17}$$

where
$$\gamma = \frac{kA_1A_2K_p}{J} \tag{1-18}$$

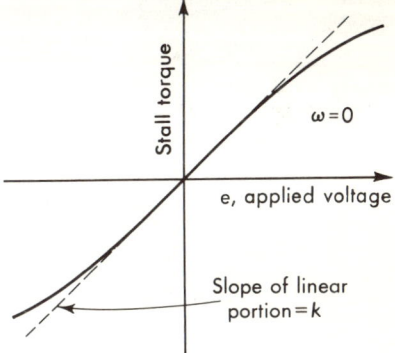

Fig. 1-10 Stall torque vs. voltage for a typical control motor.

In form, Eq. (1-17) is the same as Eq. (1-5): a linear differential equation with constant coefficients. The order of this equation, equal to that of the highest derivative contained in it, is 2. Such an equation is usually referred to as being of second order.

1-4 Damping Ratio and Undamped Natural Frequency.

Before the solution of the two differential equations Eqs. (1-5) and (1-17) is considered, two important symbols are noted:

ζ = damping ratio
ω_n = undamped natural frequency

Any linear second-order differential equation with constant coefficients can be written in terms of these two quantities. Such a second-order system is written in the form

$$\frac{d^2c}{dt^2} + 2\zeta\omega_n \frac{dc}{dt} + \omega_n^2 c = \omega_n^2 r \tag{1-19}$$

When Eq. (1-19) is compared with Eq. (1-17), the following relations hold:

$$2\zeta\omega_n = \frac{m}{J} \quad \text{and} \quad \omega_n^2 = \gamma \tag{1-20}$$

Solving these relations for ω_n and ζ yields

$$\omega_n = \sqrt{k} = \sqrt{\frac{kA_1A_2K_b}{J}} \tag{1-21}$$

$$\zeta = \frac{m}{2J\sqrt{k}} = \frac{m}{2} \frac{1}{\sqrt{kA_1A_2K_bJ}} \tag{1-22}$$

When Eq. (1-19) is compared with Eq. (1-5) for the mechanical system, the following relations result:

$$2\zeta\omega_n = \frac{B}{M} \quad \text{and} \quad \omega_n^2 = \frac{K}{M} \tag{1-23}$$

Solving these relations for ω_n and ζ yields

$$\omega_n = \sqrt{\frac{K}{M}} \tag{1-24}$$

$$\zeta = \frac{B}{2\sqrt{KM}} \tag{1-25}$$

Even in higher-order systems, where there are more than two roots of the characteristic equation, it may be that the nature of the system response depends essentially on two "least damped roots." That is, the response of an "equivalent" second-order system is utilized to approximate the response of the higher-order system. The two parameters ζ and ω_n, which describe the second-order equation, are discussed in further detail throughout the text.

Equation (1-17) represents the dynamic equation of the position servo (Fig. 1-8). This equation is expressed in terms of ω_n and ζ as follows:

$$\frac{d^2c}{dt^2} + 2\zeta\omega_n \frac{dc}{dt} + \omega_n^2 c = \omega_n^2 r \tag{1-26}$$

ω_n and ζ are given by Eqs. (1-21) and (1-22). The quantity $c(t)$ is the dependent variable or response, and $r(t)$ is the externally applied driving function.

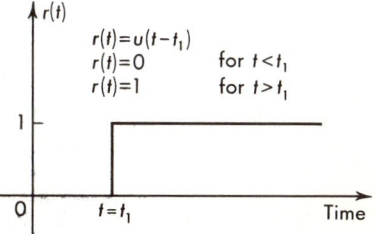

Fig. 1-11 Displaced unit step function in position.

1-5 Laplace-transform Solution of a Differential Equation.[13,27]*

The solution of Eq. (1-26), $c(t)$, can be found by use of Laplace transforms, as discussed in Appendix I. The driving function, a step function $u(t - t_1)$, is obtained from Fig. 1-11 for $t_1 = 0$. The system, described by Eq. (1-26), is at rest, that is, $c(0^+) = 0 = \left.\frac{dc}{dt}\right|_{t=0^+}$, before application of the step function.

Both sides of Eq. (1-26) are Laplace-transformed as follows:

$$s^2 C + 2\zeta\omega_n s C + \omega_n^2 C = \omega_n^2 \frac{1}{s} \tag{1-27}$$

where s is the Laplace-transform variable and $1/s$ is the Laplace transform of the unit step function (cf. Table I-1). Since initial conditions are zero, $s^2 C = \mathcal{L}(d^2c/dt^2)$ and $sC = \mathcal{L}(dc/dt)$. The transform of the output variable $c(t)$ is represented by capitalized C to indicate that the function of time $c(t)$ and the function of s, $C(s)$, which results after transforming the equation are "associated" functions.

Equation (1-27) is solved algebraically for C,

$$C = \frac{\omega_n^2}{s(s^2 + 2\zeta\omega_n s + \omega_n^2)} \tag{1-28}$$

* Superscript numerals are keyed to the Bibliography at the back of the book.

and is reduced with partial fractions* to

$$C = \frac{1}{s} - \frac{s + 2\zeta\omega_n}{(s + \zeta\omega_n)^2 + \omega_n^2(1 - \zeta^2)} \tag{1-29}$$

and is inverse Laplace-transformed, using Table I-1, with the result, for $t > 0$,

$$c(t) = 1 - \frac{e^{-\zeta\omega_n t}}{\sqrt{1 - \zeta^2}} \sin(\omega_n \sqrt{1 - \zeta^2}\, t + \phi) \tag{1-30}$$

where $\phi = \cos^{-1} \zeta$.

After the input knob is subjected to a unit step in rotation, $r(t) = u(t)$, curves of Fig. 1-12 plot output-shaft response $c(t)$ versus the dimensionless parameter $\omega_n t$, Eq. (1-30). All curves except that for $\zeta = 0$ decay

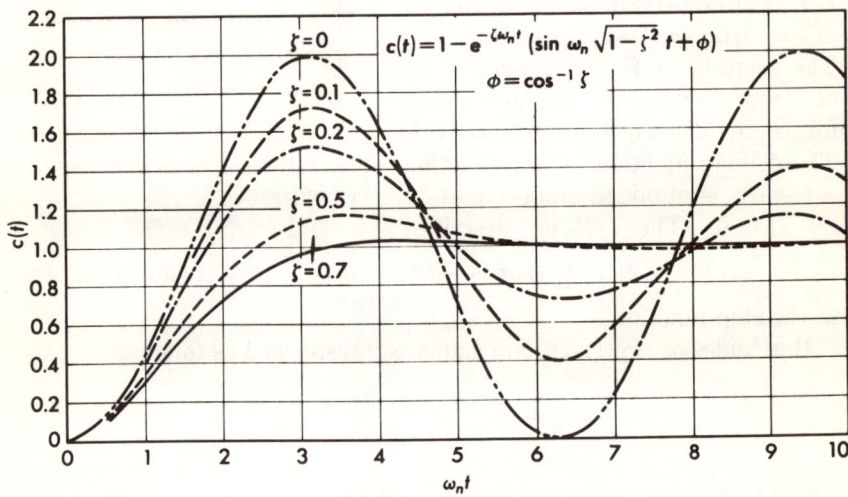

Fig. 1-12 Transient response curves for various values of ζ.

to a steady-state value of 1. Curves for small ζ damp more slowly than for larger ζ. As might be expected, the greater the damping (ζ large), the less oscillatory is the character of the response. From Eq. (1-22)

$$\zeta = \frac{m}{2\sqrt{kA_1 A_2 K_b J}} \tag{1-31}$$

* See Appendix I for an explanation of partial-fraction expansions.

The damping ratio ζ is a function of several system parameters, and it can be varied by changing any of the quantities in Eq. (1-31). For example, the amplifier gain can be decreased to increase the damping ratio.

As the system becomes more oscillatory (smaller ζ), the response $c(t)$ overshoots the steady-state value (in this case 1) by a greater amount with each cycle. Per cent overshoot, a measure of the output rise above the steady state, is defined by

$$\% \text{ overshoot} = 100 \times \frac{\text{max value reached in first overshoot} - \text{steady-state value}}{\text{steady-state value}} \quad (1\text{-}32)$$

For the second-order system of Eq. (1-26), per cent overshoot is plotted as a function of ζ in Fig. 1-13.

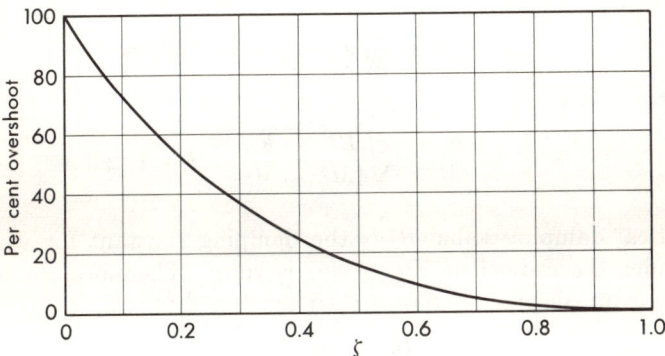

Fig. 1-13 Per cent overshoot versus ζ for a second-order system.

The response speed of a second-order system is related to ω_n and ζ. Since Fig. 1-12 is plotted against $\omega_n t$, the time required for these curves to reach 90 per cent of their steady-state value, termed the rise time, depends upon ω_n. For larger ω_n the system will have a smaller rise time. For a second-order system of this type, the undamped natural frequency is given by [refer to Eq. (1-21)]

$$\omega_n = \sqrt{\frac{kA_1A_2K_b}{J}} \quad (1\text{-}33)$$

Variation of ω_n is possible by changing servo components. Notice, however, that increasing the A_1A_2 product causes ω_n to increase and ζ

to decrease. Because of this relation between ζ and ω_n, other methods of design must be considered.

It is possible to indicate another definition for ζ and ω_n, one which will serve to explain the name given these two quantities. The quantity ζ is the ratio of the damping constant ($\zeta\omega_n$) that exists in a second-order system to the critical damping constant (ω_n). Higher-order system response is often dominated by two "least damped complex roots." For this reason, results obtained for the second-order system are often approximately true for higher-order systems.

Critical damping in a second-order system occurs when there are two equal real roots of the characteristic equation (1-26). This corresponds to $\zeta = 1$, since, in Eq. (1-29), substitution of $\zeta = 1$ results in two identical roots,

$$s_i = -\zeta\omega_n = -\omega_n \tag{1-34}$$

For the mechanical system of Fig. 1-6, the characteristic equation is

$$s^2 + \frac{B}{M}s + \frac{K}{M} = 0 \tag{1-35}$$

and the roots are

$$s_i = -\frac{B}{2M} \pm \sqrt{\frac{B^2}{4M^2} - \frac{K}{M}} \qquad i = 1, 2 \tag{1-36}$$

The critical damping value B_c is the damping constant for which the term under the radical of Eq. (1-36) is zero. Therefore, B_c is found from the expression

$$\frac{B_c^2}{4M^2} - \frac{K}{M} = 0$$

and solving,

$$B_c = 2\sqrt{KM} \tag{1-37}$$

For this value of B there exists a double real root $-B_c/2M$. The damping ratio is

$$\zeta = \frac{B/M}{B_c/M} = \frac{B}{2\sqrt{KM}} \tag{1-38}$$

The undamped natural frequency is the frequency of oscillation that occurs for zero damping. The roots of Eq. (1-35), for $B = 0$, are

$$s_i = \pm\sqrt{\frac{-K}{M}} = \pm j\sqrt{\frac{K}{M}} \qquad i = 1, 2 \tag{1-39}$$

where $j = \sqrt{-1}$. Imaginary roots result in an impulse response of the form

$$x = A_1 \cos \sqrt{\frac{K}{M}} t + A_2 \sin \sqrt{\frac{K}{M}} t \qquad (1\text{-}40)$$

The undamped natural frequency is

$$\omega_n = \sqrt{\frac{K}{M}} \qquad (1\text{-}41)$$

Substituting from Eqs. (1-38) and (1-41) into Eq. (1-35), the following is obtained:

$$s^2 + 2\left(\frac{1}{2}\frac{B}{M}\sqrt{\frac{M}{K}}\right)\sqrt{\frac{K}{M}} s + \frac{K}{M} = 0$$

$$s^2 + 2\zeta\omega_n s + \omega_n^2 = 0 \qquad (1\text{-}42)$$

This last equation is identical with the original definition of Sec. 1-4. Commonly, the damping ratio ζ is less than unity, and, if so, the roots of Eq. (1-42) can be written

$$s_i = -\zeta\omega_n \pm j\omega_n\sqrt{1-\zeta^2}$$
$$i = 1, 2 \qquad (1\text{-}43)$$

where $j = \sqrt{-1}$. Thus for ζ in the range $0 < \zeta < 1$ the roots are complex.

The roots given by Eq. (1-43) are located on a plot shown in Fig. 1-14. This plot, termed the "s plane," has (1) the real coordinate on the horizontal axis and (2) the imaginary on the vertical axis. Complex roots always appear in conjugate pairs, and the number of roots is the same as the order of the equation.

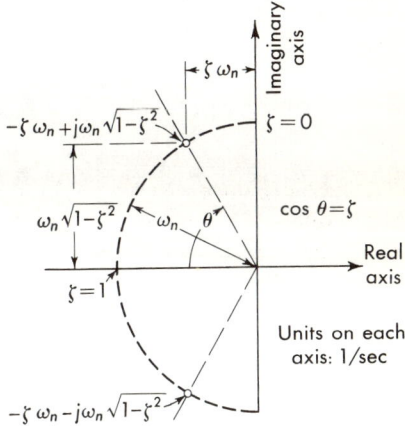

Fig. 1-14 Roots of a second-order system in the s plane.

Since Eq. (1-30) is multiplied by a negative exponential term $e^{-\zeta\omega_n t}$, the transient component decays with time. As t increases, $e^{-\zeta\omega_n t}$ decreases for $\zeta\omega_n > 0$.

1-6 Transfer Functions and Block Diagrams. The simplification resulting from the use of Laplace transforms is further increased when transfer functions and block diagrams are introduced. Consider a linear constant-coefficient differential equation that relates the output of a system to its input:

$$a_n \frac{d^n c}{dt^n} + a_{n-1} \frac{d^{n-1} c}{dt^{n-1}} + \cdots + a_1 \frac{dc}{dt} + a_0 c = b_m \frac{d^m r}{dt} + \cdots$$

$$+ b_1 \frac{dr}{dt} + b_0 r \quad (1\text{-}44)$$

where the a_n and b_n are constants, $c(t)$ is the output or response, and $r(t)$ is the input or driving function. Any element whose performance is characterized by such an equation is said to be linear. One important property of such elements is that if r_1 produces an output c_1 and r_2 produces an output c_2, then an input

$$\alpha_1 r_1 + \alpha_2 r_2 \quad (1\text{-}45)$$

produces an output

$$\alpha_1 c_1 + \alpha_2 c_2 \quad (1\text{-}46)$$

Transforming Eq. (1-44), assuming zero initial conditions;

$$(a_n s^n + a_{n-1} s^{n-1} + \cdots + a_1 s + a_0)C = (b_m s^m + \cdots + b_1 s + b_0)R \quad (1\text{-}47)$$

The ratio of the Laplace transform of the output C to the Laplace transform of the input R is the "transfer function"

$$G(s) = \frac{C}{R} = \frac{b_m s^m + b_{m-1} s^{m-1} + \cdots + b_1 s + b_0}{a_n s^n + a_{n-1} s^{n-1} + \cdots + a_1 s + a_0} \quad (1\text{-}48)$$

The transfer function $G(s)$ is a property of the element only. For linear constant-coefficient systems it is independent of the driving function and initial conditions. $G(s)$ is a rational algebraic function of s, and for lumped parameter systems the constants a_n and b_n depend only upon the elements of the system.

When the transfer function of an element or a system is known, the transform of the response can be found from the equation

$$C(s) = G(s)R(s) \quad (1\text{-}49)$$

This equation can be represented by a "block diagram," as shown in Fig. 1-15. The diagram signifies that the "output" is equal to the product of the "input" and the "box function." Hence the block diagram of Fig. 1-15 represents Eq. (1-49).

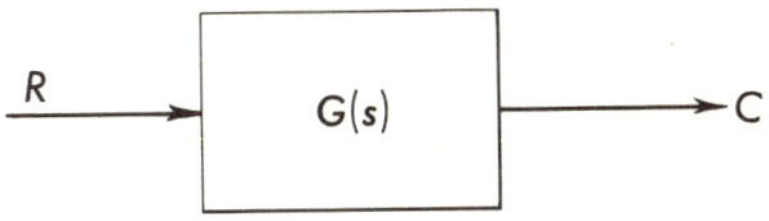

If two such systems are in series and the first physical system is not loaded by the second physical system (Fig. 1-16a), the output of the first system is the input to the second, and the following equations must hold:

Fig. 1-15 Block diagram of a linear system.

$$R_1 = G_1(s)R \qquad C = G_2(s)R_1 \qquad (1\text{-}50)$$

These equations can be combined algebraically,

$$C = G_1(s)G_2(s)R = G_1G_2R \qquad (1\text{-}51)$$

This equation shows that the two systems in series can be transposed (Fig. 1-16b) and can be represented by a single box whose transfer function is the product G_1G_2 as shown in Fig. 1-16c. This method of combination may be extended to any number of systems in series.

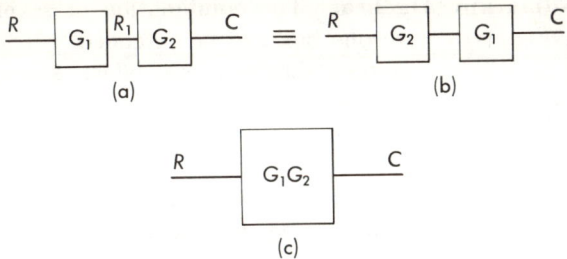

Fig. 1-16 Block-diagram algebra: two blocks in series.

As an example of transfer functions and block diagrams, consider the transfer function for the mechanical system of Fig. 1-6. The ratio of the transformed position output X to the transformed force input F, as found from Eq. (1-5), is

$$\left(s^2 + \frac{B}{M}s + \frac{K}{M}\right)X = \frac{F}{M} \qquad (1\text{-}52)$$

and the transfer function is

$$G_1(s) = \frac{X}{F} = \frac{1/M}{s^2 + (B/M)s + (K/M)} \quad (1\text{-}53)$$

The block diagram is shown in Fig. 1-17.

Fig. 1-17 Block diagram of a mechanical suspension.

Fig. 1-18 Block diagram of a position servo.

In a similar manner the transformed equation for the position servo is, from Eq. (1-26),

$$(s^2 + 2\zeta\omega_n s + \omega_n{}^2)C = \omega_n{}^2 R \quad (1\text{-}54)$$

and the overall transfer function is

$$G_2(s) = \frac{C}{R} = \frac{\omega_n{}^2}{s^2 + 2\zeta\omega_n s + \omega_n{}^2} = \frac{\gamma}{s^2 + (m/J)s + \gamma} \quad (1\text{-}55)$$

The block diagram is shown in Fig. 1-18.

1-7 Block-diagram Algebra. The combination of several systems in series (cf. Fig. 1-16) into one box can be extended. Rearranging of system diagrams to effect simplification is termed "block-diagram algebra." The identities presented in this section are frequently used in control-system design. Since transformed quantities are used, the $R(s)$, $C(s)$, and $G(s)$ can be discarded and simply R, C, and G used.

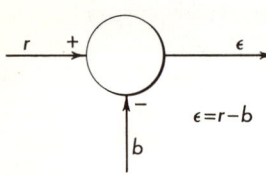

Fig. 1-19 Subtractor symbol.

Basic to the subject of feedback control systems is the error detector or subtractor, the symbol for which is shown in Fig. 1-19. The input signal (voltage, torque, position, etc.), labeled r, minus the feedback signal (same units as r so a subtraction may be effected) equals the actuating signal ε. Practical forms of subtractors are discussed in Chap. 7.

In Fig. 1-20 are summarized the most important block-diagram identities. Since each of these pairs can be interchanged in a complex system, they aid in system reduction. These identities can be verified by showing that the output from two equivalent diagrams is the same. For example,

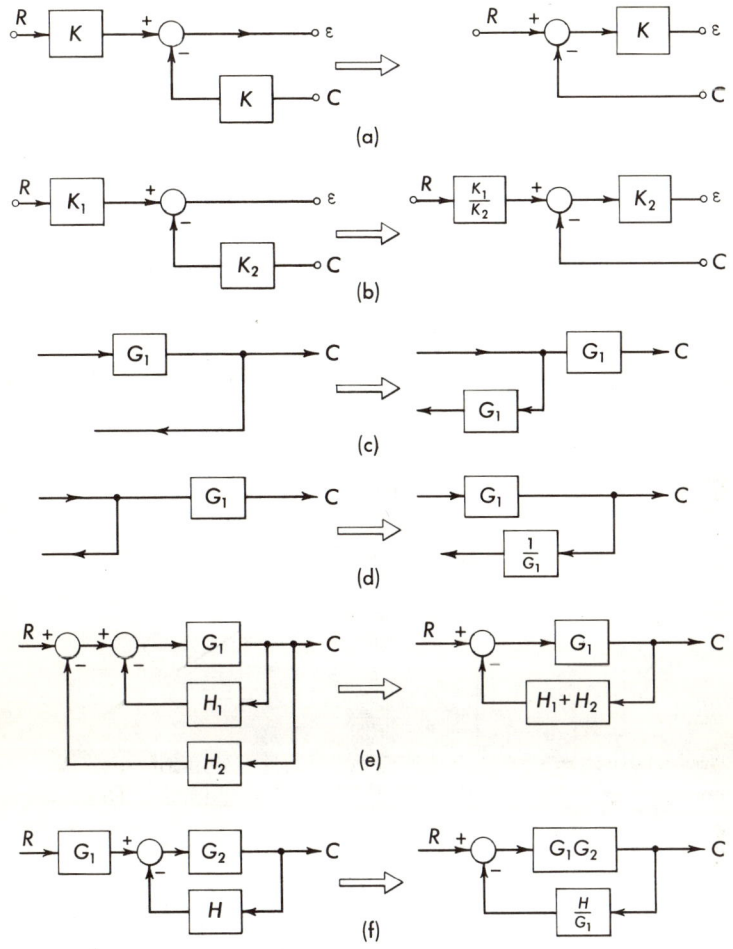

Fig. 1-20 Several block-diagram identities.

compare the output from both block diagrams of Fig. 1-20b. The diagram on the left yields

$$\varepsilon = K_1 R - K_2 C \tag{1-56}$$

The diagram on the right yields

$$\varepsilon = \left(\frac{K_1}{K_2} R - C\right) K_2 = K_1 R - K_2 C \tag{1-57}$$

which is identical with Eq. (1-56).

Control System Design

The identity shown in Fig. 1-21 is one of the most important. The actuating signal ε is written

$$\varepsilon = R - HC = \frac{C}{G} \qquad (1\text{-}58)$$

The input-output or closed-loop transfer function is found from Eq. (1-58),

$$\frac{C}{R} = \frac{G}{1 + GH} \qquad (1\text{-}59)$$

and the actuating signal to input signal relation is

$$\frac{\varepsilon}{R} = \frac{1}{1 + GH} \qquad (1\text{-}60)$$

It must be remembered that both G and H are functions of s. The variables C, R, and ε are Laplace-transformed variables and are functions of s.

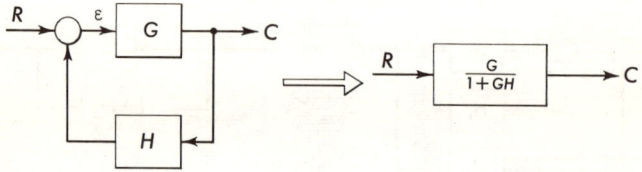

Fig. 1-21 An important block-diagram identity.

Equation (1-59) is the closed-loop transfer function of a control system with degenerative feedback in terms of open-loop quantities. This equation states that the closed-loop transfer function C/R is the forward-loop transfer function G divided by 1 plus the open-loop transfer function GH. The expression (1-59) is frequently needed in feedback-control-system design, and it should be memorized.

1-8 Signal-flow Graphs. The block diagram yields the mathematical relationships between the system input and output. Use of signal-flow graphs* permits a study of the flow of signals through the system and indicates all feedback paths in the system. The block-diagram study of control systems cannot yield the same information as the signal-flow graph.

* These were introduced by S. J. Mason, Feedback Theory—Some Properties of Signal Flow Graphs, *Proc. IRE*, vol. 41, no. 9, pp. 1144–1156, September, 1953.

The signal-flow graph for a system comprises a network of nodes which are connected by paths, termed "branches." The direction of signal flow along the branch is indicated by the arrow.

A system is described by a set of equations

$$x_i = \sum_{h=1}^{N} t_{ki} x_h \qquad i = 1, 2, \ldots, N$$

where x_i are the variables and t_{ki} are the transmissions which relate the variable x_h to the variable x_i. The variables are the nodes and the t_{ki} are associated with the branches. As an example, the equation

$$x_1 = t_{21} x_2 \qquad (1\text{-}61)$$

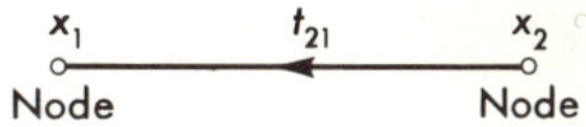

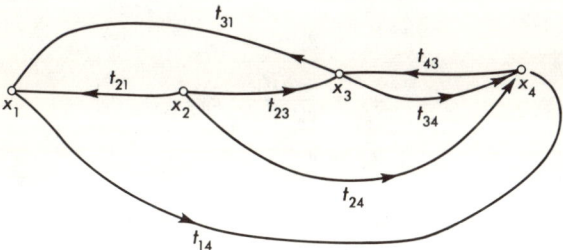

Fig. 1-22 Signal-flow graph for $x_1 = t_{21} x_2$.

is represented by the signal-flow graph of Fig. 1-22. In this diagram the nodes x_1 and x_2 are the variables and the branch from x_2 to x_1 indicates the relationship between these two nodes. The flow of the signal along the branch is only in one direction.

Fig. 1-23 Signal-flow graph for Eq. (1-62).

As another example, consider the signal-flow graph for the following system of equations:

$$\begin{aligned} x_1 &= t_{21} x_2 + t_{31} x_3 \\ x_3 &= t_{23} x_2 + t_{43} x_4 \\ x_4 &= t_{14} x_1 + t_{24} x_2 + t_{34} x_3 \end{aligned} \qquad (1\text{-}62)$$

The complete signal-flow graph is included in Fig. 1-23. The basic properties of signal-flow graphs can be summarized as follows:

1. The system variables are represented by nodes and are arranged from left to right.
2. The branch from x_i to x_h indicates the dependence of the x_h upon x_i but not the reverse.
3. As the signal x_i travels along a branch, it is multiplied by the transmission function so that $t_{ih}x_i$ arrives at node x_h.

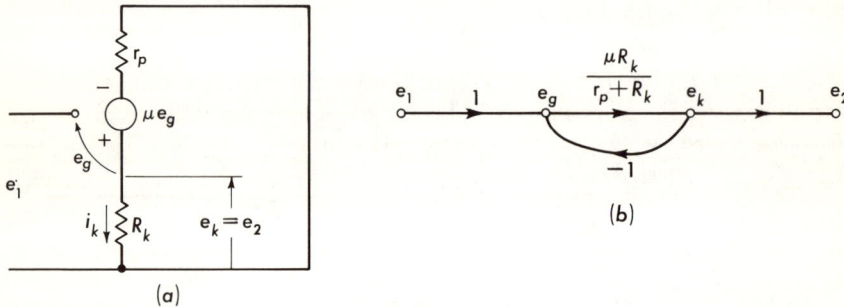

Fig. 1-24 Equivalent circuit and signal-flow diagram for a cathode follower.

As an example of the construction of a signal-flow diagram for a physical circuit consider the equivalent circuit for a cathode follower, shown in Fig. 1-24a, which satisfies the following equations:

$$e_g = e_1 - e_k$$
$$e_g = \frac{\mu R_k}{r_p + R_k} e_g \tag{1-63}$$
$$e_2 = e_k$$

The signal-flow diagram is shown in Fig. 1-24b. As a second example, the signal-flow graph is drawn for the block diagram of the simple feedback control system of Fig. 1-25a in Fig. 1-25b.

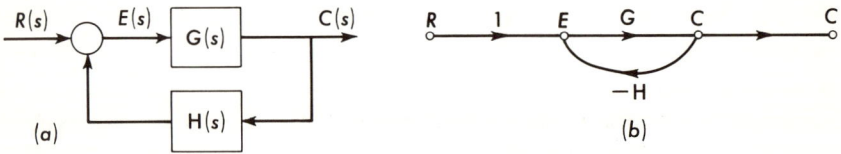

Fig. 1-25 Block diagram and signal-flow diagram for a position control system.

1-9 Transfer Functions and Convolution. Suppose the driving function in Eq. (1-44) is a unit impulse $\delta(t)$; thus

$$r(t) = \delta(t) \tag{1-64}$$

This function is defined as one which has zero value everywhere except at $t = 0$, where the magnitude is "infinite," and is such that

$$\int_{-\epsilon}^{+\epsilon} \delta(t) \, dt = 1 \tag{1-65}$$

where ϵ is a small quantity. In addition, the impulse function selects the zero value of the function as follows:

$$\int_{-\epsilon}^{+\epsilon} f(t) \delta(t) \, dt = f(t) \bigg|_{t=0} = f(0)$$

The function can be constructed from a pulse, as shown in Fig. 1-26. The pulse has height $1/a$ and width a. The strength of the pulse, equal to the area under the curve, is $(1/a)(a) = 1$; the impulse is formed by taking the limit as a approaches zero $(a \to 0)$.

The Laplace transform of an impulse is

$$\mathcal{L}[\delta(t)] = \int_0^\infty e^{-st} \delta(t) \, dt = [e^{-st}]_{t=0} = 1 \tag{1-66}$$

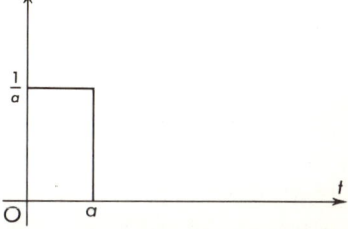

Fig. 1-26 An impulse formed by taking the limit as a approaches zero.

where in this instance, the integral includes the impulse located at the origin. This transform pair is included in Table I-1.

The transformed response $C(s)$ of the Eq. (1-44) system to a unit impulse input is functionally the same as the transfer function, since $R(s) = 1$. Hence

$$G(s) = \frac{C(s)}{1} = \frac{b_m s^m + b_{m-1} s^{m-1} + \cdots + b_1 s + b_0}{a_n s^n + a_{n-1} s^{n-1} + \cdots + a_1 s + a_0} \tag{1-67}$$

is a function having the same form as the Laplace transform of the impulse response of the system.

The inverse transform of the transfer function is known as the "weighting function." If an impulse is applied to a system, as shown in Fig. 1-27, the output is the weighting function $w(t)$. Any function $r(t)$ can be represented as the limit of the summation of a series of pulses (Fig. 1-28). If $w(t)$ is the response (at $\tau = t$) of a system when $\delta(\tau)$ is applied, then $\alpha w(t - n \Delta \tau)$ is the response (at $\tau = t$) of a system when the delayed impulse $\alpha \delta(t - n \Delta \tau)$ is applied (at $\tau = n \Delta \tau$), where α is a constant. The response of the system to a pulse which is located at $\tau = n\,\Delta\tau$ and is of height $r(n\,\Delta\tau)$ and width $\Delta\tau$ is

$$r(n\,\Delta\tau)w(t - n\,\Delta\tau)\,\Delta\tau \qquad (1\text{-}68)$$

The total response at time $\tau = t$ is the sum of the responses due to each impulse,

$$c(t) \cong \sum_{n=0}^{N} r(n\,\Delta\tau)w(t - n\,\Delta\tau)\,\Delta\tau = \sum_{n=0}^{N} r(\tau_n)w(t - \tau_n)\,\Delta\tau \qquad (1\text{-}69)$$

Letting $\Delta\tau \to 0$ and $N \to \infty$, so that $N\,\Delta\tau = t$ in Eq. (1-69), yields, in the limit, the integral

$$c(t) = \int_0^t r(\tau)w(t - \tau)\,d\tau \qquad (1\text{-}70)$$

Hence the output of a linear system with constant coefficients is found by "convolving"* the input with the weighting function. Equation (1-67)

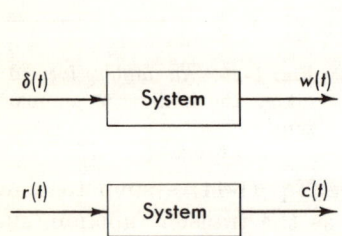

Fig. 1-27 Impulse response and weighting function.

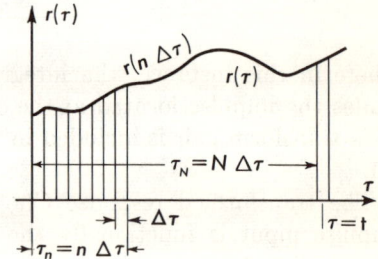

Fig. 1-28 A function represented by a sum of impulses.

shows that the weighting function can be found by taking the inverse Laplace transform of the transfer function

$$\mathcal{L}^{-1}G(s) = w(t) \qquad (1\text{-}71)$$

It is easily derived through the substitution $t - \tau = \tau'$, so that

$$c(t) = \int_0^t r(t - \tau')w(\tau')\,d\tau' \qquad (1\text{-}72)$$

---

* This is the process of carrying out the convolution integral of Eq. (1-70).

Taking the Laplace transform of both sides of Eq. (1-70) yields

$$C(s) = R(s)G(s) \tag{1-73}$$

This equation has the same form as Eq. (1-49). Reference to Appendix I yields further information on the convolution integral.

In Chap. 2 transfer functions and block diagrams of numerous elements are determined. Chapter 7 is concerned with transfer functions of numerous control-system components utilized in practice.

1-10 Control-system Conventions. Symbols used in the field of feedback controls have been standardized. Figure 1-29 shows a block

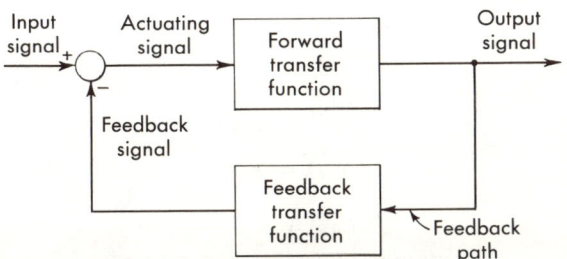

Fig. 1-29 Control-system conventions.

diagram with defined terminology that is used in this text. The symbols and the terms they represent are as follows:

$$\begin{aligned}
\text{Input signal} &= r(t) \text{ or } R(s) & &\text{reference} \\
\text{Output signal} &= c(t) \text{ or } C(s) & &\text{controlled variable} \\
\text{Actuating signal} &= \epsilon(t) \text{ or } \mathcal{E}(s) & &\text{error (if } H_1, H_2, \ldots = 1) \\
\text{Error signal} &= \epsilon(t) \text{ or } \mathcal{E}(s) & & \\
\text{Feedback signal} &= b(t) \text{ or } B(s) & &\text{if different from } c
\end{aligned}$$

Forward transfer functions $G_1(s), G_2(s), \ldots$

Feedback transfer functions $H_1(s), H_2(s), \ldots$

1-11 Stability and the Location of the Roots of the Characteristic Equation. Stability of a linear control system is defined in Sec. 1-1. This definition can be associated with mathematical criteria for determining whether or not it is satisfied in a control-system design.

System stability depends only upon the system and not upon the driving function. Hence, if a system is unstable, any excitation causes the system to achieve eventually an unbounded output. But if the system is stable, any bounded excitation will cause a bounded response.

The weighting function, obtained by taking the inverse transform of the transfer function, depends only upon the system. If the weighting function is that of a stable system, i.e., if it decays as time approaches infinity, all exponential exponents will be negative:

$$c(t) = Ae^{-\alpha_1 t} + e^{-\alpha_2 t}(B \cos \omega_r t + C \sin \omega_r t) + De^{-\alpha_3 t} \quad (1\text{-}74)$$

Thus, as $t \to \infty$, $c(t)$ approaches zero. The Laplace-transformed form of Eq. (1-74) yields the same information relative to the location of the roots of the characteristic equation. Thus if, considering a simpler case,

$$C(s) = \frac{E_1}{s + a} \quad (1\text{-}75)$$

then

$$c(t) = E_1 e^{-at} \quad (1\text{-}76)$$

and if

$$C(s) = \frac{E_2}{s - b} \quad (1\text{-}77)$$

the time response is

$$c(t) = E_2 e^{bt} \quad (1\text{-}78)$$

When the root of the transfer function lies in the left half plane, the exponent is negative. In Eq. (1-75) the root lies at $s_i = -a$, and if $a > 0$, the weighting function is that of a stable system. In Eq. (1-77) the root lies at $s_i = b$, and if $b > 0$, the weighting function is that of an unstable system, i.e. the right-hand member of Eq. (1-78) increases as time increases.

The location of typical roots in the s plane and the corresponding weighting function are shown in Fig. 1-30. Single roots, $-\alpha_1$, $-\alpha_2$ and α_1, $\alpha_2 > 0$, along the negative real axis yield a solution of thé form

$$A_3 e^{-\alpha_1 t} + B_3 e^{-\alpha_2 t} \quad (1\text{-}79)$$

A single complex conjugate pair of roots $\pm j\omega$ along the imaginary or j axis yields a transient term

$$A_2 \cos \omega t + B_2 \sin \omega t \quad (1\text{-}80)$$

Introduction to Feedback Control Systems

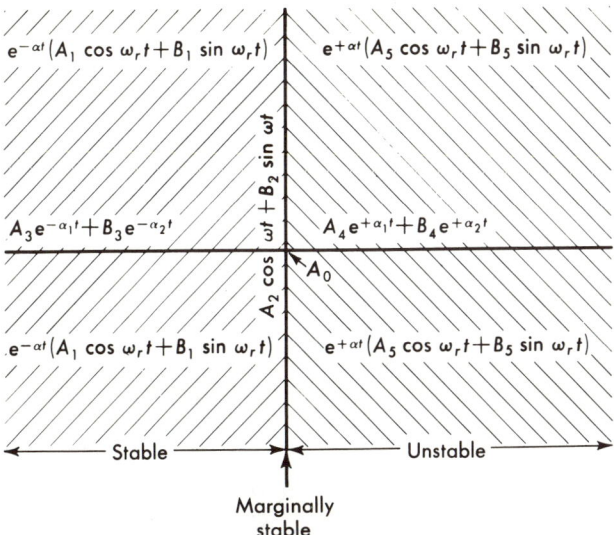

Fig. 1-30 Root plane and location of roots.

Single roots along the positive real axis, α_1, α_2 and α_1, $\alpha_2 > 0$, yield terms of the form

$$A_4 e^{+\alpha_1 t} + B_4 e^{+\alpha_2 t} \tag{1-81}$$

One simple pair of complex roots, $-\alpha \pm j\omega (\alpha > 0)$, in the left half plane gives rise to a transient term of the form

$$e^{-\alpha t}(A_1 \cos \omega_r t + B_1 \sin \omega_r t) \tag{1-82}$$

A simple pair of complex roots, $\alpha \pm j\omega$ ($\alpha > 0$), in the right half plane yields a term of the form

$$e^{+\alpha t}(A_5 \cos \omega_r t + B_5 \sin \omega_r t) \tag{1-83}$$

Double roots, α ($\alpha > 0$), of the characteristic equation occurring at one point on the real axis in the left half of the s plane result in a term of the form

$$(A_6 + B_6 t)e^{-\alpha t} \tag{1-84}$$

A single root at the origin gives rise to a term of the form

$$y_{t_1} = \text{const} = A_0 \tag{1-85}$$

A double root at the origin yields a transient term of the form

$$y_{t_2} = A_7 + B_7 t \qquad (1\text{-}86)$$

Double complex roots, $\pm j\omega$, which lie along the imaginary or j axis give rise to terms of the form

$$y_{t_3} = (a_8 + B_8 t)\cos \omega t + (C_8 + D_8 t)\sin \omega t \qquad (1\text{-}87)$$

In writing these expressions, the constants A_i, B_i, C_i, and D_i used in each expression are arbitrary. The nature of stability can now be stated in terms of the location of the roots of the transfer function denominator $(1 + GH = 0)$ or the characteristic equation as follows:

A system is stable..................	If *all* roots lie in the left half plane
A system is unstable...............	If *any* of the roots lie in the right half plane or if *any* multiple (double, triple, etc.) complex pairs of roots lie along the j axis or *any* multiple real roots lie at the origin
A system is marginally (or limitedly) stable	If any single pair of conjugate roots lies along the j axis and/or a single root lies at the origin and all other roots lie in the left half plane
A system is conditionally stable....	If all roots lie in the left half plane only for some particular condition of the system parameters. Often the system is stable in only one range of some parameter, for example, loop gain constant

Roots which lie in the right half plane result in an oscillating transient response that increases with time and renders a control system unsuitable for practical use. Single conjugate pairs of roots which lie on the imaginary axis other than at the origin result in an undamped sinusoidal (an oscillatory) response term. If all other roots lie in the left half plane except possibly a single root A at the origin, this system may be considered an oscillator providing the limiting case between stable and unstable systems.

1-12 Conclusions. The essentials of closed-loop system design are exemplified in the solution of the second-order system. In the design of a control system no direct-synthesis procedure is available. The control designer uses a combination of synthesis and analysis. Typically, the designer is faced with several components which must be used (amplifier, motor, etc.). These have differential equations from which can be determined transfer functions with the use of which the system is constructed. When the feedback loop is closed, the system may be unstable, may have too large a steady-state error, may have too small a damping ratio ζ, or may have too small a natural frequency ω_n. If so, the engineer must make changes to bring the system performance into specification agreement.

Depending upon the engineer's experience, the initial change will be more or less optimal. In any case it becomes necessary to analyze the changed system to determine any improvement. Having analyzed the first alteration, the engineer makes another change, if necessary, and analyzes the emerging system. This process is continued until the system meets performance specifications.

To be a competent control-system designer requires the capacity to analyze systems quickly and correctly. Fortunately, a formalized procedure is available for analysis of linear systems. This procedure can be outlined as follows:

1. Obtain the system differential equations and hence the transfer functions for elements of the system.
2. Solve for the steady-state component of the differential equations for certain types of inputs (from this steady-state errors can be determined).
3. Determine the nature of stability from the transfer function or characteristic equation.

Each step must be formalized to enable the engineer to complete the analysis quickly and accurately.

PROBLEMS

1-1. List five examples of open-loop control systems and draw block diagrams for each.

1-2. List five examples of closed-loop control systems and draw block diagrams for each.

1-3. Find the Laplace transform of the following functions of time:

(a) ωt
(b) $\sin \omega t$
(c) e^{-at}
(d) $\tfrac{1}{2}at^2$
(e) $u(t)$, unit step function
(f) $\delta(t)$, unit impulse

1-4. State which of the following equations are linear differential equations with constant coefficients:

(a) $\dfrac{d^3y}{dt^3} + 3\left(\dfrac{dy}{dt}\right)^2 + 4y = f(t)$

(b) $\dfrac{d^2y}{dt^2} + 2y\dfrac{dy}{dt} + 3y = \sin \omega t$

(c) $\dfrac{d^3y}{dt^3} + 3t\dfrac{d^2y}{dt^2} + 4\dfrac{dy}{dt} + 1 = e^{j\omega t}$

(d) $\dfrac{d^3y}{dt^3} + 3\dfrac{d^2y}{dt^2} + 2\dfrac{dy}{dt} + 4 = t^2 \sin \omega t$

1-5. Find the solution for the following differential equations. Assume that at $t = 0$, $y = y(0) = y_0$, and that dy/dt and all higher derivatives up to and including those equal to the order of the equation minus 1 are zero.

(a) $\dfrac{d^3y}{dt^3} + 3\dfrac{d^2y}{dt^2} + 10\dfrac{dy}{dt} = 0$ 　　　　　　　　　　　　Ans. $y = y_0$

(b) $\dfrac{d^2y}{dt^2} + 10 = 0$ 　　　　　　　　　　　　　　　　　Ans. $y = y_0 - 5t^2$

(c) $\dfrac{d^2y}{dt^2} + 7\dfrac{dy}{dt} + 15y = 0$ 　　　　　(d) $\dfrac{d^3y}{dt^3} - 3\dfrac{d^2y}{dt^2} + y = 0$

(e) $\dfrac{d^3y}{dt^3} + 8y = 0$

1-6. Plot on the s plane the roots of the characteristic equations of Prob. 1-5.

1-7. Find the solutions of the following differential equations, subject to zero initial conditions:

(a) $\dfrac{d^2y}{dt^2} - 2\dfrac{dy}{dt} + 4y = \sin \omega t$

(b) $\dfrac{d^2y}{dt^2} - 2\dfrac{dy}{dt} + 4y = e^{-2t}$

　　　　　　　　　Ans. $y(t) = \tfrac{1}{12}[e^{-2t} + e^{t}(\sqrt{3} \sin \sqrt{3}\,t - \cos \sqrt{3}\,t)]$

(c) $\dfrac{d^3y}{dt^3} + \dfrac{d^2y}{dt^2} + 4\dfrac{dy}{dt} + 4y = e^{-t}$

1-8. Compare the Laplace-transform and classical methods of solution as to ease of solution by solving the following differential equation by both procedures. Assume zero initial conditions.

$$\dfrac{d^2c}{dt^2} + 2\zeta\omega_n\dfrac{dc}{dt} + \omega_n^2 c = \tfrac{1}{2}\alpha t^2$$

Take $\zeta = 0.10$, $\omega_n = 10$ rad/sec, and $\alpha = 1$.

1-9. Plot the frequency of oscillation of the transient component of Eq. (1-30) versus ζ when $\omega_n = 1$.

1-10. Derive an analytic expression which relates per cent overshoot to damping ratio ζ for a second-order system. Plot the corresponding curve to verify Fig. 1-13.

　　　　　　　　　　　　　　　　　Ans. Overshoot $= e^{-\zeta\pi/\sqrt{1-\zeta^2}}$

1-11. (a) Find the complete solution $c(t)$ when the input to the position servo of Fig. 1-8 is the displaced ramp function which is shown in Fig. 1P-11 and is represented mathematically in the figure. Assume zero initial conditions; at $t = 0$, $c = 0$ and $dc/d\omega_n t = 0$.

(b) For values of $\zeta = 0.1, 0.3, 0.6,$ and 2.0 plot the solution in the following forms:
 (1) $c(t)$ versus $\omega_n t$.
 (2) $dc/d\omega_n t$ versus $\omega_n t$.

Introduction to Feedback Control Systems   33

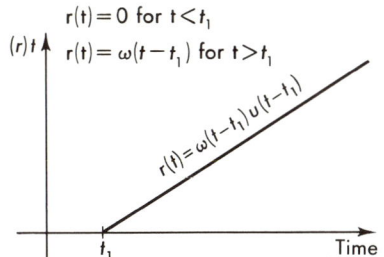

Fig. 1P-11

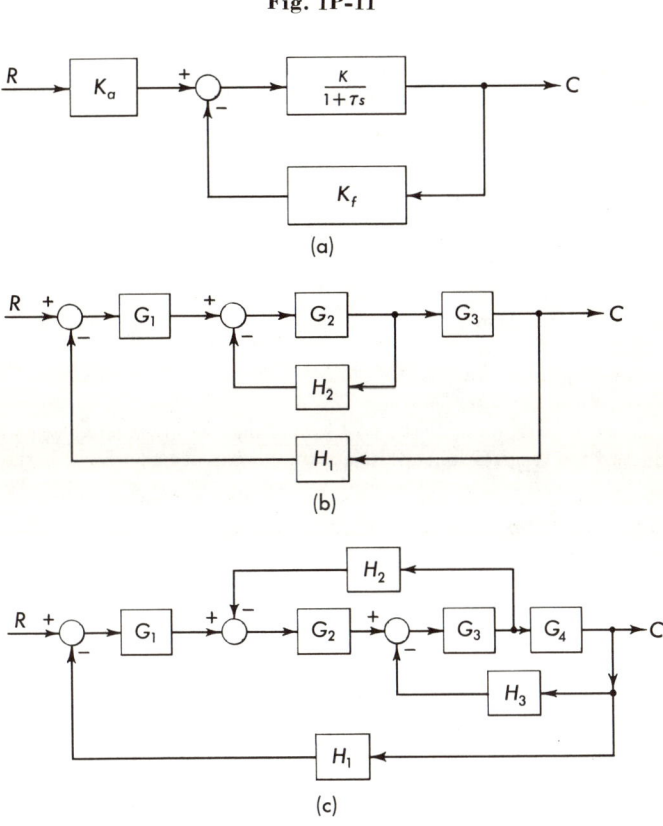

Fig. 1P-15

1-12. After completing the design of a simple position servo of the form shown in Fig. 1-8, the designer finds that the system had 79 per cent overshoot in response to a unit step input. The open-loop transfer function is

$$\frac{C}{R-C} = \frac{C}{\epsilon} = \frac{AK_m}{s(\tau_m s + 1)}$$

By what factor should he multiply the amplifier gain A so that the overshoot is reduced to 50 per cent? K_m and τ_m are motor constants. *Ans.* 0.111

1-13. The open-loop transfer function for a position servo is given by

$$\frac{C}{R-C} = \frac{K}{s(1+0.5s)(1+0.2s)}$$

$K = 1.6$, and $H(s) = 1$. Find the response of this system $c(t)$ for a unit step input $r(t) = u(t)$. The initial conditions are zero.

1-14. Determine the damping ratio, per cent overshoot, and frequency of oscillatory roots for the system of Prob. 1-13. *Ans.* $\zeta = 0.37$, $\omega = 1.55$

1-15. Reduce the block diagrams of Fig. 1P-15 and derive expressions for the closed-loop transfer functions.

1-16. Write the signal-flow graphs for the systems of Prob. 1-15.

1-17. The transfer function of a system is

$$G(s) = \frac{10}{s+3}$$

Use the convolution integral of Eq. (1-70) to find the response (output) for zero initial conditions when the input is

(a) A unit step function $u(t)$
(b) A unit ramp function $tu(t)$
(c) A unit-amplitude sinusoid $u(t) \sin t$

1-18. Verify the results of Prob. 1-17 by finding the inverse Laplace transform of the product of the transfer function $G(s)$ and the Laplace transform of the input.

chapter 2

OBTAINING THE SYSTEM DIFFERENTIAL EQUATION

2-1 Introduction. The first step in synthesizing a feedback control system is to obtain differential equations for the fixed system elements (those elements which must be "lived with"), such as mechanical components, networks, error-sensing devices, and amplifiers. This chapter deals with the writing of linear differential equations for electrical, mechanical, or electromechanical systems. The approach to the problem is identical for any of these systems. The method outlined in this chapter reduces linear systems analysis to a step-by-step procedure.

2-2 Loop Analysis of Electrical Networks.[16,39] Electrical network analysis rests upon two basic laws, namely, Kirchhoff's laws:

1. *The algebraic sum of voltage drops around a closed loop equals zero. Alternatively, the sum of voltage drops equals the sum of voltage rises around a closed loop.*
2. *The algebraic sum of currents flowing out of a circuit node equals zero.*

A common method of analysis based on the first law is called the "Maxwell mesh" or "loop" method. It consists of assuming that currents, termed "loop currents," flow in each loop of a multiloop network, algebraically summing the voltage drops around each loop, and setting each equation equal to zero. As an example of the loop analysis of an electric circuit, consider the problem of finding the output voltage e_o for the circuit of Fig. 2-1. The steps in the solution of this problem are summarized as follows:

1. Assume loop currents in a clockwise direction (i_1 and i_2 in Fig. 2-1). Although the correct solution is obtained if some other direction is assumed, certain checks are available if the currents are taken in the same direction. Be certain that (1) a current flows through every element and that (2) the number of currents assumed is sufficient.

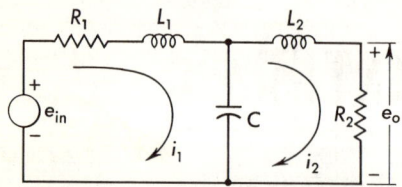

Fig. 2-1 Electric circuit example.

One method to assure that a sufficient number of currents have been assumed in a network is indicated in Fig. 2-2. This method is applicable to networks that can be drawn with no wires crossing. (Such networks are termed "planar networks.") The currents are so assumed that (a) through every element there is a current and (b) no element crosses a

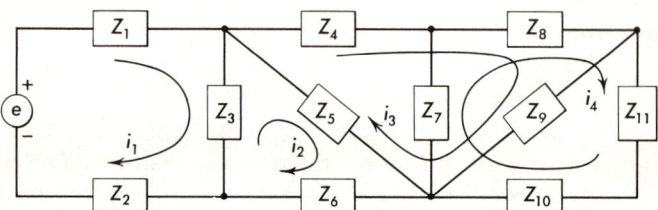

Fig. 2-2 Insufficient number of loop currents.

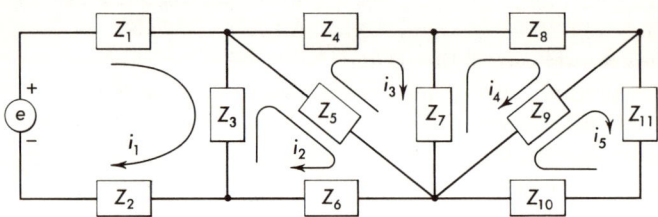

Fig. 2-3 Sufficient number of loop currents.

loop. Condition a is satisfied in Fig. 2-2, since there is current in each element. Condition b is violated, however, in the case of both currents i_3 and i_4. A correct assumption of the currents is shown in Fig. 2-3.

2. Around each loop write an equation obtained from Kirchhoff's first law (the sum of voltage drops around the loop equals the sum of voltage rises around the loop). This will result in the same number n of integro-differential equations as there are assumed currents.

3. Laplace-transform the differential equations.
4. Solve the resulting n algebraic equations for the desired current or currents. In solving these equations s is treated as an algebraic quantity.

Equations for the circuit of Fig. 2-1 are now found by following the above procedure. Currents i_1 and i_2 are assumed in a clockwise direction. The sum of voltage drops around each loop is set equal to the voltage rises. In the first loop the input voltage e_{in} is a rise, as indicated by polarity markings on Fig. 2-4. Voltage drops in the first loop occur

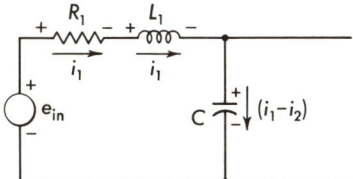
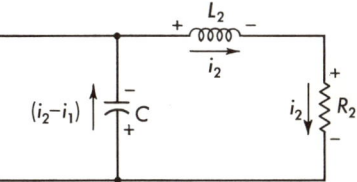

Fig. 2-4 First loop of electric circuit example.

Fig. 2-5 Second loop of electric circuit example.

across passive elements R, L, and C in the loop. Equating the sum of voltage rises to the sum of voltage drops yields

$$e_{in} = R_1 i_1 + L_1 \frac{di_1}{dt} + \frac{1}{C} \int_0^t (i_1 - i_2)\, dt \tag{2-1}$$

Laplace-transforming with zero initial conditions

$$E_{in} = \left(R_1 + L_1 s + \frac{1}{sC}\right) I_1 + \left(\frac{-1}{sC}\right) I_2 \tag{2-2}$$

Variables are capitalized to indicate that they are functions of s. For linear-system applications zero initial conditions are usually assumed, since the nature of control-system stability depends on the structure of the system and is independent of driving functions or initial conditions.

An equation for each loop must be written. In the second loop, shown in Fig. 2-5, the voltage rises (driving functions) are zero. Equating this to the sum of the voltage drops gives

$$0 = L_2 \frac{di_2}{dt} + R_2 i_2 + \frac{1}{C} \int_0^t (i_2 - i_1)\, dt \tag{2-3}$$

or in terms of the Laplace-transform operator s,

$$0 = \left(-\frac{1}{sC}\right) I_1 + \left(L_2 s + R_2 + \frac{1}{sC}\right) I_2 \tag{2-4}$$

In a shorthand notation Eqs. (2-2) and (2-4) are written in convenient form for solution:

$$E_{in} = Z_{11}I_1 + Z_{12}I_2 \tag{2-5a}$$

$$0 = Z_{21}I_1 + Z_{22}I_2 \tag{2-5b}$$

where Z_{11} is the sum of all operational impedances around the first loop and in this example is

$$Z_{11} = L_1 s + R_1 + \frac{1}{sC} \tag{2-6}$$

Z_{12} is the sum of all operational impedances common to loops 1 and 2 and in this example is

$$Z_{12} = -\frac{1}{sC} \tag{2-7}$$

For electrical networks that contain only resistors R, capacitors C, inductors L, and transformers M, the network is assumed to be linear and bilateral. Linearity, as defined in Chap. 1, infers that transforms of the voltage drops across these elements are of the form

$$\begin{aligned} E_R &= RI & E_C &= \mathcal{L}\frac{1}{C}\int_0^t i\,dt = \frac{1}{sC}I \\ E_L &= \mathcal{L}L\frac{di}{dt} = LsI & E_M &= \pm\,\mathcal{L}M\frac{di}{dt} = \pm\,MsI \end{aligned} \tag{2-8}$$

The dependent variable i, with its derivatives and integrals, appears only to the first power. Bilateral elements are those whose voltage drop is the same in magnitude for a current flowing in either direction through the element. All common elements R, L, C, and M are bilateral. A diode, shown in the schematic of Fig. 2-6a, is an example of a nonbilateral element. When a positive-polarity voltage is applied, the current is to the right as shown by the curve of Fig. 2-6b and the diode acts like a low-value resistor. When the same voltage is applied in the opposite direction, only a small current exists. The diode now behaves like a resistor of large value.

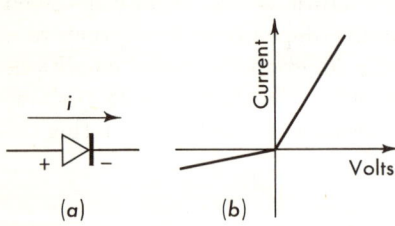

Fig. 2-6 Schematic and I-E curve of a diode, a nonbilateral element.

Obtaining the System Differential Equation

Returning to Eq. (2-5), Z_{22} is the sum of all operational impedances around loop 2 and is given by

$$Z_{22} = L_2 s + R_2 + \frac{1}{sC} \tag{2-9}$$

Z_{21} is the sum of all operational impedances common to loops 1 and 2. Since the elements are bilateral, the voltage drop across the common element C is independent of the direction of current through the element and

$$Z_{21} = Z_{12} = -\frac{1}{sC} \tag{2-10}$$

This property of linear bilateral networks proves useful in checking for errors.

Simple equations like Eq. (2-5) are easily solved by substitution. Thus, substituting for I_2 from Eq. (2-5b) into Eq. (2-5a) yields

$$E_{in} = Z_{11}\left(-\frac{Z_{22}}{Z_{21}}\right)I_2 + Z_{12}I_2$$

and

$$E_2 = I_2 R_2 = R_2 \frac{-E_{in}}{(Z_{11}Z_{22}/Z_{21}) - Z_{12}} \tag{2-11}$$

Solution of more complicated algebraic equations can be accomplished by utilizing Cramer's rule (the solution appears as the ratio of two determinants). The denominator determinant is formed from coefficients of the dependent variables. The equations are written in a symmetric fashion, as in Eq. (2-5). The denominator determinant is written directly from the network:

$$\Delta = \begin{vmatrix} Z_{11} & Z_{12} \\ Z_{21} & Z_{22} \end{vmatrix} \tag{2-12}$$

The symbol Δ is used to denote the determinant. The numerator determinant is formed by replacing the column in Δ, corresponding to the quantity to be found, by the column of driving functions.

In this example, I_2 is to be found. Replace the I_2 column which contains Z_{12} and Z_{22} with the driving-function column, which contains E_{in} and 0. The ratio of these two determinants, shown in Eq. (2-13), yields the solution

$$I_2 = \frac{\begin{vmatrix} Z_{11} & E_{in} \\ Z_{21} & 0 \end{vmatrix}}{\begin{vmatrix} Z_{11} & Z_{12} \\ Z_{21} & Z_{22} \end{vmatrix}} \tag{2-13}$$

A review of the theory of determinants is included in Appendix III. For the example of Eq. (2-12), the determinants are expanded:

$$I_2 = \frac{-Z_{21}E_{in}}{Z_{11}Z_{22} - Z_{21}Z_{12}} \qquad (2\text{-}14)$$

The transfer function E_o/E_{in} is found from Eq. (2-14):

$$E_o = R_2 I_2 = \frac{-R_2 Z_{21} E_{in}}{Z_{11}Z_{22} - Z_{12}Z_{21}} \qquad (2\text{-}15)$$

Replacing Z_{12}, Z_{21}, Z_{11}, Z_{22} by the values given in Eqs. (2-6), (2-7), (2-9), and (2-10),

$$\frac{E_o}{E_{in}} = \frac{-R_2(-1/sC)}{(L_1 s + R_1 + 1/sC)(L_2 s + R_2 + 1/sC) - (1/sC)^2} \qquad (2\text{-}16)$$

This transfer function reduces, after some algebraic manipulations, to

$$G_1(s) = \frac{E_o}{E_{in}}$$
$$= \frac{R_2}{(L_1 L_2 C)s^3 + (R_1 L_2 C + R_2 L_1 C)s^2 + (L_1 + L_2 + R_1 R_2 C)s + R_1 + R_2} \qquad (2\text{-}17)$$

The block diagram for this network is shown in Fig. 2-7, where $G_1(s)$, given by Eq. (2-17), is the transfer function.

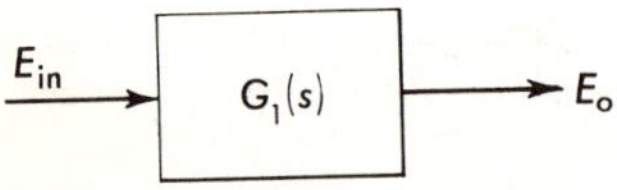

Fig. 2-7 Block diagram of electric circuit example.

The n-loop network is treated in the same manner as the example just considered. If there are n independent loops, there are n loop currents. Consequently, there are n independent equations from which the n unknown loop currents are determined. These equations are written as

$$\begin{aligned}
I_1 Z_{11} + I_2 Z_{12} + \cdots + I_j Z_{1j} + \cdots + I_n Z_{1n} &= E_1 \\
I_1 Z_{21} + I_2 Z_{22} + \cdots + I_j Z_{2j} + \cdots + I_n Z_{2n} &= E_2 \\
\cdots \cdots \cdots \cdots \cdots \cdots \cdots \cdots \cdots \cdots \cdots \cdots & \\
I_1 Z_{n1} + I_2 Z_{n2} + \cdots + I_j Z_{nj} + \cdots + I_n Z_{nn} &= E_n
\end{aligned} \qquad (2\text{-}18)$$

The quantities E_1, E_2, ..., E_n are the transforms of driving voltages or voltage sources in the loops. The quantities Z_{ij} are impedance operators and are algebraic functions of the Laplace-transform operator s.

Obtaining the System Differential Equation

The self-impedance of the loop, in the absence of mutual inductance, is

$$Z_{ii} = sL_{ii} + R_{ii} + \frac{1}{sC_{ii}} = sL_{ii} + R_{ii} + \frac{s_{ii}}{s} \qquad (2\text{-}19)$$

where L_{ii} is the self-inductance of the loop presented to the ith current flowing around the ith loop, R_{ii} is the self-resistance, and $s_{ii} = 1/c_{ii}$ is the self-elastance (reciprocal capacitance). The common-impedance terms are

$$Z_{ij} = -\left(sL_{ij} + R_{ij} + \frac{1}{sC_{ij}}\right) \qquad (2\text{-}20)$$

The quantities L_{ij}, R_{ij}, and C_{ij} are the common inductance, common resistance, and common capacitance, respectively, between the ith and jth loops. For example, capacitance in the circuit of Fig. 2-1 is common to loops 1 and 2.

If mutual inductance is present, a term $\pm sM_{ij}$ will appear in the Z_{ij} term [Eq. (2-20)] as follows:

$$Z_{ij} = -\left(sL_{ij} + R_{ij} + \frac{1}{sC_{ij}} \pm sM_{ij}\right) \qquad (2\text{-}21)$$

The algebraic sign of the mutual term is discussed in the next section. Driving functions which appear on the right side of Eqs. (2-18) are derived from active voltage sources E_i within the current loops.

Equation (2-18) is solved by determinants; the denominator determinant is

$$\Delta = \begin{vmatrix} Z_{11} & Z_{12} & Z_{13} & \cdots & Z_{1n} \\ Z_{21} & Z_{22} & Z_{23} & \cdots & Z_{2n} \\ Z_{31} & \cdots & & & \\ \cdots & \cdots & \cdots & \cdots & \cdots \\ Z_{n1} & Z_{n2} & \cdots & & Z_{nn} \end{vmatrix} \qquad (2\text{-}22)$$

where, in choosing subscripts for the impedance, the first corresponds to row and the second to column as follows:

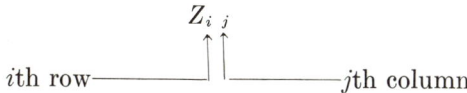

If all Z_{ij} are not equal to all Z_{ji} for each pair ij, an error is indicated—for all elements are bilateral—and the work should be checked to correct it.

42   Control System Design

For planar circuits containing only linear, bilateral, passive elements, all Z_{ij} are equal to all Z_{ji} and are negative when the current directions are assumed in the manner shown.

The numerator determinant with both the ith row and jth column crossed out is written

$$\Delta_{ij} = \begin{vmatrix} Z_{11} & Z_{12} & Z_{13} & \cdots & Z_{1j} & \cdots & Z_{1n} \\ Z_{21} & Z_{22} & Z_{23} & \cdots & Z_{2j} & \cdots & Z_{2n} \\ Z_{31} & Z_{32} & Z_{33} & \cdots & Z_{3j} & \cdots & Z_{3n} \\ \vdots & & & & & & \vdots \\ \cancel{Z_{i1}} & \cancel{Z_{i2}} & \cancel{Z_{i3}} & & \cancel{Z_{ij}} & & \cancel{Z_{ij}} \\ \vdots & & & & & & \vdots \\ Z_{n1} & Z_{n2} & Z_{n3} & \cdots & Z_{nj} & \cdots & Z_{nn} \end{vmatrix} \quad (2\text{-}23)$$

This determinant is of one degree lower than the original Δ. To find the solution for any particular current, e.g., the current in the jth loop i_j, the jth column of the determinant of Eq. (2-22) is replaced with the column of driving voltages. The solution can be written formally:

$$I_j = \frac{1}{\Delta} \sum_{i=1}^{n} (-1)^{(i+j)} E_i \Delta_{ij} \quad (2\text{-}24)$$

This equation is simply the mathematical expression for a determinant expansion.

An interesting example of the use of Eq. (2-24) is in finding the input impedance $Z_{jj}{}^{\text{in}}$ to a network. When all voltage sources are shorted and a voltage E_j is applied in only the jth loop, the input impedance to the jth loop is the ratio of E_j to the resulting current I_j. Since all voltage sources except E_j are zero, the summation of Eq. (2-24) reduces to one term:

$$I_j = \frac{1}{\Delta} (-1)^{j+j} \Delta_{jj} = E_j \frac{\Delta_{jj}}{\Delta} \quad (2\text{-}25)$$

and the input impedance is

$$Z_{jj}{}^{\text{in}} = \frac{E_j}{I_j} = \frac{\Delta}{\Delta_{jj}} \quad (2\text{-}26)$$

The input impedance for the current of Fig. 2-1 is

$$Z_{11}{}^{\text{in}} = \frac{\Delta}{\Delta_{11}} = Z_{11} - \frac{Z_{12}{}^2}{Z_{22}}$$

$$= \left(L_1 s + R_1 + \frac{1}{sC}\right) - \frac{(1/sC)^2}{L_2 s + R_2 + 1/sC} \quad (2\text{-}27)$$

2-3 Loop Analysis of Electrical Networks with Mutual Inductance. If a transformer is present in the current loop, a similar analysis is used except that a voltage is induced in the transformer primary winding as a consequence of current change in the secondary winding, and vice versa. This voltage inducement is termed "mutual inductance," and the term which represents the mutual inductance $\pm sM_{ij}$ is included in Eq. (2-21). The choice of plus or minus sign depends upon the connection or polarity of the transformer.

As an example, consider the circuit of Fig. 2-8, where a dot is positioned on each transformer. The rule associated with this convention is as

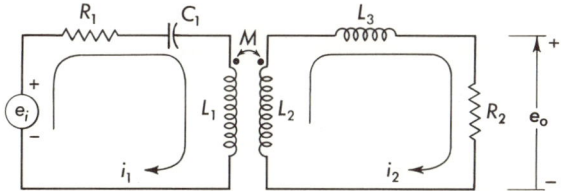

Fig. 2-8 Mutual inductance in loop circuits.

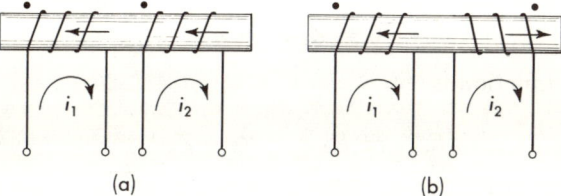

Fig. 2-9 Transformer polarity markings.

follows: If both assumed currents I_1 and I_2 flow into or out of the dotted ends of the coils, the corresponding sign of the mutual-inductance term, treated as a voltage drop, is the plus sign. If one assumed current flows into a dotted end and the other flows out of a dotted end, the corresponding sign is the minus sign.

The dots, or polarity markings, on a transformer indicate the winding sense or direction of the unit. For example, consider the two transformers of Fig. 2-9, where assumed currents I_1 and I_2 flow through transformers with different winding senses. In Fig. 2-9a the assumed currents produce fluxes along the axis of the transformer which aid each other. In this case, when the fluxes aid, the mutual inductance has the same sign as the self-inductance; thus,

$$L_1 s I_1 + M s I_2 \tag{2-28}$$

The dots, as shown in Fig. 2-9a, appear on the same ends. In Fig. 2-9b the fluxes produced by the assumed currents I_1 and I_2 are opposed and the mutual inductance and self-inductance have opposite signs; thus,

$$L_1 s I_1 - M s I_2 \qquad (2\text{-}29)$$

The dots appear on opposite ends of the coils for Fig. 2-9b.

In Fig. 2-8, the current in loop 1 flows into the dotted end and I_2 flows out of the dotted end. The sign of the mutual-inductance term is negative, and the operational impedances for the circuit are

$$Z_{11} = sL_1 + R_1 + \frac{1}{sC_1}$$
$$Z_{12} = Z_{21} = -Ms \qquad (2\text{-}30)$$
$$Z_{22} = (L_2 + L_3)s + R_2$$

The transform of the driving function is E_{in}, and the equations are

$$E_{\text{in}} = Z_{11} I_1 + Z_{12} I_2$$
$$0 = Z_{21} I_1 + Z_{22} I_2 \qquad (2\text{-}31)$$

The transfer function is

$$\frac{E_o}{E_{\text{in}}} = \frac{I_2 R_2}{E_{\text{in}}} = \frac{R_2(-Z_{21})}{Z_{11} Z_{22} - Z_{12}{}^2} \qquad (2\text{-}32)$$

When Eqs. (2-30) are inserted into Eq. (2-32) and the result is simplified, the transfer function results as

$$G_2(s) = \frac{E_o}{E_{\text{in}}}$$
$$= \frac{R_2 M C_1 s^2}{C_1[L_1(L_2 + L_3) - M^2]s^3 + [R_1(L_2 + L_3) + R_2 L_1]C_1 s^2 + (L_2 + L_3 + R_1 R_2 C_1)s + R_2} \qquad (2\text{-}33)$$

This is the transfer function between E_{in} and E_o. The block diagram is shown in Fig. 2-10.

2-4 Nodal Analysis of Electrical Networks. The nodal method of analysis is based on use of Kirchhoff's second law, namely, the algebraic sum of currents flowing out of a circuit node is equal to zero. The steps

Obtaining the System Differential Equation 45

necessary to solve an electrical network by use of nodal analysis are outlined as follows:

1. Assume potentials at all nodes in the circuit. It is convenient to choose one point or node in the circuit as at ground potential (i.e., zero volts) and assume potentials V_i at all other nodes. The node potentials, measured above ground, are the dependent variables in the node solution just as the currents are the dependent variables in the loop solution. There is no problem of finding the correct number of node potentials necessary to describe the circuit as there is in the case of loop analysis.

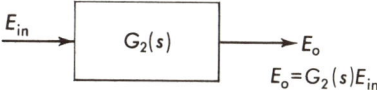

Fig. 2-10 Block diagram of the electrical network of Fig. 2-8.

2. Utilize Kirchhoff's second law; i.e., write the equation expressing the algebraic sum of currents flowing into the node point as equal to zero. This will result in the same number n of integrodifferential equations as there are assumed nodal potentials measured above ground potential. No equation is written for the node chosen at ground potential.
3. Laplace-transform the differential equations.
4. Solve the resulting n algebraic equations for the desired voltage.

The above procedure is used to find the transfer function E_o/E_{in} for the circuit of Fig. 2-11. The first step in the solution is to choose the

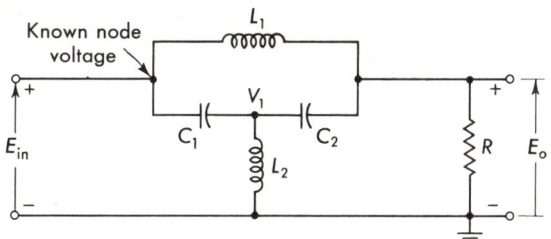

Fig. 2-11 Example of the nodal analysis method.

zero-voltage node and assume as unknowns the potentials of the other nodes. The lower lead on the circuit of Fig. 2-11 is conveniently taken as at zero potential. E_o is the potential at the output node, and V_1 is the potential at the center point (shown in Fig. 2-11). Since the input nodal pair is driven with a known voltage source E_{in}, only two unknown node potentials (E_o and V_1) exist in the problems; so only two equations must be written. An equation is not written at E_{in}, since E_{in} is a known quantity. The next step consists of writing Kirchhoff's current equation about these two node points.

Consider first the relation that exists between current and voltage across three electrical elements (R, C, and L). The transform of the current through a resistor R is

$$I_R = \mathcal{L}\frac{v_1 - v_2}{R} = \frac{V_1 - V_2}{R} \tag{2-34}$$

where the current is from node 1 to node 2, as shown in Fig. 2-12a. The current through a capacitor C is

$$I_C = \mathcal{L}C\frac{d}{dt}(v_1 - v_2) = sC(V_1 - V_2) \tag{2-35}$$

where the current is from node 1 to node 2, as shown in Fig. 2-12b. The current from node 1 to node 2 through an inductor L is

$$I_L = \mathcal{L}\frac{1}{L}\int_0^t (v_1 - v_2)\, dt = \frac{1}{sL}(V_1 - V_2) \tag{2-36}$$

The equation for the currents leaving any node, say, node 1, is formed by summing currents of the form of Eqs. (2-34) to (2-36). Capitalized quantities represent Laplace-transformed functions.

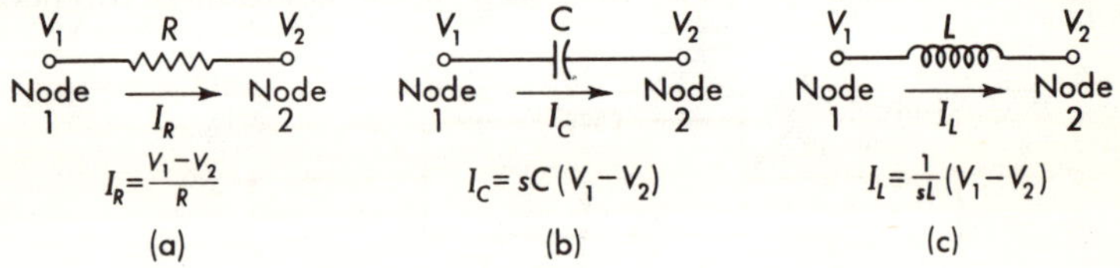

Fig. 2-12 Current through R, L, and C elements.

Figure 2-13 shows the circuit necessary to write the first equation for the circuit of Fig. 2-11. The algebraic sum of currents flowing from the node (Fig. 2-13) must equal zero; thus

$$sC_2(V_1 - E_o) + sC_1(V_1 - E_{in}) + \frac{1}{sL_2}V_1 = 0 \tag{2-37}$$

The first equation is obtained by rearranging Eq. (2-37). Since E_{in} is the driving function, it is taken to the right side as

$$V_1\left[s(C_2 + C_1) + \frac{1}{sL_2}\right] + E_o(-sC_2) = (sC_1)E_{in} \tag{2-38}$$

Obtaining the System Differential Equation 47

The second equation, written at the output node, is represented by the circuit of Fig. 2-14. Again equating the algebraic sum of the currents leaving the node to zero, we have

$$\frac{1}{sL_1}(E_o - E_{in}) + sC_2(E_o - V_1) + \frac{E_o}{R} = 0 \qquad (2\text{-}39)$$

The driving function E_{in} is shifted to the right side of the equation, and

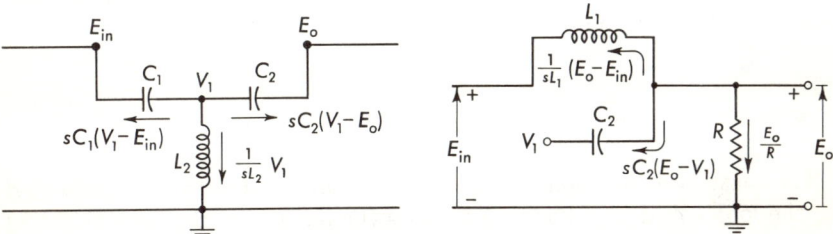

Fig. 2-13 First portion of node circuit.   Fig. 2-14 Second portion of node circuit.

the expression is rearranged as follows:

$$V_1(-sC_2) + E_o\left(sC_2 + \frac{1}{R} + \frac{1}{sL_1}\right) = \frac{1}{sL_1}E_{in} \qquad (2\text{-}40)$$

Equations (2-38) and (2-40) are the two necessary equations in the transformed variable which define the system. These equations can be rewritten in a shorthand notation as follows:

$$\begin{aligned} V_1 Y_{11} + E_o Y_{12} &= I_1 \\ V_1 Y_{21} + E_o Y_{22} &= I_2 \end{aligned} \qquad (2\text{-}41)$$

where Y_{11} is the sum of all operational admittances which tie to node 1 and in this example is

$$Y_{11} = s(C_1 + C_2) + \frac{1}{sL_2} \qquad (2\text{-}42)$$

Y_{12} is the sum of all operational admittances connecting nodes 1 and 2 and in this example is

$$Y_{12} = -sC_2 \qquad (2\text{-}43)$$

Since the network is formed from linear bilateral elements,

$$Y_{21} = Y_{12} = -sC_2 \tag{2-44}$$

Y_{22} is the sum of all operational admittances connected to node 2 and is

$$Y_{22} = \left(sC + \frac{1}{R} + \frac{1}{sL_1}\right) \tag{2-45}$$

The driving function for the nodal solution is a current. In this case, using transformed quantities,

$$I_1 = (sC_1)E_{in} \quad \text{and} \quad I_2 = \frac{1}{sL_1}E_{in} \tag{2-46}$$

The method of solution of Eqs. (2-41) is identical with that presented for the loop solution. For simple cases the output voltage can be found by substitution as in Eq. (2-11). A more systematic method of solution is based on expressing $E_o(s)$ as the ratio of two determinants; thus,

$$E_o = \frac{\begin{vmatrix} Y_{11} & I_1 \\ Y_{21} & I_2 \end{vmatrix}}{\begin{vmatrix} Y_{11} & Y_{12} \\ Y_{21} & Y_{22} \end{vmatrix}} = \frac{I_2 Y_{11} - I_1 Y_{21}}{Y_{11} Y_{22} - Y_{12}^2} \tag{2-47}$$

The transfer function E_o/E_{in} is found by substituting values into Eq. (2-47) from preceding equations:

$$\frac{E_o}{E_{in}} = \frac{[s(C_1 + C_2) + 1/sL_2](1/sL_1) - (sC_1)(-sC_2)}{[s(C_1 + C_2) + 1/sL_2](sC_2 + 1/R + 1/sL_1) - (sC_2)^2} \tag{2-48}$$

Clearing fractions yields

$$G_3(s) = \frac{E_o}{E_{in}}$$

$$= \frac{s^4(L_1 L_2 C_1 C_2) + s^2(C_1 + C_2)L_2 + 1}{s^4(L_1 L_2 C_1 C_2) + [(1/R)(C_1 + C_2)L_1 L_2]s^3 + s^2[(C_1 + C_2)L_2 + C_2 L_1] + s(L_1/R) + 1} \tag{2-49}$$

The block diagram for the transfer function of Eq. (2-49) is shown in Fig. 2-15.

Obtaining the System Differential Equation

If there are $n+1$ independent nodes, there are n independent unknown nodal potentials when one node is taken at zero potential. Consequently, there exist n independent equations relating the transforms

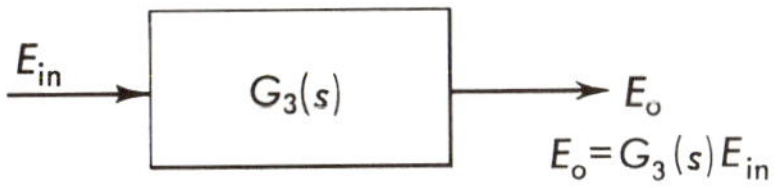

```
     E_in  ┌────────┐  E_o
    ─────▶│  G₃(s) │─────▶
           └────────┘    E_o = G₃(s) E_in
```

Fig. 2-15 Block diagram for the circuit of Fig. 2-11.

of these nodal potentials to transforms of driving currents. This set of equations may be written

$$V_1 Y_{11} + V_2 Y_{12} + \cdots + V_n Y_{1n} = I_1$$
$$V_1 Y_{21} + V_2 Y_{22} + \cdots + V_n Y_{2n} = I_2$$
$$\cdots \cdots \cdots \cdots \cdots \cdots \cdots \cdots \cdots$$
$$V_1 Y_{n1} + V_2 Y_{n2} + \cdots + V_n Y_{nn} = I_n$$

(2-50)

The driving functions $I_1, I_2, \ldots, I_n$ are transforms of active current sources or current driving functions. The general operational self-admittance term is written as

$$Y_{ii} = sC_{ii} + \frac{1}{R_{ii}} + \frac{1}{sL_{ii}} \qquad (2\text{-}51)$$

and the general common-admittance term is written

$$Y_{ij} = -\left(sC_{ij} + \frac{1}{R_{ij}} + \frac{1}{sL_{ij}}\right) \qquad (2\text{-}52)$$

Mutual inductance $\pm M_{ij}$ is purposely left out of Eq. (2-52). Mutual inductance can be treated in a nodal solution,* but it is usually more convenient to effect solution by the loop method. When circuits, which are more easily treated on the nodal basis (because there are fewer independent nodes than independent loops), involve mutual inductance, a Π equivalent circuit for the transformer can be used. The conversion to a Π equivalent circuit is shown in Fig. 2-16. This circuit is most suited for the nodal solution. A T equivalent circuit, also shown in Fig. 2-16, can be used for the loop solution.

* See Ref. 13, pp. 40–43.

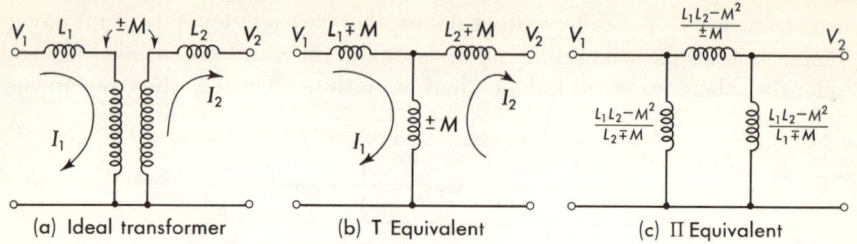

(a) Ideal transformer (b) T Equivalent (c) Π Equivalent

Fig. 2-16 Two equivalent circuits for a transformer.

Equations (2-50) are solved by determinants in a fashion similar to that utilized in loop analysis (Sec. 2-2):

$$V_j = \frac{1}{\Delta^N} \begin{vmatrix} Y_{11} & Y_{12} & Y_{13} & \cdots & I_1 & \cdots & Y_{1n} \\ Y_{21} & Y_{22} & Y_{23} & \cdots & I_2 & \cdots & Y_{2n} \\ \cdots & \cdots & \cdots & \cdots & \cdots & \cdots & \cdots \\ Y_{n1} & Y_{n2} & Y_{n3} & \cdots & I_n & \cdots & Y_{nn} \end{vmatrix} \quad (2\text{-}53)$$

The symbol Δ^N is used to indicate the determinant of the coefficients:

$$\Delta^N = \begin{vmatrix} Y_{11} & Y_{12} & \cdots & Y_{1j} & \cdots & Y_{1n} \\ Y_{21} & Y_{22} & \cdots & Y_{2j} & \cdots & Y_{2n} \\ \cdots & \cdots & \cdots & \cdots & \cdots & \cdots \\ Y_{n1} & Y_{n2} & \cdots & Y_{nj} & \cdots & Y_{nn} \end{vmatrix} \quad (2\text{-}54)$$

The solution for V_j can be written formally as

$$V_j = \frac{1}{\Delta^N} \sum_{i=1}^{n} (-1)^{i+j} I_i \Delta_{ij}^N \quad (2\text{-}55)$$

where Δ_{ij}^N is the determinant Δ^N with the ith row and the jth column crossed out.

The input admittance Y_{jj}^{in} into the node is found by opening all current sources except the one that drives the nodal pair comprised of the jth node and "ground" node and finding the ratio

$$Y_{jj}^{\text{in}} = \frac{I_j}{V_n} = \frac{\Delta^N}{\Delta_{jj}^N} \quad (2\text{-}56)$$

As another example of the nodal method, consider the circuit in Fig. 2-17 for which the input operational admittance Y_{11}^{in} is desired. Currents flowing out of a node are positive (on the left-hand side of the equation), and currents flowing into the node are treated as sources

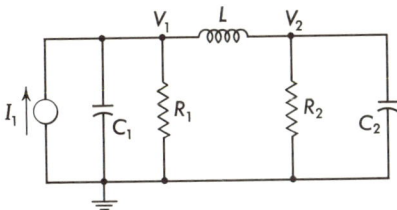

Fig. 2-17 Circuit diagram showing node locations.

(positive on the right-hand side of the equation). The self- and common operational admittances are

$$Y_{11} = \frac{1}{Ls} + \frac{1}{R_1} + C_1 s$$

$$Y_{12} = Y_{21} = -\frac{1}{Ls} \qquad (2\text{-}57)$$

$$Y_{22} = \frac{1}{Ls} + \frac{1}{R_2} + C_2 s$$

The equations are

$$I_1 = V_1 Y_{11} + V_2 Y_{12}$$

$$0 = V_1 Y_{21} + V_2 Y_{22} \qquad (2\text{-}58)$$

The input admittance is given by

$$Y_{11}{}^{in} = \frac{\Delta^N}{\Delta_{11}{}^N} = \frac{\begin{vmatrix} Y_{11} & Y_{12} \\ Y_{21} & Y_{22} \end{vmatrix}}{Y_{22}} = Y_{11} - \frac{Y_{12}{}^2}{Y_{22}} \qquad (2\text{-}59)$$

When Eqs. (2-57) are used, Eq. (2-59) reduces to

$$Y_{11}{}^{in} = \frac{s^3(LC_1C_2R_1R_2) + s^2[L(C_1R_1 + C_2R_2)] + s(R_1R_2C_1C_2 + L) + R_1 + R_2}{R_1(R_2LC_2s^2 + Ls + R_2)} \qquad (2\text{-}60)$$

2-5 Loop Analysis of Active Networks. Equations (2-18) and (2-50) of the preceding sections can be generalized to include active elements, i.e., vacuum tubes and transistors. Such elements are not completely linear, but over a limited range of operation they can be treated as linear.

52   Control System Design

Over the linear range, which is established with appropriate biasing, these elements can be replaced with an equivalent circuit.

A triode is usually represented by an equivalent loop circuit* (Fig. 2-18). Based upon usual assumptions, an equivalent voltage generator in series with a plate resistance is substituted for the vacuum tube. The points a, b, and c in each of the diagrams of Fig. 2-18 are interchangeable. μ denotes the amplification factor, and r_p the plate resistance.

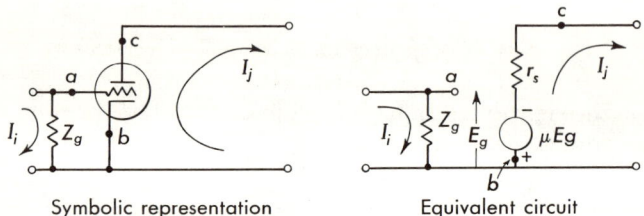

Symbolic representation   Equivalent circuit

Fig. 2-18 Equivalent circuit for a vacuum-tube triode (loop basis).

Suppose the vacuum tube is located in a network between the ith and jth meshes, as shown in Fig. 2-18. The loop equation for the jth mesh is written

$$I_1 Z_{j1} + I_2 Z_{j2} + \cdots + I_i Z_{ji} + \cdots + I_m Z_{jm} = V_j - \mu E_g \quad (2\text{-}61)$$

where impedances have the same significance as in Eqs. (2-19) and (2-20). Since $E_g = I_i Z_g$, Eq. (2-61) is written

$$I_1 Z_{j1} + I_2 Z_{j2} + \cdots + I_i(Z_{ji} + \mu Z_g) + \cdots + I_m Z_{jm} = V_j \quad (2\text{-}62)$$

If no other loops contain this tube or any other tube, the network determinant is of the form

$$\Delta = \begin{vmatrix} Z_{11} & Z_{12} & \cdots & Z_{1i} & \cdots & Z_{1m} \\ Z_{21} & Z_{22} & \cdots & Z_{2i} & \cdots & Z_{2m} \\ Z_{31} & Z_{32} & \cdots & Z_{3i} & \cdots & Z_{3m} \\ \cdots & \cdots & \cdots & \cdots & \cdots & \cdots \\ Z_{j1} & Z_{j2} & \cdots & Z_{ji} + \mu Z_g & \cdots & Z_{jm} \\ \cdots & \cdots & \cdots & \cdots & \cdots & \cdots \\ Z_{m1} & Z_{m2} & \cdots & Z_{mi} & \cdots & Z_{mm} \end{vmatrix} \quad (2\text{-}63)$$

* Numerous books discuss this equivalent circuit; see, for example, chap. 3 of Ref. 32.

This determinant is not symmetrical about its principal diagonal, contrary to the case for passive networks [Eq. (2-22)]. In general, addition of a bilateral element results in a nonreciprocal network; that is, $Z_{ij} \neq Z_{ji}$.

If Δ^0 denotes the value of Δ when $\mu = 0$, the determinant of Eq. (2-63) can be written as

$$\Delta = \Delta^0 + \mu Z_g \Delta_{ji} \tag{2-64}$$

where Δ_{ji} is the determinant resulting from canceling the jth row and ith column.

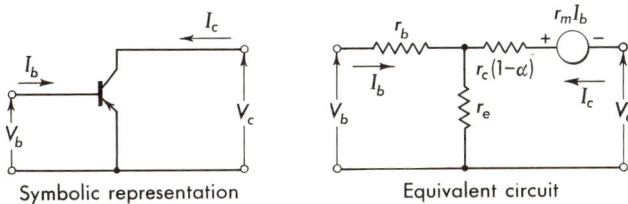

Fig. 2-19 Equivalent circuit for a transistor used in the common-emitter connection.

Inclusion of a transistor in the loop analysis modifies the network determinant in a similar fashion. One equivalent circuit[33] for a grounded-emitter stage is shown in Fig. 2-19. The resistances shown in this T network have the following physical significance:

r_e = emitter resistance or forward resistance of emitter-base diode; small in value
r_c = collector resistance or reverse resistance of collector-base diode; high in value
r_b = base resistance, which represents an interaction effect between emitter and collector
r_m = mutual resistance, which provides gain in the circuit

Consider the simple one-stage amplifier shown in Fig. 2-20. From the equivalent circuit, also shown in Fig. 2-20, the following loop equations can be written

$$E_{\text{in}} = (R_g + r_b + r_e)I_1 + (-r_e)I_2 \tag{2-65}$$

$$-r_m I_b = (-r_e)I_1 + [r_e + r_c(1 - \alpha) + R_L]I_2 \tag{2-66}$$

Because only resistance occurs in this network, these equations could be written in the time domain. They are identical when written in the

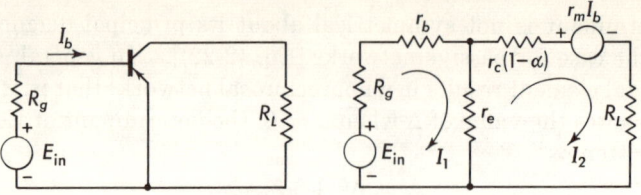

Fig. 2-20 Single-stage transistor amplifier.

transform domain, since s does not appear. Since I_b, the base current, is equal to I_1, Eq. (2-66) becomes

$$0 = (r_m - r_e)I_1 + [r_e + r_c(1 - \alpha) + R_L]I_2 \qquad (2\text{-}67)$$

Solution for any current (for example, I_2) proceeds in the normal fashion:

$$I_2 = \frac{-e_{in}(r_m - r_e)}{\begin{vmatrix} R_g + r_b + r_e & -r_e \\ r_m - r_e & r_e + r_c(1-\alpha) + R_L \end{vmatrix}} \qquad (2\text{-}68)$$

Addition of a transistor results in a nonreciprocal network; that is, $Z_{ij} \neq Z_{ji}$. The effect of inserting a transistor is similar to that of inserting a vacuum tube. There is, however, at least one important difference: circuit loading. The input impedance to a vacuum tube is high, hence it produces considerable circuit isolation. This is usually not the case with a transistor,* and for that reason the entire circuit must be analyzed as a unit, since each stage is not readily isolated from the rest of the circuit.

2-6 Node Analysis of Active Networks. A vacuum tube and a transistor enter into a node-basis analysis in much the same way that they are included in a loop analysis.

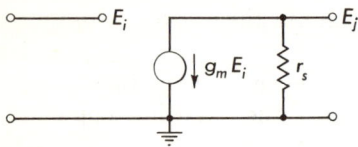

Fig. 2-21 Equivalent circuit for a vacuum-tube triode (node basis).

The equivalent node circuit for a vacuum tube is shown in Fig. 2-21. The cathode is taken as the reference node and is grounded. Potentials are assumed for all nodes; E_i is the transform of the assumed grid potential and E_j is the transform of the output node potential at the tube terminals. The operational equation for the

---

* See Ref. 9, chap. 8, for further transistor design information.

Obtaining the System Differential Equation 55

jth node pair is

$$I_j = E_1 Y_{j1} + E_2 Y_{j2} + \cdots + E_i(Y_{ji} + g_m) + \cdots + E_n Y_{jn} \quad (2\text{-}69)$$

With only one tube in the circuit, the unilateral term enters the network determinant at one point:

$$\Delta^N = \begin{vmatrix} Y_{11} & Y_{12} & \cdots & Y_{1i} & \cdots & Y_{1n} \\ Y_{21} & Y_{22} & \cdots & Y_{2i} & \cdots & Y_{2n} \\ \cdots & \cdots & \cdots & \cdots & \cdots & \cdots \\ Y_{j1} & Y_{j2} & \cdots & Y_{ji}+g_m & \cdots & Y_{jm} \\ \cdots & \cdots & \cdots & \cdots & \cdots & \cdots \\ Y_{n1} & Y_{n2} & \cdots & Y_{n1} & \cdots & Y_{nn} \end{vmatrix} \quad (2\text{-}70)$$

It is sometimes inconvenient to use the cathode as the reference node, since g_m would occur in several Δ^N terms. The determinant can always be rearranged, however, without changing its value, and g_m can be made to occur only once.

The equivalent circuit for a grounded-emitter transistor is shown in Fig. 2-22. Node analysis of such a circuit proceeds in a fashion similar to that of the vacuum-tube solution. Node potentials are assumed for all points in the circuit (and for nodes in the transistor equivalent circuit). The emitter lead is chosen as the reference potential, and E_i, E_j, and E_k are transforms of the assumed node potentials for the remaining transistor nodes. The equation for node j is written

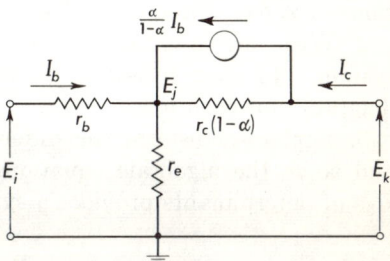

Fig. 2-22 Transistor equivalent node circuit with one current source.

$$\frac{\alpha}{1-\alpha} I_b = E_i \left(-\frac{1}{r_b} \right) + E_j \left[\frac{1}{r_b} + \frac{1}{r_e} + \frac{1}{r_c(1-\alpha)} \right] + E_k \left[-\frac{1}{r_c(1-\alpha)} \right] \quad (2\text{-}71)$$

Since $I_b = (E_i - E_j)/r_b$, Eq. (2-71) can be simplified to

$$0 = E_i \frac{-1}{(1-\alpha)r_b} + E_j \left[\frac{1}{1-\alpha} \left(\frac{1}{r_b} + \frac{1}{r_c} \right) + \frac{1}{r_e} \right] + E_k \frac{-1}{(1-\alpha)r_e} \quad (2\text{-}72)$$

The current source in Fig. 2-22 depends upon the assumed node potentials. The term on the left side of Eq. (2-71) is brought to the right side and included as part of Eq. (2-72). Such a term produces a nonreciprocal network; that is, $Y_{ij} \neq Y_{ji}$. The remaining analysis proceeds in a similar fashion to that of the preceding section.

2-7 Mechanical Networks; Linear Motion.[40] The analysis of mechanical networks is similar to the analysis of electrical networks and is accomplished by utilizing Newton's law for mechanical systems: *The algebraic sum of forces acting on a body (mass) is equal to the product of the mass and its acceleration.* The method of analysis of mechanical systems is outlined as follows:

1. Assume displacements and directions for each mass in the system. Displacements x are dependent variables, and forces f are driving functions. In assigning directions to the dependent variables, choose one direction, e.g., upward or to the right, as positive. If possible, choose all variables in this same direction. Also assume that the positive directions of x, dx/dt, and d^2x/dt^2 are the same.

2. Write an equation, utilizing Newton's law, for each mass in the system. In order to isolate the forces on each mass, draw a "free-body diagram" for each mass.

3. Laplace-transform the differential or integrodifferential equations and solve the algebraic equations for the desired linear displacement. Use of determinants provides a systematic method of solution.

The above procedure is applied to the system of Fig. 2-23. It is desired to find the transfer function X_1/F. The first step in the solution

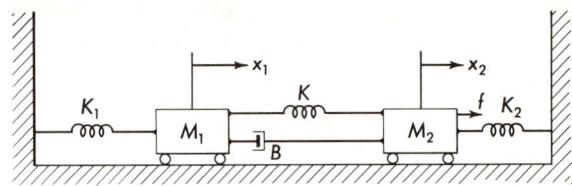

Fig. 2-23 Linear-motion mechanical system.

is to assign positive directions to M_1 and M_2 displacements and label these displacements x_1 and x_2. Spring constants K_i and damping constant B are described in Chap. 1. The restoring force of a linear spring is

$$f_K = K(x_1 - x_2) \tag{2-73}$$

The force produced by the spring is opposite in direction to x_1, as shown in Fig. 2-24a. The damping force produced by a dashpot is

$$f_B = B\left(\frac{dx_1}{dt} - \frac{dx_2}{dt}\right) \tag{2-74}$$

This force is shown in Fig. 2-24b.

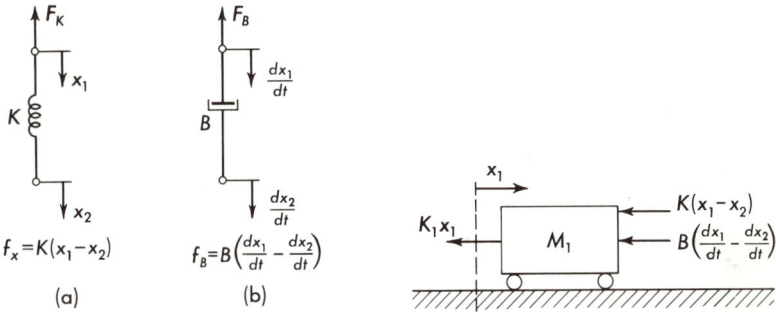

Fig. 2-24 Spring and damping forces.   **Fig. 2-25** Free-body diagram for M_1.

Utilizing Newton's law, an equation is written about each mass in the system. The free-body diagram for mass M_1 in Fig. 2-23 is shown in Fig. 2-25. As M_1 accelerates in the positive direction $(+x_1)$, the forces of the springs and dashpot oppose the motion. The sum of forces must equal the mass times the acceleration, or

$$M_1 \frac{d^2 x_1}{dt^2} = -K_1 x_1 - K(x_1 - x_2) - B\left(\frac{dx_1}{dt} - \frac{dx_2}{dt}\right) \tag{2-75}$$

When this equation is rearranged and Laplace-transformed (with zero initial conditions), the following system equation results:

$$[M_1 s^2 + Bs + (K_1 + K)]X_1 + (-Bs - K)X_2 = 0 \tag{2-76}$$

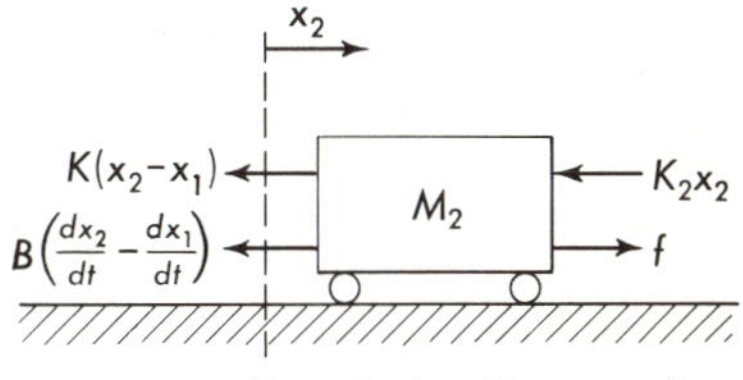

The transforms of the dependent variables are capitalized to indicate that they are functions of s.

Fig. 2-26 Free-body diagram for M_2.

The free-body diagram for M_2 is shown in Fig. 2-26. As the mass accelerates in the positive direction ($x_2 > 0$ to the right), the springs and dashpot oppose the motion. The applied force f is in the same direction as x_2; so it carries a positive sign:

$$M \frac{d^2 x_2}{dt^2} = +f - K(x_2 - x_1) - B\left(\frac{dx_2}{dt} - \frac{dx_1}{dt}\right) - K_2 x_2 \qquad (2\text{-}77)$$

When Eq. (2-77) is transformed with zero initial conditions, the rearranged force equation on M_2 becomes

$$(-Bs - K)X_1 + [M_2 s^2 + Bs + (K + K_2)]X_2 = F \qquad (2\text{-}78)$$

Again the transforms of the variables are capitalized to indicate that they are functions of s.

Equations (2-76) and (2-78) are two equations necessary for the analysis of Fig. 2-22. For purposes of simplification the two equations can be written as

$$a_{11} X_1 + a_{12} X_2 = 0$$
$$a_{21} X_1 + a_{22} X_2 = F \qquad (2\text{-}79)$$

where $\quad a_{11} = [M_1 s^2 + Bs + (K_1 + K)] \qquad a_{12} = a_{21} = -(Bs + K)$

and $\qquad a_{22} = [M_2 s^2 + Bs + (K_2 + K)] \qquad (2\text{-}80)$

Solution by determinants yields

$$X_1 = \frac{\begin{vmatrix} 0 & a_{12} \\ F & a_{22} \end{vmatrix}}{\begin{vmatrix} a_{11} & a_{12} \\ a_{21} & a_{22} \end{vmatrix}} = \frac{-a_{12} F}{a_{11} a_{22} - a_{12}^2} \qquad (2\text{-}81)$$

The transfer function in terms of system constants is found by inserting expressions for a_{11}, a_{22}, a_{12}, and a_{21}. Notice that since the mechanical system is linear and bilateral, $a_{12} = a_{21}$. The transfer function is

$$\frac{X_1}{F} = G_4(s)$$
$$= \frac{Bs + K}{M_1 M_2 s^4 + [B(M_1 + M_2)]s^3 + [M_1(K + K_2) + M_2(K + K_1)]s^2 + [B(K_1 + K_2)]s + [K_1 K_2 + K(K_1 + K_2)]} \qquad (2\text{-}82)$$

The block diagram for this system is shown in Fig. 2-27. It is important to note that the method of obtaining the transfer function is similar for both mechanical and electrical systems.

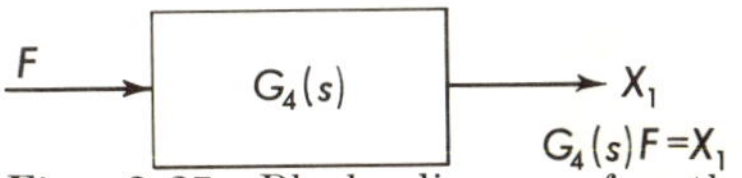

Analysis of an n-mass mechanical system follows a pattern similar to the above example. To obtain an electrical network analogy, the general equations are written with velocity $v = dx/dt$ as the dependent variable instead of displacement; whence, in the transformed equations, $V_i(s)$ appear as algebraic variables. By analogy with Eqs. (2-50) a set of force equations is written:

Fig. 2-27 Block diagram for the mechanical system of Fig. 2-23.

$$g_{11}V_1 + g_{12}V_2 + \cdots + g_{1n}V_n = F_1$$
$$g_{21}V_1 + g_{22}V_2 + \cdots + g_{2n}V_n = F_2$$
$$\cdots \cdots \cdots \cdots \cdots \cdots \cdots \cdots \cdots \cdots$$
$$g_{n1}V_1 + g_{n2}V_2 + \cdots + g_{nn}V_n = F_n$$

(2-83)

Forces F_i are driving functions, and quantities V_i are velocities. Since the variables are Laplace-transformed, they are capitalized. The mechanical self-admittances g_{ii} are of the form

$$g_{ii} = sM_{ii} + B_{ii} + \frac{K_{ii}}{s} \tag{2-84}$$

and the mechanical common admittances g_{ij} are of the form

$$g_{ij} = -\left(B_{ij} + \frac{K_{ij}}{s}\right) \tag{2-85}$$

Note the absence of sM_{ij} in the common-admittance term of Eq. (2-85). When all the variables are referred to one fixed coordinate system (inertial space), forces are transmitted to a particular mass only through springs and dashpots. Equations (2-83) are solved by determinants in a fashion similar to that utilized in loop analysis (Sec. 2-2):

$$V_j = \frac{1}{\Delta^M} \begin{vmatrix} g_{11} & g_{12} & \cdots & F_1 & \cdots & g_{1n} \\ g_{21} & g_{22} & \cdots & F_2 & \cdots & g_{2n} \\ \cdots & \cdots & \cdots & \cdots & \cdots & \cdots \\ g_{n1} & g_{n2} & \cdots & F_n & \cdots & g_{nn} \end{vmatrix} \tag{2-86}$$

where Δ^M is the determinant of the coefficients g_{ij}.

As another example, consider the mechanical system of Fig. 2-28. The positive direction is taken to the right for both x_1 and x_2. Newton's

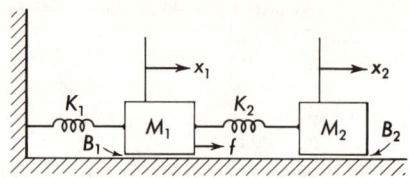

Fig. 2-28   Vibration-damper example.

force law equations for this translational system are

$$M_1 \frac{d^2x_1}{dt^2} + Kx_1 + K_2(x_1 - x_2) + B_1 \frac{dx_1}{dt} = f$$

$$M_2 \frac{d^2x_2}{dt^2} + K_2(x_2 - x_1) + B_2 \frac{dx_2}{dt} = 0 \qquad (2\text{-}87)$$

Equations (2-87) are put into the form of Eqs. (2-83) by writing them in terms of velocity rather than displacements and then rearranging:

$$M_1 \frac{dv_1}{dt} + (K_1 + K_2) \int_0^t v_1\, dt + Bv_1 - K_2 \int_0^t v_2\, dt = f$$

$$-K_2 \int_0^t v_1\, dt + M_2 \frac{dv_2}{dt} + K_2 \int_0^t v_2\, dt + B_2 v_2 = 0 \qquad (2\text{-}88)$$

These equations are Laplace-transformed with the result

$$g_{11} V_1 + g_{12} V_2 = F_1$$
$$g_{21} V_1 + g_{22} V_2 = F_2 \qquad (2\text{-}89)$$

where the transforms of the variables and driving functions are capitalized to indicate that they are functions of s.   Coefficients are

$$g_{11} = M_1 s + B_1 + \frac{K_1 + K_2}{s} \qquad g_{12} = g_{21} = \frac{-K_2}{s}$$

$$g_{22} = M_2 s + B_2 + \frac{K_2}{s} \qquad f_1 = F \qquad f_2 = 0 \qquad (2\text{-}90)$$

Figure 2-28 shows an interesting example of a vibration damper. For simplicity, taking $B_1 = 0 = B_2$ and finding the transfer function sX_1/F:

$$\frac{sX_1}{F} = \frac{Fg_{22}}{F\begin{vmatrix} g_{11} & g_{12} \\ g_{21} & g_{22} \end{vmatrix}} = \frac{g_{22}}{g_{11}g_{22} - g_{12}{}^2} \quad (2\text{-}91)$$

Inserting values from Eqs. (2-90) and simplifying yields

$$\frac{X_1}{F} = \frac{M_2 s^2 + K}{(M_1 M_2)s^4 + [M_1 K_2 + M_2(K_1 + K_2)]s^2 + K_1 K_2} \quad (2\text{-}92)$$

If f is a sinusoidal driving function (see Sec. 5-1), the output x_1 is also sinusoidal. The ratio of the complex-number representations of the sinusoidal input and output is found by replacing s by $j\omega$ in Eq. (2-92), with the result

$$\frac{X_1(j\omega)}{F(j\omega)} = \frac{K_2 - M_2\omega^2}{K_1 K_2 + \omega^4 M_1 M_2 - [M_1 K_2 + M_2(K_1 + K_2)]\omega^2} \quad (2\text{-}93)$$

The system is termed a vibration damper; for, when the angular input frequency ω is $\sqrt{K_2/M_2}$, the displacement x_1 is zero, even though a sinusoidal force is applied to the mass M_1.

2-8 Mechanical Systems; Rotational Motion. The method of obtaining differential equations for angular motion is essentially the same as for linear motion. Newton's law for angular motion can be stated: *The algebraic sum of torques acting upon an inertia is equal to the moment of inertia times the angular acceleration.* The procedure for analysis is outlined as follows:

1. Assume angles θ_i of rotation for each inertia. Take one direction, say, counterclockwise, as positive for θ_i, $d\theta_i/dt$, and $d^2\theta_i/dt^2$. Angular displacements are dependent variables, and torques are the driving functions.
2. Draw a free-body diagram and write an equation utilizing Newton's torque law for each inertia in the system.
3. Laplace-transform the differential or integrodifferential equations and, as in the preceding cases, solve the equations by taking a ratio of determinants.

The above procedure is applied to the rotational system of Fig. 2-29. The shafts connecting the inertias act as restoring springs. Free-body diagrams for this system are included in Fig. 2-30. As the moment of inertia J_1 is accelerated in the $\theta_1 > 0$ direction, the springs (shafts)

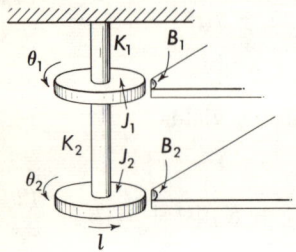

Fig. 2-29 An angular mechanical system.

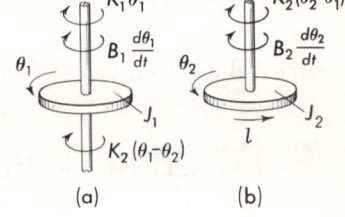

Fig. 2-30 Free-body diagram for a rotational system.

oppose the motion. Damping also causes a torque which opposes the motion. The equation for the inertia is

$$J_1 \frac{d^2\theta_1}{dt^2} = -K_1\theta_1 - K_2(\theta_1 - \theta_2) - B_1 \frac{d\theta_1}{dt} \qquad (2\text{-}94)$$

When Eq. (2-94) is transformed and rearranged, there results

$$(J_1 s^2 + B_1 s + K_1 + K_2)\Theta_1 + (-K_2)\Theta_2 = 0 \qquad (2\text{-}95)$$

In a similar fashion, the transform of the differential equation for the second moment of inertia is obtained from the free-body diagram of Fig. 2-30b. This second equation is written, in simplified form,

$$(-K)\Theta_1 + (J_2 s^2 + B_2 s + K_2)\Theta_2 = L \qquad (2\text{-}96)$$

The solution for the transform of the dependent variable, say, Θ_2, is found from Cramer's rule (the solution appears as the ratio of two determinants) as follows:

$$\Theta_2 = \frac{\begin{vmatrix} b_{11} & 0 \\ b_{21} & L \end{vmatrix}}{\begin{vmatrix} b_{11} & b_{12} \\ b_{21} & b_{22} \end{vmatrix}} = \frac{Lb_{11}}{b_{11}b_{22} - b_{12}^2} \qquad (2\text{-}97)$$

where
$$b_{11} = J_1 s^2 + B_1 s + K_1 + K_2$$
$$b_{22} = J_2 s^2 + B_2 s + K_2 \qquad (2\text{-}98)$$
$$b_{12} = b_{21} = -K_2$$

As in all preceding cases, the transfer function solution yields a ratio of polynomials in s. The transfer function is found by substituting from Eqs. (2-98) into Eq. (2-97).

The analysis of a rotational system containing n moments of inertias is generalized with the following equations:

$$\begin{aligned} h_{11}\Omega_1 + h_{12}\Omega_2 + \cdots + h_{1n}\Omega_n &= L_1 \\ h_{21}\Omega_1 + h_{22}\Omega_2 + \cdots + h_{2n}\Omega_n &= L_2 \\ \cdots\cdots\cdots\cdots\cdots\cdots\cdots\cdots\cdots \\ h_{n1}\Omega_1 + h_{n2}\Omega_2 + \cdots + h_{nn}\Omega_n &= L_n \end{aligned} \qquad (2\text{-}99)$$

The dependent variable is the angular velocity $\omega = d\theta/dt$, $(\Omega_i = \mathcal{L}\omega_i)$, and the driving functions are the torques L_i. Self-admittances h_{ii} are of the form

$$h_{ii} = sJ_{ii} + B_{ii} + \frac{K_{ii}}{s} \qquad (2\text{-}100)$$

and the common operational mechanical admittances h_{ij} are of the form

$$h_{ij} = -\left(B_{ij} + \frac{K_{ij}}{s}\right) \qquad (2\text{-}101)$$

Equations (2-99) are solved by determinants in a fashion similar to that utilized in loop analysis (Sec. 2-2):

$$\omega_j = \frac{1}{\Delta^M} \begin{vmatrix} h_{11} & h_{12} & \cdots & L_1 & \cdots & h_{1n} \\ h_{21} & h_{22} & \cdots & L_2 & \cdots & h_{2n} \\ \cdots & \cdots & \cdots & \cdots & \cdots & \cdots \\ h_{n1} & h_{n2} & \cdots & L_n & \cdots & h_{nn} \end{vmatrix} \qquad (2\text{-}102)$$

where Δ^M is the determinant of the coefficient h_{ij}.

2-9 Mechanical Coupling; Gear Trains. A gear train is a common method of coupling mechanical systems. Consider the mechanical system of Fig. 2-31, where a torque l drives a load through a gear train. The source has a moment of inertia J_1 and damping B_1; the load has a moment of inertia J_2, damping B_2, and spring constant K_2. Angles θ_1 and θ_2 are positions of the input and output shafts, respectively. Transformed equations for the system are

$$J_1 s^2 \Theta_1 = L - L_1 - B_1 s \Theta_1 \qquad (2\text{-}103)$$

and
$$J_2 s^2 \Theta_2 = L_2 - B_2 s \Theta - K_2 \Theta_2 \qquad (2\text{-}104)$$

Angles θ_1 and θ_2 are related by the number of gear teeth; therefore, only one dependent variable exists in this problem. The relation between θ_1 and θ_2 and between l_1 and l_2 must be found. Let r_1 be the radius of

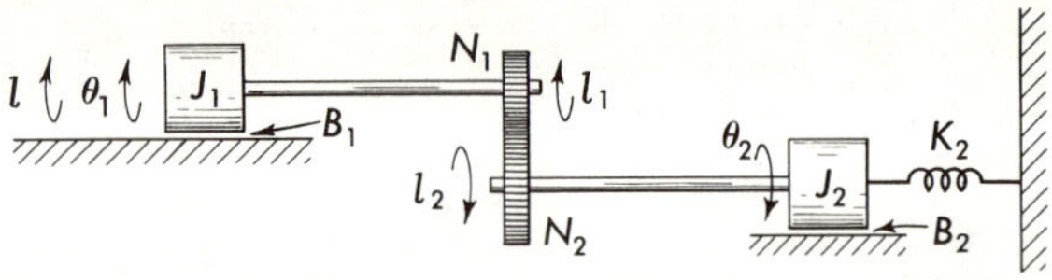

Fig. 2-31 Gear train used to couple mechanical systems.

the first gear and r_2 be the radius of the second gear; then, since the linear distance traveled in a given time interval along the surface of each gear by a point in contact with both at some instant must be the same, $\theta_1 r_1 = \theta_2 r_2$. The number of teeth on the gear surface is proportional to the gear radius; therefore,

$$\theta_1 N_1 = \theta_2 N_2 \qquad (2\text{-}105)$$

where N_1 is the number of teeth on the input gear and N_2 is the number of teeth on the load gear. The equation may be written

$$\theta_2 = \frac{N_1}{N_2} \theta_1 \qquad (2\text{-}106)$$

Since the rates at which work is done by the two gears are equal,

$$l_1 \frac{d\theta_1}{dt} = l_2 \frac{d\theta_2}{dt} = l_2 \frac{d}{dt} \theta_1 \frac{N_1}{N_2} = l_2 \frac{N_1}{N_2} \frac{d\theta_1}{dt}$$

and canceling $d\theta_1/dt$,

$$l_1 = \frac{N_1}{N_2} l_2 \qquad (2\text{-}107)$$

If N_1/N_2 is less than 1, the angular displacement of the load is less than the angular displacement of the source (if $\theta_1 = \theta_2 = 0$ at $t = 0$), whereas the torque applied to the load is greater than the torque applied by the source. Solution of Eq. (2-104) for L_2 by substituting $\theta_2 = (N_1/N_2)\theta_1$ yields

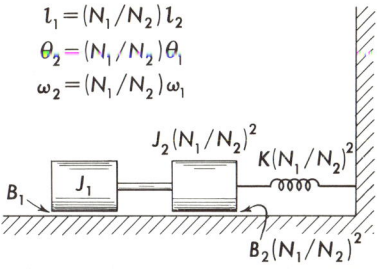

$$L_2 = \frac{L_1}{N_1/N_2}$$

$$= (J_2 s^2 + B_2 s + K_2)\frac{N_1}{N_2}\Theta_1 \quad (2\text{-}108)$$

Fig. 2-32 Reflecting inertia, spring, and damping constants across a gear train.

The complete expression for transform of torque L in terms of Θ_1 is found by substituting from Eq. (2-108) into Eqs. (2-103) and (2-104); thus

$$L = (J_1 s^2 + B_1 s)\Theta_1 + L_2\frac{N_1}{N_2}$$

$$L = \left[J_1 + J_2\left(\frac{N_1}{N_2}\right)^2\right]s^2 + \left[B_1 + B_2\left(\frac{N_1}{N_2}\right)^2\right]s + K_2\left(\frac{N_1}{N_2}\right)^2 \Theta_1 \quad (2\text{-}109)$$

Figure 2-32 summarizes the effect of a gear train by referring the moment of inertia, damping constant, and spring constant to the source side of the gear train.

It is helpful to note the similarity in relationships for an ideal transformer and for a gear train. In a transformer with a turns ratio of $N_1:N_2$ (step-down) the transforms of the voltages and currents and the operational impedances are related by

$$E_1 = \frac{N_1}{N_2}E_2 \qquad I_1 = \frac{N_2}{N_1}I_2$$

$$Z_1 = \left(\frac{N_1}{N_2}\right)^2 Z_2 = \left(\frac{N_1}{N_2}\right)^2 \left(Ls + R + \frac{1}{sC}\right) \quad (2\text{-}110)$$

Symbols are defined in Fig. 2-33. In a gear train, with a gear ratio of $N_1:N_2$, the transforms of the torques and velocities and the operational

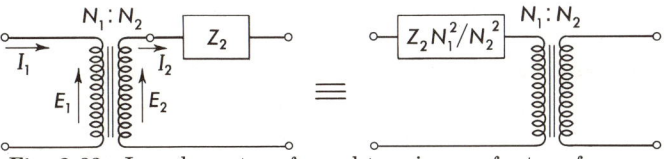

Fig. 2-33 Impedance transformed to primary of a transformer.

mechanical impedances are related by

$$L_1 = \frac{N_1}{N_2} L_2$$

$$\omega_1 = \frac{N_2}{N_1} \omega_2 \tag{2-111}$$

$$H_1 = \left(\frac{N_1}{N_2}\right)^2 h_2 = \left(\frac{N_1}{N_2}\right)^2 \left(J_2 s + B_2 + \frac{K_2}{s}\right)$$

2-10 Electromechanical Networks. Numerous electromechanical devices are encountered in engineering and scientific applications. Solenoids, actuators, motors, generators, gyroscopes, accelerometers, and loudspeakers are just a few of many. Although the performances of these devices exhibit nonlinear characteristics, the corresponding characteristic curves are linearized as a first approximation. Steps for solving such a system are as follows:

1. Determine or plot a characteristic curve which describes the physical behavior of the electromechanical coupling.
2. Linearize this characteristic curve (or curves) over the region of operation desired and write an analytical equation expressing it.
3. Therewith write the remaining system equations of performance and solve in a manner similar to that used for electrical and mechanical circuits.

In many control systems it is necessary to find the transfer function for a combination of electrical and mechanical components. As an electromechanical system example, consider the loudspeaker shown in the photograph of Fig. 2-34a and in the schematic diagram of Fig. 2-34b. The transfer function X/E is desired. The problem is solved by consideration of the coupling between the electrical and mechanical circuits.

As the mass moves in the coil, a back voltage is generated as follows:

$$e_b = k_1 \frac{dx}{dt} = k_1 v \tag{2-112}$$

k_1 is assumed to be constant, at least through the range of operation considered (which is usually quite small). It is a function of the number of turns of wire wound on the coil and the magnetic field strength.

As current is applied to the coil, a field which interacts with the field of the permanent magnet is created; and a force is developed on the

Obtaining the System Differential Equation   67

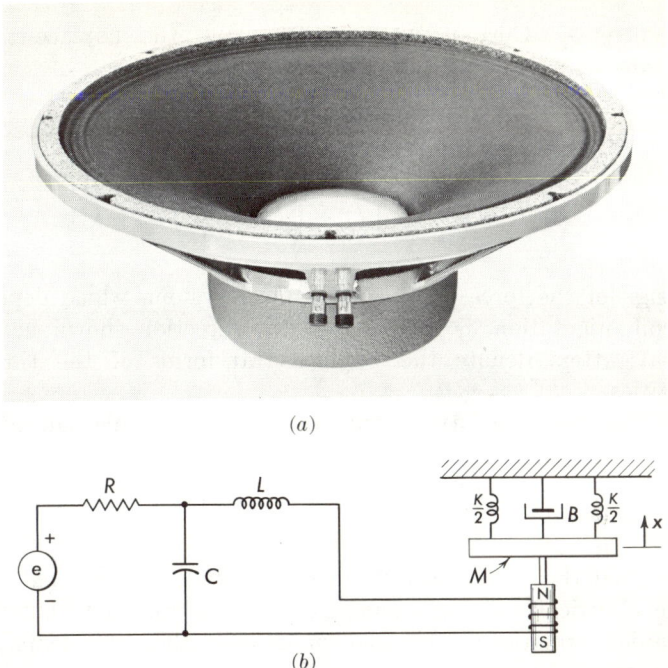

Fig. 2-34 (a) Photograph of a loudspeaker. *(Courtesy of J. B. Lansing Sound, Inc., Los Angeles, Calif.)* (b) An equivalent circuit for a loudspeaker.

mass. This force, proportional to the applied current, is expressed as follows:

$$F_e = k_2 i \qquad (2\text{-}113)$$

Again, k_2 is assumed to be constant and depends upon the number of turns on the coil and the magnetic field strength. In a consistent set of units, the two constants are numerically equal ($k_1 = k_2$).

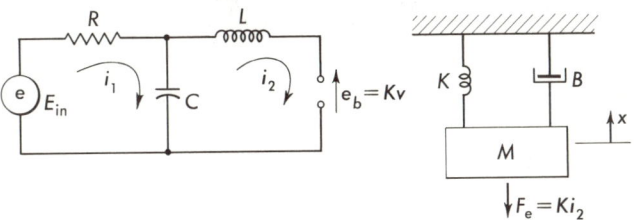

Fig. 2-35 Two parts of equivalent circuit for a loudspeaker.

The circuit is separated into two parts, as shown in Fig. 2-35, and solved. The electric circuit is solved with loop analysis by using two loop currents. Currents are assumed, and the Laplace-transformed equations, as written for zero initial conditions, are

$$E_{in} = \left(R + \frac{1}{sC}\right)I_1 + \left(-\frac{1}{sC}\right)I_2$$

$$0 = \left(-\frac{1}{sC}\right)I_1 + \left(Ls + \frac{1}{sC}\right)I_2 + k_1 V \quad (2\text{-}114)$$

The sign of the force on the mechanical system, which depends upon the coil orientation, is assumed in the direction shown in Fig. 2-35. Capital letters denote the Laplace transforms of the time-variable quantities.

The Laplace transform of the equation for the mechanical circuit is

$$0 = (0)I_1 + (k_2)I_2 + \left(Ms + B + \frac{K}{s}\right)V \quad (2\text{-}115)$$

Notice that the voice coil introduces a voltage proportional to velocity in the electrical equations and a force proportional to current in the mechanical equations. The determinant of the coefficients is written from Eqs. (2-114) and (2-115) as follows:

$$\Delta = \begin{vmatrix} R + \frac{1}{sC} & -\frac{1}{sC} & 0 \\ -\frac{1}{sC} & Ls + \frac{1}{sC} & k_1 \\ 0 & k_2 & Ms + B + \frac{K}{s} \end{vmatrix} \quad (2\text{-}116)$$

Expansion gives

$$\Delta = \left(R + \frac{1}{sC}\right)\left[\left(Ls + \frac{1}{sC}\right)\left(Ms + B + \frac{K}{s}\right) - k_1 k_2\right]$$
$$- \frac{1}{s^2 C^2}\left(Ms + B + \frac{K}{s}\right) \quad (2\text{-}117)$$

The transfer function is

$$\frac{X}{E_{in}} = \frac{V/s}{E_{in}} = \frac{1}{s\Delta}\left(-\frac{1}{sC}k_2\right) = -\frac{k_2}{s^2 C \Delta} \quad (2\text{-}118)$$

where Δ is given by Eq. (2-117). Substituting for Δ and clearing fractions gives a ratio of polynomials in s. The time solution for a particular driving function E_{in} is found by substituting for E_{in} in Eq. (2-118), solving for X, and inverting.

2-11 Analogies. Equations of motion are derived similarly for all lumped-parameter linear systems, whether the systems are electrical, mechanical, or electromechanical. The procedures, the use of basic laws, and the solutions are much the same. Because there are system similarities, it is possible to construct electrical analogies, both nodal and loop, for mechanical circuits. The electrical analogue is found by deriving an electrical network having a differential equation of the same form as that of the mechanical system. Although the analogous electric circuit is often determined intuitively, the following procedure is suggested:

1. Write the mechanical system equations.
2. Substitute, as discussed below, electrical quantities using electrical-network constants and variables.
3. Interpret these equations to yield the analogue network.

a. Nodal Analogue. To obtain a nodal analogue for a mechanical equation, compare the operational admittance of the mechanical circuit [Eqs. (2-83)],

$$\left(sM_{ii} + B_{ii} + \frac{K_{ii}}{s}\right)V_i \cdots = F_i \qquad (2\text{-}119)$$

or in Eqs. (2-99),

$$\left(sJ_{ii} + B_{ii} + \frac{K_{ii}}{s}\right)\Omega_i \cdots = L_i \qquad (2\text{-}120)$$

with the operational admittance in Eqs. (2-50),

$$\left(sC_{ii} + \frac{1}{R_{ii}} + \frac{1}{sL_{ii}}\right)V_i \cdots = I_i \qquad (2\text{-}121)$$

Comparison of the coefficients of these equations indicates the analogous quantities listed in the following table:

Mass M or moment of inertia J.......... Analogous to capacitance C
Damping constant B.................... Analogous to electrical conductance $1/R$
Spring constant K..................... Analogous to inverse inductance $1/L$
Forces or torques....................... Analogous to current
Velocity............................... Analogous to voltage

Summarizing,

$$\left.\begin{array}{c}M\\J\end{array}\right\} \sim C \quad B \sim \frac{1}{R} \quad K \sim \frac{1}{J} \quad \left.\begin{array}{c}F\\L\end{array}\right\} \sim I \quad \left.\begin{array}{c}V\\\Omega\end{array}\right\} \sim E \quad (2\text{-}122)$$

b. Loop Analogue. Comparison of Eqs. (2-119) and (2-120) with Eq. (2-18) indicates

$$\left(sL_{ii} + R_{ii} + \frac{1}{sC_{ii}}\right)I_i \cdots = E_i \quad (2\text{-}123)$$

which yields the analogies listed in the following table:

Mass M or moment of inertia J . . . . Analogous to inductance L
Damping constant B . . . . Analogous to resistance R
Spring constant K . . . . Analogous to inverse capacitance $1/C$
Forces or torques. Analogous to voltages
Velocity. . . . . Analogous to current

Summarizing,

$$\left.\begin{array}{c}M\\J\end{array}\right\} \sim L \quad B \sim R \quad K \sim \frac{1}{C} \quad \left.\begin{array}{c}F\\L\end{array}\right\} \sim E \cdot \left.\begin{array}{c}V\\\Omega\end{array}\right\} \sim I \quad (2\text{-}124)$$

As might be noted from Sec. 2-11a and b, there is a possibility of more than one electrical analogue existing for a given mechanical circuit.

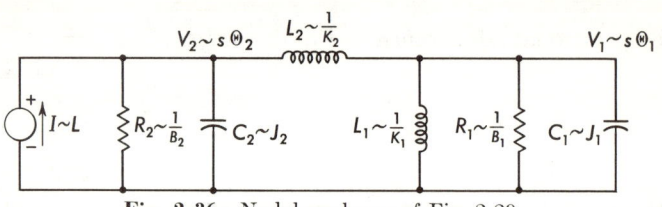

Fig. 2-36 Nodal analogue of Fig. 2-29.

As an example, an analogue for the mechanical system of Fig. 2-29 is found. Figure 2-36 shows the nodal analogue which is found from Eqs. (2-122). Equations (2-124) are used to find the loop analogue of Fig. 2-37. Usually electric networks have "duals."* The two networks

* Additional information on duality of electric networks is found in Ref. 13, chap. 2, sec. 15.

Obtaining the System Differential Equation 71

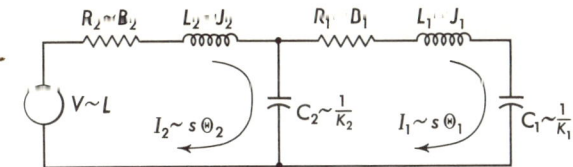

Fig. 2-37 Loop analogue of Fig. 2-29.

thus derived are duals, since node pair voltages in Fig. 2-36 are analogous to loop currents in Fig. 2-37.

2-12 Separable and Nonseparable Networks. Block diagrams and transfer functions are found for most of the examples in this chapter. The input to the block is "multiplied" or operated upon by the quantity in the block to equal the output.

In general it is desirable, from the point of view of simplicity, to write transfer functions for small sections of the system. A common error

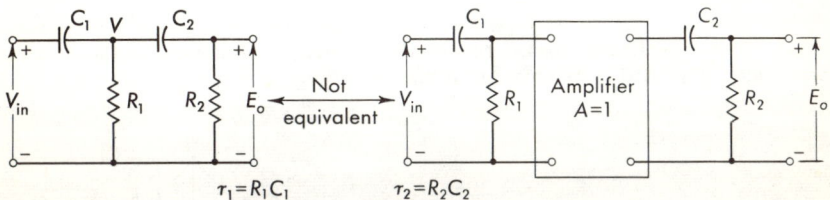

Fig. 2-38 Nonseparable networks.

in writing transfer functions results when networks or components are analyzed separately although, in actuality, the network must be considered as an entity. For example, consider the network of Fig. 2-38. The student is often tempted to write this network as two blocks, each with the same overall transfer function. This leads to

$$\frac{E_o}{E_{in}} = \cancel{\frac{(\tau_1 s)(\tau_2 s)}{(\tau_1 s + 1)(\tau_2 s + 1)}} \quad \cancel{\frac{\tau_1 \tau_2 s^2}{\tau_1 \tau_2 s^2 + (\tau_1 + \tau_2)s + 1}} \quad (2\text{-}125)$$

The transfer function for this network is correctly found by solving two nodal equations:

$$V\left[s(C_1 + C_2) + \frac{1}{R_1}\right] + E_o(-sC_2) = sC_1 V_{in}$$

$$V(-sC_2) + E_o\left(sC_2 + \frac{1}{R_2}\right) = 0 \quad (2\text{-}126)$$

The transfer function is found by solving these equations:

$$\frac{E_o}{V_{in}} = \frac{s^2\tau_1\tau_2}{s^2\tau_1\tau_2 + s(\tau_1 + a\tau_2) + 1} \qquad (2\text{-}127)$$

where $\tau_1 = R_1C_1$, $\tau_2 = R_2C_2$, and $a = 1 + R_1/R_2$. Notice in this case that the equations, though somewhat similar in form, actually differ by the inclusion of the factor a in Eq. (2-127).

Generally, two networks can be separated if the second network does not load the first. To enable separation of the networks, the magnitude of the output impedance of the first network must be much smaller than the magnitude of the input impedance of the second network (at least by a factor of 10). A vacuum tube provides good isolation, since the complex-number input impedance of a tube is usually high. Because of the loading effects of transistors (as pointed out earlier) similar isolation is not always provided.

PROBLEMS

2-1. Write the Laplace transforms of the differential equations for the electrical networks of Fig. 2P-1. Assume zero initial conditions.

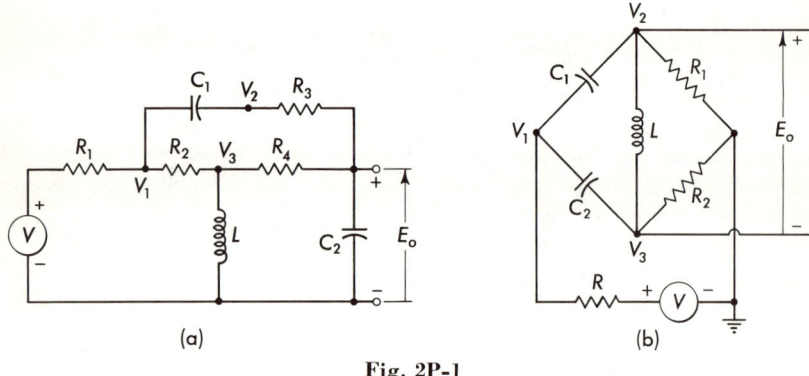

Fig. 2P-1

2-2. Write Laplace-transformed equations for the networks of Fig. 2P-2. Assume zero initial conditions.
Ans. (a)

$$I_1\left(R_0 + \frac{1}{sC_1}\right) + I_2\left(-\frac{1}{sC_1}\right) + I_3(0) = V$$

$$I_1\left(-\frac{1}{sC_1}\right) + I_2\left(sL_1 + R_1 + \frac{1}{sC_1}\right) + I_3(-sM) = 0$$

$$I_1(0) + I_2(-sM) + I_3\left(sL_2 + R_2 + \frac{1}{sC_2}\right) = 0$$

Obtaining the System Differential Equation 73

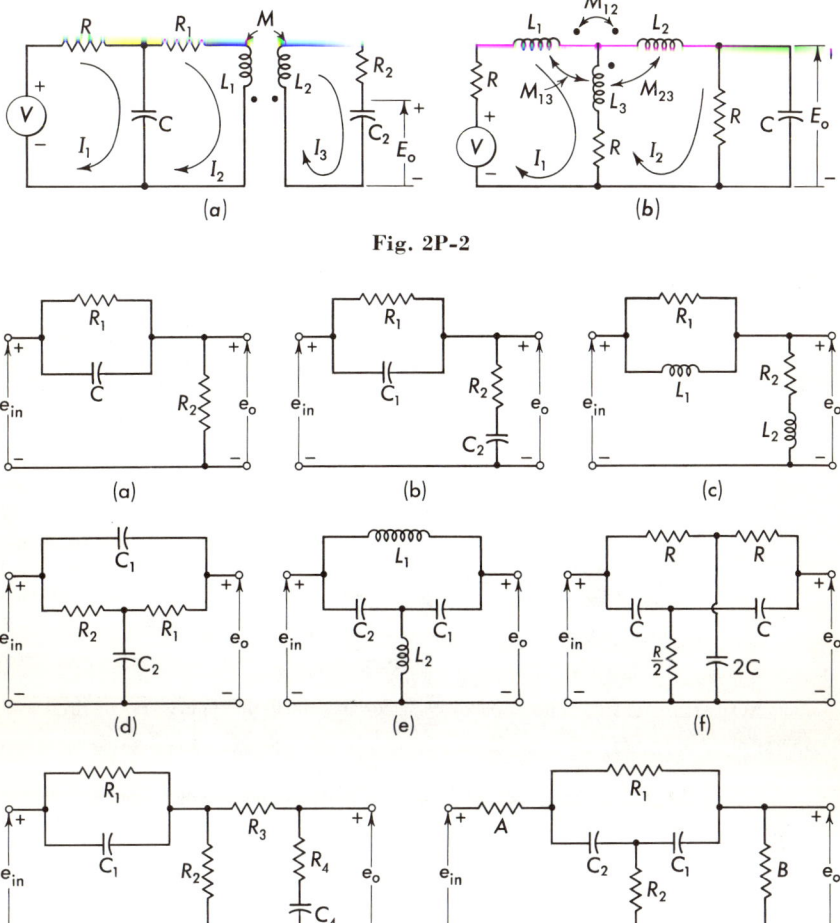

Fig. 2P-2

Fig. 2P-3

2-3. Write transfer functions for the electrical networks of Fig. 2P-3. All networks are driven from zero source impedance and into infinite load impedance. Assume zero initial conditions.

Ans. (a)

$$\frac{E_o}{E_{in}} = \frac{s + 1/a\tau_a}{s + 1/\tau_a} \quad \text{where } a = 1 + \frac{R_1}{R_2}, \ \tau_a = \frac{R_1 R_2 C}{R_1 + R_2}$$

Ans. (d)

$$\frac{E_o}{E_{in}} = \frac{s^2/\omega_0^2 + rns/\omega_0 + 1}{s^2/\omega_0 + ns/\omega_0 + 1}$$

$$\omega_0 = R_1 R_2 C_1 C_2 \qquad r = \frac{C_1(R_1 + R_2)}{C_1(R_1 + R_2) + C_2 R_2} \qquad n = \frac{C_1(R_1 + R_2) + C_2 R_2}{\omega_0}$$

Ans. (f)
$$\frac{E_o}{E_{in}} = \frac{\tau^2 s^2 + 1}{\tau^2 s^2 + 4\tau s + 1} = RC$$

2-4. Find the transfer function E_o/E_{in} and block diagram for the network of Prob. 2-2b.

2-5. Write Laplace-transformed differential equations for the mechanical networks of Fig. 2P-5. Assume zero initial conditions.

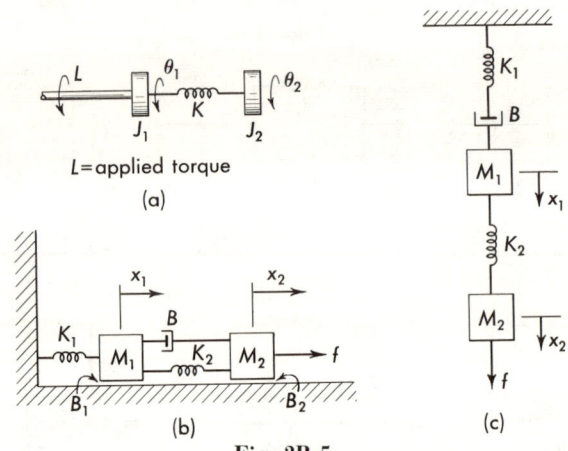

Fig. 2P-5

2-6. Write transfer functions X_2/F or Θ_2/L for the mechanical networks of Prob. 2-5.
Ans. (c)
$$\frac{X_1}{V} = \frac{(K_2/s)(Bs + 2K)}{(M_1 s^2 + Bs + K)(M_2 s^2 + Bs + 2K + K_2) - (Bs + 2K)^2}$$

2-7. Write the Laplace-transformed differential equations for the mechanical system shown in Fig. 2P-7. $v(t)$ is a velocity input driving function. Assume zero initial conditions. Find the transfer function
$$\frac{X_1}{V} = G_1(s)$$

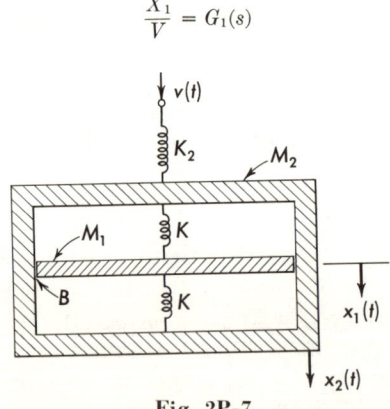

Fig. 2P-7

2-8. Find the transfer function for the fluid coupling shown in Fig. 2P-8,

$$\frac{\Theta_3}{L} = G_2(s)$$

Assume zero initial conditions.

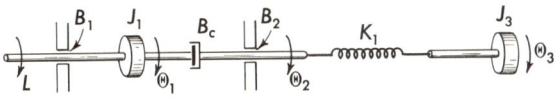

Fig. 2P-8

2-9. Find the transfer function for the electromechanical system of Fig. 2P-9. Assume linear torque-speed curves for the motor. The potentiometer rotates through 10 turns. e_{in} is a zero impedance source.

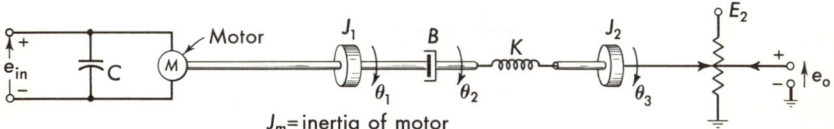

Fig. 2P-9

2-10. A torsion rod is frequently used to measure the effective moment of inertia of gear trains and inertia loads. Such a rod is fitted as shown in Fig. 2P-10a. When the system is disturbed, the time required for 20 cycles of oscillation is found to be 10 sec. The torsion rod is then coupled to a shaft of a gear train and load moment of

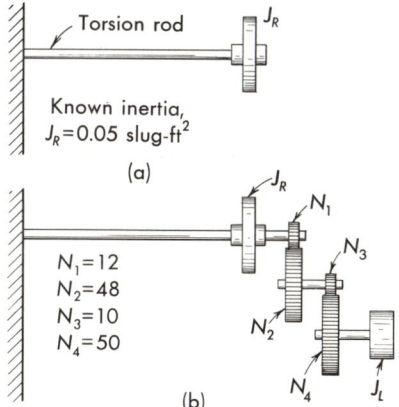

Fig. 2P-10

inertia as shown in Fig. 2P-10b. When the system is shocked, the time required for 10 cycles of oscillation is found to be 18 sec. What is the effective moment of inertia of the combined gear train and load moment of inertia? Assume $\zeta \ll 1$ for both cases and assume all shafts rigid with the exception of the torsion rod.

Ans. 0.648 slug-ft²

2-11. Find the transfer function E_o/Y of the system of Fig. 2P-11 used to measure high-frequency displacements of a shake table. The potentiometer measures the difference in displacement between M_1 and M_2. The driving function is a position

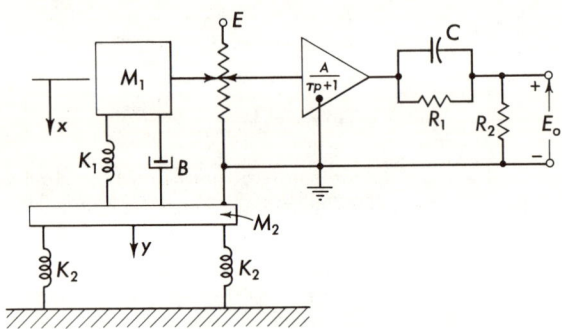

Fig. 2P-11

$y(t)$. Ignore the acceleration of gravity, and assume infinite input and zero output impedance of the amplifier. The maximum travel of the linear motion potentiometer is d. Assume zero initial conditions.

Ans.

$$\frac{E_o}{Y} = \frac{-AER_2Ms^2(sRC+1)}{l(R_1+R_2)(M_1s^2+Bs+K_1)(\tau s+1)\left(sC\frac{R_1R_2}{R_1+R_2}+1\right)}$$

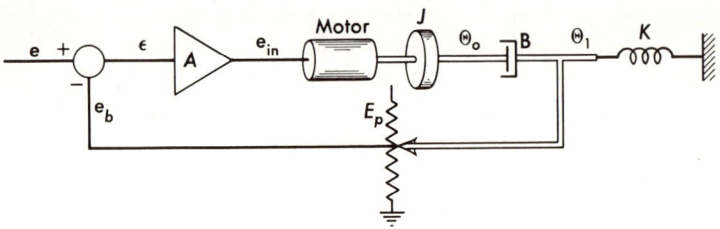

Fig. 2P-12

2-12. Find the transfer function Θ_1/E of the system of Fig. 2P-12.

$ke_{in} = Ms\Theta_0 + L$

θ_{max} = maximum travel of potentiometer

ω_0 = no-load speed

L_0 = stall torque

E_o = rated control volts

2-13. Find the transfer function $X(s)/F(s)$ for the system in Fig. 2P-13.

$$\text{Ans.} \quad \frac{X}{F} = \frac{-l_1/l_2}{Ms^2 + Bs + K}$$

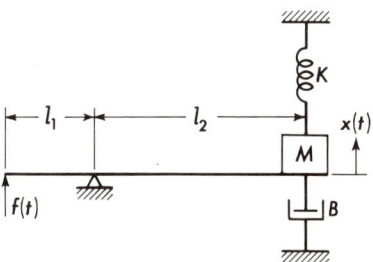

Fig. 2P-13

2-14. Find the transfer function $G(s) = Y(s)/E(s)$ for the system in Fig. 2P-14. Assume back-emf constant k_1 and force constant k_2 as discussed in Sec. 2-10.

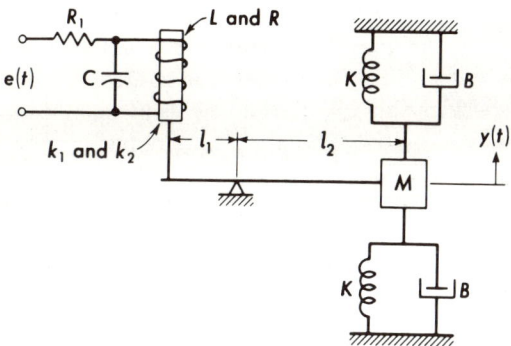

Fig. 2P-14

2-15. Find the transfer function E_2/E_1 for the system in Fig. 2P-15.

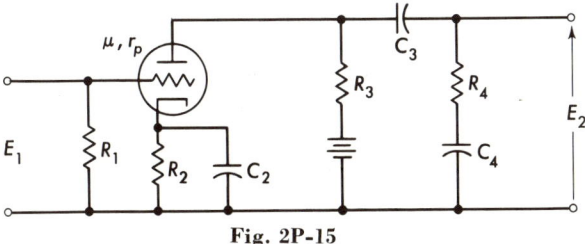

Fig. 2P-15

chapter 3

STEADY-STATE ERRORS

3-1 The Steady-state Component. Solution of a linear differential equation having constant coefficients can be divided into two parts: the transient and the steady-state components. In this chapter the steady-state solutions are found for particular types of inputs. In the case of feedback control systems it is desirable to find the steady-state error which results for a particular type of input. The steady-state error resulting from the difference between the steady-state values of the input and the output variables is one of the measures of control-system performance.

The design specification for a feedback control system may include noise reduction, sensitivity, stability, low-frequency errors, accuracy, steady-state errors, and many others.* Starting points in basic systems design are usually the steady-state error and the stability requirements. Stability information is derived from the transient solution or weighting function (i.e., the response to an applied impulse). Chapters 4 and 5 treat the stability problem. Steady-state errors that result for various types of inputs and load disturbances are discussed in this chapter.

3-2 Steady-state Errors Due to Input Disturbances. To determine the nature of system stability, the characteristic equation alone is necessary. However, when investigating steady-state errors (the steady-state solutions of differential equations), it is necessary to know the type of driving function applied to the system. Three types of input driving function are considered in this text; they are usually adequate for basic control design applications. They are a step function in position, a ramp function in position (a step function in velocity), and a

---

* These terms are defined at various points in the text.

step function in acceleration, respectively denoted analytically as

$$r = A \tag{3-1}$$

$$r = vt \tag{3-2}$$

$$r = \tfrac{1}{2}at^2 \tag{3-3}$$

where A, v, and a are constants. The functions are applied at $t = 0$ and are valid for positive t only. The system is assumed to be stable when deriving the steady-state errors.

Consider the steady-state error that results if a step function of position is applied. The system under consideration is shown in the block diagram of Fig. 3-1: $G(s)$, the forward loop gain, is normally written in the following "standard form":

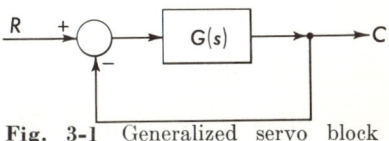

Fig. 3-1 Generalized servo block diagram for $H = 1$.

$$G(s) = K \frac{(1 + \tau_1 s)(1 + \tau_3 s) \cdots}{s^n (1 + \tau_2 s)(1 + \tau_4 s)}$$

where K is a constant independent of s and $G(s)$ is a function of s. As a first case, take the feedback transfer function $H(s)$ equal to unity. In this case the transform of the steady-state error, the difference between the desired output R and the actual output C, is

$$\varepsilon = R - C \tag{3-4}$$

and is the actuating signal. Reference to Eq. (1-63), where $H = 1$, yields the transform of the error ε in terms of input R,

$$\varepsilon = \frac{1}{1 + G(s)} R \tag{3-5}$$

Equation (3-5) may be written alternatively as

$$[1 + G(s)]\varepsilon = R \tag{3-6}$$

If r is equal to the constant A, as given by Eq. (3-1), the steady-state component is also a constant if the system is stable. From a physical point of view, if the input r to the position system of Chap. 1 is a constant, then, after all transient terms die out, the output c is a constant.

All output derivatives, velocity, acceleration, etc., are zero. In the steady state the error will also be constant. (This constant can be zero.)

The final-value theorem, discussed in Appendix I, affords a simple means of finding steady-state errors. This theorem is written

$$\lim_{t \to \infty} y(t) = \lim_{s \to 0} sY(s) \tag{3-7}$$

provided

1. $Y(s) = \mathcal{L}y(t)$.
2. $y(t)$ is the response of a stable system [i.e., all poles of $Y(s)$ lie in the left half plane].

Notice that the value of the time function at infinity is found directly from the transformed function at $s = 0$.

As an example, suppose it is desired to find the steady-state ($t \to \infty$) value for the function $\epsilon(t)$ if the transformed function has the form

$$\varepsilon(s) = \frac{1}{1 + G(s)} R(s) \tag{3-8}$$

and the system transfer function has the form

$$G = \frac{\frac{5}{3}K(s+3)(s+2)}{(s+10)(s^2+s+1)} \tag{3-9}$$

where, for simplicity in this equation and elsewhere in the book, G is written for $G(s)$. In this case, the limit of G as s approaches zero is the constant K_j, or, stated mathematically,

$$\lim_{s \to 0} G = K \tag{3-10}$$

Since the Laplace transform of a step function of amplitude A is $R = A/s$, the steady-state error for a unit step function applied to a closed-loop system is

$$\lim_{t \to \infty} \epsilon(t) = \lim_{s \to 0} s \frac{A/s}{1+G} = \frac{A}{1 + \lim_{s \to 0} G} \tag{3-11}$$

When Eq. (3-10) is substituted into Eq. (3-11),

$$\lim_{t \to \infty} \epsilon(t) = \frac{A}{1+K} \tag{3-12}$$

Equation (3-12) is presented pictorially in Fig. 3-2. When a step in position is applied to the system, the output tends to follow this step; however, a steady-state position difference ϵ_0 between r and c results. This difference can be reduced by increasing the system gain K. Although the error can be made small, it will always be present, since infinite gain ($K \rightarrow \infty$) is not practical.

This example shows how the steady-state error is found in a feedback control system. Most control systems can be classified by a few types according to the form of the open-loop transfer function G. The steady-state errors resulting from step inputs in position, velocity, and acceleration are found for each type of system. As a result, error constants are derived, and the steady-state system errors are readily found.

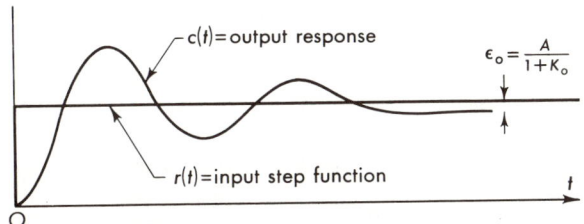

Fig. 3-2   Step-function response for a type 0 system.

3-3 Classification of Feedback Control Systems According to Type. In order to classify a system by its type, consider the block diagram given in Fig. 3-1. The quantity G in this diagram is an operational transfer function appearing in the forward loop. The feedback function H is taken equal to 1. The transfer function relating error to R is given by the expression

$$\varepsilon = \frac{R}{1+G} \tag{3-13}$$

The quantity G, termed the "open-loop transfer function," is found by multiplying all the transfer function operators cascaded in the loop. It is of the form typified by the transfer function

$$G = \frac{K(\tau_1 s + 1)(\tau_3 s + 1)}{s^n(\tau_2 s + 1)(\tau_4 s + 1)} \tag{3-14}$$

That is, G is the ratio of two factored polynomials. All values of τ are time constants for various components. If one should go through a system and compute the transfer functions for each of the items cascaded in a loop, the product is the open-loop transfer function. When these transfer functions are multiplied and all terms are collected, there will usually be a factor $(s - 0)^n = s^n$ in the denominator.

This factor in the denominator is due to poles at the origin of the open-loop transfer function. Values of s ($s_i = -1/\tau_i$, i even) that make the denominator of G zero are termed "poles." Values of s ($s_i = -1/\tau_i$, i odd) that make the numerator zero are termed "zeros."

Owing to the form of this transfer function, the system has a definite steady-state response to various types of input depending upon the value of s. Hence, the steady-state error for various types of input is specified by knowing the value of n [the exponent of s in the denominator of Eq. (3-14)]. It is convenient, then, to classify feedback control systems in terms of n in Eq. (3-14). This is done here before discussing other steady-state conditions in order to simplify terminology.

Type 0 system, one for which $n = 0$ in Eq. (3-14)
Type 1 system, one for which $n = 1$ in Eq. (3-14)
Type 2 system, one for which $n = 2$ in Eq. (3-14)
Type 3 system, one for which $n = 3$ in Eq. (3-14)

3-4 Static-error Constants.[38] Based upon a system which has unity feedback ($H = 1$), static-error constants are defined as

$$\text{Position-error constant} = K_0 = \lim_{s \to 0} G(s) \quad (3\text{-}15)$$

$$\text{Velocity-error constant} = K_v = \lim_{s \to 0} sG(s) \quad (3\text{-}16)$$

$$\text{Acceleration-error constant} = K_a = \lim_{s \to 0} s^2 G(s) \quad (3\text{-}17)$$

$G(s)$ is the forward-loop transfer function as shown in Fig. 3-1. For the case of $H = 1$, the inverse transform of the actuating signal

$$\varepsilon(s) = R - C$$

is the error.

Each of the inputs of Eqs. (3-1) to (3-3) is applied to a system whose transfer function is given by Eq. (3-14). In this manner the significance of the various error constants becomes apparent.

Position-error Constant. The transform of the error, again corresponding to the transform of the actuating signal for the case $H = 1$, is given by

$$\varepsilon(s) = \frac{R(s)}{1 + G} \quad (3\text{-}18)$$

To find the steady-state or final value of $\epsilon(t)$ when $r(t)$ is a step function in position, take the limit as t approaches infinity. From the final-value theorem,

$$\epsilon_0 = \lim_{t \to \infty} \epsilon(t) = \lim_{s \to 0} s \frac{R(s)}{1 + G(s)} \tag{3-19}$$

But $R(s) = A/s$ for a step function. This term cancels with the s in the numerator:

$$\epsilon_0 = \lim_{t \to \infty} \epsilon(t) = \frac{A}{1 + \lim_{s \to 0} G(s)} = \frac{A}{1 + K_0} \tag{3-20}$$

where K_0 is the position constant. For a type 0 system ($n = 0$)

$$K_0 = K \tag{3-21}$$

K_0 is the position constant and, as indicated in Eq. (3-21), is of finite value for $n = 0$.

For type 1 system (or higher, $n \geq 1$)

$$\frac{\epsilon_0}{A} = \lim_{s \to 0} \frac{1}{1 + \dfrac{K(\tau_1 s + 1)(\tau_2 s + 1)}{s^n(\tau_2 s + 1)(\tau_4 s + 1)}} = 0 \tag{3-22}$$

The steady-state error in response to a step function for systems of type 1 or greater is equal to zero. The time response to step function input for systems with $n \geq 1$ has zero steady-state error. Therefore, the difference between r and c (the error) is zero. The time response for $n = 0$ systems is that of the example given in Fig. 3-2. The ratio of position error to magnitude of the applied step function is $1/(1 + K_0)$.

Velocity-error Constant. Consider next the response to systems with a ramp input (for $r = vt$). Again the transfer function is written and the final-value theorem used,

$$\lim_{t \to 0} \epsilon(t) = \lim_{s \to 0} s \frac{v/s^2}{1 + G(s)} \tag{3-23}$$

where $R(s) = v/s^2$ is the Laplace transform of $r(t) = vt$. Equation (3-23) is simplified as follows:

$$\lim_{t \to \infty} \frac{\epsilon(t)}{v} = \lim_{s \to 0} \frac{1}{s + sG(s)} = \frac{1}{\lim_{s \to 0} sG} = \frac{1}{K_v} \tag{3-24}$$

This limit does not exist for $n = 0$; that is, a type 0 system has infinite output, since, setting $n = 0$, we have

$$\frac{\epsilon_v}{v} = \frac{1}{\lim_{s \to 0} \dfrac{K(\tau_1 s + 1)(\tau_3 s + 1)}{(\tau_2 s + 1)(\tau_4 s + 1)}} = \frac{1}{\lim_{s \to 0} s} \to \infty \quad (3\text{-}25)$$

For a type 1 system, $n = 1$ and Eq. (3-24) has a constant, finite value given by

$$\frac{\epsilon_v}{v} = \frac{1}{\lim_{s \to 0} \dfrac{K(\tau_1 s + 1)(\tau_3 s + 1)}{s(\tau_2 s + 1)(\tau_4 s + 1)}} = \frac{1}{K} \quad (3\text{-}26)$$

The expression ϵ_v/v is the ratio of position error to input velocity and is defined by $1/K_v$, as follows:

$$\frac{1}{K_v} = \frac{\epsilon_v}{v} \quad \text{thus } K_v = \frac{v}{\epsilon_v} \quad (3\text{-}27)$$

Comparison of Eq. (3-27) with (3-26) indicates that the velocity constant K_v is given by the limit of sG as s approaches zero, or

$$K_v = \lim_{s \to 0} sG \quad (3\text{-}28)$$

where $K_v = 0$ for $n = 0$, $K_v = K$ for $n = 1$, and $K_v = \infty$ for $n \geq 2$.

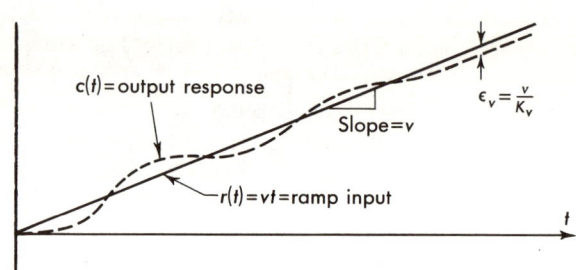

Fig. 3-3 Ramp-function response for a type 1 system.

If a type 1 system is subjected to a ramp input, the output has the form of Fig. 3-3. The position error ϵ_v is the difference between input r and output c. As an example, suppose an antenna is to track a target with a maximum rate of 20 rad/min = $\frac{1}{3}$ rad/sec with a maximum position error of $\frac{1}{2}° = 0.0087$ rad. The velocity constant obtained from this specification is equal to the tracking rate ($\frac{1}{3}$ rad/sec) divided by the tracking position error (0.0087 rad), or

$$K_v = \frac{v}{\epsilon_v} = \frac{\frac{1}{3}}{0.0087} = 38.2 \text{ sec}^{-1} \tag{3-29}$$

A type 1 system with a K_v of at least 38.2 is required to meet this specification.

For a type 2 system (or higher, $n \geq 2$)

$$\lim_{s \to \infty} \frac{s(\tau_1 s + 1)(\tau_3 s + 1)}{s^2(\tau_2 s + 1)(\tau_4 s + 1)} = \infty \tag{3-30}$$

and the steady-state position error is zero. Correspondingly, $K_v = \infty$ for $n \geq 2$.

In summary, if a ramp function is applied to a type 0 system, the error becomes infinite, which means that a type 0 system cannot follow a ramp function. A type 1 system follows with the same velocity but has a finite position error. This case is shown in Fig. 3-3. For systems of type 2 and higher, the steady-state error is zero, and these systems follow with no steady-state error.

Acceleration-error Constant. To determine the acceleration-error constant K_a consider the input to be a step function of acceleration, or

$$r(t) = \tfrac{1}{2}at^2 \tag{3-31}$$

The Laplace transform of this driving function is listed in Table I-1 as

$$R(s) = \frac{a}{s^3} \tag{3-32}$$

The steady-state position error due to an acceleration input is

$$\epsilon_a = \lim_{t \to \infty} \epsilon(t) = \lim_{s \to 0} s \frac{a/s^3}{1+G} = \frac{a}{\lim_{s \to 0} s^2 G(s)} \tag{3-33}$$

Replacing the expression for G from Eq. (3-14) yields

$$\frac{\epsilon_a}{a} = \frac{1}{\lim_{s \to 0} s^2 \frac{K(\tau_1 s + 1)(\tau_3 s + 1)}{s^n (\tau_2 s + 1)(\tau_4 s + 1)}} \tag{3-34}$$

The steady-state position error due to an acceleration input is symbolized by ϵ_a. This error is infinite for both type 0 and 1 systems, since $\lim_{s \to 0} s^2 G = 0$ appears in the denominator. For type 2 systems, the error is a finite constant given by

$$\frac{\epsilon_a}{a} = \frac{1}{K} = \frac{1}{K_a} \quad \text{thus} \quad \frac{a}{\epsilon_a} = K_a \tag{3-35}$$

The acceleration-error constant K_a is the ratio of the amplitude of the step in acceleration to the steady-state position error. In the steady state, type 0 and 1 systems cannot follow an acceleration input without

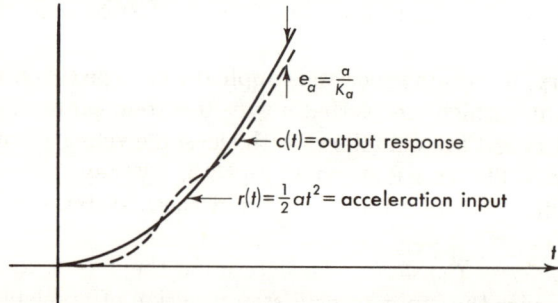

Fig. 3-4 Acceleration response for a type 2 system.

error and a type 2 system follows with a finite steady-state position error. The type 2 case is shown in the sketch of Fig. 3-4.

From the above consideration, it is seen that a control system has zero steady-state error in response to a unit step function only if $n \geq 1$. A type 0 system with velocity inputs (ramp inputs) has a steady-state output velocity different from the constant input velocity, and in the limit the position error becomes infinitely large. For a type 1 system, the steady-state output velocity is equal to the constant input velocity; however, there is a constant position error between steady-state output and constant-velocity input. For types 2, 3, and higher systems, the system operates in the steady state with the input and the output in exact correspondence.

When type 0 and 1 systems are excited with a parabolic function, the acceleration is not reproduced in the steady state and the error increases continuously to infinity. For type 2 systems, the acceleration is reproduced in the steady state, but with a constant position error. For systems of type 3 or higher, the position error is zero and there is an exact correspondence between input and output in the steady state. Information regarding static-error coefficients is summarized in Table 3-1.

Table 3-1 Summary of Error Constants

n	K_0	ϵ_0	K_v	ϵ_v	K_a	ϵ_a
0	Finite constant	$\dfrac{A}{1+K_0}$	0	∞	0	∞
1	∞	0	Finite constant	$\dfrac{v}{K_v}$	0	∞
2	∞	0	∞	0	Finite constant	$\dfrac{a}{K_a}$

Of the several aforementioned constants, the velocity constant is most commonly encountered. Written in terms of an equation, the velocity constant is equal to the ratio of constant input velocity to the steady-state position error. When the transfer function is typically written as follows:

$$G = \frac{K_v(\tau_1 s + 1)(\tau_3 s + 1)}{s(\tau_2 s + 1)(\tau_4 s + 1)} \quad (3\text{-}36)$$

the limit of sG as s approaches zero is K_v. Hence, the constant multiplier of the transfer function, when the latter is written in the standard form shown here, is the velocity constant when there is a first-order pole at the origin due to a factor of s in the denominator.

From a consideration of Table 3-1, it might appear without further investigation that the higher the system type, the "better" the control system. For systems of type 2 there is zero error for either a step in position or a step in velocity and only a finite error for a step in acceleration. However, when the system stability is under consideration, those systems with a pole of order 2 or higher at the origin (a factor of s^n occurring in the denominator, $N \geq 2$), corresponding to type 2 and higher, are increasingly difficult to stabilize. Systems with a pole of order 1 or 0 at the origin are more easily stabilized. As with most other engineering problems, the design of feedback control systems usually involves a compromise between the value of the steady-state error that results in a system and the value of the settling time that may be obtained with a particular type of system.

The compromise between steady-state error and settling time can be seen from a physical consideration of the mechanical system of Fig. 3-5. When a soft spring supports the mass, the steady-state deflection of the mass under its own weight is large. Vibrations of the system are sluggish in nature, i.e., the settling time is large. If it is of second order, there will be, correspondingly, a low resonant frequency but a relatively large damping ratio. The response of this system is similar to that of a low-gain feedback system. When a tight (hard) spring supports the mass, the steady-state deflection (steady-state error) is small, and the system responds more rapidly as manifested by a shorter settling time. For the second-order system, the natural frequency is higher and the damping ratio is smaller; and the system is relatively less stable. A similar phenomenon occurs in a high-gain control system.

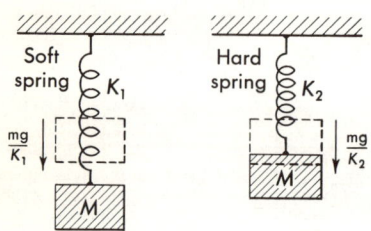

Fig. 3-5 Comparison of a soft and hard spring.

3-5 Generalized Steady-state Errors. If the block diagram has the more general form of Fig. 3-6, that is, if $H \neq 1$ and an input transfer function G_2 is added, steady-state errors take on a different significance. Input to the summing point is $G_2 R$, and the feedback to the summing

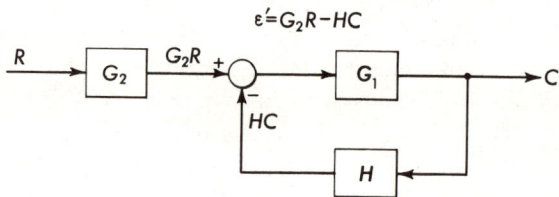

Fig. 3-6 Generalized block diagram.

point is HC. The actuating signal $\epsilon'(t)$ is no longer the difference between the input r and the output c, but its transform is given by

$$\epsilon' = G_2 R - HC \tag{3-37}$$

In this form the transfer function is written as

$$\frac{\epsilon'}{R} = \frac{G_2}{1 + G_1 H} \tag{3-38}$$

Steady-state Errors

For step inputs of position ($r = A$), the steady-state actuating signal is given by

$$\epsilon'_0 = \lim_{s \to 0} s \frac{G_2(A/s)}{1 + G_1H} = \lim_{s \to 0} \frac{AG_2}{1 + G_1H} \tag{3-39}$$

For step inputs of velocity ($r = vt$), the actuating signal is found from the transfer function as follows:

$$\frac{\epsilon'_v}{v} = \lim_{s \to 0} s \frac{G_2}{1 + G_1H} \frac{1}{s^2} = \lim_{s \to 0} \frac{G_2}{sG_1H} \tag{3-40}$$

Similarly, for step inputs of acceleration ($r = \frac{1}{2}at^2$), the actuating signal is found from

$$\frac{\epsilon'_a}{a} = \lim_{s \to 0} \frac{G_2}{s^2 G_1 H} \tag{3-41}$$

Although the steady-state actuating signal does not correspond to the steady-state error ($r - c$), its magnitude may be a measure of system performance. Quite often the transfer functions of the input and feedback blocks (G_2 and H, respectively, of Fig. 3-6) are constants (transducers) which change the input variable of mechanical position to an output variable of electrical voltage, or pressure to angle, etc. In this case direct subtraction of input and output is meaningless. For example, if the input r is a pressure and the output c a position, the difference $r - c$ is meaningless. Before subtraction is possible, an appropriate constant (transducer), with transfer function $K = $ position/pressure, must be included so the actuating signal can be obtained.

For cases where $H \neq 1$, useful information can still be obtained from the error constants provided the feedback function satisfies the condition

$$H(s) = \text{const} \tag{3-42}$$

That is, if $H(s)$ is a constant, error constants can be used to determine the steady-state error in the system. For example, with a potentiometer (converting shaft position to voltage) for the output transducer, H is a constant and can be included with G in all the equations of this section. In this case the voltage error is the difference $R - HC$ in which R and HC have like dimensions and can be meaningfully compared.

In the more general case, when H is a function of s, the final-value theorem is used to compute the steady-state actuating signal.

90 Control System Design

3-6 Steady-state Errors Due to Output Load Disturbances.
Steady-state errors also result from disturbances applied at the control output. Consider, for example, the simple position servo of Chap. 1. The input is a position $r(t)$, and the output $c(t)$ is positioned by a motor driven with an amplifier whose input is the actuating signal (equal to the error $\epsilon = r - c$). This system, with $H = 1$, is shown in the block diagram of Fig. 3-7, where K_s is the constant transfer function of the input and output potentiometers and A is the transfer function of the amplifier.

The motor transfer function $K_m/s(\tau s + 1)$ is applicable only when an inertia load is driven with voltage input signal. When a constant torque is also applied to the output shaft of this motor, the differential equation

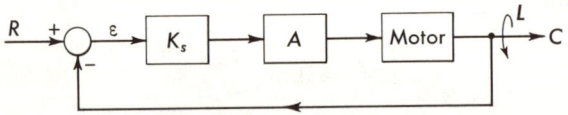

Fig. 3-7 Position control system subjected to output torque.

characterizing the motor performance must be reconsidered. The transform of this equation is

$$L_m = K_1 E_{in} - K_2 sC \qquad (3\text{-}43)$$

where Eq. (3-43) is derived directly from the linearized motor speed-torque curves presented in Chap. 1. K_1 and K_2 are constants determinable from the linearized speed-torque curves. For the present application, the motor torque L_m is balanced by an inertial torque $J\, d^2c/dt^2$ and the output torque L. In terms of transforms,

$$L_m = Js^2 C - L \qquad (3\text{-}44)$$

where L denotes the Laplace-transformed disturbing torque on the output shaft. The shaft angle C, the disturbing torque L, and the motor torques are all taken in the same direction. Equating Eqs. (3-43) and (3-44),

$$C = \frac{K_1 E_{in} + L}{s(Js + K_2)} \qquad (3\text{-}45)$$

The voltage input to the motor is related to the error as follows:

$$E_{in} = K_s A \epsilon = K_s A (R - C) \qquad (3\text{-}46)$$

When E_{in} from Eq. (3-46) is substituted in Eq. (3-45), the differential equation of motion of the system is

$$(Js^2 + K_2 s + K_1 K_s A)C = K_1 K_s A R + L \qquad (3\text{-}47)$$

Equation (3-47) is that of a system with two inputs: the reference signal R and the output torque L. If L is now considered as the control-system input, the transfer function C/L for $R = 0$ is written as follows:

$$\frac{C}{L} = \frac{1}{Js^2 + K_2 s + K_1 K_s A} \qquad (3\text{-}48)$$

Equation (3-48) represents the transfer function for a linear system whose driving function is the torque. Both transient and steady-state effects result from the application of such a torque. If the form of the torque is known, the output c can be found by ordinary methods, i.e., by use of Laplace transforms or by classical procedures.

Since the system was at rest before the application of torque, it is correct to take $r = 0$. Whether $r = 0$ or any other constant, the response due to application of the disturbing torque to the system at rest would be given by Eq. (3-48). It is simpler, however, to work with one driving component ($r = 0$). In such case, $c = -\epsilon(t)$. The transfer function expressed by Eq. (3-48) is the output in response to an applied output-shaft torque, and since $c = -\epsilon(t)$, it can be written, with Laplace transforms, for a step torque of magnitude L_0 as

$$\varepsilon = \frac{-L_0/s}{Js^2 + K_2 s + K_1 K_s A} \qquad (3\text{-}49)$$

The steady-state error ϵ_0 as found by use of the final-value theorem is

$$\lim_{t \to \infty} \epsilon(t) = \lim_{s \to 0} s \, \frac{-L_0/s}{Js^2 + K_2 s + K_1 K_s A} = \frac{-L_0}{K_1 K_s A} \qquad (3\text{-}50)$$

where L_0/s is the Laplace transform of the step function in torque applied to the output shaft.

Physically, a position servo can be considered as similar to a rotational spring. If an error exists, the motor will develop a torque to cancel the error. In particular, if a torque is applied to the output shaft, an error will result ($\epsilon = -c$). The error will produce a voltage at the motor terminals and hence develop a torque to cancel the applied torque. Similarly, if a rotational spring system is torqued, as shown in Fig. 3-8,

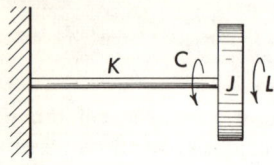

Fig. 3-8 Torque gradient for a feedback control system.

the spring will rotate through an angle C until the torque KC developed by the spring equals the torque applied. The spring constant K is equal to the torque per unit angle. For the feedback system, the torque per unit angle is $L_0/\epsilon_0 = K_1 K_s A$. If the torque is any function other than a constant, the error resulting from the application of the load torque is obtained in the conventional manner.

This example can be extended to the general control system shown in Fig. 3-9, where K_2 is the constant portion of G_2 and is removed from $K_2 G_2$, since the torque is transferred through G_2 only. The output

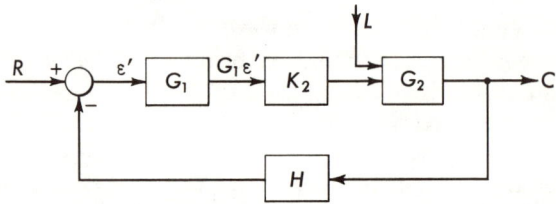

Fig. 3-9 General block diagram subjected to output disturbances.

disturbance, indicated by L in Fig. 3-9, enters directly into the final block G_2. Equations for the system of Fig. 3-9 are

$$(K_2 G_1 \varepsilon' + L) G_2 = C \qquad (3\text{-}51)$$

$$\varepsilon' = R - HC \qquad (3\text{-}52)$$

The two arrows going into the G_2 block indicate that the sum $(L + K_2 G_1 \varepsilon')$ is the input to G_2. Combining Eqs. (3-51) and (3-52),

$$G_2[K_2 G_1 (R - HC) + L] = C \qquad (3\text{-}53)$$

which when rearranged becomes

$$(1 + K_2 G_1 G_2 H) C = G_2 (L + K_2 G_1 R) \qquad (3\text{-}54)$$

If the system is originally at rest, no generality is lost by taking $r = 0$, and the transform of the actuating signal ε' is

$$\varepsilon' = -HC \qquad (3\text{-}55)$$

Equation (3-55) is substituted into Eq. (3-54), and the differential equation relating the transform of the actuating signal ε' to that of the applied disturbing torque L is written

$$(1 + K_2 G_1 G_2 H)\varepsilon' = -HG_2 L \qquad (3\text{-}56)$$

If $r = 0$, the block diagram of Fig. 3-9 may be redrawn (Fig. 3-10) to correspond, in an input-output form, to Eq. (3-56). The normal control system is given by Fig. 3-9. The fictitious system in Fig. 3-10 is useful for analysis purposes. The subtraction, indicated in Fig. 3-9, is replaced by a block with a -1 gain incorporated in the H block. If the standard subtractor (output subtracted from input) is used in Fig. 3-10, another -1 gain block must be included in the feedback path. The error in response

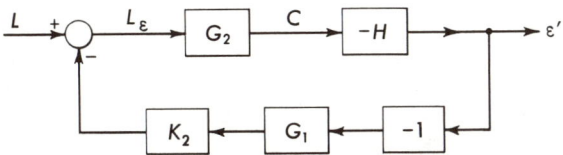

Fig. 3-10 Inverted block diagram with output torque as input.

to any type of disturbing torque L at the output of the original system is determined from either Eq. (3-56) or the block diagram of Fig. 3-10. As an example, consider a torque which is increasing with time, $l = at$, applied to the output. Substituting $\mathcal{L}at = L = a/2s^2$ into Eq. (3-56),

$$\lim_{t \to \infty}(t) = \lim_{s \to 0} \varepsilon(s) = \lim_{s \to 0} \frac{-a/2s^2}{Js^2 + K_2 s + K_1 K_s A} = \infty$$

3-7 Control-system Specifications. Since specifications for a closed-loop system must be interpreted by the control-system designer, a review of the more commonly encountered requirements is presented. Specifications as generally given are divided into those with quantities in the frequency domain (i.e., stated in terms of frequency) and those with quantities in the time domain (i.e., stated in terms of time response). Often a combination of the two kinds is given.

Frequency-domain Specifications. A control-system specification is commonly written in the same sense as an electronic amplifier or filter. Hi-fi enthusiasts may recall that their amplifiers are often specified and compared on the basis of their bandwidth, say, flat from 20 to 20,000 cps. Most filters, i.e., bandpass, low-pass, band-reject, etc., are specified in terms of the amplitude response vs. frequency. The *bandwidth* (Fig. 3-11) is the frequency range between which the amplitude response does not drop 3 db (0.707 amplitude ratio) below the amplitude at the center of the passband. The value of a ratio A_2/A_1 in decibels is defined as $20 \log_{10} A_2/A_1$, where A_2/A_1 commonly denotes an amplitude ratio.

The bandwidth indicates to some degree the speed of response of the system. In filter theory the bandwidth determines the ability of the system to reproduce the shape of the input signal. For example, if a

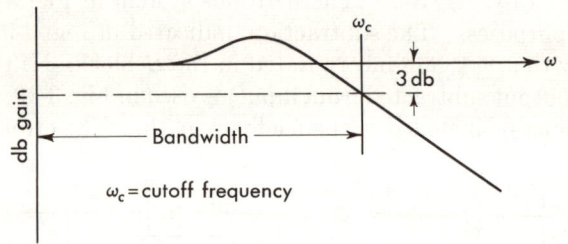

Fig. 3-11 Definition of bandwidth.

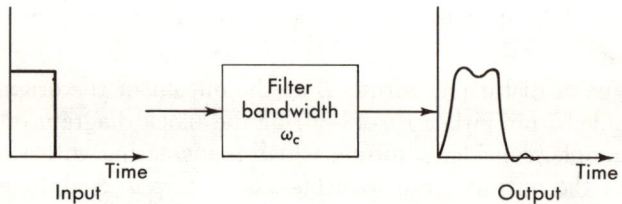

Fig. 3-12 Effect of bandwidth on pulse shape.

square pulse is applied to the input of a low-pass filter which has a bandwidth limit (indicated by giving the finite cutoff frequency* ω_c), the output is not a square pulse but might be a distorted waveform as shown in Fig. 3-12. In some specifications not only the bandwidth but also the characteristics of the cutoff are given. For example, the rate of cutoff beyond the cutoff frequency may be specified as 12 db/octave. An octave is a factor-of-2 change in frequency.

The d-c performance of a control system is frequently specified. It is included as a frequency specification, since direct current is the limiting case of zero frequency. The various error constants, and especially the velocity constant K_v, provide a simple method of d-c system specification.

---

* The cutoff frequency is the limit of the passband. It is the frequency where the amplitude is down 3 db, as shown in Fig. 3-11.

A variety of other frequency-domain specifications may be encountered in practical control-system design. For example, if a control system is used in an aircraft which has a structural resonance at a particular frequency due to the construction, it may be necessary to design a system which blocks the transmission of signals in this band of frequencies but passes both higher and lower frequencies. Otherwise, the aircraft may be subjected to severe vibration resulting in erratic flight. If noise which is limited to a particular frequency band is present, a special band-pass characteristic must be specified to block the undesirable frequencies.

Time-domain Specifications. Often the desired characteristics of the system performance are specified in terms of time-domain quantities, i.e., by the time response to a step function or a ramp function. In general, the transient component is required to satisfy certain time-domain requirements exactly. It is, however, impossible to find the transient component until most of the design has been completed. As an aid in bridging the gap between specifications given in the time domain and useful design quantities (such as damping ratio ζ and natural resonant frequency ω_n) the second-order solution may prove useful. Care must be exercised in evaluating more complex systems, since the extension of the second-order system analysis to these systems can be misleading. But at least the second-order system is a starting point for practical design.

Second-order system response to a step function input is considered in Chap. 1. From the transient component of the second-order system response it is to be observed that the design based on the calculation of time-domain specifications can be rather cumbersome because of the amount of numerical work involved. Charts which relate various time-domain specifications, such as per cent overshoot and damping ratio, are presented in Chap. 1. For higher-order systems, beyond the second, these charts are used only as a guide and a starting point.

A typical response to a unit step function input for a particular damping ratio is shown in Fig. 3-13. The curve is conveniently described in terms of three quantities: the per cent overshoot, the rise time, and the settling time (which is related to the damping ratio).

The *overshoot*, in per cent of the final value, measures the amount the output overshoots the steady-state response for a unit step input. For Fig. 3-13, the overshoot is $A_1 \times 100$ per cent, since the steady-state value is unity.

The *rise time* is defined as the time required for the response to a unit step function input to rise from 10 to 90 per cent of the final value. An alternative and substantially equivalent definition may be used: the reciprocal of the slope of the response at the instant of the response is half the final value. The first definition is generally preferable.

The *settling time* is defined as the time required for the response to cross the steady-state value again after overshooting the first time. This is shown in Fig. 3-13. It is also defined as the time required for the response to a unit step function to decrease and to reach, and thereafter stay within, a specified percentage of its final value.

The damping ratio, discussed in Chap. 1, is defined by Eq. (1-22). This quantity may be calculated from the time response when the ratio of

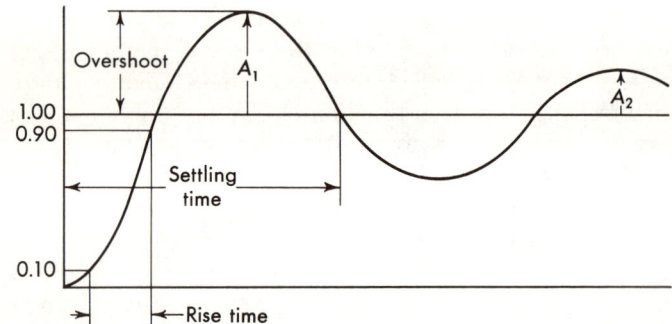

Fig. 3-13 Time-domain specifications shown on a typical response curve.

two successive positive overshoots is known: A_1 and A_2 shown in Fig. 3-13. The damping ratio ζ for a second-order system is given by

$$\zeta = \frac{(\tfrac{1}{2}\pi) \ln (A_1/A_2)}{1 + [(\tfrac{1}{2}\pi) \ln (A_1/A_2)]^2} \tag{3-57}$$

Besides the difficulty of relating these given time-response quantities directly to frequency-response information or frequency-domain data, another difficulty arises in specifying control systems on the basis of time-domain information: No common definition has been accepted for time-domain quantities. For example, the rise time can be defined as (1) the time for response to increase from 10 to 90 per cent of steady-state output, (2) the time for the response to increase from 5 to 95 per cent of steady-state output, or (3) the value of the reciprocal of the slope at 50 per cent of steady-state output. Definitions of quantities given in this chapter correspond as closely as possible to present control-system definitions.

PROBLEMS

3-1. Find position, velocity, and acceleration coefficients for the following forward-loop transfer functions ($H = 1$ for all systems):

(a) $G = \dfrac{K}{(s+10)(s+100)}$

(b) $G = \dfrac{K}{s(s+10)(s+100)}$ Ans. $K_0 = \infty$, $K_v = 10^{-3} K$, $K_a = 0$

(c) $G = \dfrac{Ks}{(s+1)(s+10)(s+100)}$

(d) $G = \dfrac{K}{s^2(s^2 + 2\zeta\omega_n s + \omega_n^2)}$

(e) $G = \dfrac{100}{(s^2 + 2\zeta\omega_n s + \omega_n^2)}$ Ans. $K_0 = \dfrac{100}{\omega_n^2}$, $K_v = 0 = K_a$

(f) $G = \dfrac{100(s+10)(s+50)}{s^3(s^2 + 2\zeta\omega_n s + \omega_n^2)}$

3-2. Find the actuating signal for the system shown in Fig. 3P-2. Find the position, velocity, and acceleration constants and determine their significance in this general case.

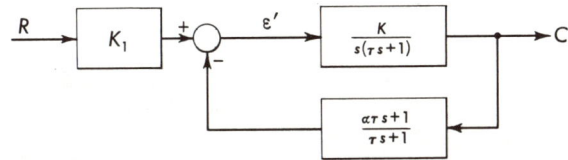

Fig. 3P-2

3-3. Determine the transform of the actuating signal and the difference $\varepsilon(s) = R - C$ for the problem of Fig. 2P-3. Which of the two is the better measure of control-system performance? How are the two related to the error constants?

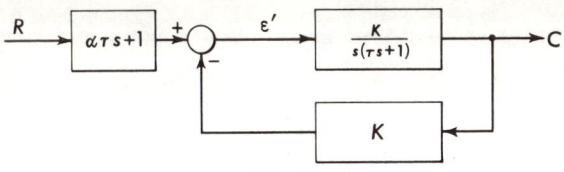

Fig. 3P-3

3-4. Derive the relation between the damping ratio ζ and the amplitudes of two successive overshoots A_1 and A_2 of the system of Fig. 3-13. Ans. Eq. (3-57)

3-5. For the systems of Prob. 3-1 find the steady-state errors to a unit step input and to a unit ramp input.

3-6. For the system of Fig. 3P-6, find the steady-state error for
(a) A unit step position input
(b) A unit step velocity input
(c) A unit step acceleration input

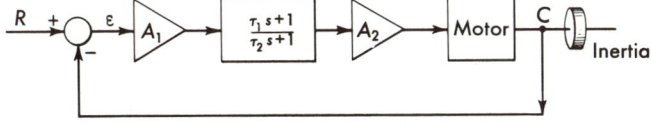

Fig. 3P-6

98   Control System Design

3-7. For the open-loop transfer function

$$G = \frac{K(s + s_1)}{s^2(s + s_2)(s + s_4)}$$

(a) State the type of system.

(b) Find K_0, K_v, and K_a.   *Ans.* $K_0 = \infty = K_v$, $K_a = Ks_1/s_2s_4$

3-8. For the system of Fig. 3P-6, find the steady-state error for a constant output torque applied to the system. Assume zero input.   *Ans.* $\varepsilon_o = L_o/A_1A_2K_m$

3-9. Repeat Prob. 3-8 with an output load torque given by the equation

$$L = bt$$

3-10. Repeat Prob. 3-8 with the transfer function of the second amplifier changed from a constant A_2 to $K(1/s)$.   *Ans.* $\varepsilon_{ss} = 0$

3-11. Repeat Prob. 3-8 with the transfer function of the second amplifier changed from a constant A_2 to Ks.   *Ans.* $\varepsilon_{ss} = \infty$

3-12. For the second-order system of Fig. 1-12, with $\zeta = 0.1$, find the following: (a) Per cent overshoot; (b) rise time; (c) settling time.

3-13. For the system of Prob. 2-12

(a) Find K_p, K_v, and K_a.

(b) What type of system is this?

(c) What is the error (steady-state) if E is a constant?

3-14. In a position servo, addition of "rate feedback" increases the damping. The block diagram for the system is shown in Fig. 3P-14.

(a) Derive the damping ratio and resonant frequency and show how they depend on τ.

(b) Calculate the steady-state error in response to a unit ramp function input.

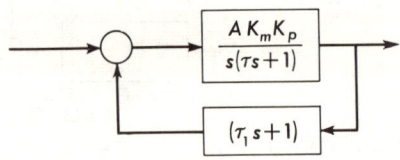

Fig. 3P-14

chapter 4

THE ROOT-LOCUS METHOD[11,12]

4-1 Introduction. As indicated in Chap. 1, the design of closed-loop systems requires repetitive analysis. The engineer must determine the system effects for particular inputs. After several trials and adjustments, a fairly optimum system can result. It is important for the control-system engineer to be able to analyze the result of each successive trial rapidly, and the importance of the root-locus method lies in the rapidity of analysis. The root-locus method depicts the entire system character (transient behavior), and effects from loop gain changes, component time constants, and stabilizing-network configurations can be quickly analyzed.

4-2 The Second-order System. In Chap. 1 the second-order system (Fig. 4-1) is studied and the complete time solution found for this system (cf. Secs. 1-4 and 1-5). The value of amplifier gain constant A necessary to provide system stability is an important quantity that must be determined in the design.

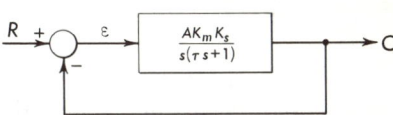

Fig. 4-1 Second-order position servo.

The motor-load transfer function $K_m s(\tau s + 1)$, the time constant τ, and motor constant K_m are fixed when the motor is chosen. The motor* is chosen from a knowledge of load power requirements. How does the transient response, which is directly related to the location of the roots of the system characteristic equation, change as the variable quantities A and K_s are altered? For the system of Fig. 4-1 the answer can be obtained analytically without difficulty.

---

* See Chap. 7 for further information on motors.

100   Control System Design

The transfer function relating output to input is easily found to be

$$\frac{C}{R} = \frac{G}{1+G} = \frac{AK_mK_s/s(\tau s+1)}{1+AK_mK_s/s(\tau s+1)} \qquad (4\text{-}1)$$

where s is the Laplace-transform variable. Equation (4-1) is simplified to

$$\frac{C}{R} = \frac{AK_mK_s}{\tau s^2 + s + AK_mK_s} \qquad (4\text{-}2)$$

The impulse response is found from Eq. (4-2) by setting $R(s) = 1$ (the Laplace transform of a unit impulse is 1) and inverting; thus,

$$c(t) = \mathcal{L}^{-1}\frac{AK_mK_s}{\tau s^2 + s + AK_mK_s} \qquad (4\text{-}3)$$

This is evaluated by first finding the roots of the denominator from

$$s^2 + \frac{1}{\tau}s + \frac{AK_mK_s}{\tau} = 0 \qquad (4\text{-}4)$$

The location of the two roots of this equation determines the transient behavior; hence, it yields information with regard to the degree of system stability. If the roots lie in the right half of the s plane, the response is of the form

$$c(t) = e^{+\alpha t}(A \sin \omega t + B \cos \omega t) \qquad (4\text{-}5)$$

If the roots should lie on the imaginary axis, the response is of the form

$$c(t) = A \sin \omega t + B \cos \omega t \qquad (4\text{-}6)$$

In either case, the system would be unsatisfactory, since Eq. (4-5) characterizes an oscillatory transient response of increasing peak amplitude and Eq. (4-6) typifies that of an undamped sinusoidal response. Locations of the roots determine the nature of the system stability. The damping ratio, the undamped natural frequency, and the frequency of oscillation can be found from knowledge of the root locations.

The two roots of Eq. (4-4) are found from the binomial expression

$$s_i = -\frac{1}{2\tau} \pm \sqrt{\frac{1}{4\tau^2} - \frac{AK_mK_s}{\tau}} \qquad i = 1, 2 \qquad (4\text{-}7)$$

For definiteness consider the following numerical values: $\tau = 0.01$ and $K_m = 5.0$. The values of s_i as AK_s varies are shown in Table 4-1. The plot of these root variations is shown in Fig. 4-2. As the gain AK_s is varied, the roots move continuously along the two branches of a locus, i.e., the locus of the roots. The values of gain constant that produce particular roots are indicated along the plot. The "root locus," or the plot of the values of the roots as a function of the gain constant, affords

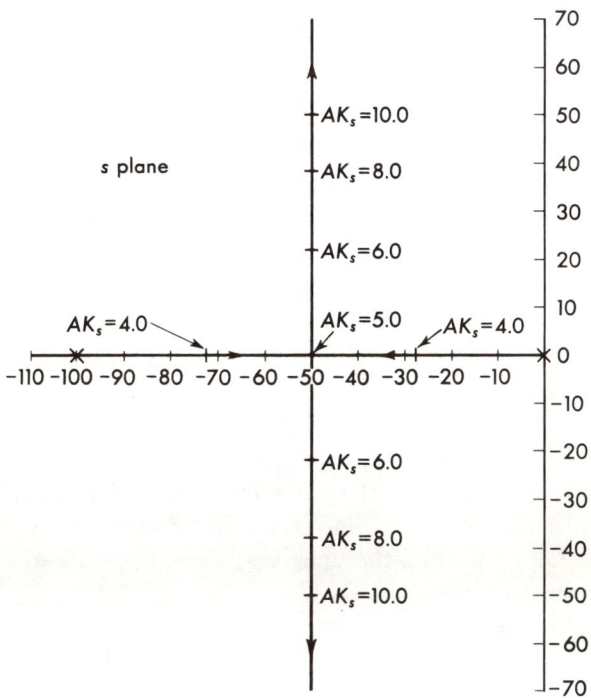

Fig. 4-2. Plot of roots of second-order system.

knowledge of the form of the transient solution for all values of the gain constant.

For example, if $AK_s = 8.00$, the roots are

$$s_i = -50 \pm j38.7 \tag{4-8}$$

and the impulse response (transient component) is

$$c(t) = e^{-50t}(A \cos 38.7t + B \sin 38.7t) \tag{4-9}$$

where A and B are constants that depend upon the initial conditions.

Table 4-1 Values of Roots of Second-order System as Gain Is Varied

AK_s	Root value at indicated location	
	s_1	s_2
0.1	−99.5	−0.5
1.0	−94.7	−5.3
2.0	−88.7	−11.3
3.0	−81.6	−18.4
4.0	−72.4	−27.6
5.0	−50.0	−50.0
6.0	$-50 + j22.4$	$-50 - j22.4$
7.0	$-50 + j31.6$	$-50 - j31.6$
8.0	$-50 + j38.7$	$-50 - j38.7$
9.0	$-50 + j44.7$	$-50 - j44.7$
10.0	$-50 + j50$	$-50 - j50$
50.0	$-50 + j150$	$-50 - j150$
100.0	$-50 + j217.9$	$-50 - j217.9$

System stability, however, depends upon the location of the roots of the characteristic equation of the system and not upon the driving function or the initial conditions. Since the locus of the roots of this second-order system does not cross the imaginary axis, the system is stable for all values of amplifier gain constant.

The plot of the root locations as the system gain constant is varied is calculated analytically as above only to demonstrate the nature of the root-locus method. In practice the root locus can be effected by graphical means, which enables the engineer to plot the positions of the roots as the gain constant is varied.

4-3 The Locus of Roots. The root-locus method is based upon knowledge of the locations of the roots of the system with zero gain constant, i.e., with the feedback loop open. In most cases the locations are easily determined from the open-loop transfer function GH. The function $G(s)$ is the forward-loop transfer function, and $H(s)$ is the feedback-loop transfer function. These quantities, included in the block diagram of Fig. 4-3, are for the

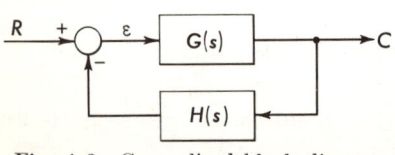

Fig. 4-3 Generalized block diagram.

example of Fig. 4-1

$$K = \frac{AK_mK_s}{\tau} \qquad G(s) = \frac{K}{s(s + 1/\tau)} \qquad H(s) = 1 \qquad (4\text{-}10)$$

In subsequent expressions, the "function of s" is understood; that is, G is written for $G(s)$ and H for $H(s)$.

Consider the following expression, which relates output to input and is derived from the block diagram of Fig. 4-3:

$$\frac{C}{R} = \frac{G}{1 + GH} \qquad (4\text{-}11)$$

where $1 + GH = 0$ is the characteristic equation. The nature of the stability of a system depends upon the impulse response (transient component). The location of the roots of $1 + GH = 0$ determines the degree of system stability.

In the example for a gain of 8.00 the roots of the characteristic equation are

$$s_1 = -50 + j38.7 \qquad s_2 = -50 - j38.7 \qquad (4\text{-}12)$$

A complex number written in rectangular form $a + jb$ is expressed in polar form as follows:

$$\sqrt{a^2 + b^2}\, e^{j\phi} = \sqrt{a^2 + b^2}\, \underline{/\phi} \qquad (4\text{-}13)$$

where $\phi = \tan^{-1}(b/a)$. The roots of Eq. (4-12) can also be expressed in polar form:

$$s_1 = \sqrt{50^2 + 38.7^2}\, e^{j\,\tan^{-1}[38.7/(-50)]} = 63.2 e^{j142.3°}$$
$$= 63.2\underline{/142.3°} \qquad (4\text{-}14)$$
$$s_2 = 63.2 e^{-j142.3°} = 63.2\underline{/-142.3°}$$

In polar form these roots are located by a magnitude (63.2) and by angles ($\pm 142.3°$). In general, the roots of the characteristic equation are complex numbers (real numbers are special cases of complex numbers) and can be written in polar form:

$$s_i = A_i e^{j\phi_i} \qquad i = 1, 2 \qquad (4\text{-}15)$$

where A_i is the magnitude and ϕ_i is the angle of the root which is expressed in polar form. In a similar fashion, any term of the form $s + a$ can be written in polar form from knowledge of its magnitude and phase angle, thus, as $Ae^{j\phi}$.

The open-loop transfer function GH can be expressed as a ratio of factored polynomials; thus, for example,

$$GH = \frac{K_1(s\tau_1 + 1)(s\tau_3 + 1)}{s^n(s\tau_2 + 1)(s\tau_4 + 1)} \tag{4-16}$$

Equation (4-16) can be rewritten in the following form, which should always be used in root-locus analysis:

$$GH = \frac{K_1\tau_1\tau_3}{\tau_2\tau_4} \frac{(s + 1/\tau_1)(s + 1/\tau_3)}{s^n(s + 1/\tau_2)(s + 1/\tau_4)} \tag{4-17}$$

Each factor in the GH function is considered as a complex number and is written in polar form as, typically,

$$s + \frac{1}{\tau_1} = A_1 e^{j\phi_1} \tag{4-18}$$

Hence, the entire GH function is a complex quantity and can be written in polar form as

$$GH = \frac{K(A_1 e^{j\phi_1})(A_3 e^{j\phi_3})}{(A_0{}^n e^{jn\phi_0})(A_2 e^{j\phi_2})(A_4 e^{j\phi_4})} \tag{4-19}$$

$$GH = \frac{KA_1A_3}{A_0{}^n A_2 A_4} e^{j[(\phi_1 + \phi_3 + \cdots) - (n\phi_0 + \phi_2 + \cdots)]} = Ae^{j\phi} \tag{4-20}$$

The algebraic equation from which the roots are determined,

$$1 + GH \equiv 1 + Ae^{j\phi} = 0 \quad \text{thus } GH = -1 \tag{4-21}$$

furnishes the two expressions

$$\text{Angle of } GH \equiv \arg GH \equiv \phi = (2k + 1)180° \tag{4-22}$$

where $k = 0, \pm 1, \pm 2, \pm 3, \ldots$, and

$$\text{Magnitude of } GH \equiv |GH| \equiv A = 1 \tag{4-23}$$

Equations (4-22) and (4-23) result from noting that

$$GH = -1 = 1\epsilon^{j(2k+1)180°}$$

that is, the angle or the argument of GH is an odd multiple of 180° and the magnitude of GH is 1. These equations furnish the basis for the root-locus method.

The equations yielding construction of the root-locus criterion follow from comparison of Eq. (4-20) with Eqs. (4-22) and (4-23). The angle equation yields

$$(\phi_1 + \phi_3) - (n\phi_0 + \phi_2 + \phi_4) = (2k + 1)180° \qquad (4\text{-}24)$$

where $k = 0, \pm 1, \pm 2, \ldots$. The magnitude equation yields

$$\frac{KA_1A_3}{A_0{}^n A_2 A_4} = 1 \qquad (4\text{-}25)$$

The root locus is plotted by finding the locus of all points s_i in the s plane which satisfy Eq. (4-24). When the locus is completely plotted, Eq. (4-25) is used to scale it in terms of the values of the gain constant K that correspond to particular values of the roots on the locus.

To avoid confusion, the various singularities are defined as follows:

A *zero* is a value of s for which the numerator of GH is zero.
A *pole* is a value of s for which the denominator of GH is zero.
A *root* is a value of s which satisfies the equation $1 + GH = 0$.

Notice that if s_j is a pole of GH, it is also a pole of $1 + GH$, since the addition of 1 to infinity yields infinity.

Consider the example of Fig. 4-1, cited earlier in this chapter. In this example the transfer function is

$$GH = \frac{AK_m K_s}{\tau} \frac{1}{s(s + 1/\tau)} = 500 A K_s \frac{1}{s(s + 100)} \qquad (4\text{-}26)$$

Since Eq. (4-24) (the angle equation) is first used to make the plot, only the GH function is examined. The system has a pole at the origin, a pole at $\tau = -100$, and no finite zeros. These poles are first put on the

106 Control System Design

diagram of Fig. 4-4. A × designates a pole, a ○ designates a zero, and a dot designates a root.

Next, a point s_1 is estimated as a possible root of $1 + GH = 0$. If s_1 is a root, the sum of the angles of GH as evaluated for $s = s_1$ must be an odd multiple of 180°. In the present instance the angle (argument) of GH is

$$\arg GH = -(\theta_0 + \theta_2) = -\left(\tan^{-1}\frac{\beta_1}{-\alpha_1} + \tan^{-1}\frac{\beta_1}{100 - \alpha_1}\right) \quad (4\text{-}27)$$

The angles subtended at the poles of GH—θ_0 and θ_2—are shown in Fig. 4-4. The sign is negative because the positive angles subtended by line segments drawn from the poles of the denominator to a point on the root

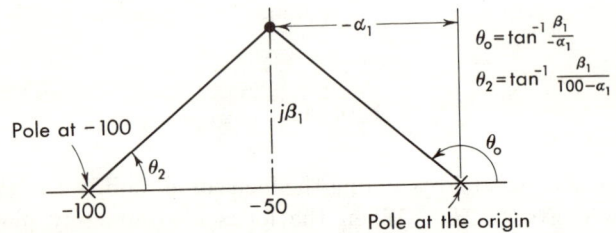

Fig. 4-4 Location of the poles of $500AK_s/s(s + 100)$.

locus are entered as negative in the calculation of the argument of GH. The locus of all points at which the angles subtended to the zeros and poles algebraically sum to an odd multiple of 180° is the root locus. The reader can verify from the plot of Fig. 4-2 that the sum of the angles subtended to the poles from any root is $-180°$. (This particular transfer function has no finite zeros.)

Figure 4-5 shows the zero and pole locations for the transfer function

$$GH = \frac{K(s + \alpha_1 + j\beta_1)(s + \alpha_1 - j\beta_1)}{s(s + 1/\tau_2)(s + 1/\tau_4)} \quad (4\text{-}28)$$

The characteristic function $1 + GH$ for this system is solved graphically for its roots by locating a trial point s_1 in the s plane and moving it until the angle equation (4-22) is satisfied. Application of Eqs. (4-22) and (4-23) to the situation depicted in Fig. 4-5 indicates that if s_1 is a root of the characteristic equation, the algebraic sum of all the angles (positive for zeros and negative for poles) must equal an odd multiple of 180°

and the magnitude must equal unity, or, written in equation form,

$$\Sigma\theta_i = \theta_1 + \theta_3 + (-\theta_0 - \theta_2 - \theta_4) = (2k + 1)180° \qquad (4\text{-}29)$$

$$A = \frac{K|s_1 + \alpha_1 + j\beta_1| \, |s_1 + \alpha_1 - j\beta_1|}{|s_1| \, |s_1 + 1/\tau_2| \, |s_1 + 1/\tau_4|} = 1 \qquad (4\text{-}30)$$

Each angle in $\Sigma\theta_i$ is an angle of the vector $s_1 + 1/\tau_i$ measured with respect to a zero-angle polar axis. Each term in A is equal to the distance from the respective pole or zero to the point s_i. The root-locus

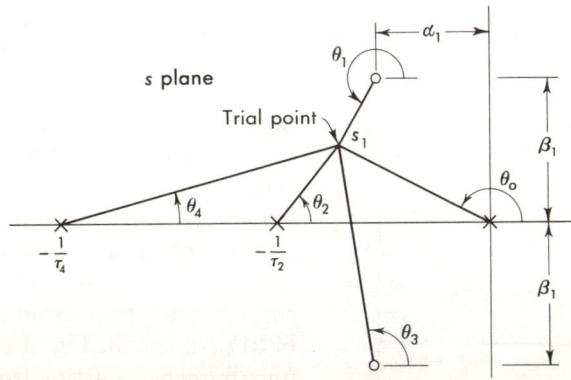

Fig. 4-5 Location of zeros and poles for

$$\frac{K[(s + \alpha_1)^2 + \beta_1^2]}{s(s + 1/\tau_2)(s + 1/\tau_4)}$$

diagram is plotted by using the condition of Eq. (4-29): thus a point s_i on the locus is effected by moving s_i about in the s plane until the angle equation is satisfied.

The system gain constant K for a particular operating point along this locus is found by application of Eq. (4-30). Usually the procedure of first finding the root locus and then finding the gain corresponding to particular points on the locus is easier than trying to satisfy both the angle and the magnitude conditions simultaneously. Often, for the purposes of certain problems, it is unnecessary to plot all branches of the locus and/or to determine K.

The essence of the root-locus construction has been outlined; however, a set of useful rules which reduces the time consumed in a completely trial-and-error procedure is presented in the following section.

108 Control System Design

4-4 Rules for Rapid Construction of Root-locus Diagrams. The rules which are stated and demonstrated in this section enable the engineer to sketch a locus diagram rapidly. Proofs for the rules are postponed to Sec. 4-7. The following open-loop transfer function is used to demonstrate the method:

$$GH = \frac{K(s+10)}{s(s+5)(s+2+j15)(s+2-j15)} \qquad (4\text{-}31)$$

The zeros and poles are shown in Fig. 4-6. Notice that the scales on the real and imaginary axes of the s plane are identical.

Rule 1. Continuous curves, which comprise the branches of the locus, start at each pole of GH, at which $K = 0$. The branches of the locus, which are single-valued functions of gain, terminate on the zeros of GH, at which $K = \infty$. The zeros and poles of GH are first located on the s plane by a $\times$ for each pole and a $\bigcirc$ for each zero. According to Rule 1, a branch of the locus starts at a pole and terminates at a zero. Each branch is a continuous curve between a pole and a zero and is a single-valued function of gain along the curve. For the example of Eq. (4-31), shown in Fig. 4-6, there are four branches starting from the four poles located at

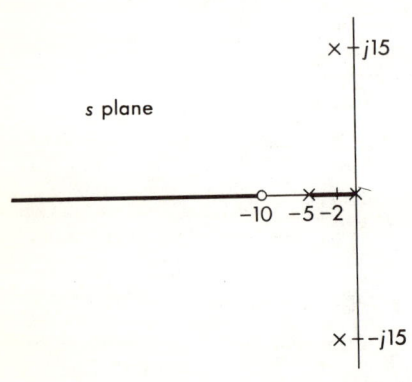

Fig. 4-6 Example of Rules 1 and 2 applied to the system

$$GH = \frac{K(s+10)}{s(s+5)[(s+2)^2 + 15^2]}$$

$s = 0 \qquad s = -5$

$s = -2 + j15 \qquad s = -2 - j15$

Since there is only one finite zero, located at $s = -10$, three of the branches must terminate on zeros at infinity. The rule, then, can be expanded to read: The branches of the locus start at the poles and terminate on either finite zeros or on $n_P - n_Z$ zeros located at infinity. n_P is the number of poles and n_Z is the number of finite zeros of GH. The number of separate branches equals the number of finite poles if, as is usually the case, $n_P \geq n_Z$.

The gain constant K is usually positive, corresponding to subtractive feedback, and varies from zero to infinity. On occasion it may be necessary to work with negative K, corresponding to additive feedback. In this case, the correct angle equation is that the angles sum to $2k(180°)$,

$k = 0, \pm 1, \pm 2, \ldots$. Since this is not often the case in feedback control systems, only positive K is subsequently considered.

Rule 2. *The locus includes all points along the real axis to the left of an odd number of poles plus zeros.* To determine if a branch of the locus exists in a given region along the real axis, count the number of real zeros plus real poles which lie to the right of the trial point. If there is an odd number, this line segment is included in a branch of the locus. For the example of Fig. 4-6, the branches of the locus include the line segments for the real axis from the origin to the point $s = -5$ and from $s = -10$ to $-$ infinity. These regions along the real axis are shown darkened in Fig. 4-6. Complex zeros or poles, such as $s = -2 \pm j15$, are ignored in applying this rule.

Rule 3. *As K approaches ∞, the branches of the locus become asymptotic to straight lines with angles*

$$\frac{(2k+1)180°}{n_P - n_Z} \quad \text{for } k = 0, \pm 1, \pm 2, \ldots \text{ until all angles not differing by multiples of } 2\pi \text{ are obtained} \quad (4\text{-}32)$$

where n_P is the number of poles and n_Z is the number of zeros

If the number of finite poles exceeds the number of finite zeros, some of the branches terminate on zeros which are located at infinity. Rule 1 states that the number of branches which terminate at infinity is equal to the number of finite poles minus the number of zeros. As the gain becomes large, these branches tend to become asymptotic to straight lines whose angles with the polar axis are given by Eq. (4-32).

For the example of Fig. 4-6, $n_P = 4$ and $n_Z = 1$ and the asymptotic angles are computed as follows:

$$\theta_k = \frac{(2k+1)180°}{3}$$

$$\begin{aligned} k &= 0 & \theta_k &= 60° \\ k &= 1 & \theta_k &= 180° \\ k &= 2 & \theta_k &= 300° \end{aligned} \quad (4\text{-}33)$$

Since only three branches go to infinity, all the needed angles are computed, and they are plotted in Fig. 4-7. No error would result if different consecutive values of k had been chosen, since the angles would differ by multiples of 2π after any three consecutive k's were inserted into Eq. (4-33). The above rule gives the asymptotic directions or angles; the point from which these asymptote lines radiate is given by the next rule.

110   Control System Design

Rule 4. *The starting point on the real axis from which the asymptotic lines radiate is given by*

$$CG = \frac{\Sigma \text{ poles} - \Sigma \text{ zeros}}{n_P - n_Z} \qquad (4\text{-}34)$$

This point is termed the center of gravity of the roots. The angular directions to which the branches of the locus are asymptotic are given by

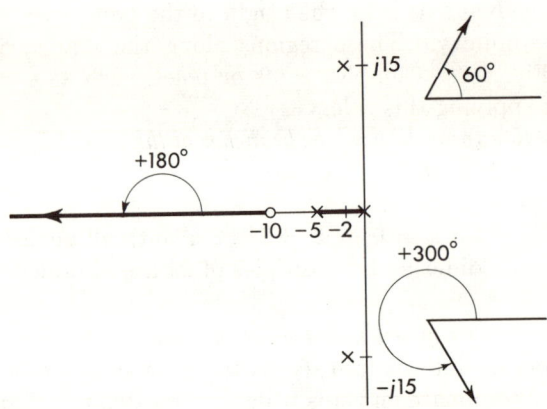

Fig. 4-7 Asymptotic directions determined by Rule 3.

Rule 3. The asymptotic lines start at the center of gravity of the zero-pole configuration. This point is found from Eq. (4-34). For the example of Fig. 4-6, the center of gravity is found as follows:

$$\Sigma \text{ poles} = (-5) + (-2 + j15) + (-2 - j15) + 0 = -9$$
$$\Sigma \text{ zeros} = -10$$
$$n_P = 4$$
$$n_Z = 1$$

and
$$CG = \frac{(-9) - (-10)}{4 - 1} = \frac{1}{3} \qquad (4\text{-}35)$$

The asymptotic lines, characterized by Eq. (4-33), start from the center of gravity given by Eq. (4-35). These lines are placed on the s plane as shown in Fig. 4-8.

Since both complex zeros and complex poles always appear in conjugate pairs—i.e., they are at equal vertical distances from the real axis—the center of gravity always lies on the real axis.

Rule 5. *The breakaway point s_b is found from the equation*

$$\frac{-1}{|s_b + 1/\tau_2|} + \frac{-1}{|s_b + 1/\tau_4|} + \frac{+1}{|s_b + 1/\tau_1|} = \frac{-1}{|s_b + 1/\tau_6|} + \frac{1}{|s_b + 1/\tau_3|} \quad (4\text{-}36)$$

where $|s_b + 1/\tau_i|$ is the magnitude of the distance from the assumed breakaway point s_b to the real axis zero or pole at $s = -1/\tau_i$. The distances are read graphically. From Rule 2 the regions where the locus exists along the real axis are determined. If an acceptable region on the real axis lies between a single-order zero and a single-order pole ("single order" means one zero or one pole located at a point), the direction of increasing k in the locus is usually from the pole to the zero along the real axis. If an acceptable region along the real axis lies between two single-order poles, the two branches which start from these poles usually approach each other along the axis. As the gain is increased, these branches merge, corresponding to a double-order root at the point at which they coalesce (come together). As the gain is further increased, the two branches break away from the real axis and form the two branches on which a complex conjugate pair of roots lie.

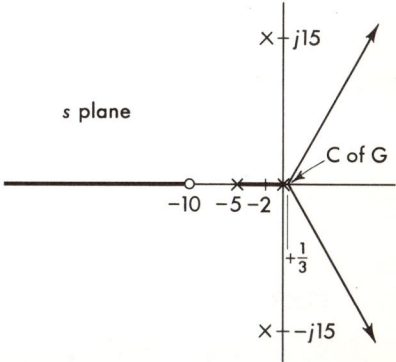

Fig. 4-8 Location of center of gravity of roots by Rule 4.

The point s_b, at which the branches come together and break away, is found as follows:

1. Estimate a point s_b.
2. Measure the distances between the estimated point and the real-axis zeros and poles.
3. Equate the sum of the reciprocals of the distances from s_b to the zeros and poles to the left of s_b to the sum of the reciprocals of the distances to the right of s_b. These are added with a negative sign for the poles and a positive sign for the zeros.
4. If the sum to the left does not equal the sum to the right, estimate another value of s_b and repeat.

The above procedure should be performed by trial and error. If, in an attempt to find an analytical solution, fractions are cleared in Eq.

(4-36), a polynomial in s_b will result. Solving for s_b requires finding the roots of a polynomial, an arduous task that the root-locus method obviates. In many cases, it is not even necessary to find s_b, an intelligent guess being suitable. For difficult cases (zeros and poles close to the real axis) it may be necessary to find s_b to aid in sketching the locus.

The zero-pole configuration of Fig. 4-6 indicates that a locus exists between 0 and -5. To find the point at which the two roots break away from the axis, estimate a point $s_b = -2.5$, as shown in Fig. 4-9.

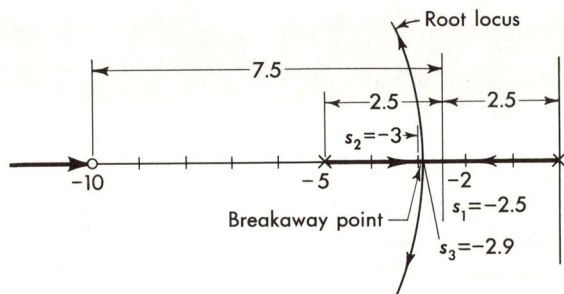

Fig. 4-9 Location of the breakaway point from the real axis by Rule 5.

The sum of the reciprocals [cf. Eq. (4-36)] to the left of s_1 equated to the sum of the reciprocals to the right of s_1 results in the numerical expression

$$-\frac{1}{2.5} + \frac{1}{7.5} \stackrel{?}{=} -\frac{1}{2.5} \qquad (4\text{-}37)$$

This equation is not quite satisfied, since

$$-4.00 + 0.133 = -0.267 \neq -0.400 \qquad (4\text{-}38)$$

As a second estimate, take $s_{b_2} = -3.0$ and obtain

$$-\frac{1}{2.0} + \frac{1}{7.0} \stackrel{?}{=} -\frac{1}{3.0}$$
$$-0.500 + 0.143 = -0.357 \neq -0.333 \qquad (4\text{-}39)$$

As a third estimate, take $s_{b_3} = -2.9$ and obtain

$$-\frac{1}{2.1} + \frac{1}{7.1} \stackrel{?}{=} -\frac{1}{2.9}$$
$$-0.476 + 0.141 = -0.335 \approx -0.345 \qquad (4\text{-}40)$$

Probably this value of $s_{b_3} = -2.9$ would be satisfactory. The process can be repeated, however, until the desired accuracy is achieved. The result is shown in Fig. 4-9.

In this example, and in all cases where complex poles or zeros are located relatively far from the real axis, the complex poles or zeros can be ignored in the computation of the breakaway point. When the complex poles or zeros are near the real axis, their effect must be included. Also, if only one real-axis pole or zero exists, the net change in angle contributed by complex poles or zeros must be included. Figure 4-10 shows a method of computing the net change in angle contributed by complex poles. The angles

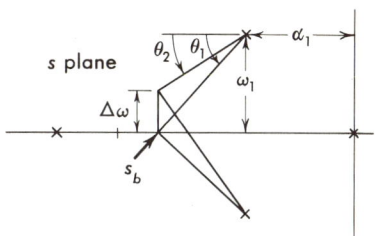

Fig. 4-10 Net change in angle from complex poles.

$$\tan \theta_1 = \frac{\omega_1}{s_b - \alpha_1} \quad \text{and} \quad \tan \theta_2 = \frac{\omega_1 - \Delta\omega}{s_b - \alpha_1} \quad (4\text{-}41)$$

where s_b is the estimated breakaway point. α_1 is the real component of the complex pole, ω_1 is the imaginary component of the complex pole, and $\Delta\omega$ is the vertical distance from the real axis. For small $\Delta\omega$, θ_1 and θ_2 are approximately equal. The difference is expressed with the small-angle approximation

$$\theta_1 - \theta_2 \approx \tan(\theta_1 - \theta_2) = \frac{\tan \theta_1 - \tan \theta_2}{1 + \tan \theta_1 \tan \theta_2} \quad (4\text{-}42)$$

The total angular contribution is given by

$$2(\theta_1 - \theta_2) = \frac{2[\omega_1/(s_b - \alpha_1) - (\omega_1 - \Delta\omega)/(s_b - \alpha_1)]}{1 + \omega_1(\omega_1 - \Delta\omega)/(s_b - \alpha_1)^2}$$

$$\approx \frac{2\Delta\omega(s_b - \alpha_1)}{(s_b - \alpha_1)^2 + \omega_1^2} \quad (4\text{-}43)$$

As an example, consider the zero-pole configuration of Fig. 4-11, which has a transfer function

$$GH = \frac{s^2 + 4}{s(s + 10)} \quad (4\text{-}44)$$

Two roots leave the poles (located at the origin and at $s = -10$) and

114   Control System Design

approach each other. The breakaway point s_b is computed from the equation

$$\frac{-\Delta\omega}{|s_b + 10|} = \frac{-\Delta\omega}{|s_b|} + \frac{2\Delta\omega s_b}{|s_b^2 + 2^2|} \qquad (4\text{-}45)$$

Estimate $s_b = -2$

$$-\frac{1}{8} \stackrel{?}{=} -\frac{1}{2} + \frac{(2)(2)}{4 + 4}$$

$$-0.125 \neq -0.5 + 0.5 \qquad (4\text{-}46)$$

Estimate $s_b = -1.7$

$$-\frac{1}{8.3} \stackrel{?}{=} -\frac{1}{1.7} + \frac{(2)(1.7)}{(1.7)^2 + 4}$$

$$-0.1205 \neq -0.588 + 0.494 = -0.094 \qquad (4\text{-}47)$$

Estimate $s_b = -1.65$

$$-\frac{1}{8.35} \stackrel{?}{=} -\frac{1}{1.65} + \frac{(2)(1.65)}{(1.65)^2 + 4}$$

$$-0.1197 \approx -0.606 + 0.492 = -0.114 \qquad (4\text{-}48)$$

The breakaway point is at $s = -1.65$, and the root-locus sketch is approximately that shown in Fig. 4-11. This rule is also applicable in the same manner for locating the point where two complex roots strike the real axis as the gain is increased.

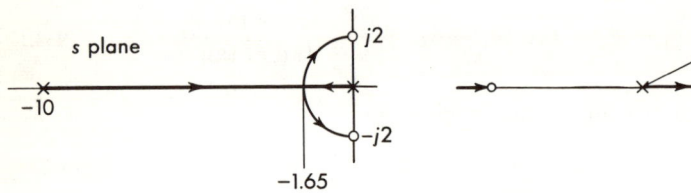

Fig. 4-11 Breakaway from real axis with complex zeros included.

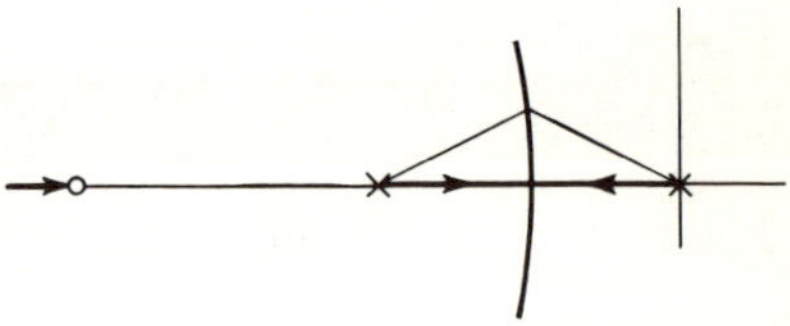

Fig. 4-12 Measurement of breakaway point by gain-variation method.

The breakaway point (also the arrival point) can also be found by considering the gain variation along the locus. The gain is zero at each pole, and it increases, as shown in Fig. 4-12, with distance from each pole. If the gain* is measured at points along the axis between the poles, it will have a maximum value at the breakaway (or a minimum value at an arrival point between two zeros). This fact affords a useful

---

* Gain measurement is considered in Sec. 4-6.

method for finding breakaway points at other locations. For example, consider the locus, shown in Fig. 4-13, for the function

$$GH = \frac{K}{s(s + 1)(s^2 + s + 10)} \qquad (4\text{-}49)$$

One branch leaves each pole, as shown. The two real-axis branches coalesce at the point $s = -\frac{1}{2}$ and leave the real axis at angles of $\pm 90°$. These branches are in line with the branches leaving the two complex conjugate poles at $s = -\frac{1}{2} \pm j3.13$. Again the two branches coalesce

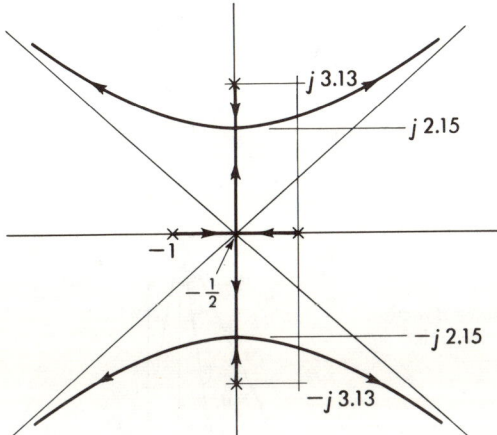

Fig. 4-13 Breakaway point along a vertical line.

and break away as shown in the figure. The points along the line $s = -\frac{1}{2}$ at which the breakaway point occurs are easily found by measuring the gain along this line and determining the point where this gain is a maximum (in this case, at $s = -\frac{1}{2} \pm j2.15$).

Rule 6. Two roots leave or strike the axis at the breakaway point at angles of $\pm 90°$. The two roots of Figs. 4-9 and 4-11 approach each other along the real axis, coalesce, and break away. The angles which the locus makes with the real axis are $\pm 90°$. This is shown in Figs. 4-9 and 4-11.

Rule 7. The angle of departure of the branches from complex poles and the angle of arrival of the branches at complex zeros are found by summing the angles to all other zeros and poles. The desired angle is found by subtracting 180° from this sum. Knowledge of the initial angle of departure of the branches from the complex poles is helpful in sketching root-locus diagrams. The zero-pole configuration of Fig. 4-6 is redrawn in Fig. 4-14 for the purpose of finding the angle of departure from the complex poles at $s = -2 \pm j15$. The angles subtended by the poles and zeros to the pole in question are added (positive for zeros and negative for poles):

$$-(\theta_0 + \theta_2 + \theta_4) + \theta_1 \qquad (4\text{-}50)$$

A protractor is used to measure these angles:

$$-(97.6° + 90° + 78.7°) + 62° = -204.3° \qquad (4\text{-}51)$$

When 180° is subtracted from this angle, an angle of departure of

$$-204.3° - 180° = -384.3°$$

is found. This angle is shown in Fig. 4-14.

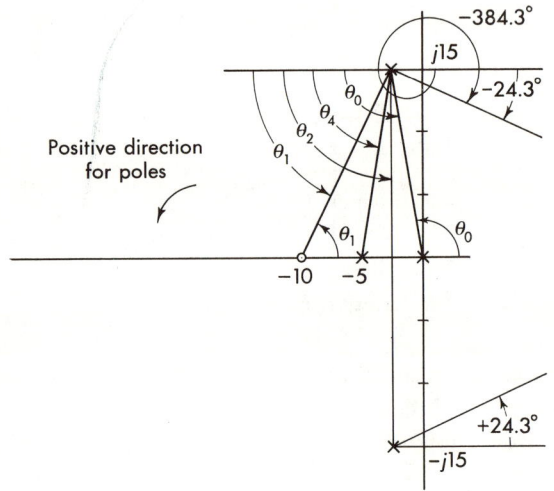

Fig. 4-14 Angle of departure from complex poles.

The Spirule, which is discussed later, is convenient for finding the angle of departure. Its application to this rule is presented in Sec. 4-9.

The angles at which the branches arrive at complex zeros can be found from an extension of this rule. As an example, consider the zero-pole configuration of Fig. 4-15. The angles subtended at $s = j2$ are added:

$$\theta_3 - \theta_0 - \theta_2 = 90° - 90° - 11.3° = -11.3° \qquad (4\text{-}52)$$

The angle of arrival is found by subtracting 180° from this angle:

$$-11.3° - 180° = -191.3° \tag{4-53}$$

The method of finding the departure angle is similar to that of finding the arrival angle except for the direction of positive angles. For a pole, the positive direction is counterclockwise, as shown in Fig. 4-14. For a zero, the positive angle is clockwise, as shown in Fig. 4-15.

The above seven rules are fundamental to the rapid sketching of root-locus diagrams. In general, when a new problem is approached, the rules are used in the given order. Practice should enable the student to apply the rules efficiently and rapidly.

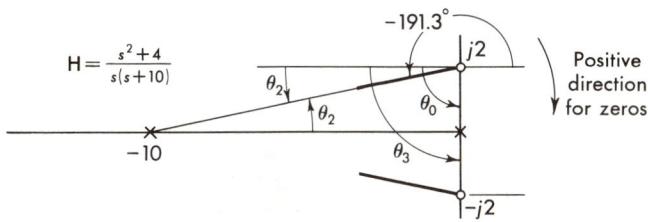

Fig. 4-15 Angle of arrival at complex zeros.

The root-locus procedure is based upon the location of the poles and zeros of GH in the s plane. These points do not move. They are merely the terminal points for the branches of the locus of the roots of

$$1 + GH = 0$$

If a branch of the locus crosses the **imaginary axis** and goes into the right half plane at some gain constant K, **the system** becomes unstable at that value of K. The degree of stability **is determined** largely by the roots near the imaginary axis. Even when **more exact** plotting aids (i.e., Spirule) are used, the root locus is first sketched with the aid of the preceding rules. The sketch gives an idea of the form of the locus, and it is therefore helpful in making a more accurate plot.

4-5 Example of the Application of the Above Rules. As an aid in understanding the above seven rules, consider the following open-loop transfer function:

$$GH = \frac{K}{s(s+3)(s^2+6s+64)} \tag{4-54}$$

118   Control System Design

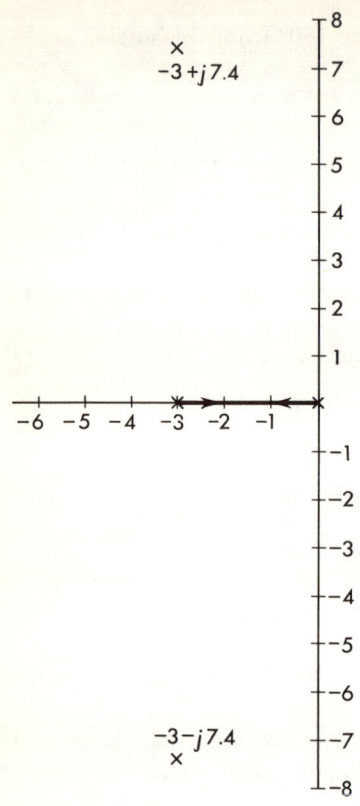

Fig. 4-16 Pole location for the transfer function

$$GH = \frac{K}{s(s+3)(s^2+6s+64)}$$

The sketches of Figs. 4-16 and 4-17 demonstrate the use of the rules to sketch the root locus for the system with open-loop transfer functions as in Eq. (4-54).

The procedure is outlined as follows:

1. Location of zeros and poles
   Zeros: None
   Poles: $0, -3, -3+j7.4, -3-j7.4$
2. A branch of the locus exists along the real axis between $s = 0$ and $s = -3$.
3. Asymptotic angles:

$$\theta = \frac{2k+1}{4-0} 180° = \begin{cases} +45° & k=0 \\ +135° & k=1 \\ +225° & k=2 \\ +315° & k=3 \end{cases}$$

(4-55)

4. Center of gravity, starting point of asymptotes:

$$CG = \frac{0-3-3+j7.4-3-j7.4}{4-0}$$

$$= \frac{-9}{4} = -2.25 \quad (4\text{-}56)$$

5. Breakaway point from real axis. Since the complex poles are at some distance from the axis, the exact value of the breakaway point s_b will be well approximated if taken midway between the two real axis poles or at -1.5.

6. The breakaway angle from the real axis is $\pm 90°$.

7. Breakaway angle from complex poles:

$$-\theta_0 - 2\theta_2 = -112.1° - 180° = -292.1°$$

Subtracting 180°,

$$-292.1° - 180° = -472.1° \quad (4\text{-}57)$$

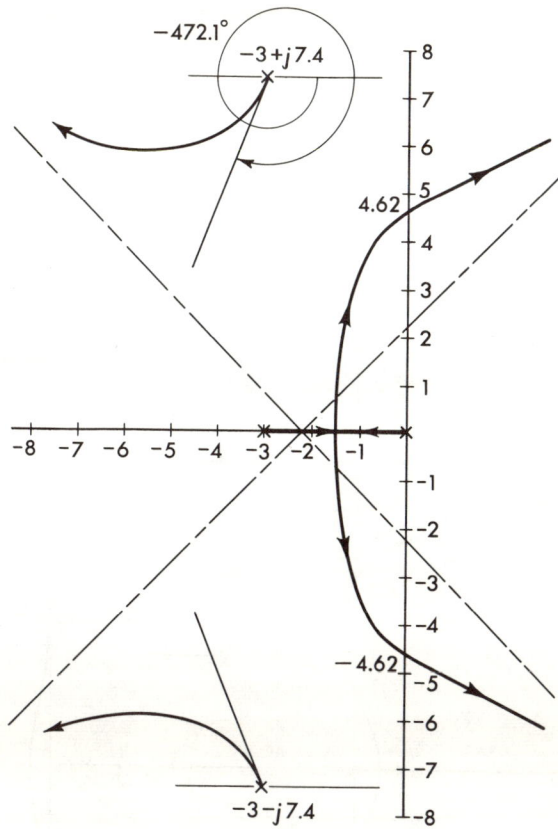

Fig. 4-17 Root-locus sketch for the system

$$GH = \frac{K}{s(s+3)(s^2+6s+64)}$$

For many problems all the rules are not needed. In those cases, of course, only the rules which are applicable are chosen. A collection of commonly encountered root-locus diagrams is included in Fig. 4-18.

4-6 Measurement of Gain. The rules of Sec. 4-4 center about the angle criterion [Eq. (4-22)]

$$\arg . GH = (2k+1)180° \tag{4-58}$$

After the locus has been sketched and certain points located more accurately with the Spirule (Sec. 4-8), the values of gain which occur at certain

120 Control System Design

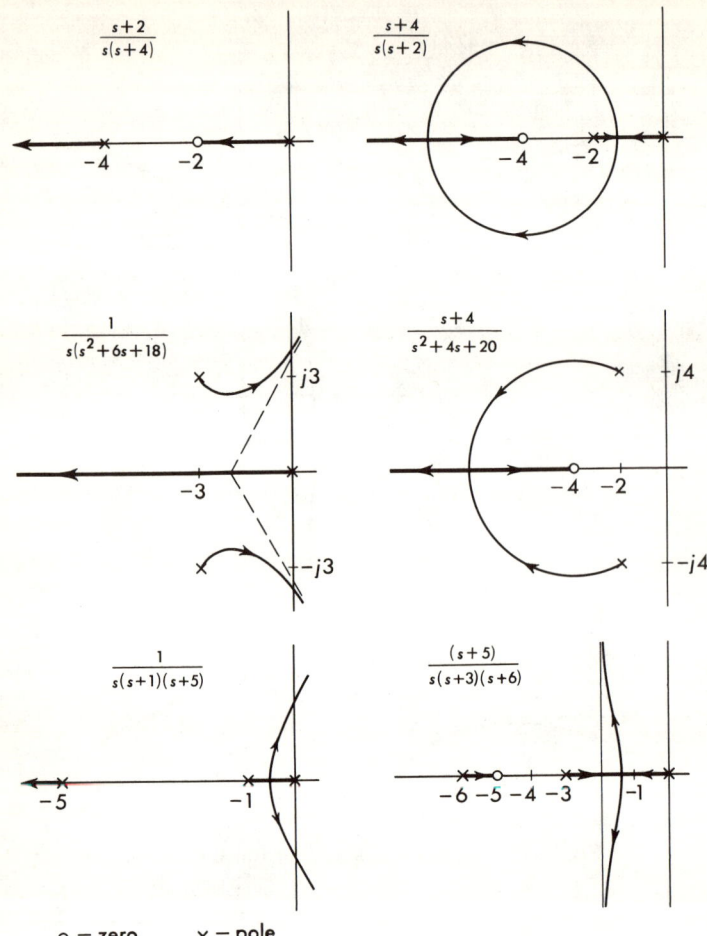

o = zero   x = pole

Fig. 4-18 Collection of root-locus diagrams.

points along the locus must be found. The gain K is evaluated from Eq. (4-23):

$$|GH| = 1 \qquad (4\text{-}59)$$

where $|GH|$ is the product of the distances from zeros to the point at which the gain is to be evaluated divided by the product of the distances from the poles to the point:

$$K = \frac{\text{product of pole distances}}{\text{product of zero distances}} \qquad (4\text{-}60)$$

If there are no zeros, the denominator of Eq. (4-60) is 1. The zero-pole distances are measured directly from the root-locus plot.

The example of Sec. 4-5, which is plotted in Fig. 4-17, is used to demonstrate the measurement of gain. From Eq. (4-60)

$$\frac{1}{|GH|} = |s|\,|s+3|\,|s+3+j7.4|\,|s+3-j7.4| \qquad (4\text{-}61)$$

The gain at the point where the locus crosses the imaginary axis is required. The magnitudes in Eq. (4-61) are read from Fig. 4-17, and computations give the gain as 1,245. In this example and in general practice the zero-pole distances can be read directly with a pair of dividers such as found in most drafting sets. The distances can also be measured with the Spirule.

4-7 Proof of the Root-locus Construction Rules. The rules are verified in the same order as they are presented in Sec. 4-4.

Rule 1. Continuous curves, which comprise the branches of the locus, start at each pole of GH, at which $K = 0$. The branches of the locus, which are single-valued functions of gain, terminate on the zeros of GH, at which $K = \infty$.

The proof of this rule follows from a consideration of the characteristic equation. The GH function of Eq. (4-17) is substituted into $1 + GH$, giving

$$\frac{s^n(s + 1/\tau_2)(s + 1/\tau_4) + K(s + 1/\tau_1)(s + 1/\tau_3)}{s(s + 1/\tau_2)(s + 1/\tau_4)} = 0 \qquad (4\text{-}62)$$

where K replaces the product $K_1\tau_1\tau_3/\tau_2\tau_4$. The zeros of GH are represented by an odd subscript on τ, and the poles of GH by an even subscript on τ. The system stability is determined from the equation formed from the numerator of $1 + GH$,

$$s^n\left(s + \frac{1}{\tau_2}\right)\left(s + \frac{1}{\tau_4}\right) + K\left(s + \frac{1}{\tau_1}\right)\left(s + \frac{1}{\tau_3}\right) = 0 \qquad (4\text{-}63)$$

For $K = 0$ the second part of Eq. (4-63) is zero and the roots for $K = 0$ are given by

$$s^n\left(s + \frac{1}{\tau_2}\right)\left(s + \frac{1}{\tau_4}\right) = 0 \qquad (4\text{-}64)$$

The roots of $1 + GH = 0$ are equal to the poles of GH for $K = 0$, or the branches of the locus start on the poles of GH.

If K is allowed to become large, the first portion of Eq. (4-63) becomes much smaller than the second, and as K approaches infinity in the limit (i.e., for $K = \infty$) the roots of $1 + GH = 0$ are given by

$$\left(s + \frac{1}{\tau_1}\right)\left(s + \frac{1}{\tau_3}\right) = 0 \tag{4-65}$$

and the branches of the locus terminate on the zeros of GH.

When the number of finite zeros is less than the number of poles, the rule is best proved by considering the value of GH for large values of s,

$$GH = K\frac{s^{n_Z}}{s^{n_P}} = \frac{K}{s^{n_P - n_Z}} \tag{4-66}$$

where n_P is the number of poles and n_Z is the number of zeros of GH. For large s, there exist $n_P - n_Z$ zeros at infinity, thus a $(n_P - n_Z)$-order zero at infinity. It follows that $n_P - n_Z$ branches of the locus terminate on these zeros at infinity.

The locus is a single-valued function of K, since for a given K the roots of $1 + GH = 0$ are uniquely specified.

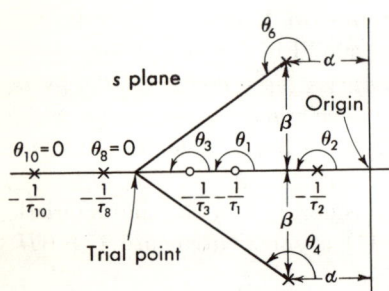

Fig. 4-19 Proof of Rule 2.

Rule 2. *The locus includes all points along the real axis to the left of an odd number of poles plus zeros.*

The validity of this rule is based upon the angle equation

$$\arg GH = \underline{/GH} = (2k + 1)180° \tag{4-67}$$

where $k = 0, \pm 1, \pm 2, \pm 3, \ldots$. In Fig. 4-19 the zero-pole configuration is shown for the function

$$GH = \frac{K(s + 1/\tau_1)(s + 1/\tau_3)}{(s + 1/\tau_2)(s + 1/\tau_8)(s + 1/\tau_{10})[(s + \alpha)^2 + \beta^2]} \tag{4-68}$$

The angle equation (4-67) is written as an algebraic sum of angles subtended by the poles and zeros:

$$\theta_1 + \theta_3 - (\theta_2 + \theta_4 + \theta_6 + \theta_8 + \theta_{10}) = (2k + 1)180° \tag{4-69}$$

where $k = 0, \pm 1, \pm 2, \pm 3, \ldots$.

The sum of the angles subtended by the complex poles, $\theta_4 + \theta_6$ in Fig. 4-19, is 360°, since the locations of complex conjugate zeros and poles for systems described by linear differential equations with constant coefficients are symmetrical with respect to the real axis of s. Any poles or zeros to the left of the trial point contribute nothing ($\theta_8 = 0$ and $\theta_{10} = 0$). The angles to the right of the trial point contribute either $+180°$ or $-180°$:

$$\theta_1 - \theta_3 - \theta_2 = 180° + 180° = 180°$$

At any point where an odd number of zeros plus poles exists to the right of the trial point, the sum is $(2k + 1)180°$, and a branch of the locus exists.

Rule 3. *As K approaches ∞, the branches of the locus become asymptotic to straight lines with angles*

$$\frac{2k+1}{n_P - n_Z} 180° \qquad k = 0, \pm 1, \pm 2, \pm 3, \ldots \qquad (4\text{-}70)$$

where n_P is the number of poles and n_Z is the number of zeros.

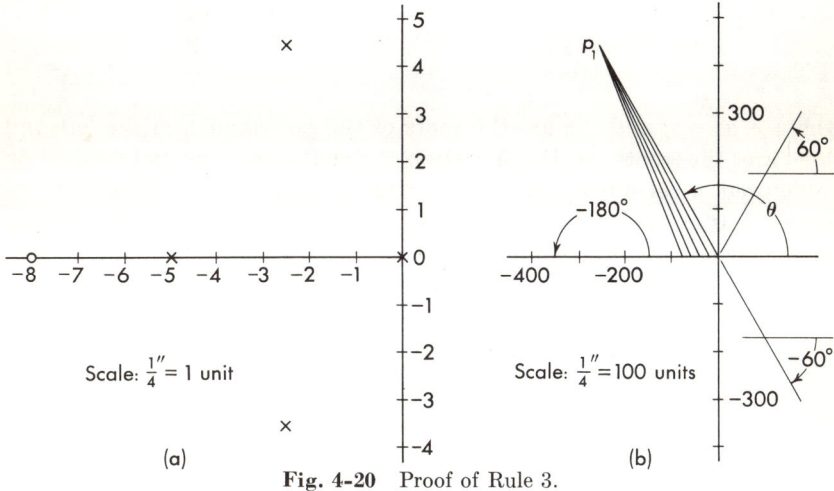

Fig. 4-20 Proof of Rule 3.

This rule is proved by reference to Fig. 4-20a and b. The scale of the axes in Fig. 4-20b is increased over that of Fig. 4-20a. The scale change reduces the clustering of zeros and poles almost to a point. Figure 4-20b shows the zero-pole configuration for large gain constant K which corresponds to roots of large magnitude. The angle subtended to a point s_1 by a line drawn from any of the zeros or poles is essentially the same.

124   Control System Design

The sign of the angles subtended by lines drawn from zeros is opposite to the sign of angles subtended by lines drawn from the poles. The sum of the angles $\Sigma \theta_i$ for Fig. 4-20b is $-(n_P - n_Z)\theta$, where n_P equals the number of poles and n_Z equals the number of zeros. If s_1 is a point on the locus (i.e., if s_1 is a root of $1 + GH = 0$), the sum of the angles must equal $(2k + 1)180°$, $k = 0, \pm 1, \pm 2, \pm 3, \ldots$, and

$$-(n_P - n_Z)\theta = (2k + 1)180°$$

$$0 = \frac{(2k + 1)180°}{n_P - n_Z} \qquad (4\text{-}71)$$

where θ is the asymptotic angle and the negative sign is included in $2k + 1$, since usually $n_P > n_Z$.

Rule 4. *The starting point for the asymptotic lines is the center of gravity of the zero-pole configuration.*

Proof of this rule stems from the theory of polynomial equations. A factored polynomial can be expanded as in the following third-order polynomial equation:

$$(s + a)(s + b)(s + c) = s^3 + (a + b + c)s^2 + (ab + ac + bc)s + abc = 0 \qquad (4\text{-}72)$$

where $-a$, $-b$, and $-c$ are the roots of the polynomial on the left and thus are the roots of the equation. Even when the polynomial is unfactored, a certain amount of information can be obtained about the roots. For example, the negative of the sum of the roots, $a + b + c$, is the coefficient of the second highest power of s in the polynomial, and $(-1)^n$ times the product of the roots is the constant term. Although Eq. (4-72) is only a third-order equation, the same statement can be shown to be true for an equation of any order.

The application to this rule follows from a consideration of the characteristic equation

$$1 + GH = 0 \qquad (4\text{-}73)$$

GH is a ratio of polynomials,

$$GH = \frac{KN(s)}{D(s)} = \frac{s^l + \cdots + a_l}{s^{l+r} + b_1 s^{l+r-1} + \cdots + b_{l+r}} \qquad (4\text{-}74)$$

where l and r are integers, a_i and b_i are constants, and N and D denote the numerator and denominator polynomials, respectively. Substituting

Eq. (4-74) into Eq. (4-73) gives

$$1 + GH = \frac{D(s) + KN(s)}{D(s)} \tag{4-75}$$

The equation of the root locus is found by setting the numerator of the right-hand member of Eq. (4-75) equal to zero, whence

$$-K = \frac{D(s)}{N(s)} = \frac{s^{l+r} + b_1 s^{l+r-1} + \cdots + b_{l+r}}{s^l + a_1 s^{l-1} + \cdots + a_l} \tag{4-76}$$

A polynomial, which can be compared with Eq. (4-72), results when the denominator is divided into the numerator:

$$s^r + (b_1 - a_1)s^{r-1} + \cdots = -K \tag{4-77}$$

When rearranged, Eq. (4-77) becomes

$$s^r + (b_1 - a_1)s^{r-1} + \cdots + K = 0 \tag{4-78}$$

where $b_1 - a_1$ is the negative of the sum of the roots of Eq. (4-78) and K is $(-1)^n$ times the product of the roots. $b_1 - a_1$ is a constant which is independent of gain as long as r is greater than 2. This is no limitation, since the rule is of value only when $r \geq 2$, that is, when $n_P - n_Z \geq 2$. When $r = 1$, the asymptote lies along the 180° line and the CG is of no interest. When $r = 0$, no asymptotes exist, because all branches terminate on finite zeros.

Since $b_1 - a_1$ is independent of the gain constant K, the sum of the roots remains at a fixed value for any value of gain. Because the roots occur in complex conjugate pairs, the sum of the roots of Eq. (4-78) is a real number; that is, $b_1 - a_1$ is real.

For large values of the gain constant and for large values of s, after division of Eq. (4-78) by K, the other terms in s in Eq. (4-78) are negligible compared with s^r/K. The result is that

$$\frac{s^r}{K} + 1 \cong 0 \quad \text{whence } s \cong \sqrt[r]{-K} = |K|^{1/r} \epsilon^{j(\pi/r + 2\pi k/r)}$$
$$k = 1, \ldots, r - 1 \tag{4-79}$$

The r roots of this equation are located on a circle and are equidistant from each other. As K increases, the roots move out along the straight lines determined by Rule 3 and indicated in Eq. (4-79).

The sum of the roots is a constant and its location is independent of gain constant K. The asymptotic lines emanate from a single point, termed the "center of gravity," as if all roots were located at that point. Hence, there are r roots effectively located at the part s_{CG}. The sum of the roots $-(b_1 - a_1)$ is equated to the sum of r roots, all of equal value s_{CG}, with the result that

$$(r)(s_{CG}) = -(b_1 - a_1)$$

Hence, the asymptotic lines radiate from the point on the real axis given by

$$\text{CG} = -\frac{b_1 - a_1}{r} \tag{4-80}$$

This point is termed the "center of gravity of the roots." In terms of the original polynomial,

$b_1 = $ negative of the sum of poles
$a_1 = $ negative of the sum of zeros
$r = $ difference between degree of denominator and numerator, $n_P - n_Z$

Equation (4-80) is written in words as

$$\text{CG} = \frac{\Sigma \text{ poles} - \Sigma \text{ zeros}}{n \text{ poles} - n \text{ zeros}} \tag{4-81}$$

Rule 5. *The breakaway point s_b is found from the equation*

$$\frac{-1}{|s_b + 1/\tau_2|} + \frac{-1}{|s_b + 1/\tau_4|} + \frac{+1}{|s_b + 1/\tau_1|} = \frac{-1}{|s_b + 1/\tau_6|} + \frac{1}{|s_b + 1/\tau_3|}$$

where $|s_b + 1/\tau_i|$ is the magnitude of the distance from the assumed breakaway point s_b to the real-axis zero or pole at $s = -1/\tau_i$.

The proof of this rule is demonstrated by Fig. 4-21. If the trial point s_b is moved a small vertical distance $\Delta\omega$ off the real axis, the net change in angle contributed from all singularities must be zero. From the example shown in Fig. 4-21, the following expression is written:

$$-(\theta_6 + \theta_4 + 180° - \theta_2) = (2k + 1)180°$$

$$-(\theta_6 + \theta_4) = -\theta_2 \tag{4-82}$$

$$\frac{-\Delta\omega}{|s_b + 1/\tau_6|} + \frac{-\Delta\omega}{|s_b + 1/\tau_4|} = \frac{-\Delta\omega}{|s_b + 1/\tau_2|}$$

where θ has been replaced by tan θ. The point s_b is determined from Eq. (4-82). It should be emphasized that a graphical, trial-and-error solution should be used when determining the breakaway point.

If complex zeros or poles occur in the GH function, the net change in angle is included with Eq. (4-43). The proof is the same.

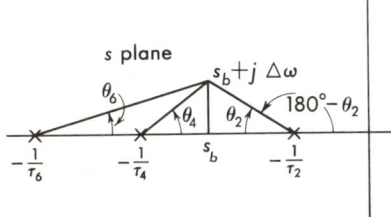

Fig. 4-21 Proof of Rule 5.

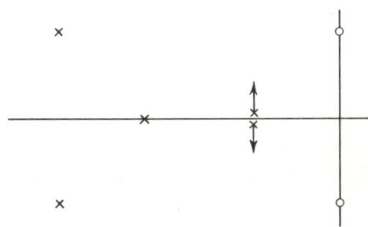

Fig. 4-22 Locus leaves double real-axis poles at $\pm 90°$.

Rule 6. Two roots leave or strike the axis at the breakaway point at an angle of $\pm 90°$.

Before verifying this rule, it is important to show that the branches leave a double real-axis pole at $\pm 90°$. Suppose a pair of poles is at a point along the real axis where an even number of poles plus zeros exists, as shown in Fig. 4-22. According to Rule 2, branches do not exist to the right or left of the double poles. To find the angle of departure from these real-axis poles, draw a small circle around them, as shown in Fig. 4-23. The angles of vectors drawn from the zeros and poles to a point s on the circle are algebraically summed as usual. As the radius of the circle is made small, the contribution to this sum from all real-axis zeros and poles is zero (for those to the left) or $k360°$ (for those to the right).

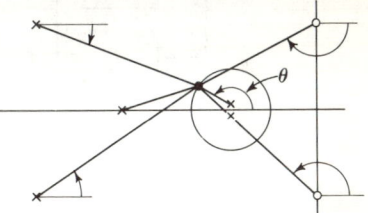

Fig. 4-23 Proof that the locus leaves a double real-axis pair of poles at $\pm 90°$.

If there is an odd number of zeros plus poles to the right, the branch does not leave the axis at that point but goes along the real axis.

The contribution from all complex zeros and poles cancels, since these always appear as complex conjugate pairs. The entire angular contribution, then, comes from the pair of poles within the small circle. The branches exist at points on the small circle where the sum of these angles is 180°, or

$$2\theta = 180° \quad \text{where } \theta = 90° \tag{4-83}$$

Hence, the branches leave a pair of real-axis poles at $\pm 90°$.

Rule 6 is verified by breaking the system, as indicated in Fig. 4-24a. The overall transfer function of the original system (Fig. 4-24a) is given by

$$\frac{K_1 G_1}{1 + K_1 G_1 H_1} \tag{4-84}$$

and for the modified multiple-loop system of Fig. 4-24b

$$\frac{K_1 G_1 H_1 / (1 + K_1 G_1 H_1)}{1 + K_1 G_1 K' H_1 / (1 + K_1 G_1 H_1)} = \frac{K_1 G_1}{1 + (1 + K') K_1 G_1 H_1} \tag{4-85}$$

The denominators of Eqs. (4-84) and (4-85) have the same form. The constant in the latter equation is made up of a product of two constants:

$$K_1 (1 + K') \tag{4-86}$$

As K_1 is varied, with $K' = 0$, the root locus is plotted in the normal fashion. The value of K_1 is so set that two roots of $1 + K_1 G_1 H_1$ coalesce.

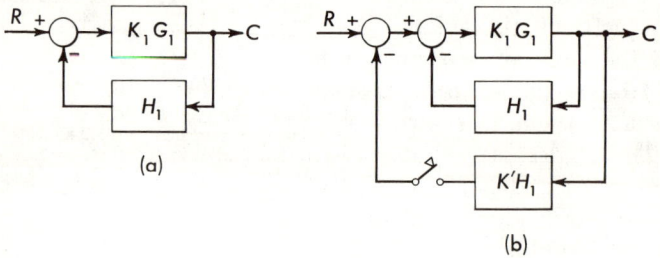

Fig. 4-24 Block diagrams used to prove Rule 6.

With this value of gain, the outer loop is closed and the roots of $1 + K_1 G_1 H_1$ become poles of the new function that is to be plotted, as K' varies from zero:

$$1 + \frac{K_1 G_1}{1 + K_1 G_1 H_1} K' H_1 \tag{4-87}$$

The two roots of $1 + K_1 G_1 H_1 = 0$ which are at the same point are now two identical poles. As shown above, when K' is increased, two branches leave these roots at $\pm 90°$.

If the complete locus had been plotted for the original equation $1 + K_1G_1H_1 = 0$, the locus would have the same shape as for the second equation $1 + (1 + K')K_1G_1H_1 = 0$. It must be concluded, then, that the branches break away from the real axis at $\pm 90°$.

Rule 7. *The angle of departure of the branches from complex poles and the angle of arrival of the branches at complex zeros are found by summing the angles to all other zeros and poles. The desired angle is found by subtracting $180°$ from the sum.*

Rule 7 defines the initial angle of departure of the branches leaving complex poles. The angle subtended by the complex conjugate pole to a test point located a small distance Δs from the complex conjugate pole

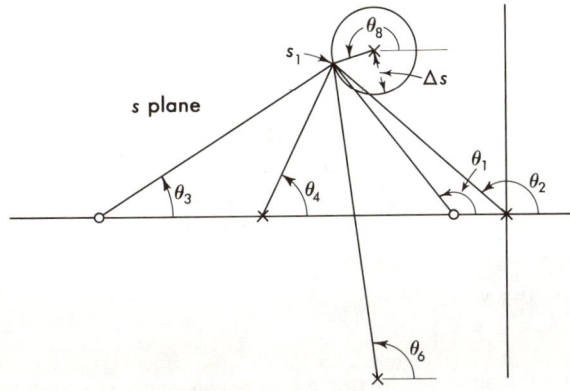

Fig. 4-25 Proof of Rule 7.

is the angle of departure of the branch from the pole. This angle is determined by drawing a small circle about the pole as shown in Fig. 4-25. The angles subtended by all the singularities, including the complex conjugate poles under consideration, are summed at a point s_1 along the small circle:

$$+\theta_1 + \theta_3 - (\theta_2 + \theta_4 + \theta_6 + \theta_8) = (2k + 1)180° \qquad (4\text{-}88)$$

Equation (4-88) is solved for θ_8:

$$\theta_8 = [\theta_1 + \theta_3 - (\theta_2 + \theta_4 + \theta_6)] - (2k + 1)180° \qquad (4\text{-}89)$$

As the small circle about the point is reduced to zero ($\Delta s \to 0$), the angles θ_i are simply the angles subtended between vectors drawn from the various poles and zeros to the particular complex pole in question. The angles of vectors from all the poles and zeros to the complex pole are summed in the usual fashion (the angles of vectors drawn from poles are taken negative and those of vectors drawn for zeros are taken positive). The angle of departure θ_8 differs from this sum by 180°. In Sec. 4-9 a simpler method is given for finding the angle of departure by use of the Spirule.

When the rule is applied to find the angle of arrival at a complex zero, only a change in sign is required. This is seen by reference to Eq. (4-89). If θ_8 was the angle of arrival at a zero, the opposite sign is taken in the sum.

4-8 The Spirule. Use of the root-locus technique requires the measurement and algebraic summation of angles and the measurement and multiplication (or division) of lengths.

These graphical constructions can be performed by using a protractor to measure the angles of the vectors to the trial point and a ruler to determine the magnitudes of these vectors for which the gain constant K is computed. If the locus is constructed on graph paper, the magnitude may also be measured with dividers and the lengths determined by comparison with the scales of the paper. For greater accuracy the coordinates of the trial point can be obtained from the known pole and zero positions and the real and imaginary parts of the various vectors can be computed. A desk calculator and a table of tangents permit the determination, to any degree of accuracy, of the locations of particular points on the locus.

Most control problems do not require the accuracy possible with the latter method. In fact, the real advantage in the root-locus technique rests in the fact that the root-locus plots can be sketched rapidly; hence, the control system can be analyzed quickly. For this reason and to facilitate the needed computation, a device called the "Spirule"* has been developed. The Spirule performs three functions:

1. Adds the angles of directed lines from the zeros to a point on the locus and subtracts the angles of directed lines from the poles to a point on the locus.
2. Effects multiplication of the distances from a point to the poles and division by the product of the distances to the zeros.
3. Gives the damping ratio ζ for applicable points on the locus.

The Spirule, which is shown in the sketch of Fig. 4-26, comprises a transparent protractor for addition of the angles and a logarithmic spiral

* The Spirule was developed by Mr. Walter Evans and is available from the Spirule Company, 9728 El Venado, Whittier, Calif.

for multiplication of vector lengths. It consists of a circular disk and an arm which is held to the disk with an eyelet that provides light friction between the disk and arm. A pivot, in the form of a pin or a movable insert, is provided on the eyelet to locate the Spirule over the trial point.

The instrument is used to locate points on a locus which has been previously sketched. A user finds the angle of GH at a trial point s_1 by linear addition of angles and finds the magnitude at a point on the locus

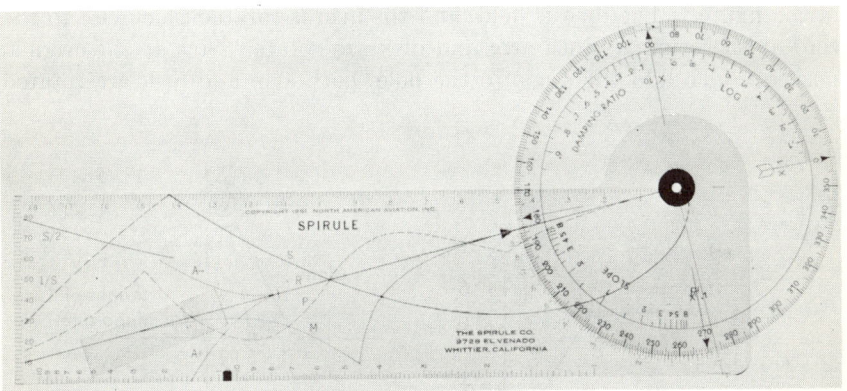

Fig. 4-26 The Spirule.

by adding the logarithms of the vectors and then taking the antilog. Both computations are effected directly by use of the Spirule.

4-9 Use of the Spirule to Sum Angles. The first requirement in plotting root-locus diagrams is to choose equal scales on the plot. It is convenient to choose a scale on the chart that corresponds to an even multiple of the scale on the top edge of the arm. (A distance of 0.1 on the scale equals $\frac{1}{2}$ in.)

To find vector angles from the point s_1 with the Spirule, the recommended sequence is:

1. Align the R arrow on the arm with the 0° mark on the disk.
2. Place the center of the disk over the point where the angles are to be measured and set the R arrow pointing to the left.
3. *For a zero* hold the disk and rotate the arm from the horizontal until the R arrow on the arm passes through the zero. Read the positive angle on the disk opposite the R arrow.
4. *For a pole* rotate both the arm and the disk so that the R arrow on the arm passes through the pole. Hold the disk and rotate the R arrow back to horizontal. Read the negative angle from the circular disk.

5. When adding angles, proceed as in 1 and 2. Add zeros by applying 3 successively, then subtract the poles by applying 4 successively. The net angle must equal $(2k + 1)180°$ to satisfy the angle condition of Eq. (4-29).

As an example, in Fig. 4-27, consider use of the Spirule to check the angle condition at point s_1. The Spirule is placed at the point s_1, as shown in the figure. The disk is held, and the arm is rotated clockwise to the zero at the origin. Both arm and disk are rotated back to the original position. To add the angle to the pole, both arm and disk are rotated

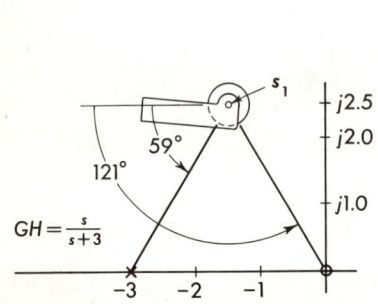 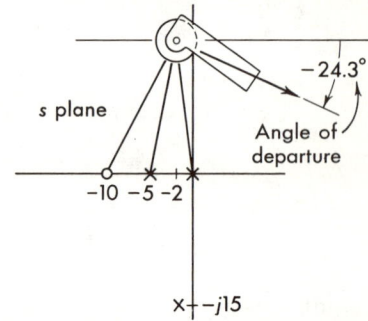

Fig. 4-27 Use of the Spirule for plotting root-locus diagrams.

Fig. 4-28 Use of the Spirule to find the angle of departure from complex poles.

until the R arrow of the arm passes through the pole. The disk is held, and the arm is rotated counterclockwise back to the horizontal. The Spirule reads 62°. The angle condition is not satisfied. Use the Spirule to see if the angle condition is satisfied when the zero at the origin is replaced by a pole. (Answer: It is.)

As a second example, consider again the problem of finding the breakaway angle from the complex pole of Fig. 4-14. Place the Spirule on the point $-2 + j15$. The 0° line is aligned with the R arrow of the arm. Follow the procedure outlined above. Hold the disk, and move the R arrow of the arm to the zero. Return the disk, and successively subtract the angles at the poles located at 0, -5, and $-2 - j15$. The Spirule should read 156°. To find the angle of departure, set the Spirule, with the arm remaining in position, so that the 0° mark on the disk is horizontal to the left. The Spirule is then in the position shown in Fig. 4-28. The angle of departure is read to the right as $-24.3°$.

4-10 The Spirule Used to Find Lengths. The logarithmic curve on the arm of the Spirule permits the multiplication of lengths. The angle between the arm and the curve is proportional to the logarithm of the distance from the center of the rule to the log curve.

To multiply lengths:

1. Align the upper side of the arm with the 0° mark on the disk.
2. Place the center of the disk over the point where the lengths are to be measured.
3. *For a zero* rotate both disk and arm until the log curve passes through the zero. Hold the disk and move the arm until the top of the arm passes through the zero. The reciprocal of the magnitude $|s + 1/\tau_1|$ is read on the log-spiral curve opposite the arrow on the disk. If the ×1 arrow is pointing to the log curve, the number along the curve is the reciprocal magnitude. If the ×0.1 arrow is pointing to the log curve, the reciprocal magnitude is 0.1 times the number read along the log spiral.
4. *For a pole* put the top of the arm (i.e., rule) on the pole. Hold the disk and move the arm until the log curve passes through the pole. The length of the vector, $|s + 1/\tau_2|$, is read opposite the arrow. The length is found by multiplying the number found on the log curve by the ×1, the ×10, or the ×0.1 number which is on the disk and which is pointing to the log curve.
5. *The net length*, or the gain constant K, is found by applying Rule 3 successively for the zeros and Rule 4 for the poles. The gain constant K is the number found on the log curve times the × number found on the disk, both multiplied by the scale factor, or

$$K = (\text{scale factor})(\text{number on log curve})(\times \text{ number}) \quad (4\text{-}90)$$

6. *Scale factor.* The scale of the problem does not necessarily agree with the scale on the Spirule. The scale factor for Eq. (4-90) is

$$\text{Scale factor} = \sigma^{n_P - n_Z} \quad (4\text{-}91)$$

where σ is the number on the problem scale read opposite the 1.0 on the Spirule arm. n_P is the number of poles, and n_Z is the number of zeros.

As an example, consider the GH function of Fig. 4-29. To find the gain at s_1, place the Spirule at the position shown in the figure. Place the arm on the pole at the origin, hold the disk, and rotate the arm until the curve is on the pole. The ×1 arrow on the disk is pointing to 0.29.

Repeat again from the pole at $s = -3$. The ×0.1 arrow is now pointing to 0.84. Place the pivot of the Spirule at the origin; the number read opposite the 1 on the arm is 10. The scale factor is $(10)^2 = 100$. The gain, which is equal to the product of the two lengths, is

$$(0.84)(\times 0.1)(100) = 8.4$$

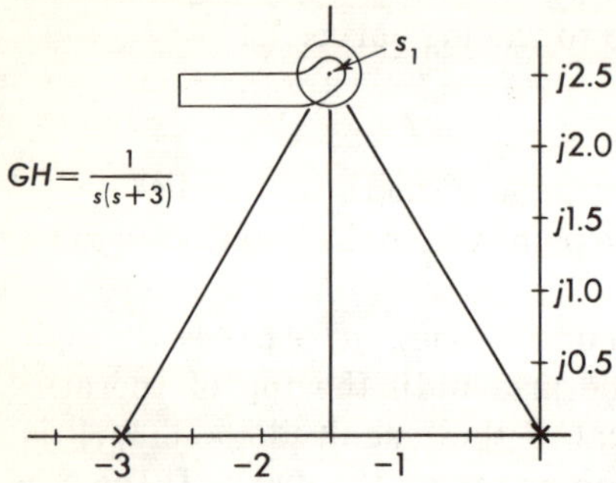

$GH = \dfrac{1}{s(s+3)}$

Fig. 4-29 Use of the Spirule to measure gain.

The dashed portion of the log curve is an extension of the logarithmic spiral for vectors whose lengths are greater than 1. Since the curve in this region has been reversed in sense (so the curve does not run off the rule), step 3 is used for poles and step 4 for zeros; i.e., the rules are reversed.

4-11 The Spirule Used to Find Damping Ratio. The damping ratio is derived in Chap. 1 for the second-order characteristic equation

$$s^2 + 2\zeta\omega_n s + \omega_n^2 = 0 \tag{4-92}$$

Any pair of complex conjugate roots is located as in Fig. 4-30. The horizontal distance from the axis is $\zeta\omega_n$, and the vertical distance is $\omega_n \sqrt{1 - \zeta^2}$. This is verified by finding the roots of Eq. (4-92),

$$s_1 = -\zeta\omega_n \pm j\omega_n \sqrt{1 - \zeta^2} \tag{4-93}$$

The angle θ is computed from Fig. 4-30 to be

$$\cos \theta = \frac{\zeta\omega_n}{\omega_n} = \zeta \tag{4-94}$$

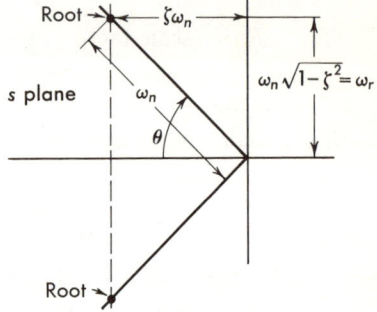

Fig. 4-30 Use of the Spirule to find damping ratio.

The damping ratio is read directly from the s plane with the Spirule. Set the arrow on the rule opposite 180° on the disk, and place the center of the disk at the origin of the s plane so that 180° on the disk lies along the negative real axis. Move the arm to the root where the damping ratio is to be determined. The damping ratio ζ is read from the intersection of the arm and the scale on the disk.

4-12 Root-locus Plots with Variable Other than Gain. All examples presented thus far in this chapter deal with location of the roots of a closed-loop system as the loop gain is varied. Often it is necessary to optimize a system with respect to another variable, for example. The root-locus method is basically a method of determining the position of the roots of a polynomial equation as a quantity K is varied whenever the polynomial equation can be written in the form

$$1 + K \frac{N(s)}{D(s)} = 0 \qquad (4\text{-}95)$$

In general, any problem, control or otherwise, that can be put into the form of Eq. (4-95) can be solved with the aid of the root-locus method.

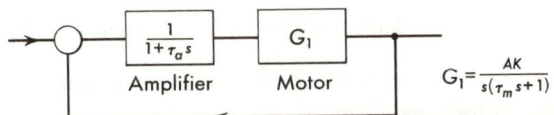

Fig. 4-31 Effect of time constant τ_a on control-system stability.

The parameter K must not be a function of s, and it must appear as a multiplicable quantity. The problem is simplified if the polynomials $N(s)$ and $D(s)$ are factored.

As an example, consider the problem of determining the effect of amplifier time constant τ_a upon the control system pictured in Fig. 4-31. The problem has been solved and the gain AK set for a particular value of τ_a and τ_m. To determine the effect upon the system of variations of τ_a, the block diagram is altered as shown in Fig. 4-32. The $1/(1 + \tau_a s)$ term is represented by two blocks. The G_1 block, which represents the product of the amplifier constant A and motor transfer function,

$$G_1 = \frac{AK}{s(\tau_m s + 1)} \qquad (4\text{-}96)$$

is divided as shown in Fig. 4-32b. The two loops are interchanged as shown in Fig. 4-32c. The inner loop is combined into one block, $1/(1 + G_1)$, and the outer loop is closed around the entire system with the final diagram taking the form of Fig. 4-32e. Use is made of block-

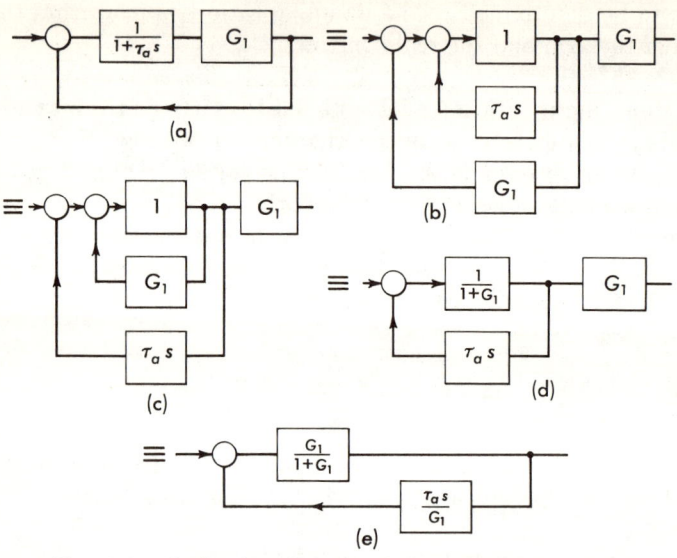

Fig. 4-32 Reduction of the block diagram of Fig. 4-31.

diagram algebra presented in Chap. 1 (Sec. 1-8) for these manipulations. The GH function for this diagram is

$$GH = \frac{\tau_a s}{1 + G_1} = \frac{\tau_a s}{1 + AK/s(\tau_m s + 1)} \qquad (4\text{-}97)$$

Clearing fractions in the right-hand member gives

$$GH = \tau_a \frac{s^2(s + 1/\tau_m)}{s^2 + (1/\tau_m)s + AK/\tau_m} \qquad (4\text{-}98)$$

Equation (4-98) satisfies the requirements for use in a root-locus construction with τ_a as the variable. The poles and zeros are fixed for given values of A, K, and τ_m. The system has three zeros—a single zero located at $-1/\tau_m$ and a double zero at the origin—and two poles located at

$$-\frac{1}{2\tau_m} \pm j\sqrt{\frac{AK}{\tau_m} - \left(\frac{1}{2\tau_m}\right)^2} \qquad (4\text{-}99)$$

The locus, which is a plot of the system roots as the amplifier time constant is increased, is shown in Fig. 4-33. The form of the locus is different from that found in a conventional control problem. In this problem there are more zeros than poles. Rather than a branch ending at infinity on an infinite zero, here the branch starts at infinity on an infinite pole and comes in from $s = \infty \epsilon^{-j\pi}$ along the negative real axis to terminate on a finite zero. All the root-locus rules are applicable, except that the variable is the time constant τ_m rather than the gain constant K.

The unusual form of this root locus, i.e., more zeros than poles, can be avoided if the following inverse function is plotted:

$$1 + \frac{1}{K}\frac{1}{GH} = 1 + K'\frac{1}{GH} \qquad (4\text{-}100)$$

In this form the zeros and poles are interchanged and the more usual plot is obtained.

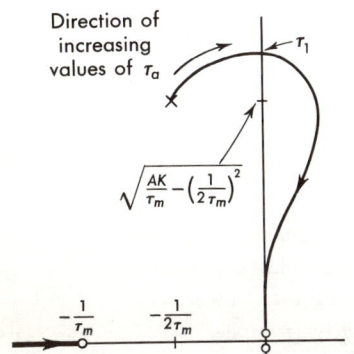

Fig. 4-33 Locus of third-order system as time constant is varied.

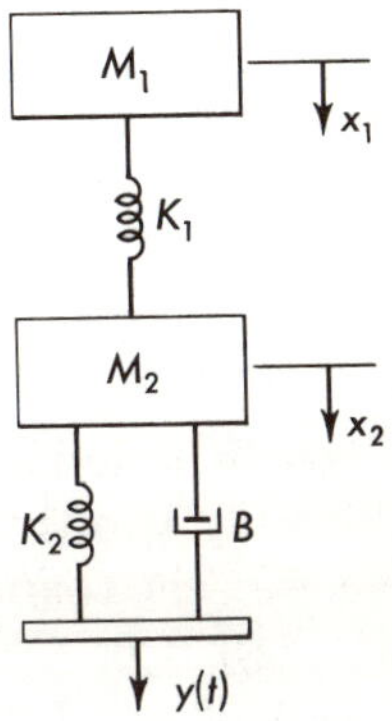

Fig. 4-34 Shock mount which is optimized with root-locus method.

In either case, the significance of a root in the right half plane is still the same. For the system of Fig. 4-33 the amplifier time constant can become no larger than τ_1, at which value a branch of the locus crosses into the right half plane.

Electrical and mechanical circuit problems are also amenable to solution with the root-locus method. Consider the shock mount problem indicated in Fig. 4-34. This system is driven with a position input $y(t)$, and the following values are used:

$$W_1 = 50 \text{ lb } (M_1 = 1.55 \text{ slugs})$$
$$W_2 = 20 \text{ lb } (M_2 = 0.62 \text{ slug})$$
$$K_1 = 100 \text{ lb/in.} = 1{,}200 \text{ lb/ft}$$
$$K_2 = 70 \text{ lb/in.} = 840 \text{ lb/ft}$$

138 Control System Design

The value of B, in pound-seconds per foot, is to be optimized for maximum possible damping on the upper mass. From a practical point of view, it is interesting to ascertain if an optimum does exist. If $B = 0$, the system is undamped. If $B = \infty$, M_2 is fastened rigidly to the base and again the system is undamped. Since energy is dissipated for values between $B = 0$ and $B = \infty$, it is reasonable to assume that an optimum value of damping does exist.

The problem is solved by arranging the characteristic equation so that B is the parameter of a root-locus plot. The equations of motion for the system are

$$(M_2 s^2 + Bs + K_1 + K_2)X_2 + (-K_1)X_1 = (Bs + K_2)Y$$
$$(-K_1)X_2 + (M_1 s^2 + K_1)X_1 = 0 \quad (4\text{-}101)$$

Solution of these equations for X_1 yields

$$\frac{X_1}{Y} = \frac{(K_1 K_2 / M_1 M_2)(1 + sB/K_2)}{s^4 + (B/M_2)s^3 + [(K_1 + K_2)/M_2 + K_1/M_1]s^2 \\ + (BK_1/M_1 M_2)s + K_1 K_2/M_1 M_2} \quad (4\text{-}102)$$

The characteristic function, which is in the denominator of Eq. (4-102), contains B in only two terms. To solve the shock mount problem with B as a parameter, the characteristic equation must be written in the form $1 + B[N(s)/C(s)] = 0$. With this form as a goal, the characteristic equation is rearranged as follows:

$$\left[s^4 + \left(\frac{K_1 + K_2}{M_2} + \frac{K_1}{M_1} \right)s^2 + \frac{K_1 K_2}{M_1 M_2} \right] + \frac{Bs}{M_2}\left(s^2 + \frac{K_1}{M_1} \right) = 0 \quad (4\text{-}103)$$

When Eq. (4-103) is divided by the first term in brackets, the correct form results:

$$1 + B \frac{N(s)}{D(s)} = 1 + \frac{(B/M_2)s(s^2 + K_1/M_1)}{s^4 + [(K_1 + K_2)/M_2 + K_1/M_1]s^2 + K_1 K_2/M_1 M_2} \\ = 0 \quad (4\text{-}104)$$

When the numerical values of the problem are substituted into Eq. (4-104), the characteristic equation reduces to

$$1 + B \frac{1.61 s(s^2 + 773)}{(s^2)^2 + (4.06 \times 10^3)s^2 + 1.05 \times 10^6} = 0 \quad (4\text{-}105)$$

which is factored (with the binomial equation) to

$$1 + B \frac{1.61s(s^2 + 773)}{(s^2 + 3{,}780)(s^2 + 274)} = 0 \qquad (4\text{-}106)$$

The left-hand member of Eq. (4-106) has two imaginary zeros at $\pm j27.8$, one zero at the origin, two imaginary poles at $\pm j16.5$, and two imaginary poles at $\pm j61.5$.

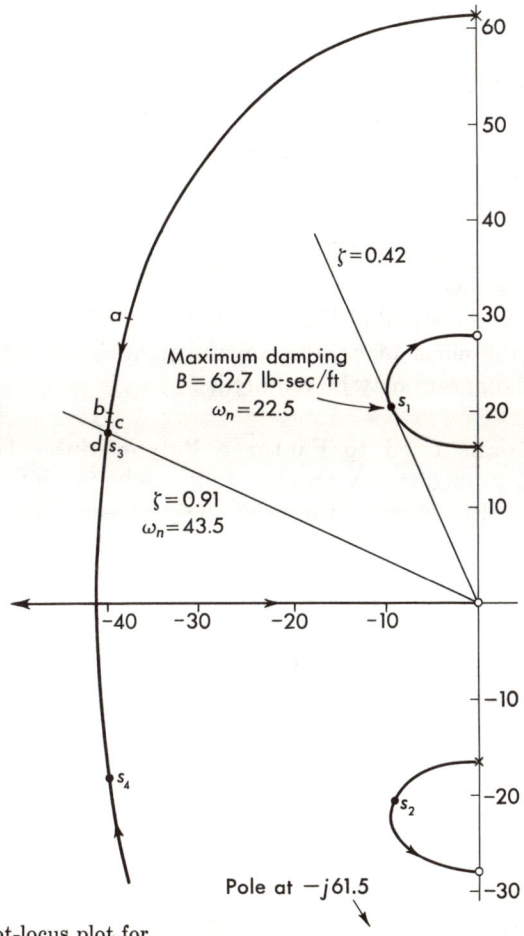

Fig. 4-35 Root-locus plot for

$$\frac{1.61Bs(s \pm j27.8)}{(s \pm j16.54)(s \pm j61.5)}$$

The root locus, shown in Fig. 4-35, is plotted in the normal manner. The variable is the damping constant B. The system has four roots; however, the least damped roots correspond to s_1 and s_2. The maximum damping is found by drawing a line tangent to the inner locus from the origin. The damping ratio at this point is $\zeta = 0.42$. The value of B which produces this root is found in the same manner as the gain is found (cf. Sec. 4-6). The product of the pole distances divided by the zero distance yields

$$B = 62.7 \text{ lb-sec/ft} \tag{4-107}$$

At this value of B the other two roots s_3 and s_4 are located at $\zeta = 0.91$ and $\omega_n = 43.5$. These roots are sufficiently damped and far removed to have negligible effect.

The method of this example can be applied whenever the parameter to be investigated can be isolated and the transformed characteristic equation written in the form

$$1 + K \frac{N(s)}{D(s)} = 0 \tag{4-108}$$

where K is the parameter and N and D are polynomials in s but independent of K. If the effect of varying several parameters is to be studied, more than one diagram may be required.

4-13 Root Locus Used to Factor a Polynomial. The root-locus method can be applied to any problem that can be put into control form, i.e., the form of Eq. (4-95). Consider the problem of finding the roots of the polynomial

$$s^4 + 5s^3 + 22s^2 + 80s + 96 = 0 \tag{4-109}$$

This equation is rewritten in the form

$$s^2(s^2 + 5s + 22) + 80(s + \tfrac{96}{80}) = 0 \tag{4-110}$$

The quadratic is factored:

$$s^2(s + \tfrac{5}{2} + j3.96)(s + \tfrac{5}{2} - j3.96) + 80(s + 1.2) = 0 \tag{4-111}$$

Equation (4-111) is rewritten in the form

$$1 + \frac{80(s + 1.2)}{s^2(s + \tfrac{5}{2} + j3.96)(s + \tfrac{5}{2} - j3.96)} = 0 \tag{4-112}$$

The equation is now in control form, with one zero at -1.2 and four poles located as follows:

Two at the origin
One at $-\frac{5}{2} + j3.96$
One at $-\frac{5}{2} - j3.96$

The root locus is plotted, as shown in Fig. 4-36, as a function of a variable K. The roots of Eq. (4-112), which are the roots of the original equation, are found from the plot of Fig. 4-36 for the value of $K = 80$. At this value of gain the four roots are found from the four branches of the locus to be

$$s_{1,2} = \pm j4$$
$$s_3 = -2 \qquad (4\text{-}113)$$
$$s_4 = -3$$

and are marked in Fig. 4-36 with a small square.

In general, the procedure for factoring an nth-order polynomial consists in splitting the original equation

$$x^n + a_1 s^{n-1} + a_2 x^{n-2} + \cdots$$
$$+ a_{n-1} x + a_n = 0 \qquad (4\text{-}114)$$

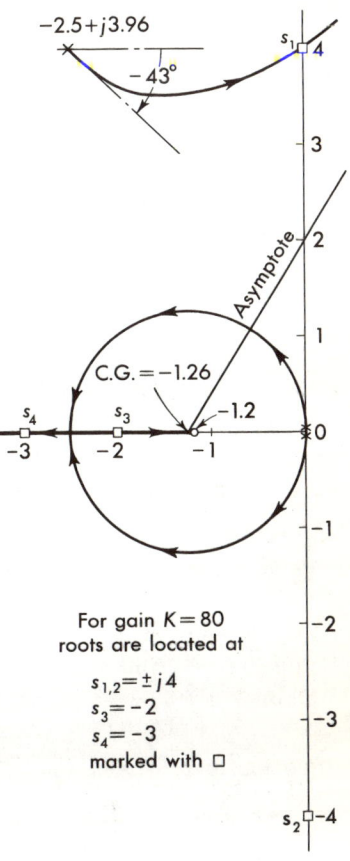

Fig. 4-36   Root locus used to find the roots of a polynomial.

as follows:

$$\left\{ \left[(x^2 + a_1 x + a_2) x^2 + a_3 \left(x^2 + \frac{a_4}{a_3} x + \frac{a_5}{a_3} \right) \right] x^2 \right. $$
$$\left. + a \left(x^2 + \frac{a_7}{a_6} x + \frac{a_8}{a_6} \right) \right\} x^2 + \cdots \qquad (4\text{-}115)$$

The quadratic terms are factored by use of the binomial expression, and the roots for each bracketed term are found from a root-locus plot. Illustratively, take $n = 6$ in Eq. (4-114). The equation

$$x^6 + a_1 x^5 + a_2 x^4 + a_3 x^3 + a_4 x^2 + a_5 x + a_6 = 0 \qquad (4\text{-}116)$$

is arranged as follows:

$$\left[(x^2 + a_1 x + a_2)x^2 + a_3\left(x + \frac{a_4}{a_3}\right)\right]x^2 + a_5 x + a_6 = 0 \quad (4\text{-}117)$$

The quadratic is then factored:

$$\left[(x + r_1)(x + r_2)x^2 + a_3\left(x + \frac{a_4}{a_3}\right)\right]x^2 + a_5 x + a_6 = 0 \quad (4\text{-}118)$$

Equation (4-118) is next rewritten and solved with two root-locus plots. The first is for the system defined by

$$1 + a_3 \frac{x + a_4/a_3}{x^2(x + r_1)(x + r_2)} = 0 \quad (4\text{-}119)$$

The roots $-r_3$, $-r_4$, $-r_5$, and $-r_6$ of this equation are found from a root-locus plot with the following conditions:

Zero at $-a_4/a_3$
Poles at $-r_1$ and $-r_2$ and two at the origin
Gain $= a_3$

Equation (4-118) is now written

$$[(x + r_3)(x + r_4)(x + r_5)(x + r_6)]x^2 + a_5 x + a_6 = 0 \quad (4\text{-}120)$$

A second root-locus plot is used to find the roots of Eq. (4-120), which is rearranged as

$$1 + \frac{a_5(x + a_6/a_5)}{x^2(x + r_3)(x + r_4)(x + r_5)(x + r_6)} = 0 \quad (4\text{-}121)$$

The root locus has one zero and six poles, four located at $-r_3$, $-r_4$, $-r_5$, and $-r_6$ and two at the origin. The six roots of Eq. (4-121), which are the six roots of the original equation (4-116), are found for a gain of a_5.

There are several methods[11] of using the root locus to factor polynomials. For example, the characteristic equation could be split into cubics instead of quadratics. The method presented, however, seems easier.

4-14 Root Locus Applied to Multiple-loop Systems. A multiple-loop control system consists of several feedback paths. Figure 4-37 indicates a block diagram for a typical multiple-loop system. The inner loop can be replaced by a single block with transfer function

$$\frac{C}{R_2} = \frac{K_2 G_2}{1 + A_1 K_2 H_1 G_2} \qquad (4\text{-}122)$$

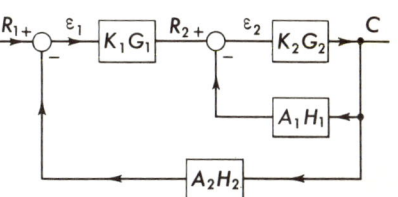

Fig. 4-37 A multiple-loop control system.

This equation is solved by root-locus methods to find the roots of the characteristic equation for a particular value of $A_1 K_2$. These roots become poles in the overall system, so the outer loop system is plotted with the equation

$$GH = K_1 A_2 G_1 H_2 \frac{K_2 G_2}{1 + A_1 K_2 H_1 G_2} \qquad (4\text{-}123)$$

Besides the roots of $1 + A_1 K_2 H_1 G_2$, which are poles of GH, the zeros and poles of G_1, H_2, and G_2 are added and a complete plot is made. The application of the root locus is identical with that for single-loop systems. The difficulty that exists for multiple-loop systems is that, as $A_1 K_2$ varies, the roots of $1 + A_1 K_2 H_1 G_2$ and hence the poles of GH vary. Hence, two plots are necessary to determine the response.

As an example, take

$$K_1 G_1 = \frac{K_1}{s} \qquad K_2 G_2 = \frac{K_2}{0.5s + 1} \qquad (4\text{-}124)$$

$$A_1 H_1 = \frac{1}{(0.5s + 1)^2} \qquad A_2 H_2 = 1$$

The root-locus sketch for the inner loop is shown in Fig. 4-38. For a particular gain K_2 the roots are located, as shown in Fig. 4-38, at $-r_1$, $-r_2$, and $-r_3$. The outer-loop locus consists of the transfer function

$$\frac{K_1 K_2 (0.5s + 1)^2}{s[(0.5s + 1)^3 + K_2]} \qquad (4\text{-}125)$$

The locus for this system is shown in Fig. 4-39, where the variable along the curve of this second locus is K_1, since K_2 is fixed.

The optimization of a multiple-loop system requires that two plots be made, one for the inner loop and one for the outer. A variation of

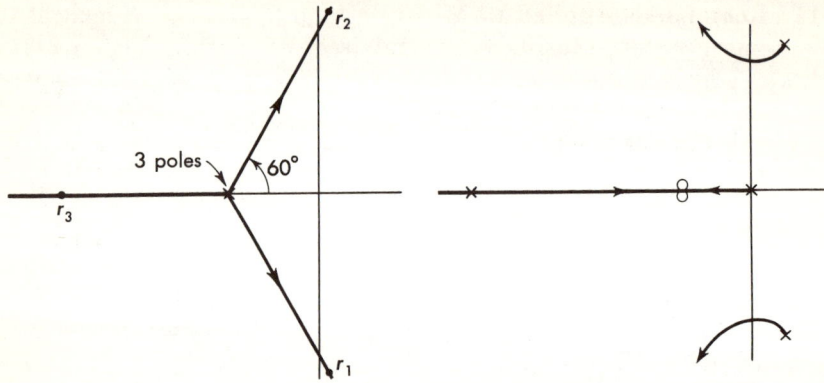

Fig. 4-38 Root-locus plot for inner loop of a multiple-loop control system.

Fig. 4-39 The second root-locus plot for a multiple-loop system.

the inner-loop gain changes the location of the poles of the outer-loop transfer function. Hence, the design of a multiple-loop system requires more plots than that of a single-loop system.

PROBLEMS

Except where noted, in the problems of this chapter $G(s)$ represents the forward transfer function in a system with unity feedback, $H = 1$.

4-1. Plot the root of the equation $G(s) = K_1/(s+1)$ as K_1 varies.

4-2. Plot the root of the equation $G(s) = K_1(s+3)$ as K_1 varies.

4-3. Plot the roots of the equation

$$G(s) = \frac{K_1}{s(s+1)(s+3.5)(s+3+j2)(s+3-j2)}$$

as K_1 varies.

4-4. Plot the roots of the equation

$$G(s) = \frac{K_1 s(s+2)}{(s+5)(s+3+j4)(s+3-j4)}$$

as K_1 varies.

4-5. $G(s) = \dfrac{K(s+2+j2)(s+2-j2)}{(s+1)(s+3)}$.

(a) Plot the roots of the equation as K_1 varies.

(b) Determine the net angle subtended by the poles and zeros from point s_0 located at $-3 + j2$. Is this point on the locus?

4-6. $G(s) = \dfrac{K_1}{s(s+1)(s+10)}$.

(a) Plot the roots of the above equation as K_1 varies.

(b) Find the distance between point $s = -\tfrac{1}{2}$ and the following points:

(1) $(-2.414 + j1.414)$
(2) $(-4 + j0)$
(3) $(2 + j0)$

4-7. For each of the following transfer functions locate the zeros and poles and make root-locus *sketches*. Discuss the stability of each case.

(a) $G(s) = \dfrac{K_1}{s(2s + 1)}$

(b) $G(s) = \dfrac{K_1(3s + 1)}{s(2s + 1)}$

(c) $G(s) = \dfrac{K_1(s + 1)}{s(2s + 1)}$

(d) $G(s) = \dfrac{K_1}{s(s + 1)(2s + 1)}$

(e) $G(s) = \dfrac{K_1}{s^2 + 10s + 100}$ *Ans.* Stable for any K_1

(f) $G(s) = \dfrac{K_1(s + 1)}{s^2 + s + 10}$ *Ans.* Stable for any K_1

(g) $G(s) = \dfrac{K_1 s}{(s + 1)(s + 10)}$ *Ans.* Stable for any K_1

(h) $G(s) = \dfrac{K_1}{s(s + 2)(s^2 + 2s + 37)}$ *Ans.* Unstable for $K_1 > 30$

4-8. For any of the systems in Prob. 4-7 that are unstable, estimate the values of gain constant K_1 for which the branches cross the imaginary axis.

4-9. In the systems characterized by the following G, find the gain when the damping ratio is $\zeta = 0.2$.

(a) $\dfrac{K_1}{s(s + 1)(0.2s + 1)}$ *Ans.* $K = 2.3$

(b) $\dfrac{K_1(s + 0.2)}{s(s + 1)(0.2s + 1)(s + 20)}$ *Ans.* $K = 250$

(c) $\dfrac{K_1(s + 2)}{s(s + 1)(0.2s + 1)(s + 200)}$ *Ans.* $K = 460$

(d) $\dfrac{K_1(s + 5)}{s(s + 1)(0.2s + 1)(s + 500)}$ *Ans.* $K = 625$

4-10. Sketch the root-locus diagrams for the following functions and discuss stability. For any of the systems that are unstable, estimate the values of the gain constant K_1 of the system when the branches cross the imaginary axis.

(a) $G = \dfrac{K_1}{s^2 + 2s + 100}$ $H = \dfrac{1}{s}$ (b) $G = \dfrac{K(1 + 3s)}{s^2 + 2s + 100}$ $H = \dfrac{1}{s^2}$

4-11. Sketch the root-locus diagram for the system characterized by

$$G = \dfrac{K_1(s + 2)}{s(s + 20)} \quad \text{and} \quad H = \dfrac{s + 4}{s^2}$$

and discuss stability. Estimate the value of the gain constant K_1 for which the system becomes unstable.

4-12. For each of the following G functions plot the root-locus diagram and determine the roots of the characteristic equation for values of K_1 equal to 1, 10, 100, and 1,000.

(a) $G = \dfrac{K_1}{s(s+1)}$

Ans. Gain	Root
1	$-\frac{1}{2} \pm j\frac{\sqrt{3}}{2}$
10	$-\frac{1}{2} \pm j\frac{\sqrt{39}}{2}$
100	$-\frac{1}{2} \pm j10$
1,000	$-\frac{1}{2} \pm j\sqrt{1000}$

(b) $G = K_1(s+1)$

(c) $G = \dfrac{K_1}{s^2(s+1)^2}$

4-13. Find the gain K_1 when the damping ratio of the system characterized by the following transfer functions is 0.1.

(a) $G = K_1 \dfrac{s^2 + 2s + 20}{(0.5s + 1)^3}$

(b) $G = \dfrac{K_1}{s(0.5s + 1)(0.05s + 1)}$

(c) $G = \dfrac{K_1 s(s^2 + 10)}{(s^2 + 5)(s^2 + 30)}$

4-14. Are there any zeros of $1 + G(s)$ in the right half plane for the following function?

$$G(s) = \dfrac{10}{s(0.1s + 1)(0.5s + 1)(s + 1)}$$

Find all four roots of this equation.

Ans. $s_1 = -9.7$; $s_2 = -4.4$; $s_3, s_4 = 0.55 \pm j2.1$

4-15. For what least value of K_1 does the following system become unstable?

$$G(s) = \dfrac{K_1}{s^2(s + 1)(s + 10)}$$

4-16. Determine the velocity error constant K_v and the resonant frequency ω_r when the gain in the system

$$G(s) = \dfrac{K_1}{s(1 + 0.5s)(1 + 0.2s)} \qquad H = 1$$

is set for a damping ratio ζ equal to 0.40. Ans. $K_v = 1.56$

4-17. Repeat Prob. 4-16 for when a series lead network is added. The system functions become

$$G = \dfrac{0.25 K_1 (1 + 0.3s)}{(1 + 0.075s) s (1 + 0.5s)(1 + 0.2s)} \qquad H = 1 \qquad Ans.\ K_v = 3.06$$

4-18. Repeat Prob. 4-16 for a system with position plus rate feedback. The system functions become

$$G = \frac{K_1}{s(1+0.5s)(1+0.2s)} \qquad H = 1 + 0.3s \qquad Ans.\ K_v = 6.0$$

4-19. Repeat Prob. 4-16 for a system with two series-parallel lead networks. The system functions become

$$G = \frac{K_1(1+1/6s)}{(1+1/60s)s(1+0.5s)(1+0.2s)} \qquad H = \frac{1+1/6s}{1+1/60s} \qquad Ans.\ K_v = 90.0$$

4-20. Use the Spirule to find the root-locus diagrams for

(a) $G = \dfrac{K}{s(\tau_1 s + 1)(\tau_2^2 s^2 + \tau_2 s + 1)(\tau_3 s + 1)}$

(b) $G = \dfrac{K}{s(\tau_1 s + 1)(\tau_2 s + 1)^2(\tau_3 s + 1)}$

where $\tau_1 = 2\tau_2$ and $\tau_1 = 3\tau_3$. Determine accurately the initial directions of the motions of the roots, the asymptotic directions, and the center of gravity.

4-21. For the system with an open-loop transfer function of the form

$$G = \frac{(s+\alpha)(s+\alpha/2)}{s^2(s+10)^2} K_1$$

sketch on one diagram a root locus for each of the following values of α: 1, 4, 6.67, 9, 12.

4-22. Determine the initial angles of departure from the complex poles and sketch the root-locus diagrams for the following transfer functions:

(a) $\dfrac{1+\tau_1 s}{\tau_2^2 s^2 + 2\zeta\tau_2 s + 1} \qquad \dfrac{\zeta}{\tau_2} < \dfrac{1}{\tau_1}$

(b) $\dfrac{1+\tau_1 s}{\tau_2^2 s^2 + 2\zeta\tau_2 s + 1} \qquad \dfrac{\zeta}{\tau_2} > \dfrac{1}{\tau_1}$

4-23. A simple position servo is expressed in block-diagram form as shown in Fig. 4P-23.
(a) Is the system stable for all values of gain A?
(b) Upon construction of the above servo, it is found to have zero damping for a gain $A = 30$. The instability is due to parasitic time lags in the synchros, amplifier,

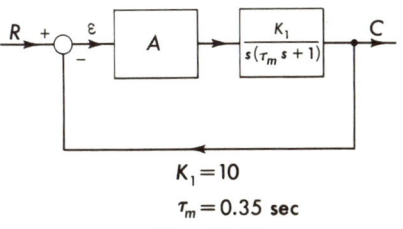

$K_1 = 10$
$T_m = 0.35$ sec
Fig. 4P-23

148   Control System Design

and motor. Approximate lumping these effects together using an additional time lag in the forward transfer function, i.e., by a factor $1/(1 + \tau_a s)$. What is the value of τ_a?
Ans. $\tau_a = 0.00336$

4-24. Use the root-locus method to find the roots of the following equations:

(a) $s^7 + 2s^6 - s^5 - 2s^4 + 4s^3 + 8s^2 - 4s - 8 = 0$

(b) $s^4 - 4s^3 - 7s^2 + 22s + 24 = 0$ *Ans. $s_1 = -1$, $s_2 = -2$, $s_3 = 3$, $s_4 = 4$*

(c) $2s^5 + 17s^4 + 78s^3 + 167s^2 + 246s + 90 = 0$

4-25. For the vibration damper of Fig. 4-34, assume the following values: $W_1 = 50$ lb, $W_2 = 20$ lb, $K_1 = 100$ lb/in., and $B = 63$ lb-sec/ft. Optimize the system, with root-locus techniques, by varying the lower spring constant K_2 to obtain the greatest damping on W_2.
Ans. $K_2 = 675$ lb/ft

4-26. One of the amplifiers in the system of Fig. 4P-26 is inverted so that a negative gain constant instead of a positive gain constant is obtained. The resulting system, without compensation, is unstable.

(a) Show, by means of a root-locus plot, why this system is unstable.

(b) The system can be made stable by appropriate compensation. What form of compensation should be used?

Note: When a negative gain is present, the angles must be assumed to $2k180°$ instead of $k180°$. The principles of locus plotting otherwise remain the same.

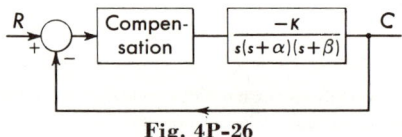

Fig. 4P-26

4-27. Sketch root-locus diagrams for the following systems:

(a) $GH = \dfrac{K}{s^4 + 16}$ (b) $GH = \dfrac{K}{s^4 - 16}$

(c) $GH = \dfrac{K}{(s^2 + 4)(s^2 - 1)}$ (d) $GH = \dfrac{K}{(s^2 + 1)(s^2 - 4)}$

chapter 5

STABILITY; THE FREQUENCY-ANALYSIS METHOD

5-1 The Impedance Concept. The concept of electrical impedance is familiar to most engineers. Early in their electrical engineering training, engineers learn to find the steady-state component solutions of differential equations with sinusoidal driving functions by means of complex-number calculations, also called the impedance method. For example, consider the circuit of Fig. 5-1, where a simple series circuit is shown. If e_1 is a sinusoidal driving voltage, that is, $e_1 = \sqrt{2}\, E_1 \sin \omega t$, the complex current amplitude through the circuit is given by

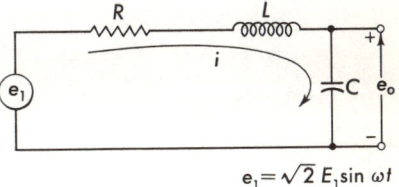

Fig. 5-1 A simple electrical circuit.

$$I(j\omega) = \frac{E_1 e^{j\phi}}{R + j\omega L + 1/j\omega C} \qquad (5\text{-}1)$$

where ωL is the "inductive reactance" and $1/\omega C$ is the "capacitive reactance" of the circuit. $e^{j\phi}$ expresses the fact that time is so selected that e_1 has zero phase angle. The units of $|I|$ are amperes (rms),* and E_1 is the rms amplitude of the applied voltage. To find the complex

---

* rms is the abbreviation for the root-mean-square value and, for sinusoidal functions, is equal to $1/\sqrt{2}$ times peak value.

amplitude of the output voltage $E_o(j\omega)$, Eq. (5-1) is multiplied by the impedance $1/j\omega C$ and the rms output voltage is written

$$E_o(j\omega) = \frac{E_1}{1 - \omega^2 LC + j\omega CR} \qquad (5\text{-}2)$$

Because of the appearance of the complex quantity j, the output voltage $E_o(j\omega)$ is a complex quantity. $E_o(j\omega)$ can be written in polar form as a function of magnitude and an angle. For the circuit of Fig. 5-1 the complex output voltage, written in polar form, is

$$E_o(j\omega) = \frac{E_1}{\sqrt{(1 - \omega^2 LC)^2 + (\omega CR)^2}\;\underline{/+ \tan^{-1} \omega CR/(1 - \omega^2 LC)}} \qquad (5\text{-}3)$$

A ratio of complex quantities is put into polar form by first writing the numerator and the denominator separately in polar form. The expression is written completely in polar form by dividing the magnitude and subtracting the angle of the denominator from that of the numerator. The symbol $\underline{/\phi}$ denotes the angle or argument in the complex quantity $e^{j\phi}$.

The phasor expression for the output voltage for Eq. (5-3) is

$$E_o(j\omega) = \frac{E_1}{\sqrt{(1 - \omega^2 LC)^2 + (\omega RC)^2}}\;\underline{\bigg/ - \tan^{-1} \frac{\omega RC}{1 - \omega^2 LC}} = A(\omega)\underline{/\phi(\omega)} \qquad (5\text{-}4)$$

This expression represents the output voltage as a function of frequency ω. The complex expression for $E_o(j\omega)$ [Eq. (5-4)] has a magnitude designated by $A(\omega)$ and an angle $\phi(\omega)$. When a variable-frequency signal generator is applied to the input, it is theoretically possible to plot the magnitude and the phase angle of the output voltage $E_o(j\omega)$ as the input frequency is varied. These plots depict the so-called "frequency response" property of the network.

It is important to notice that $E_o(j\omega)$, as given in Eq. (5-4), is independent of time. The input voltage is, however, a function of

$$e_1(t) = \sqrt{2}\,E_1 \sin \omega t \qquad (5\text{-}5)$$

where $\sqrt{2}$ times the rms voltage E_1 is the peak value. The output voltage expressed as a function of time is found from the complex expression of Eq. (5-4). The magnitude of $E_o(j\omega)$, as given in Eq. (5-4), is the rms amplitude of the output sine wave, and ϕ is the phase angle which must

Stability; the Frequency-analysis Method

$$e_o(t) = \frac{\sqrt{2}\,E_1}{\sqrt{(1 - \omega^2 LC)^2 + (\omega CR)^2}} \sin\left(\omega t - \tan^{-1} \frac{\omega CR}{1 - \omega^2 LC}\right) \quad (5\text{-}6)$$

Although a differential equation is required to describe the circuit of Fig. 5-1, the engineer who is familiar with the impedance concept can quickly solve and obtain the steady-state component solutions to these equations. If it is necessary, the steady-state time component can be obtained by appropriate interpretation of amplitude and phase angle of the complex solution [as is indicated in going from Eq. (5-4) to Eq. (5-6)].

The impedance concept can be applied with equal success and with equal reduction of labor to any linear system, whether it be electrical, mechanical, or electromechanical. To understand better the nature of the process involved, the two equations for Fig. 5-1 are written as follows:

$$e_1 = Ri = L\frac{di}{dt} + \frac{1}{C}\int_0^t i\,dt \quad (5\text{-}7)$$

and

$$C\frac{de_o}{dt} = i \quad (5\text{-}8)$$

where i is the instantaneous current in the loop of Fig. 5-1. Combining Eq. (5-8) and Eq. (5-7), the following differential equation relating e_1 to e_o results:

$$e_1 = RC\frac{de_o}{dt} + LC\frac{d^2 e_o}{dt^2} + e_o \quad (5\text{-}9)$$

The impedance concept is based upon the following principle: *If e_1 is a sinusoidal voltage, then, in the steady state, the current that flows in a linear network and hence all voltages that appear in the network are also sinusoidal quantities of the same frequency, but usually with different amplitudes and different phase angles.* The same statement applies to complex-number representations of sinusoidal quantities denoted by corresponding primed letters. Using this basic principle of linear systems, let e_1 be represented by

$$e_1 = \sqrt{2}\,E_1 e^{j\omega t} \quad (5\text{-}10)$$

where $e^{j\omega t}$ can be written

$$e^{j\omega t} = \cos \omega t + j \sin \omega t \quad (5\text{-}11)$$

Application of the above-cited principle to Eq. (5-9) indicates that e_o is a sinusoidal signal of the same frequency but of different amplitude and phase; hence

$$e_o = \sqrt{2}\, E_o e^{j\phi} e^{j\omega t} = \sqrt{2}\, \bar{E}_o e^{j\omega t} \quad (5\text{-}12)$$

where $\bar{E}_o$ is the phasor of the output voltage. When e_1 and e_o from Eqs. (5-10) and (5-12) are substituted into Eq. (5-9), the following expression is obtained:

$$\sqrt{2}\, E_1 e^{j\omega t} = RC \frac{d}{dt}\sqrt{2}\, \bar{E}_o e^{j\omega t} + LC \frac{d^2}{dt^2}\sqrt{2}\, \bar{E}_o e^{j\omega t} + \sqrt{2}\, \bar{E}_o e^{j\omega t} \quad (5\text{-}13)$$

Since $\bar{E}_o$ is independent of time, the differentiations indicated in Eq. (5-13) can be carried out, with the result

$$E_1 \sqrt{2}\, e^{j\omega t} = RC\bar{E}_o j\omega \sqrt{2}\, e^{j\omega t} + LC\bar{E}_o (j\omega)^2 \sqrt{2}\, e^{j\omega t} + \bar{E}_o \sqrt{2}\, e^{j\omega t} \quad (5\text{-}14)$$

Since the expression $\sqrt{2}\, e^{j\omega t}$ occurs in each term of Eq. (5-14), it can be canceled and Eq. (5-14) is thus simplified to

$$E_1 = E_o e^{j\phi}[jRC\omega + LC(j\omega)^2 + 1] \quad (5\text{-}15)$$

By comparing Eq. (5-9) with Eq. (5-15) we see that these two equations are similar. In Eq. (5-15) $j\omega$ appears to the same order that differentiation with respect to time d/dt appears in Eq. (5-9). This parallelism can be formalized as follows:

$$\frac{d}{dt} \to j\omega \quad \frac{d^2}{dt^2} \to (j\omega)^2 \quad \int_0^t dt \to \frac{1}{j\omega} \quad e_i \to E_i(j\omega) \quad (5\text{-}16)$$

5-2 Generalized Impedance Functions. The principle demonstrated in solving the example of Fig. 5-1 can be generalized to any linear system. When the derivative operator d^n/dt^n in the differential equation of any system is replaced by $(j\omega)^n$, an algebraic equation results. This equation can be solved for the dependent variable, and the solution yields the complex-number representation of the steady-state value of the dependent variable corresponding to a sinusoidal driving function input.

For the example of Fig. 5-1, Eq. (5-15) is solved for $E_o e^{j\phi}$, resulting in the following expression:

$$E_o e^{j\phi} = \frac{E_1}{(1 - \omega^2 LC) + j\omega RC}$$

$$= \frac{E_1}{\sqrt{(1 - \omega^2 LC)^2 + (\omega RC)^2}} \bigg/ -\tan^{-1}\frac{\omega RC}{1 - \omega^2 RC} \quad (5\text{-}17)$$

Equation (5-17) is identical with Eq. (5-4), which demonstrates the validity in this example of substituting $(j\omega)^n$ for d^n/dt^n.

As an example of the impedance method for finding the sinusoidal steady-state solution to control systems, consider the system of Chap. 1, which is also shown in the block diagram of Fig. 5-2. The differential equation relating the input position r to the output position c is written as follows:

$$\frac{d^2c}{dt^2} + 2\zeta\omega_n\frac{dc}{dt} + \omega_n^2 c = \omega_n^2 r \quad (5\text{-}18)$$

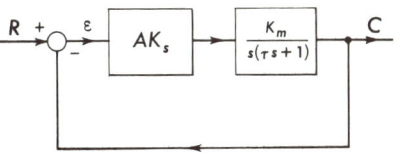

Fig. 5-2 An example of a position servo.

To find the steady-state reponse $C(j\omega)$ when $R(j\omega)$ denotes a sinusoidal input signal, substitute $j\omega$ for d/dt and $(j\omega)^2$ for d^2/dt^2 in Eq. (5.18) and $C(j\omega)$ and $R(j\omega)$ for c and r and solve for $C(j)\omega$:

$$C(j\omega) = \frac{\omega_n^2 R(j\omega)}{\omega_n^2 - \omega^2 + j2\zeta\omega_n\omega} \quad (5\text{-}19)$$

where $C(j\omega)$ and $R(j\omega)$ indicate that these are complex quantities and not functions of time. Dividing numerator and denominator by the square of the undamped natural resonant frequency ω_n^2 and substituting $u = \omega/\omega_n$ gives

$$\frac{C(ju)}{R(ju)} = \frac{1}{1 - u^2 + j2\zeta u} \quad (5\text{-}20)$$

where u is the frequency ratio ω/ω_n and ζ is the damping ratio. Equation (5-20) is written in polar form as follows:

$$\frac{C(ju)}{R(ju)} = \frac{1}{\sqrt{(1 - u^2)^2 + (2\zeta u)^2}} \bigg/ -\tan^{-1}\frac{2\zeta u}{1 - u^2} = A(u)/\underline{\phi(u)} \quad (5\text{-}21)$$

The output $C(ju)$ is the complex-number representation of a sinusoid with the same frequency u as the applied input $r(t)$. As shown in Eq. (5-21), the amplitude is multiplied by A and the phase is shifted by an angle ϕ. If the frequency of the sinusoidal driving signal is varied, the closed-loop frequency response, pictured in Fig. 5-3, is obtained. Amplitude and phase curves shown in Figs. 5-3a and b are plotted against the logarithm of the frequency ratio for values of damping ratio in the range 0.1 to 2.0. Figure 5-3a is the decibel value of the amplitude $|C(ju)/R(ju)|$ plotted against the logarithm of the frequency ratio u, and Fig. 5-3b is

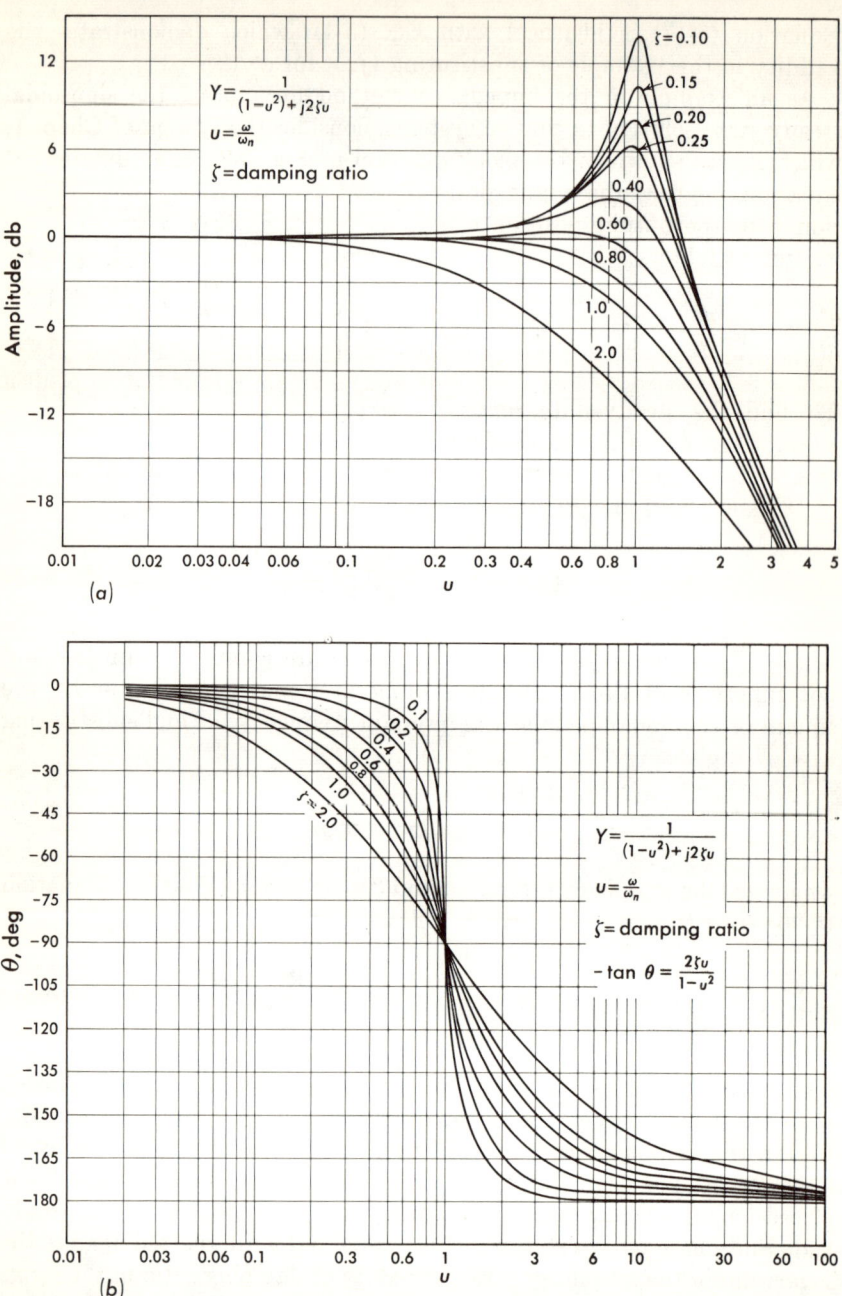

Fig. 5-3 (a) Amplitude of the second-order transfer function; (b) phase shift for the second-order transfer function.

the phase shift of $C(ju)$ with respect to $R(ju)$ plotted as a function of the logarithm of the frequency ratio.

The decibel, abbreviated db, is defined by

$$\text{db} = 20 \log_{10} \frac{A_2}{A_1} \qquad (5\text{-}22)$$

where A_2/A_1 denotes an amplitude ratio (i.e., a ratio of voltage out of an amplifier divided by voltage into the amplifier). The decibel was originally defined as a logarithm of power ratio; however, in control-system applications it is used as an amplitude ratio as in Eq. (5-22). A chart that permits conversion between decibels and magnitude of the amplitude ratio is included in Fig. 5-4. Table 5-1 presents the same information in numerical form: gain corresponds to positive db and attenuation to negative db.

The frequency where the peak of each curve of Fig. 5-3a occurs is plotted against the damping ratio in Fig. 5-5.

It is important to observe the simplicity with which the steady-state solution of differential equations is obtained when the driving function is a sinusoid. It is only necessary to replace d^n/dt^n by $(j\omega)^n$ and the time quantities by their corresponding complex-number representations. Given a transfer function in which the Laplace-transform operator s is indicated, replacing s by $j\omega$ will result in the corresponding frequency transfer function.

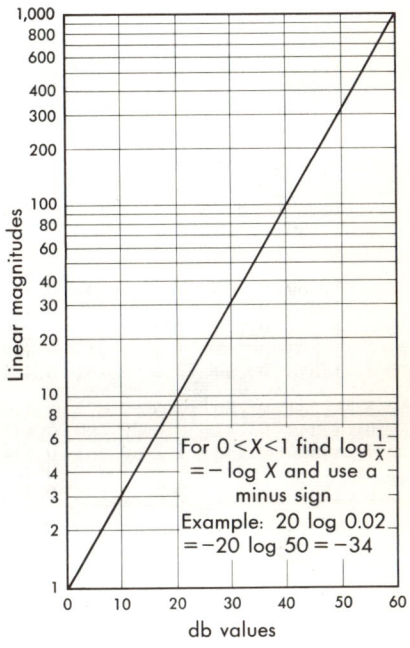

Fig. 5-4 Linear magnitudes vs. decibel values.

5-3 Plotting Impedance and Frequency Transfer Functions. The labor required to plot simple impedance and frequency transfer functions point by point for a series of frequency values is considerable; however, a method for the easy plotting of impedance and frequency transfer functions has been developed. The resulting plots are known as "Bode plots." This method avoids the difficulty of computing polar

Table 5-1 Relation between Decibels and Gain or Loss

db	Gain	Loss	db	Gain	Loss	db	Gain	Loss	db	Gain	Loss
0.0	1.000	1.000	5.0	1.778	0.562	10.0	3.162	0.316	15.0	5.623	0.178
0.1	1.012	0.988	5.1	1.799	0.556	10.1	3.199	0.313	15.1	5.689	0.176
0.2	1.023	0.977	5.2	1.820	0.549	10.2	3.236	0.309	15.2	5.754	0.174
0.3	1.035	0.966	5.3	1.841	0.543	10.3	3.273	0.306	15.3	5.821	0.172
0.4	1.047	0.955	5.4	1.862	0.537	10.4	3.311	0.302	15.4	5.888	0.170
0.5	1.059	0.944	5.5	1.884	0.531	10.5	3.350	0.298	15.5	5.957	0.168
0.6	1.072	0.933	5.6	1.906	0.525	10.6	3.388	0.295	15.6	6.026	0.166
0.7	1.084	0.922	5.7	1.928	0.519	10.7	3.428	0.292	15.7	6.096	0.164
0.8	1.096	0.912	5.8	1.950	0.513	10.8	3.467	0.288	15.8	6.166	0.162
0.9	1.109	0.902	5.9	1.972	0.507	10.9	3.508	0.285	15.9	6.237	0.160
1.0	1.122	0.891	6.0	1.995	0.501	11.0	3.548	0.282	16.0	6.310	0.158
1.1	1.135	0.881	6.1	2.018	0.496	11.1	3.589	0.279	16.1	6.383	0.157
1.2	1.148	0.871	6.2	2.042	0.490	11.2	3.631	0.275	16.2	6.457	0.155
1.3	1.162	0.861	6.3	2.065	0.484	11.3	3.673	0.272	16.3	6.531	0.153
1.4	1.175	0.851	6.4	2.089	0.479	11.4	3.715	0.269	16.4	6.607	0.151
1.5	1.189	0.841	6.5	2.113	0.473	11.5	3.758	0.266	16.5	6.684	0.150
1.6	1.202	0.832	6.6	2.138	0.468	11.6	3.802	0.263	16.6	6.761	0.148
1.7	1.216	0.822	6.7	2.163	0.462	11.7	3.846	0.260	16.7	6.839	0.146
1.8	1.230	0.813	6.8	2.188	0.457	11.8	3.890	0.257	16.8	6.918	0.145
1.9	1.245	0.803	6.9	2.214	0.452	11.9	3.935	0.254	16.9	6.998	0.143
2.0	1.259	0.794	7.0	2.239	0.447	12.0	3.981	0.251	17.0	7.079	0.141
2.1	1.274	0.785	7.1	2.265	0.441	12.1	4.027	0.248	17.1	7.161	0.140
2.2	1.288	0.776	7.2	2.291	0.436	12.2	4.074	0.245	17.2	7.244	0.138
2.3	1.303	0.767	7.3	2.317	0.432	12.3	4.121	0.243	17.3	7.328	0.136
2.4	1.318	0.759	7.4	2.344	0.427	12.4	4.169	0.240	17.4	7.413	0.135
2.5	1.334	0.750	7.5	2.371	0.422	12.5	4.217	0.237	17.5	7.499	0.133
2.6	1.349	0.741	7.6	2.399	0.417	12.6	4.266	0.234	17.6	7.586	0.132
2.7	1.365	0.733	7.7	2.427	0.412	12.7	4.315	0.232	17.7	7.674	0.130
2.8	1.380	0.725	7.8	2.455	0.407	12.8	4.365	0.229	17.8	7.762	0.129
2.9	1.396	0.716	7.9	2.483	0.403	12.9	4.416	0.226	17.9	7.852	0.127
3.0	1.413	0.708	8.0	2.512	0.398	13.0	4.467	0.224	18.0	7.943	0.126
3.1	1.429	0.700	8.1	2.541	0.394	13.1	4.519	0.221	18.1	8.035	0.124
3.2	1.445	0.692	8.2	2.571	0.389	13.2	4.571	0.219	18.2	8.128	0.123
3.3	1.462	0.684	8.3	2.601	0.384	13.3	4.624	0.216	18.3	8.222	0.122
3.4	1.476	0.676	8.4	2.630	0.380	13.4	4.677	0.214	18.4	8.318	0.120
3.5	1.496	0.668	8.5	2.661	0.376	13.5	4.732	0.211	18.5	8.414	0.119
3.6	1.514	0.660	8.6	2.692	0.371	13.6	4.786	0.209	18.6	8.511	0.117
3.7	1.531	0.653	8.7	2.723	0.367	13.7	4.842	0.207	18.7	8.610	0.116
3.8	1.549	0.646	8.8	2.754	0.363	13.8	4.898	0.204	18.8	8.710	0.115
3.9	1.567	0.638	8.9	2.786	0.359	13.9	4.955	0.202	18.9	8.810	0.113
4.0	1.585	0.634	9.0	2.818	0.355	14.0	5.012	0.200	19.0	8.913	0.112
4.1	1.603	0.624	9.1	2.851	0.351	14.1	5.070	0.197	19.1	9.016	0.111
4.2	1.622	0.617	9.2	2.884	0.347	14.2	5.129	0.195	19.2	9.120	0.110
4.3	1.641	0.609	9.3	2.918	0.343	14.3	5.188	0.193	19.3	9.226	0.108
4.4	1.660	0.602	9.4	2.951	0.339	14.4	5.248	0.191	19.4	9.333	0.107
4.5	1.679	0.596	9.5	2.985	0.335	14.5	5.309	0.188	19.5	9.441	0.106
4.6	1.698	0.589	9.6	3.020	0.331	14.6	5.370	0.186	19.6	9.550	0.105
4.7	1.718	0.582	9.7	3.055	0.327	14.7	5.432	0.184	19.7	9.661	0.103
4.8	1.738	0.575	9.8	3.090	0.324	14.8	5.495	0.182	19.8	9.772	0.102
4.9	1.758	0.569	9.9	3.126	0.320	14.9	5.559	0.180	19.9	9.886	0.101

plots (i.e., amplitude radially and positive phase angle counterclockwise); it centers about the use of the logarithmic function. Decibel gain and phase angle are plotted against logarithmic frequency.

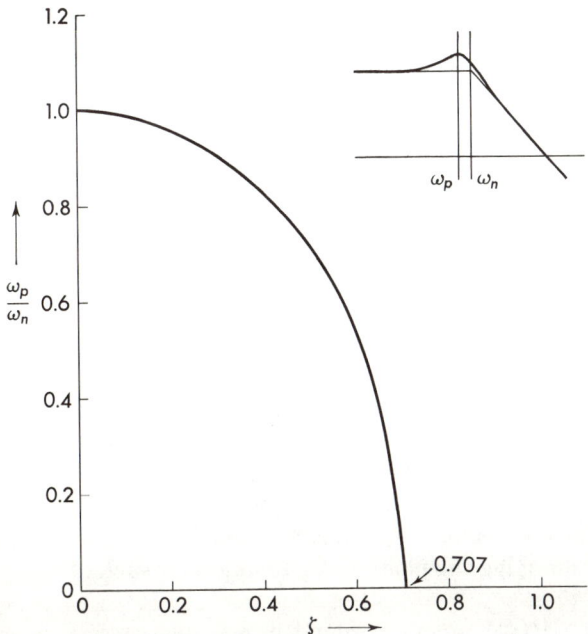

Fig. 5-5 Frequency where peak amplitude occurs for the second-order system.

To appreciate the advantages of using the logarithm of the frequency transfer function, suppose that the frequency transfer function

$$GH(j\omega) = \frac{K(j\omega\tau_1 + 1)(j\omega\tau_3 + 1)}{(j\omega)^n(j\omega\tau_2 + 1)(j\omega\tau_4 + 1)} \quad (5\text{-}23)$$

is written in polar form:

$$GH(j\omega) = \frac{K[\sqrt{(\omega\tau_1)^2 + 1}\,/\tan^{-1}\omega\tau_1][\sqrt{(\omega\tau_3)^2 + 1}\,/\tan^{-1}\omega\tau_3]}{\omega^n/n90°[\sqrt{(\omega\tau_2)^2 + 1}\,/\tan^{-1}\omega\tau_2][\sqrt{(\omega\tau_4)^2 + 1}\,/\tan^{-1}\omega\tau_4]}$$

$$GH(j\omega) = A(\omega)e^{j\phi(\omega)} \quad (5\text{-}24)$$

where $A(\omega)$ is the magnitude of $GH(j\omega)$ and $\phi(\omega)$ is the corresponding phase angle. The logarithm* to the base e of the last member of Eq.

* Throughout the text, "ln x" is used to indicate the logarithm of x to the base e.

(5-24) is
$$\ln GH(j\omega) = \ln A(\omega)e^{j\phi(\omega)} = \ln A(\omega) + \ln e^{j\phi(\omega)}$$
$$\ln GH(j\omega) = \ln A(\omega) + j\phi(\omega) \tag{5-25}$$

The logarithm of the frequency transfer function has a real part (the logarithm of the magnitude) and an imaginary part (the phase angle as expressed in radians). If Eq. (5-24) is combined with Eq. (5-25), the logarithm of the magnitude and the phase angle can be written separately in two expressions:

$$\ln A(\omega) = \ln K + \ln \sqrt{(\omega\tau_1)^2 + 1} + \ln \sqrt{(\omega\tau_3)^2 + 1} - \ln \omega^n$$
$$- \ln \sqrt{(\omega\tau_2)^2 + 1} - \ln \sqrt{(\omega\tau_4)^2 + 1} \tag{5-26}$$

$$\phi(\omega) = \tan^{-1} \omega\tau_1 + \tan^{-1} \omega\tau_3 - n90° - \tan^{-1} \omega\tau_2 - \tan^{-1} \omega\tau_4 \tag{5-26a}$$

Because the logarithm of the product of two quantities is equal to the sum of the respective logarithms, each factored portion of Eq. (5-24) is treated separately. These separate factors are plotted in decibel units, since multiplication of each side of Eq. (5-26) by 20/2.3* converts the units to decibels. The total decibel magnitude is formed simply by summing the individual values algebraically. Equation (5-26a) indicates that the phase-angle function is formed in a similar manner by algebraically summing the phase-angle function for each of the individual terms in Eq. (5-24).

Since most $GH(j\omega)$ functions found in control systems are composed of factors comprising zeros and poles, the decibel magnitude and phase vs. log frequency need be considered for only a few types of terms. Based upon the knowledge of four simple types, a complete $GH(j\omega)$ function can be built up simply by appropriately adding the magnitude and phase curves.

5-4 The Asymptotic Approximation. The most commonly encountered functions in control systems are listed below:

Frequency-invariant factors	K
Terms corresponding to simple* zeros and poles at the origin	$j\omega$ or $1/j\omega$
Linear terms corresponding to simple zeros	$j\omega\tau_1 + 1$
Linear terms corresponding to simple poles	$(j\omega\tau_2 + 1)^{-1}$
Quadratic terms corresponding to simple zeros and poles	$\left[\left(j\dfrac{\omega}{\omega_n}\right)^2 + 2\zeta\dfrac{\omega}{\omega_n}j + 1\right]^{\pm 1}$

* "Simple" means the zero or pole at a point is of order 1.

* Since $\ln x = (\log_{10} x) \ln 10 = 2.3 \log_{10} x$, it follows that $20 \log_{10} x = 20/2.3 \ln x$.

Frequency-response curves for each of these functions can be plotted point by point. Owing to the nature of the plot, however, a linear asymptotic approximation of the decibel magnitude curves permits a rapid method of plotting these factors. In all cases these curves are plotted on semilog paper. The linear scale is that of the decibel gain plotted against the logarithm of the frequency ω. Phase angle is plotted on a linear scale against the logarithm of the frequency ω.

Frequency-invariant Factors. The product of the gain constants K, which is a constant independent of frequency, is plotted from the function

$$K_{db} = 20 \log_{10} K \qquad (5\text{-}27)$$

where K represents the product of all frequency-invariant terms in the $GH(j\omega)$ function when written in the standard form displayed in Eq.

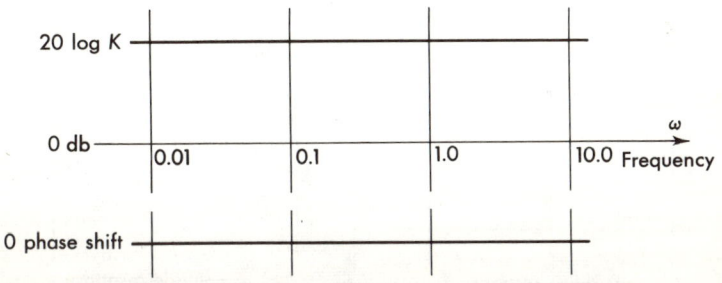

Fig. 5-6   Decibel gain and phase angle for a constant.

(5-23). Equation (5-27) is plotted in Fig. 5-6 as a constant with zero phase angle.

Zeros or Poles at the Origin. For zeros or poles at the origin,

$$(j\omega)^n \quad \text{or} \quad \frac{1}{(j\omega)^n} \qquad (5\text{-}28)$$

amplitude and phase curves are found by taking the logarithm of these functions as follows:

$$\ln (j\omega)^{\pm n} = \pm n \ln \omega \pm jn90° \qquad (5\text{-}29)$$

For simple zeros or poles the integer n is unity. The amplitude is $\pm n\, 20 \log_{10} \omega$ db, and the phase angle is $\pm n 90°$. The $\pm n$ accounts for either zeros or poles; $+n$ corresponds to zeros, and $-n$ corresponds to poles.

The phase-angle curve for a simple zero or pole is constant at $\pm n 90°$. For a single zero at the origin, the phase angle is a constant $+90°$; and for a single pole at the origin, the phase angle is a constant $-90°$, as shown in Fig. 5-7a.

The amplitude plot is a straight line with a slope of approximately $\pm n6$ db/octave, where n is the order of the zeros or poles at the origin. An octave change in frequency corresponds to a doubling or halving of the fundamental frequency. For a single zero, the straight line is of slope $+6$ db/octave and intersects the 0-db axis at the point $\omega = 1$. The curve for a single pole, shown in Fig. 5-7b, is a straight line with a slope of -6 db/octave and passes through the 0-db axis at the point $\omega = 1$.

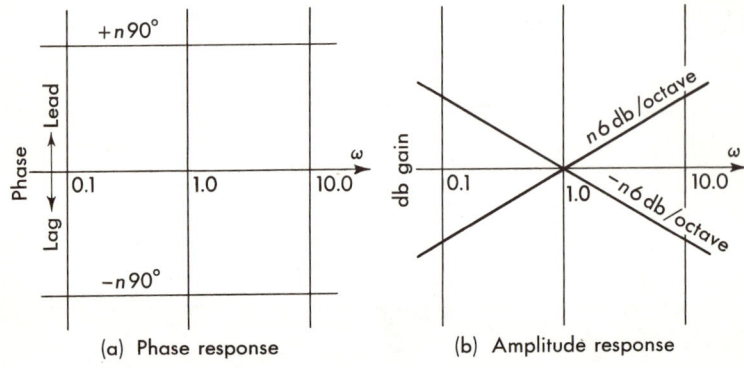

(a) Phase response (b) Amplitude response

Fig. 5-7 Phase and decibel gain for $(j\omega)^{\pm n}$.

These straight lines, shown in Fig. 5-7a and b, are not asymptotic approximations; rather, they are the actual curves. When amplitude in decibels is plotted against the logarithm of the angular frequency, the resulting amplitude curves are straight lines passing through the 0-db axis at $\omega = 1$.

The $\pm n6$ db/octave slope of the amplitude curves can be verified by considering the change in decibel amplitude Δ db for a change in frequency of one octave from ω_1 to ω_2. This is written, for $n = 1$, as

$$\Delta \text{ db} = 20(\log \omega_2 - \log \omega_1) = 20 \log \frac{\omega_2}{\omega_1} \tag{5-30}$$

Taking ω_2 equal to $2\omega_1$, since a frequency change of one octave corresponds to a doubling of the fundamental frequency,

$$\Delta \text{ db} = 20 \log_{10} 2 = (20)(0.30103) \approx 6 \text{ db/octave} \tag{5-31}$$

A straight line with a slope of 6 db/octave on the amplitude curves, shown in Fig. 5-7b, intersects the 0-db line at $\omega = 1$.

For ease in plotting, the constant portion of the transfer function K can be combined with the $\pm(j\omega)^n$ term. Usually only a single or double pole exists at the origin. As an example, consider a single pole at the origin. The constant and this pole are combined as follows:

$$\frac{K(\omega\tau_1 + 1) \cdots}{j\omega(\omega\tau_2 + 1) \cdots} = \frac{(\omega\tau_1 + 1) \cdots}{j(\omega/K)(\omega\tau_2 + 1) \cdots} \quad (5\text{-}32)$$

The term to be plotted is

$$\frac{1}{j(\omega/K)} \quad (5\text{-}33)$$

Term (5-33) has the same slope as the term corresponding to the single pole at the origin, $1/j\omega$, but the amplitude curve intersects the 0-db line at a frequency $\omega = K$. The phase angle remains a constant of $-90°$.

If a double pole is at the origin, the slope is -12 db/octave and the amplitude crosses the 0-db line at $\omega = \sqrt{K}$.

It is convenient when working with frequency methods to write the transfer function in factors of the form $\tau s + 1$, such that K includes all the constants factored from the s-dependent portion of the transfer function. This is different in form from that necessary for a root-locus construction. For the root-locus construction the terms should be put in factors of the form $s + 1/\tau$.

Simple Zeros. For simple zero factors of the form

$$j\omega\tau_1 + 1 \quad (5\text{-}34)$$

a linear asymptotic approximation is used. For $\omega\tau_1 \ll 1$,

$$20 \log_{10} |j\omega\tau_1 + 1| \approx 20 \log_{10} 1 = 0 \text{ db} \quad (5\text{-}35)$$

For small values of ω the magnitude remains at 0 db. When ω becomes much greater than 1 ($\omega\tau_1 \gg 1$),

$$20 \log_{10} |j\omega\tau_1 + 1| \approx 20 \log_{10} \omega\tau_1 \quad (5\text{-}36)$$

For large values of ω, the amplitude and phase plot corresponding to a simple zero resulting from $j\omega\tau_1 + 1$ resembles the plot for the term $j\omega\tau_1$. The slope, for large ω, is 6 db/octave. The 6 db/octave straight-line

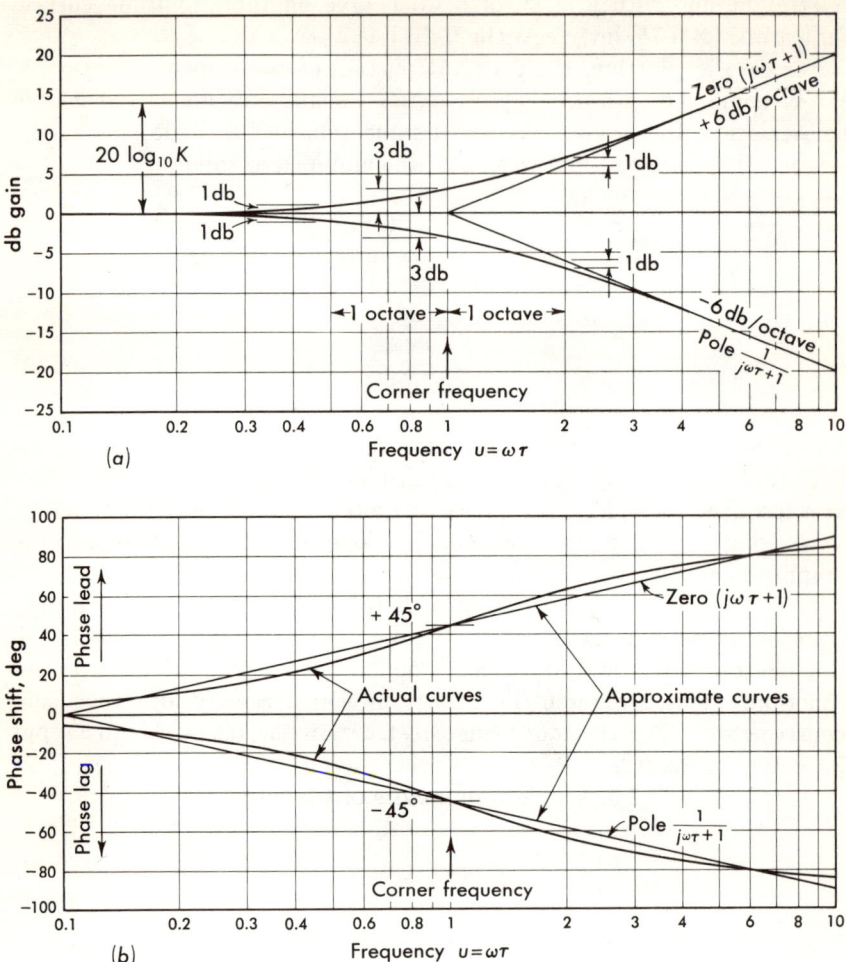

Fig. 5-8 (a) Amplitude response corresponding to factors of K, $j\omega\tau + 1$, and $1/(j\omega\tau + 1)$; (b) phase response corresponding to factors of $j\omega\tau + 1$ and $1/(j\omega\tau + 1)$.

asymptote intersects the 0-db line at $\omega\tau_1 = 1$, thus at $\omega = 1/\tau_1$. The point of intersection at $\omega = 1/\tau_1$ is termed the "corner frequency." The two straight lines, one along the 0-db and the other at $+6$ db/octave intersecting at the point $\omega = 1/\tau_1$, are the asymptotic approximations corresponding to a simple zero factor of the form $j\omega\tau_1 + 1$. This straight-line asymptote is shown in Fig. 5-8a. The procedure to follow when plotting the amplitude curve for $j\omega\tau_1 + 1$ is outlined as follows:

1. Locate the corner frequency, that is, $\omega = 1/\tau_1$.
2. Plot a 6 db/octave line through this point to the right (increasing frequency) and a straight line along 0 db to the left.
3. If more accuracy is required, find the actual curve as explained in the next paragraph and as shown on Fig. 5-8a.

The actual amplitude curve deviates only slightly from the straight-line asymptotes. As is shown in Fig. 5-8a, the curve rises 3 db above the asymptote at the corner frequency and approximately 1 db at one octave above and one octave below the corner frequency. These deviations from the asymptotic approximation are computed by evaluating the magnitude of the zero factor $j\omega\tau_1 + 1$ at the corner frequency, at twice, and at one-half the corner frequency, as follows:

At $\omega = 1/\tau_1$
$$20 \log |j1 + 1| = 20 \log \sqrt{2} = 3 \text{ db}$$
$$\text{Deviation from asymptote} = 3 \text{ db}$$

At $\omega = 2/\tau_1$
$$20 \log |j2 + 1| = 20 \log \sqrt{5} = 6.99$$
$$\text{Deviation from asymptote} = 6.99 - 6 \approx 1 \text{ db}$$

At $\omega = \tfrac{1}{2}\tau_1$
$$20 \log \left| \frac{j1}{2} + 1 \right| = 20 \log \sqrt{\tfrac{5}{4}} = 0.969$$
$$\text{Deviation from asymptote} \approx +1.0 \text{ db}$$

The phase angle for a simple zero factor is obtained from the expression

$$\phi = \tan^{-1} \omega\tau_1 \tag{5-37}$$

The frequency ω is plotted on a logarithmic scale. The arctangent curve has a value of 45° when $\omega\tau_1 = 1$, which is at the corner frequency. The phase curve starts at 0°, increases to a maximum of 90°, and is symmetric about the 45° point. The complete frequency-response curve for the zero comprises the amplitude curve shown in Fig. 5-8a and the phase-angle curve shown in Fig. 5-8b.

Since the phase curve is a familiar arctangent curve, it, too, is easily sketched. An approximate straight-line phase curve can be used to aid in sketching the phase response. The straight line passes through 0° at one-tenth of the corner frequency, 45° at the corner frequency, and 90° at 10 times the corner frequency. If a line is drawn as shown in Fig.

5-9, the maximum deviation from the actual curve is 6°. These straight-line approximations are also included in Fig. 5-8b. The procedure used to plot the phase curve is as follows:

1. For each zero locate the point which is one decade lower in frequency.
2. Plot a straight line with slope +13.2°/octave (+45°/decade) through this point to the right and continuing up for 2 decades, i.e., until +90° is reached. The curve is then flat.
3. If more accuracy is required, add the deviations shown in Fig. 5-9.

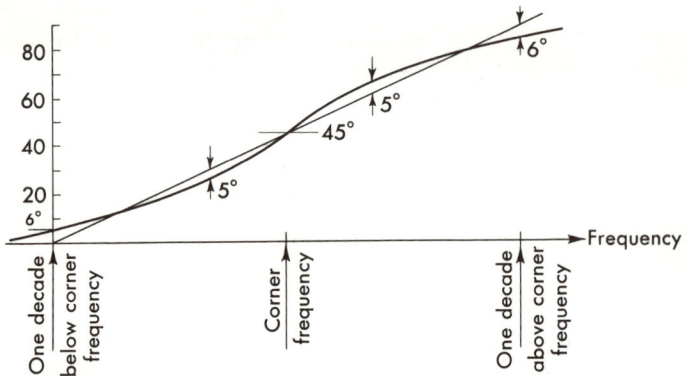

Fig. 5-9 Approximate phase curve.

Simple Poles. Simple pole factors of the form $1/(j\omega\tau_2 + 1)$ can be treated in a fashion similar to that for simple zero factors. Since the logarithm of a reciprocal quantity is equal to the negative of the logarithm of the quantity or

$$20 \log \frac{1}{j\omega\tau_2 + 1} = -20 \log (j\omega\tau_2 + 1) \qquad (5\text{-}38)$$

the curve for a simple pole factor is similar to that for a simple zero factor except that it is reflected about the 0-db line. For small frequencies, $\omega \ll 1/\tau_2$, the amplitude remains at 0 db. For large frequencies, $\omega \gg 1/\tau_2$, the asymptote is a straight line of slope -6 db/octave. This asymptote, shown in Fig. 5-8a, intersects the 0-db axis at $\omega = 1/\tau_2$. The frequency $\omega = 1/\tau_2$ is termed the "corner frequency" or the "break point" for this particular pole. As in the case of the zero, the actual amplitude curve deviates from the straight-line approximation by -3 db at the corner frequency and by -1 db at both $\omega = 1/2\tau_2$ and $\omega = 2/\tau_2$.

The phase-angle curve is similar to that for a zero, but it is reflected about the 0-rad line. Since the pole is in the denominator of GH, the sign is changed when the angle is brought into the numerator,

$$\phi = -\tan^{-1} \omega \tau_2 \tag{5-39}$$

Equation (5-39) is an arctangent curve which starts from zero phase angle and approaches a value of $-90°$ for large frequency. The $-45°$ phase-angle point occurs at the corner frequency. The same straight-line approximation that is used for zeros is applicable for poles. The phase curve and the straight-line approximation are included in Fig. 5-8b.

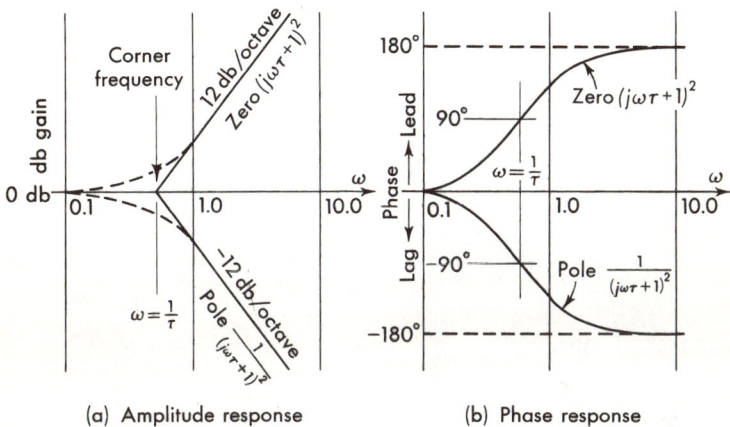

(a) Amplitude response   (b) Phase response

Fig. 5-10 Frequency response of double poles and zeros.

If the transfer function has a repeated zero or pole—for example, if there are two equal poles or two equal zeros—the amplitude curve is somewhat similar to that for a single zero or pole. However, the slope changes from 6 to 12 db/octave, and the phase angle is at $\pm 90°$ at the corner frequency (for a zero or pole, respectively) rather than at $\pm 45°$ as for a single zero or pole. The phase angle varies from zero to $\pm 180°$ rather than from zero to $\pm 90°$. The corresponding functions are shown in Fig. 5-10a and b.

Quadratic Zeros and Poles. Occasionally, quadratic pole factors of the form

$$G(j\omega) = \frac{\omega_n^2}{-\omega^2 + j2\zeta\omega_n\omega + \omega_n^2} = \frac{1}{-(\omega/\omega_n)^2 + j2\zeta\omega/\omega_n + 1} \tag{5-40}$$

occur in the $GH(j\omega)$ function. The right-hand member of Eq. (5-40) can be put into dimensionless form by taking $u = \omega/\omega_n$ with the result

$$G(ju) = \frac{1}{1 - u^2 + j2\zeta u} \qquad (5\text{-}41)$$

The magnitude and phase of the right-hand member of Eq. (5-41) are plotted in Fig. 5-3a and b. Because the amplitude and phase response for quadratic pole factors depend not only on the corner frequency but also upon the damping ratio ζ, a dimensionless chart of the form shown in Fig. 5-3a and b is used to make the plot. The amplitude and phase response are plotted by locating the corner frequency and damping ratio for the particular quadratic factor, as found by comparison of the given expression with Eq. (5-40). For example, suppose it is required to plot the magnitude and phase vs. frequency for the function

$$G(j\omega) = \frac{1}{(j\omega)^2 + 3j\omega + 10} \qquad (5\text{-}42)$$

Equation (5-42) is put into the form of Eq. (5-38) by comparison:

$$\frac{\frac{1}{10}}{(j\omega)^2/10 + \frac{3}{10}j\omega + 1} = \frac{1}{(j\omega/\omega_n)^2 + 2\zeta j\omega/\omega_n + 1} \qquad (5\text{-}43)$$

Equating like terms,

$$\omega_n = \sqrt{10} \qquad \text{and} \qquad \zeta = \frac{\omega_n}{2}\frac{3}{10} = \frac{3.0}{2\sqrt{10}} = 0.475 \qquad (5\text{-}44)$$

The corner frequency $\omega = \omega_n$ is first located on the Bode plot. The first approximation is drawn in the same manner as for a double pole located at the corner frequency ω_n; that is, a -12 db/octave line going to the right and starting from the corner frequency for the amplitude response and an approximation to the arctangent curve extending from 0 to $-180°$ for the phase response. Because the deviation from these asymptotes depends upon the damping ratio, it is necessary to refer to Fig. 5-11, which gives the deviations from the -12 db/octave amplitude for various damping ratios. The deviations from the phase approximation are given in Fig. 5-12. For control systems which contain not only single-order real zeros and poles but also single-order complex poles, use of these charts enables the engineer to make a Bode plot easily. If a complex pole exists in the transfer function, the curves of Fig. 5-3a and b are mirrored as for simple zeros and poles.

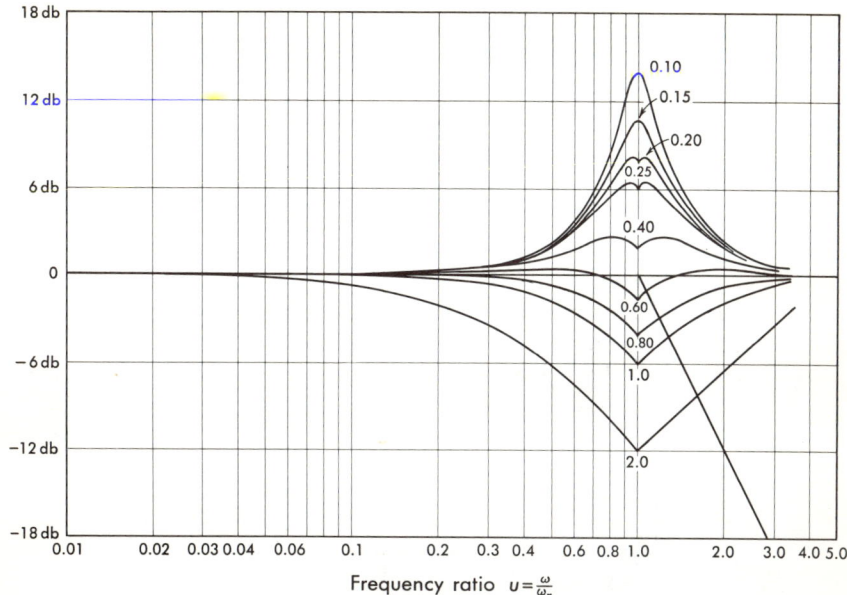

Fig. 5-11 Amplitude differences for the quadratic function $1/(1 - u^2 + j2\zeta u)$.

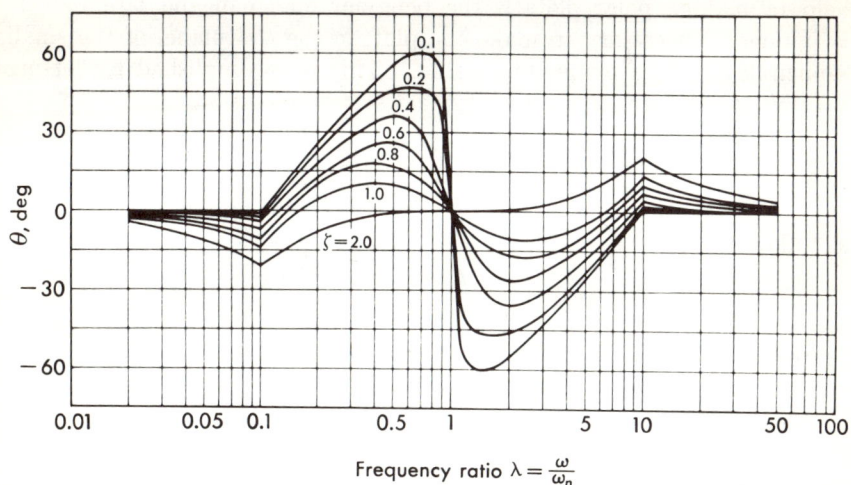

Fig. 5-12 Phase-shift deviation from a straight-line approximation for a second-order system.

168 Control System Design

5-5 Polar Plots and Amplitude vs. Phase-shift Plots. Usually it is not required to make a polar plot, since stability information can be obtained directly from decibel gain and phase vs. log frequency (Bode) plots. However, in certain cases it is necessary to make a polar plot. These plots can be made, of course, by plotting point by point; i.e., substituting various values of ω ($\omega_1, \omega_2, \omega_3, \ldots, \omega_n$) into $GH(j\omega)$, calculating $|GH|$ and arg $GH(j\omega)$, and plotting the magnitude and phase as shown in Fig. 5-13. This method is quite laborious, and it should be avoided.

The polar plot can be plotted easier directly from the Bode plots, in which the decibel gain and phase are plotted vs. frequency. After decibel gain is converted to amplitude (cf. Fig. 5-4), the amplitude and phase are read directly from the Bode plot at the various values of ω. Of great

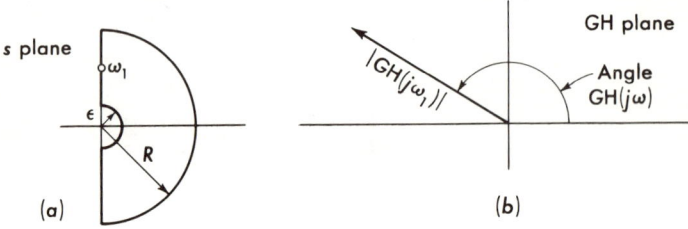

Fig. 5-13 Polar plot of $GH(j\omega_1)$.

help in making polar plots is the behavior for small and large ω. It is frequently necessary to make the plot, in the GH plane, of the small semicircle in the s plane, Fig. 5-13a. This is accomplished by letting $s = \epsilon e^{j\phi}$ and substituting into a typical system equation,

$$GH(s) = \frac{K(s + s_1)(s + s_3)}{s^2(s + s_2)(s + s_4)^2}$$

as follows:

$$GH(\epsilon e^{j\phi}) = \frac{K(\epsilon e^{j\phi} + s_1)(\epsilon e^{j\phi} + s_3)}{\epsilon^2 e^{j2\phi}(\epsilon e^{j\phi} + s_2)(\epsilon e^{j\phi} + s_4)^2}$$

Since ϵ is small,

$$GH(\epsilon e^{j\phi}) \approx \frac{K s_1 s_3}{s_2 s_4} \frac{e^{-2j\phi}}{\epsilon^2} \approx \mathcal{K} e^{-2j\phi}$$

when $\mathcal{K}$ is a large number. The small semicircle in the s plane maps into a large semicircle in the GH plane, depending on the number of poles at the origin. In Fig. 5-14 is shown the shape of the GH contour for various numbers of poles at the origin.

The same analysis can be used if complex poles exist at $\pm j\omega_0$. In this case, let $s = j\omega_0 + \epsilon e^{j\phi}$. A semicircle around this pole maps into large contours in the GH plane similar to those shown in Fig. 5-14. In a similar fashion, letting $s = Re^{j\phi}$, the large semicircle in the s plane plots into a small semicircle around the origin in the GH plane.

Another method of plotting the frequency-response data is termed the decibel gain vs. phase-shift plot. In this plot, the decibel gain is plotted

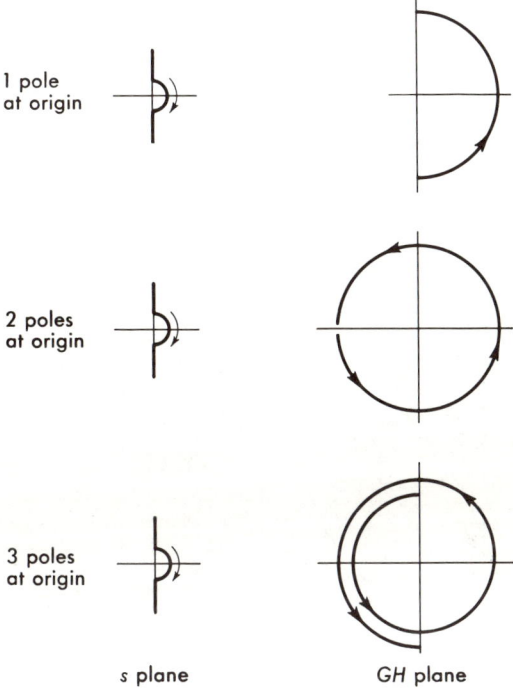

Fig. 5-14 Small circle at the origin of the s plane maps into large circle in GH plane.

on the vertical axis and the phase shift on the horizontal axis. The frequency ω is a parameter. The advantage of this plot is that, to vary the gain K, the plot is moved vertically with respect to the scale (up for increasing gain). These plots are obtained by first making the Bode plot and then transferring to the decibel vs. phase plot in much the same manner as for a polar plot. In this transfer, however, it is not necessary to convert from decibels to amplitude. As an example of this plot, the corner plots for the second-order system (Fig. 6-3a and b) are depicted in Fig. 5-15.

5-6 The s Plane and the $GH(s)$ Plane. When studying stability from the root-locus point of view as in Chap. 4, the s plane is used. The roots of $1 + GH(s) = 0$ are located on this s plane. When all the roots are in the left half plane, the transients due to an impulse input die out; thus the system is stable. If any root should lie in the right half plane, however, the corresponding exponential terms have increasing amplitudes and build up with increasing time. Pairs of complex conjugate roots of

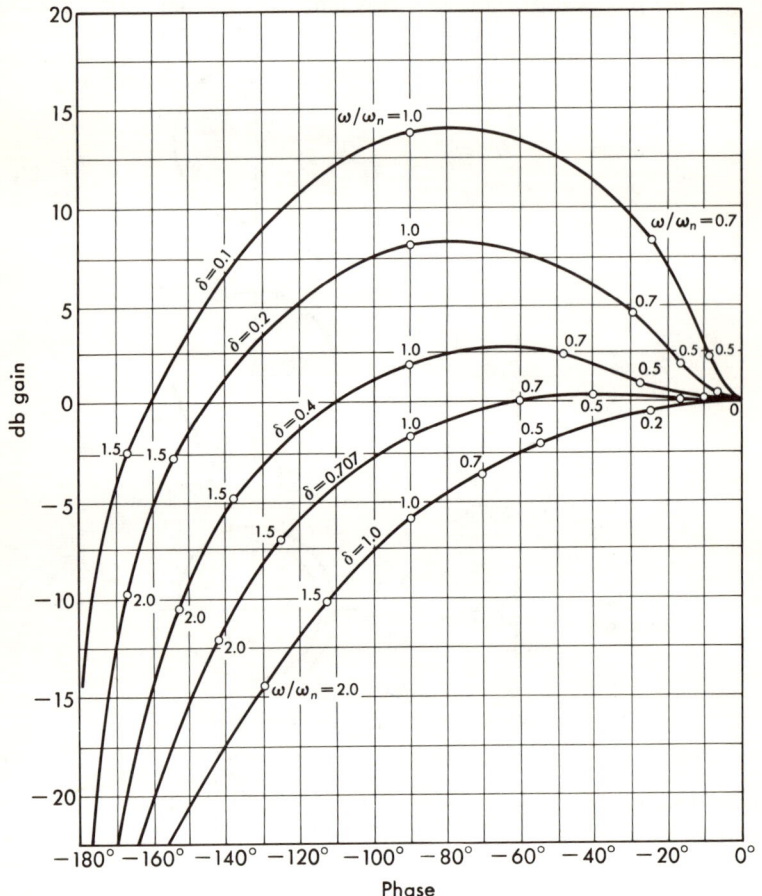

Fig. 5-15 Decibels vs. phase shift for second-order system.

order 1 that lie on the imaginary axis but not at the origin correspond to sinusoidal oscillations which neither build up nor decay. Pairs of complex conjugate roots on the imaginary axis or multiple real roots at the origin, of order 2 or higher, correspond to responses where the amplitude increases with time oscillatorily or monotonically; thus, the system is unstable. If roots on the imaginary axis are of order 1 and all other roots are in the left half plane, the resulting response is bounded and the system is termed marginally stable, i.e., considered as a sort of boundary case between stable and unstable systems.

A designer is usually directly interested in knowing if any roots lie in the right half plane; for if so, the system is unstable. To determine if any roots of $1 + GH = 0$ are in the right half plane, the contour shown in Fig. 5-16 is chosen. This contour encloses the entire right half of the s plane. The radius of the circular arc portion of the contour is made so large that all possible roots in the right half plane are included. This is usually done by considering the limiting case of infinite radius.

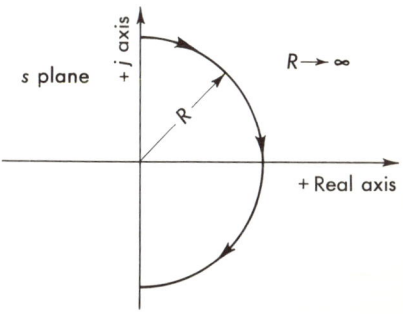

Fig. 5-16 Contour enclosing the right half of s plane.

The plot of $GH(s)$ as s runs the contour of Fig. 5-16 yields the mapping of this contour located in the s plane into a contour in the GH plane. A closed curve C_1, such as shown in the s plane of Fig. 5-17a, maps into a

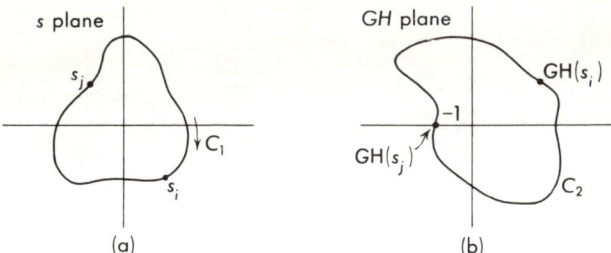

Fig. 5-17 Contour in s plane mapped into contour in GH plane.

closed curve C_2 in the GH plane, as shown in Fig. 5-17b. For single-valued functions, a one-to-one correspondence exists between a point on C_1, the curve in the s plane, and a point on C_2, the map in the GH plane. If a point is moved along C_1 in the direction of the arrow (clockwise), the mapped point moves along C_2 in a direction which depends upon the GH function.

Let s_j denote a root of $1 + GH = 0$ and suppose the contour C_1 passes through the point s_j in the s plane, as shown in Fig. 5-17a. If s_j is a root, then $GH(s_j) + 1 = 0$; thus for $s = s_j$, $GH(s_j) = -1$, that is, the contour C_2 in the GH plane passes through the point $GH = -1$.

The characteristic function for a closed-loop system is typified by

$$1 + GH = K_1 \frac{(s + s_1)(s + s_2)(s + s_3)}{(s + s_a)(s + s_b)(s + s_a)} \quad (5\text{-}45)$$

where $-s_1, -s_2, \ldots$ are the roots and $-s_a, -s_b, \ldots$ are the poles of $1 + GH$. For subsequent use it is to be noted that poles of GH are also poles of $1 + GH$, since substitution of $s = s_i$, etc., in either gives an infinite modulus. Each factor in the expression for $1 + GH$ is a complex number and hence can be represented by a vector, as shown in Fig. 5-18. The vector extends from the fixed points, s_1, s_2, s_3, and poles, s_a, s_b, s_c, to the variable point s. Suppose the variable point moves, in a clockwise direction, on a contour such as to make one complete revolution about s_2, as shown in Fig. 5-18. Then the vector $s + s_2$ makes one complete clockwise revolution because the contour encircles this root. Since all the other roots and poles are external to the contour, each of the remaining vectors makes no revolution. Since the term $s + s_2$ in Eq.

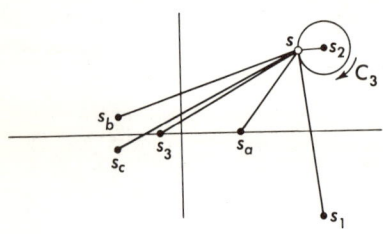

Fig. 5-18 Encirclement of a root in the s plane.

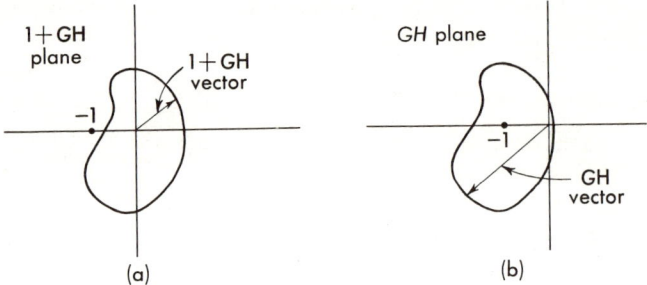

Fig. 5-19 $(1 + GH)$ plane and GH plane.

(5-45) changes phase by 360° (corresponding to one complete revolution about s_2), the $1 + GH$ incurs a change in phase of 360°. Hence, a vector representing $1 + GH$ [in the $(1 + GH)$ plane] would make one clockwise encirclement of the origin. This is shown in Fig. 5-19a. The remaining roots and poles contribute no change in the phase of $1 + GH$.

Because the roots are in the numerator of Eq. (5-45), one *clockwise* rotation about s_2 results in one *clockwise* encirclement of the origin of the $(1 + GH)$ plane, which is related to the GH plane as shown in Fig.

5-19b. In this figure, one CW (clockwise) encirclement of the origin of the $(1 + GH)$ plane corresponds to one CW encirclement of the $-1 + j0$ point in the GH plane.

Suppose the closed contour in the s plane is made to encircle both a root and a pole, as shown in Fig. 5-20a, in a CW direction. In this case the vectors drawn from both s_2 and s_a to s rotate through one complete CW revolution, or 360°. The factor $s + s_2$, corresponding to the root s_2, contributes 360° to the change in phase of $1 + GH$, since it is in the numerator of Eq. (5-45). The factor $s + s_a$, corresponding to the pole s_a, contributes $-360°$ to the change in phase, since it is in the denominator. Hence the net change in phase of $1 + GH$ is zero, and the resulting map in the GH plane does not encircle the -1 point (see Fig. 5-20b).

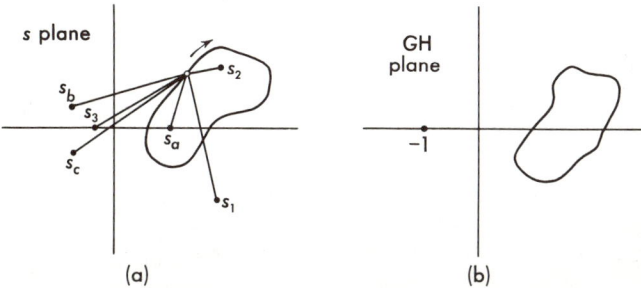

Fig. 5-20 Enclosing a root and a pole in s plane causes no net enclosure of -1 point in GH plane.

If the closed contour in the s plane is enlarged to include s_1, s_2, and s_a, the net change in phase of $1 + GH$ is 360°. A CW encirclement of a root causes a CW encirclement of the -1 point in the GH plane. A CW encirclement of a pole causes a CCW (counterclockwise) encirclement of the -1 point in the GH plane.

These results can be summarized as follows: A CW encirclement of a region in the s plane causes $n_N = n_R - n_P$ clockwise encirclements of the -1 point in the GH plane, where n_N = number of CW encirclements of the -1 point in the GH plane, n_R = number of roots located within the contour in the s plane, n_P = number of poles that are located within the contour in the s plane. Generally, n_N is positive if $n_R > n_P$ in the right half plane. In this case the -1 point is encircled in the same direction (CW) as the contour in the s plane (CW). n_N is zero if $n_R = n_P$ and the -1 point is not encircled. When n_N is negative, $n_R < n_P$ and the -1 point is encircled in the direction opposite (CCW) to the contour in the s plane (CW).

5-7 The Nyquist Stability Criterion. The frequency-analysis method of determining feedback-control-system stability can be followed by use of the Nyquist criterion. Use of this criterion provides a simple, graphical technique that allows the control-system engineer to determine linear-system stability. A rather complete mathematical derivation of the Nyquist criterion is given in Appendix V.

Roots of the characteristic equation, and thus zeros of the characteristic function $1 + GH(s)$ which lie in the right half of the s plane, evidence an unstable closed-loop system. Suppose the contour of Fig. 5-16 is made large enough to include the entire right half of the s plane. If this contour is mapped on the GH plane, the number of CW encirclements of the -1 point of the resulting map yields information permitting stability determination of the closed-loop system.

The Nyquist stability criterion can be stated as follows: The open-loop transfer function $GH(s)$ is expressed as the ratio of two factored polynomials in the variable s and written in the form

$$GH(s) = \frac{K(s\tau_1 + 1)(s\tau_3 + 1)}{(s)^n(s\tau_2 + 1)(s\tau_4 + 1)} \qquad (5\text{-}46)$$

Now as s travels a closed contour comprised of the imaginary axis from $-j\infty$ to $+j\infty$ and then the infinite right-hand semicircle from $s = Re^{j\pi/2}$ to $s = Re^{-j\pi/2}$ ($R = \infty$), the polar plot of $GH(s)$ encircles the $-1 + j0$ point in a CCW direction n_N times. n_N is given by

$$n_N = n_R - n_P \qquad (5\text{-}47)$$

where n_N is the number of clockwise encirclements of the point -1 (a negative n_N corresponds to CCW encirclements), n_P is the number of poles of GH in the right half plane, and n_R is the number of roots of $1 + GH = 0$ which lie in the right half plane. $G(s)$ is the forward-loop transfer function, and $H(s)$ is the feedback-loop transfer function. These functions are indicated in Fig. 5-21. As noted above, the poles of GH are also poles of $1 + GH$.

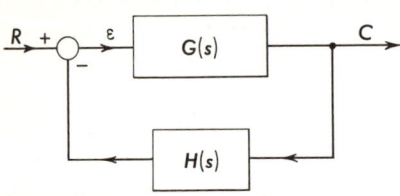

Fig. 5-21 A typical closed-loop system.

In practice, however, a somewhat simpler contour than that shown in Fig. 5-16 can be used. Since only systems having characteristic equations with constant real coefficients are considered in this text, the roots of the characteristic function must either be real or occur in complex conjugate pairs. As a result of the constant real coefficients, the map of the upper half of the $j\omega$ axis and the upper half of the infinite semicircle is the mirror image of the lower half of the $j\omega$ axis and lower half of the infinite semicircle. Further, the pole and zero structures in these two areas are likewise mirror images. Hence, it is necessary to plot the simpler closed contour comprised of only the upper portion of the imaginary axis and the upper portion of the semi-infinite circle to determine if any roots of $1 + GH(s) = 0$ lie in the right half of the s plane.

So that all the right half of the s plane is encircled, the semicircular contour must have an infinitely large radius R. Physical systems usually have zero response to an infinite frequency; hence, the large semicircle in the s plane usually maps onto the GH plane as a point at the origin. Practically, only real, positive frequencies from zero to infinity need be mapped, and this plot is obtained by applying sinusoidal inputs and varying the frequency of the signals from zero to infinity.

The Nyquist criterion gives information regarding the number of roots minus the number of poles of $1 + GH = 0$. But if GH can be expressed as the ratio of factored polynomials (and often it occurs naturally in this form), and since the poles of $1 + GH$ are identical with the poles of GH, the number of poles of GH in the right half of the s plane can be found by inspection of Eq. (5-46). Hence, under the conditions just stated the number of poles with positive real parts can easily be determined. The number of clockwise encirclements n_N of the point $-1 + j0$ in the $GH(s)$ plane is found from the GH-plane plot. Then both n_N and n_P in Eq. (5-47) are known, so that the number of roots of $1 + GH = 0$ in the right half of the s plane can be found from Eq. (5-47), with the result

$$n_R = n_N + n_P \tag{5-48}$$

As an example of the application of the Nyquist criterion to a simple system, consider the open-loop transfer function

$$GH = \frac{A}{s(\tau s + 1)} \tag{5-49}$$

This is the open-loop transfer function of a system which comprises an amplifier driving a motor as its forward loop and unity feedback. The gain constant A is the product of the potentiometer constant, the motor constant, and the amplifier amplication constant, and τ is the time constant of the motor. Replacing s by $j\omega$, as discussed in Sec. 5-1, gives the frequency transfer function $GH(j\omega)$ as

$$GH(j\omega) = \frac{A}{j\omega(j\omega\tau + 1)} \tag{5-50}$$

To apply the Nyquist criterion, it is necessary to obtain the closed contour by making a plot of $GH(j\omega)$ as the angular frequency ω is varied from zero to infinity and of $GH(s)$ as s traverses the upper half of the infinite semicircle. This semi-map is then reflected in the real axis of the GH plane. For definiteness, let τ be equal to 0.1 sec.

Use is made of the Bode asymptotic approximation (Sec. 5-4) to approximate the plot of the $GH(j\omega)$ function as the frequency is varied. The plot can be used in the following two ways:

1. The polar Nyquist plot can be made from the Bode asymptotic approximation. The stability criterion can then be applied to this polar diagram, as discussed above.
2. The stability criterion can be applied through the direct use of the Bode diagram, i.e., plot of decibel gain vs. log frequency, in a fashion not yet discussed.

Accordingly, the first of the two above-mentioned methods is used to determine the stability of the system described by Eq. (5-50).

The Bode diagram for the transfer function of Eq. (5-50) is constructed, as discussed in Sec. 5-4, by adding corresponding ordinates of the plots for

$$\frac{A}{j\omega} \quad \text{and} \quad \frac{1}{0.1j\omega + 1} \tag{5-51}$$

as shown in Fig. 5-22. The resulting amplitude plot is a straight-line segment with slope of -6 db/octave to the frequency $\omega = 10$. For $\omega > 10$, the plot is a straight-line segment with slope of -12 db/octave. The curve is plotted for $A = 1$. A change in gain constant A merely results in a vertical translation of the amplitude curve. The phase curve starts at $-90°$, corresponding to the term $s = 1/j\omega$ resulting from the pole at the origin, and continues to decrease as ω increases, to a limit of $-180°$. Its equation is $-\tan^{-1}(0.1\omega/1) - 90°$.

The polar plot is constructed from the Bode plot by plotting corresponding values of amplitude and phase for a sufficient number of frequencies and joining these points by a smooth curve as shown in Fig. 5-23. Figure 5-23a is a plot of the contour C_1 in the s plane, and Fig. 5-23b is the map, C_2, in the GH plane. The arrows on each of these curves indicate the direction of increasing values of ω. Use of the Nyquist criterion shows the system of this example (Fig. 5-23b) to be stable. Examination of the transfer function of Eq. (5-49) shows that $n_P = 0$; that is, there are no poles of GH (and hence of $1 + GH$) in the right half of the s plane. Also, the number of CW encirclements of the

−1 point is zero for any value of amplifier constant A; it follows from Eq. (5-48) that $n_R = 0$. Hence, the characteristic function of the system has no roots in the right half of the s plane for any value of A, and thus the system is stable for all values of A.

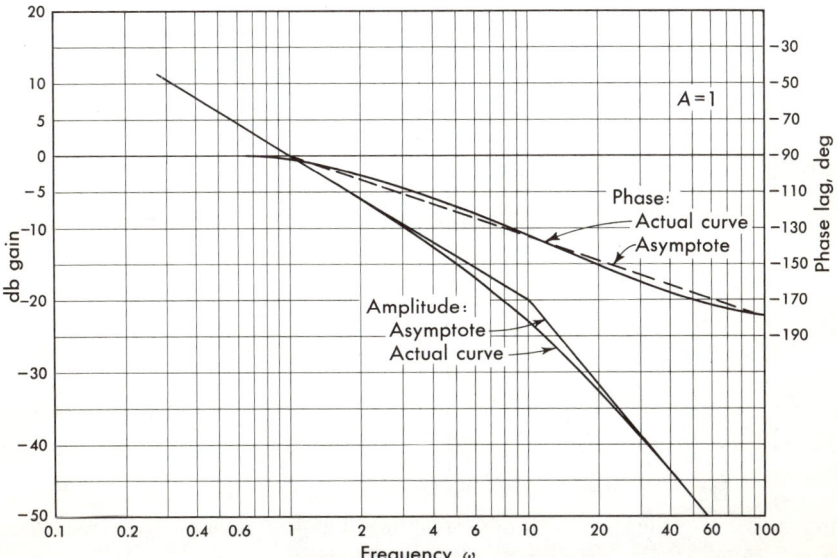

Fig. 5-22 Bode diagram (amplitude and phase) for the system

$$\frac{A}{j\omega(0.1j\omega + 1)}$$

As indicated in Chap. 4, knowledge of the open-loop transfer function $GH(s)$ can be used to determine the roots of $1 + GH = 0$. With the frequency-analysis method, the open-loop frequency response $GH(j\omega)$ is plotted, in magnitude and phase angle, as a Bode plot. The frequency is varied from zero to infinity. For example, replacing s by $j\omega$ in Eq. (5-46) gives

$$GH(j\omega) = \frac{K_1(j\omega\tau_1 + 1)(j\omega\tau_3 + 1)}{(j\omega)^n(j\omega\tau_2 + 1)(j\omega\tau_4 + 1)} \tag{5-52}$$

In general, it is convenient to put the frequency transfer function in the form typified in Eq. (5-52). If the factors are not of the form $\tau j\omega + 1$, as, for example, in

$$GH = \frac{A(j\omega + 5)(j\omega + 2)}{[(j\omega)^2 + 3j\omega + 18](j\omega + 3)} \tag{5-53}$$

178 Control System Design

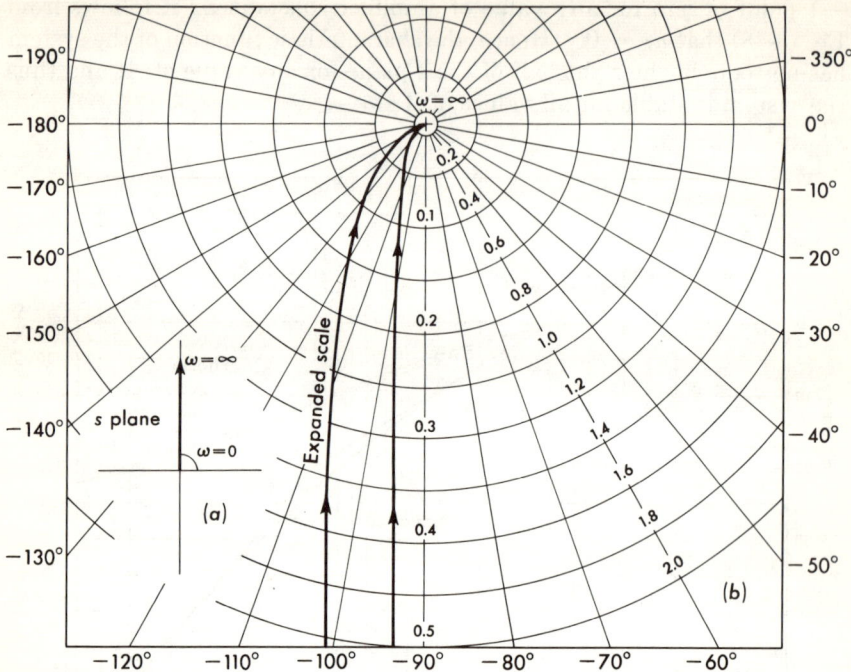

Fig. 5-23 (a) Plot of $s = j\omega$ as ω is varied from 0 to $+\infty$; (b) polar Nyquist diagram of

$$\frac{1}{j\omega(0.1j\omega + 1)}$$

the numerator and denominator of the expression can be rewritten to obtain

$$GH = \frac{A5(j\omega/5 + 1)2(j\omega/2 + 1)}{18[(j\omega)^2/18 + j\omega/6 + 1]3(j\omega/3 + 1)}$$

$$GH = \frac{10A}{54} \frac{(j\omega/5 + 1)(j\omega/2 + 1)}{[(j\omega)^2/18 + j\omega/6 + 1](j\omega/3 + 1)} \quad (5\text{-}54)$$

This latter form has the advantage of isolating the factors that multiply the gain. In addition, this form is necessary so that all poles and zeros other than those at the origin start at 0 db for low frequency. In this case the gain constant K is related to the amplifier gain as follows:

$$K = \tfrac{10}{54}A \quad (5\text{-}55)$$

Consider, as another example, a system which has the following $GH(j\omega)$ function:

$$GH(j\omega) = \frac{A}{j\omega(j\omega + 1)(\tfrac{1}{2}j\omega + 1)} \tag{5-56}$$

The Bode plot is constructed, with the asymptotic approximation, in Fig. 5-24. The polar plot is constructed by plotting points of corresponding amplitude (converting decibels to amplitude) and phase at

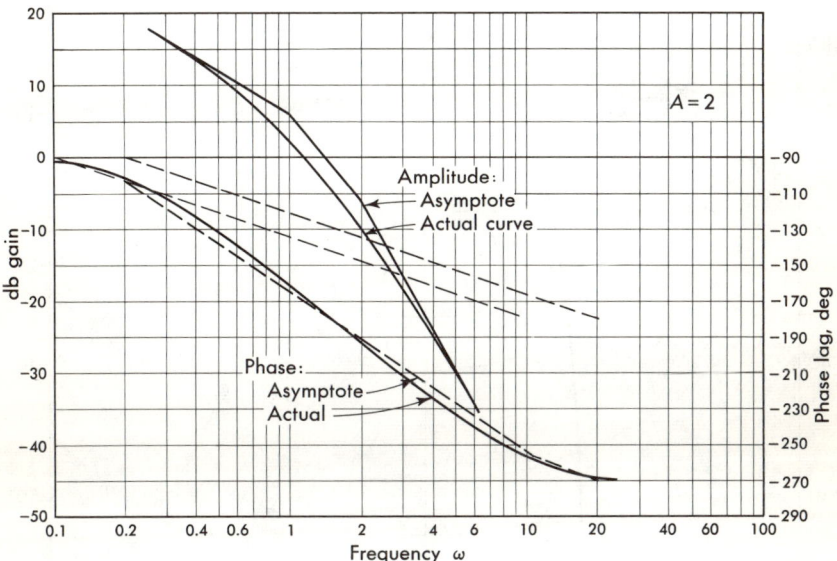

Fig. 5-24 Bode plot for
$$\frac{A}{j\omega(j\omega + 1)(\tfrac{1}{2}j\omega + 1)}$$

several values of frequencies and drawing a smooth curve through them. The plot for the $GH(j\omega)$ function of Eq. (5-56) is shown in Fig. 5-25 for two values of amplifier constant: $A = 2$ and $A = 10$. The smaller value of gain constant, $A_1 = 2$, results in a GH plot which does not encircle the point -1. Thus $n_N = 0$; and since $n_P = 0$ by inspection of Eq. (5-56), it follows that $n_R = 0$ and thus the system is stable. The larger value of gain constant, $A_2 = 10$, results in a $GH(j\omega)$ function which does encircle the point -1 as the frequency is varied from zero to infinity. Thus $n_R = n_N + n_P = n_N + 0 \neq 0$, and this system is unstable.

In summary, by utilizing an amplitude and phase plot of $GH(j\omega)$ in conjunction with the Nyquist stability criterion, the engineer is enabled

180   Control System Design

to determine whether or not any roots of the characteristic equation lie in the right half of the s plane, and thus whether or not the system is stable.

The root-locus plot and the Bode plot method in conjunction with the Nyquist criterion for determining stability are perfect complements. When the designer is in the initial stages of synthesizing a control system,

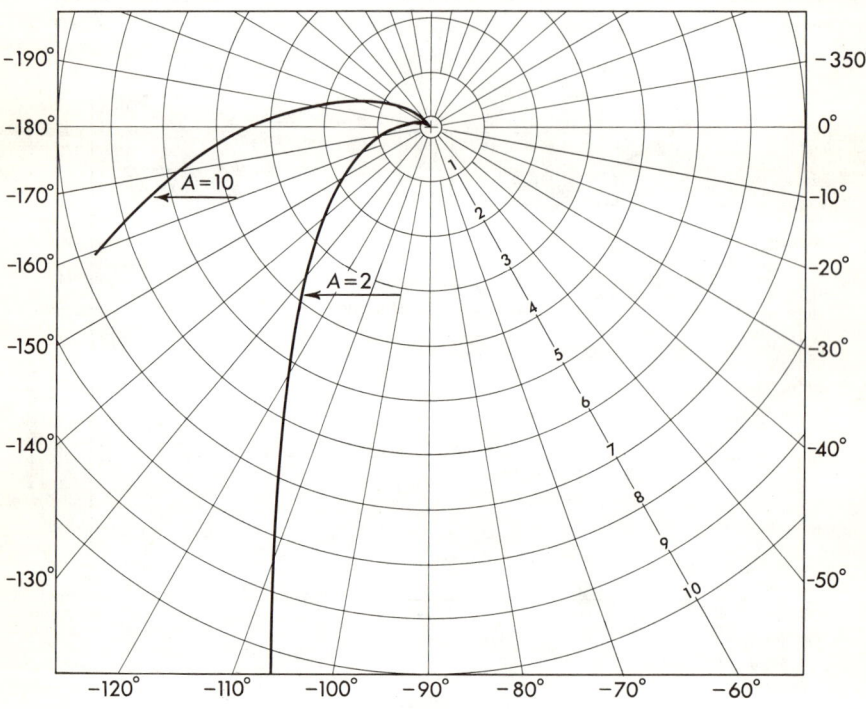

Fig. 5-25 Polar Nyquist diagram of

$$\frac{A}{j\omega(j\omega + 1)(\tfrac{1}{2}j\omega + 1)}$$

use of the root-locus plot is most convenient for rapid analysis. When the control system is built, however, an amplitude and phase response test of $GH(j\omega)$ can usually be run on most physical systems so that system stability can be studied by use of these experimentally determined curves. If a system is constructed of components whose transfer functions are not known analytically, a frequency-response test provides a means of approximating the frequency transfer function experimentally (Sec. 5-14). In summary, root-locus techniques should be used when designing control systems using analytically determined transfer func-

Stability; the Frequency-analysis Method

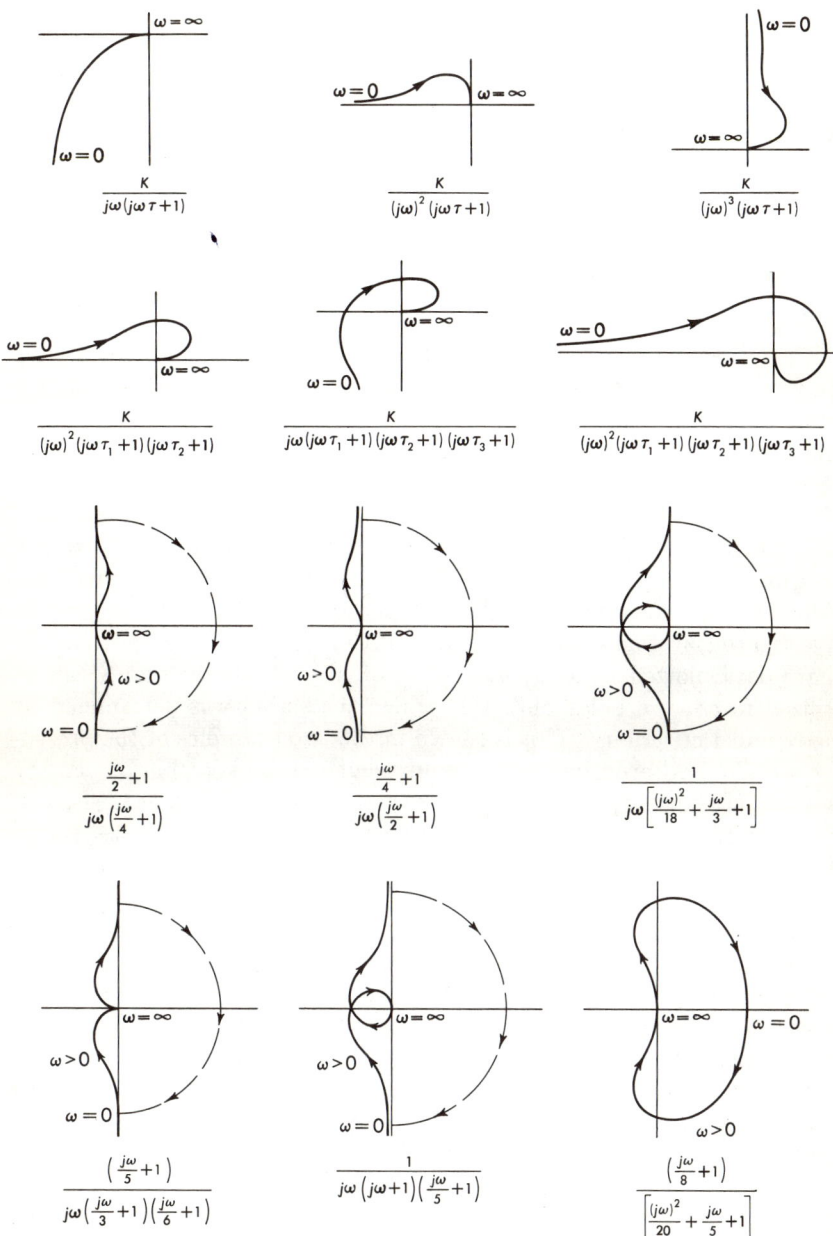

Fig. 5-26 Collection of Nyquist diagrams.

tions. The frequency-analysis method should be used when studying systems experimentally or when determining, also experimentally, frequency transfer functions which are difficult to derive analytically.

To facilitate using the frequency-analysis method, the student should be familiar with the shapes of the plots of various frequency transfer functions. For that reason, plots of several simple transfer functions are included in Fig. 5-26 in polar Nyquist form. The first six of these plot $GH(j\omega)$ for positive $j\omega$. The remaining six show complete polar plots for both $\pm j\omega$ and for the small semicircle about the pole at the origin. The small semicircle in the s plane maps into the large semicircle in the GH plane.

The number of encirclements n_N of the -1 point is found by drawing a straight line out from the -1 point in any direction. n_N equals the number of times the $GH(j\omega)$ contour crosses this line in one direction minus the number of times the contour crosses this line in the opposite direction.

5-8 Stability Criterion with the Decibel Gain and Phase vs. Frequency Diagrams. The decibel gain and phase vs. logarithmic frequency transfer function plots of control systems can be used in investigating system stability in either of two ways. The amplitude and phase data plotted as a function of frequency (Bode diagram), can be utilized to effect a polar plot from which the stability is determined by the Nyquist criterion. This is shown in the two examples of the preceding section. Alternatively, an "equivalent" stability criterion can be derived in a form applicable directly to the amplitude, in decibels, and phase vs. frequency diagrams. In certain cases it may be necessary to make the polar plot for purposes of clarifying the use of the Bode plot. This step is frequently required for conditionally stable systems.

The stability criterion utilizing the polar diagram is based on knowledge of the number of CW encirclements of the critical point $-1 + j0$ in the $GH(s)$ plane. If $n_P = 0$, as is often the case, and if the GH locus encircles or goes through this point, the system certainly has at least one root in the right half plane, or on the $j\omega$ axis, and hence is either unstable or marginally stable. The point on the Bode diagram which corresponds to the critical $-1 + j0$ point in the GH plane is located by taking the logarithm of -1; thus,

$$\log_{10}(-1) = \log 1e^{-j180°} = 0 - j180° \tag{5-57}$$

The critical point in the Bode diagram is 0 db on the amplitude curve and $k180°$, where k is any odd integer, on the phase curve, both at the same frequency. If the $GH(j\omega)$ curve encircles the $-1 + j0$ point on a polar plot, the magnitude of $GH(j\omega)$ will be greater than 1 when the phase shift is 180°. Hence, encirclement of the 0-db and 180° critical point in the Bode diagram means that the amplitude is greater than 0 db when the phase shift is 180°. Under this condition, the characteristic function has a root in the right half of the s plane and hence the system is unstable, again provided there are no poles in the right half plane. If the amplitude is exactly 0 db when the phase shift is 180°, this corresponds to the polar $GH(j\omega)$ locus passing through the point $-1 + j0$, and the root lies on the imaginary axis. Some examples of both stable and unstable systems are now considered.

5-9 Examples of Bode Construction. Suppose the stability of the system with the following transfer function is investigated:

$$GH = \frac{K}{s(s/5 + 1)(s/10 + 1)} \tag{5-58}$$

K is the gain constant of the open loop, and the GH function is represented by a pole at $s = 0$, at $s = -5$, and at $s = -10$. The Bode diagram consists of four components: one for K, one for the single pole at the origin, and one for each of the other two poles. These components are shown in Fig. 5-27.

The frequency transfer function, written as a function of ω by replacing s with $j\omega$, is

$$GH(j\omega) = \frac{K}{j\omega(j\omega/5 + 1)(j\omega/10 + 1)} \tag{5-59}$$

The corner frequencies at $\omega = 5$ and $\omega = 10$ are first located on the graph of Fig. 5-27. Notice that the frequency scale is logarithmic. Corresponding to the pole at the origin, a straight-line segment passing through the 0-db axis at $\omega = 1$ with a slope of -6 db/octave is first drawn. Adding this line to the horizontal line of $20 \log_{10} K$ for the gain constant K gives a line which goes through the 0-db axis at the value $\omega = K$. In this example, $20 \log K$ is taken as -4 db. Corresponding to the pole at -5, arising from the term $1/(j\omega/5 + 1)$, a straight-line segment, which coincides with the 0-db line, is drawn out to the corner frequency at $\omega = 5$. This line drops off at -6 db/octave, as shown in Fig. 5-27. The contribution from the term $1/(j\omega/10 + 1)$ is a line segment which is asymptotic to the 0-db line out to the corner frequency and then drops off at -6 db/octave.

184   Control System Design

The phase-shift curves corresponding to each of these four terms are inserted as shown in the figure. The phase of the gain constant is 0°. The phase of the $1/j\omega$ term is constant at $-90°$. The phase of the term $1/(j\omega/5 + 1)$ proceeds from 0 to $-90°$ at $-45°$/decade starting at $\omega = 0.5$ and intersects $-45°$ at the corner frequency $\omega = 5$. The phase of the term $1/(j\omega/10 + 1)$ changes from zero to $-90°$ at $-45°$/decade starting at $\omega = 1$ and intersects $-45°$ at the corner frequency $\omega = 10$.

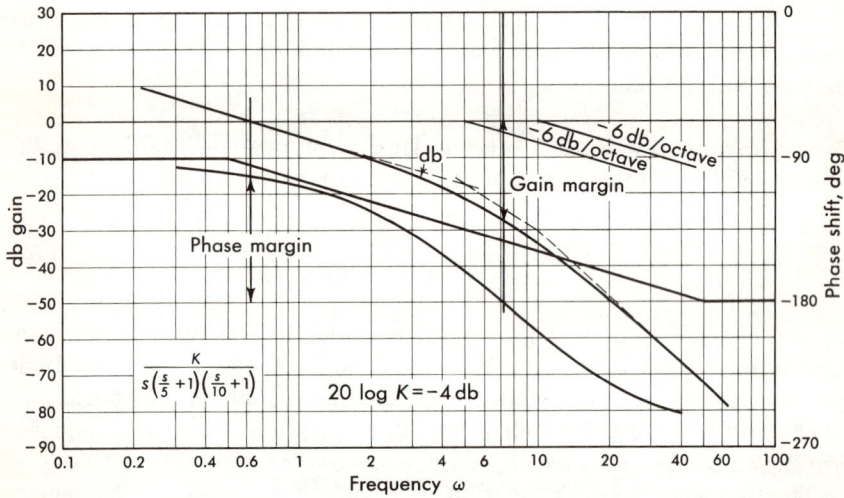

Fig. 5-27 Amplitude and phase response of
$$\frac{K}{j\omega(j\omega/5 + 1)(j\omega/10 + 1)}$$

To obtain the complete frequency response of this system, it is only necessary to (1) add the amplitude curves, thereby obtaining the asymptotic approximation shown, and (2) add the phase-angle curves, which results in the composite phase-angle curve shown in Fig. 5-27. Generally, the information obtained from the use of asymptotes is satisfactory for the determination of stability. For this particular case, when the phase angle is $-180°$, the amplitude is less than 0 db. If the gain K in the example of Fig. 5-27 is made greater, the resulting system is "less stable." In this latter case, the asymptotes would, of necessity, be "faired in" by noting that the true curve for a linear factor deviates from the asymptote by -3 db at each corner frequency and -1 db at frequencies of both half and twice the corner frequency. Another example is now given to indicate the nature of the asymptotic approximation when the system is near instability.

Suppose the stability of the system with open-loop and frequency transfer functions is investigated.

$$G(s) = \frac{K_1(s+5)}{s^2(s+10)} \qquad G(j\omega) = \frac{\overbrace{\tfrac{1}{2}K_1}^{K}(j\omega/5+1)}{(j\omega)^2(j\omega/10+1)} \qquad (5\text{-}60)$$

The terms of the numerator and denominator have been put into the form $j\omega\tau + 1$. The break points are first located on the Bode diagram. The

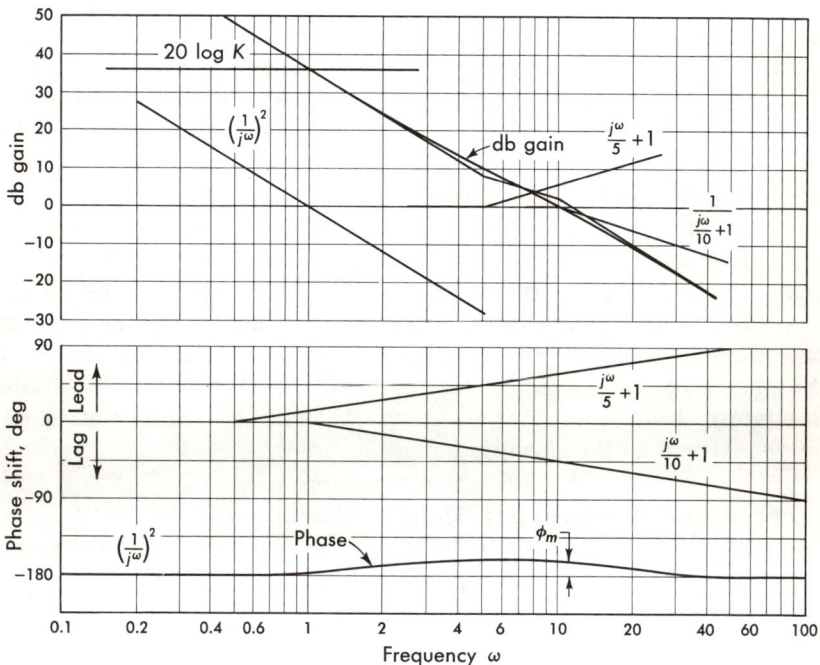

Fig. 5-28 Amplitude and phase of

$$\frac{K(j\omega/5+1)}{(j\omega)^2(j\omega/10+1)}$$

magnitude curve consists of four parts. The first is a constant, $20 \log_{10} K$. The second curve, contributed by $1/(j\omega)^2$, is a straight-line segment intersecting the 0-db axis at $\omega = 1$ with a slope -12 db/octave. The third curve is a 0-db line segment out to the first corner frequency ($\omega = 5$). From this point the amplitude increases at a rate of 6 db/octave. The fourth term, $1/(j\omega/10 + 1)$, has a curve which is a 0-db line segment out to the corner frequency, at $\omega = 10$. From this point the curve is a line segment which drops off at a rate of -6 db/octave. These curves are shown in Fig. 5-28.

The phase-angle curve for the constant K is a line of constant value of $0°$ and for the $1/(j\omega)^2$ term a line segment of constant value of $-180°$. A variation from 0 to $90°$ at a slope of $13.3°$/octave ($45°$/decade) with the $45°$ value at the corner frequency of $\omega = 5$ is contributed by $j\omega/5 + 1$. A variation from 0 to $-90°$ at a slope of $-13.3°$/octave with the corner frequency phase angle of $-45°$ results from the term $j\omega/10 + 1$. Adding the four separate amplitude curves of this system results in the total-amplitude curve shown by the heavy line on the figure. Adding the four phase-angle curves results in the composite-system phase angle shown by the heavy line. The actual amplitude curves are found from the asymptotes by raising the amplitude curve 3 db at each corner frequency for zeros and lowering -3 db at each corner frequency for poles, also adding (for zeros) or subtracting (for poles) 1 db at frequencies of one-half and twice the corner frequency.

The composite-amplitude curve is shown in Fig. 5-28 by the heavy line. Because of the double pole at the origin, this phase curve is near $-180°$ at the low-frequency end, but the term with the zero at $s = -5$ produces phase lead which keeps the phase curve of the composite system above $-180°$ until a frequency at which the values of the amplitude curve become less than 0 db is reached. To determine accurately the frequency at which unit amplitude occurs, the smooth curve rather than the asymptote must be used.

5-10 M and N Contours; Nichols Charts. The Nyquist stability criterion, and its application to determine the stability of control systems, also yields information regarding the "degree of stability" of a system. If a root of $1 + GH = 0$ exists on the imaginary axis in the s plane, the plot of $GH(j\omega)$ goes through the point $-1 + j0$; see, for example, Fig. 5-17a and b. If the root is in the right half of the s plane, the plot of $GH(j\omega)$ encircles the point $-1 + j0$. The correspondence between the contour in the s plane and the map in the $GH(j\omega)$ plane can be further extended, as shown in Fig. 5-29.

Figure 5-29b, c, and d indicates mapping of the imaginary axis of the s plane onto the $GH(j\omega)$ plane for three locations of the root s_i. If one of a pair of roots is located at the point s_1 on Fig. 5-29a, a plot somewhat as shown in Fig. 5-29b results as the frequency is varied from zero to infinity. The $GH(j\omega)$ map encircles the point $-1 + j0$, which indicates that the transient component stems from two roots in the right half of the s plane. The system oscillates with increasing peak amplitudes. If, however, one of a pair of roots is located at the point s_2, the plot in the $GH(s)$ plane resembles that of Fig. 5-29c, in which the curve goes through the point $-1 + j0$. This system exhibits oscillations which neither build up nor die out. If one of the roots is located at the point s_3, the corresponding plot of $GH(j\omega)$ neither encircles nor goes through the point $-1 + j0$, but instead misses the point, as shown in Fig. 5-29d.

The closer the root s_3 is to the imaginary axis, the closer the curve $GH(j\omega)$ is to the $-1 + j0$ point. On the basis of such fact, a relation exists between the nearness of approach to the point $-1 + j0$ in the GH plane and the nearness of the root s_3 to the imaginary axis in the s plane. Since the horizontal distance from the j axis to the root is equal to the constant $\zeta\omega_n$ in the exponential term, the damping ratio which is associated with this pair of roots is related to the nearness of

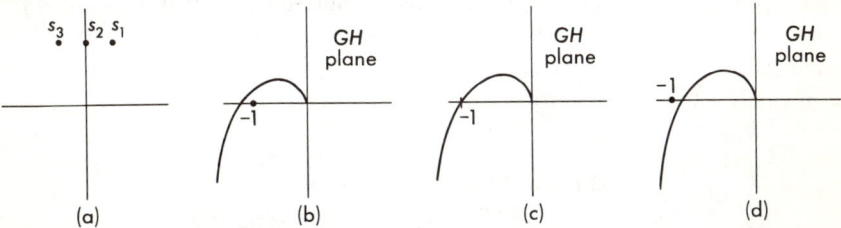

Fig. 5-29 Relation between nearness of approach to the point -1 and the root location.

approach of the map to the point $-1 + j0$ (i.e., to the "degree of instability"). The nearness of approach of the $GH(j\omega)$ plot to the point $-1 + j0$ can be related exactly to the system damping ratio for a system characterized by second-order differential equation of performance (see Appendix V). For higher-order systems, the relative stability can be similarly approximated if a pair of dominant roots of the characteristic equation exists.

The M and N contours are based upon use of the closed-loop frequency transfer function

$$\frac{C(j\omega)}{R(j\omega)} = \frac{G(j\omega)}{1 + G(j\omega)} \qquad (5\text{-}61)$$

for systems with unit feedback, $H = 1$. The values of the open-loop frequency transfer function $G(j\omega)$ and the denominator $1 + G(j\omega)$ can be indicated by vectors, which are shown in Fig. 5-30. Equation (5-61) can be written in an alternative form (polar form):

$$\frac{C(j\omega)}{R(j\omega)} = \left|\frac{G(j\omega)}{1 + G(j\omega)}\right| \bigg/ \tan^{-1} \frac{\text{Im}\,[C(j\omega)/R(j\omega)]}{\text{Re}\,[C(j\omega)/R(j\omega)]} \qquad (5\text{-}62)$$

where Im means "imaginary part of" and Re means "real part of." The $M(\omega)$ and $N(\omega)$ functions are defined in terms of the amplitude and phase of Eq. (5-62) by

$$\frac{C(j\omega)}{R(j\omega)} = M\underline{/\tan^{-1} N} \qquad (5\text{-}63)$$

where these functions are defined by

$$M = \left|\frac{G}{1+G}\right| \quad \text{and} \quad N = \frac{\text{Im}\,[C(j\omega)/R(j\omega)]}{\text{Re}\,[C(j\omega)/R(j\omega)]} \qquad (5\text{-}64)$$

The M and N curves are lines along which the magnitude is constant ($M = \text{const}$) and lines along which the phase angle is constant ($N = \text{const}$). By use of polar coordinates, the equations of M and N curves are easily shown to be circles.

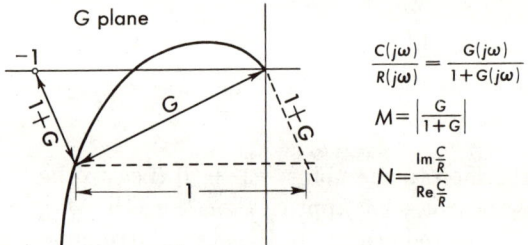

Fig. 5-30 Definition of M and N.

The M circles are shown in Fig. 5-31. The larger the value of M, the smaller the radius of the circle about the point $-1 + j0$ and the closer the center of the circle to the -1 point. The corresponding values enabling plot of the circles are found from

$$\text{Radius of } M \text{ circle} = \left|\frac{M}{M^2 - 1}\right|$$

$$\text{Abscissa coordinate of center of } M \text{ circle} = \frac{-M^2}{M^2 - 1} \qquad (5\text{-}65)$$

The derivation of these expressions is given in Appendix V. A series of M circles is superimposed on the polar plot for a particular function

$G(j\omega)$. The plot of the $G(j\omega)$ function is tangent to a particular M circle that corresponds to the maximum value of M (denoted as M_p), which occurs at a particular ω. Knowledge of the value of M_p enables an estimate of the damping ratio to be obtained.

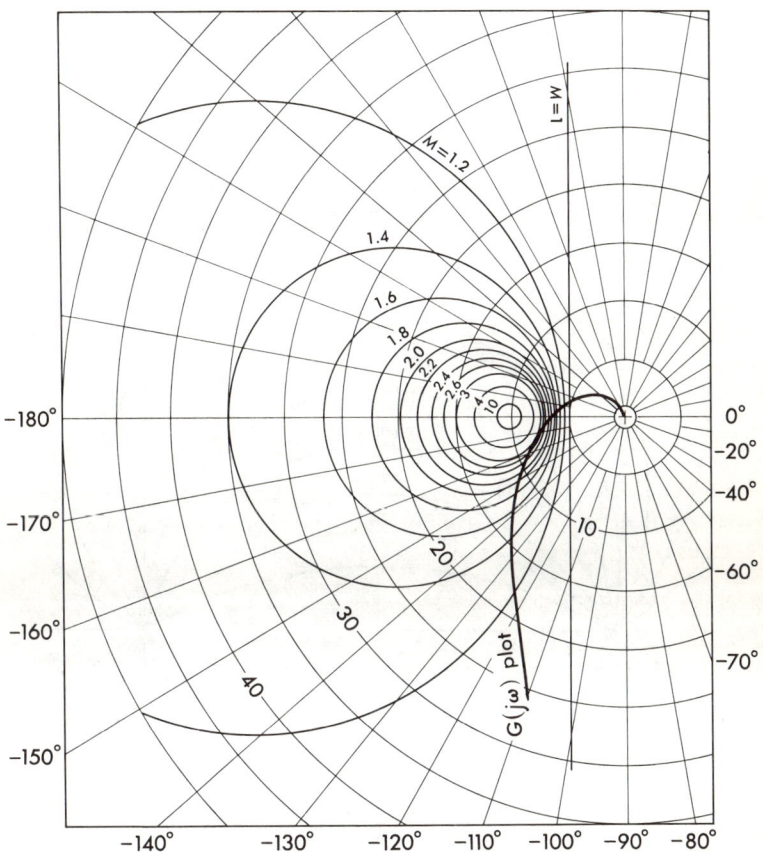

Fig. 5-31 M circles on polar plot. Radius of M circles $= |M/(M^2 - 1)|$. Abscissa coordinate of center of M circle $= -M^2/(M^2 - 1)$.

In Fig. 5-31, for example, $G(j\omega)$ is plotted as the frequency is varied from zero to infinity. On this plot the curve is tangent to the $M = 3$ curve. M_p for this system is equal to 3.

The N circles are shown in Fig. 5-32. The locus for all circle centers lies along the line Re $G(j\omega) = -\frac{1}{2}$. The circles are located along this line according to the equations

$$\text{Radius of } N \text{ circle} = \sqrt{\frac{1}{4} + \left(\frac{1}{2N}\right)^2} \quad (5\text{-}66a)$$

$$\text{Ordinate coordinate of } N \text{ circle} = \frac{1}{2N} \quad (5\text{-}66b)$$

The N circles, the equations of which are derived in Appendix V, are significant in determining the phase angle of the closed-loop frequency

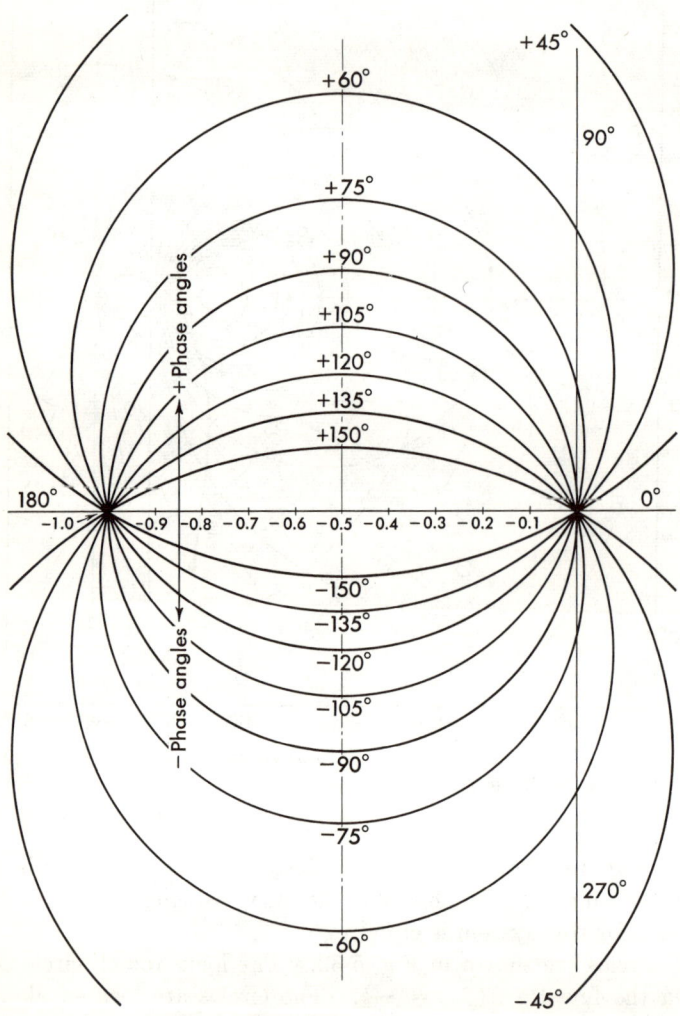

Fig. 5-32 N circles in the G plane.

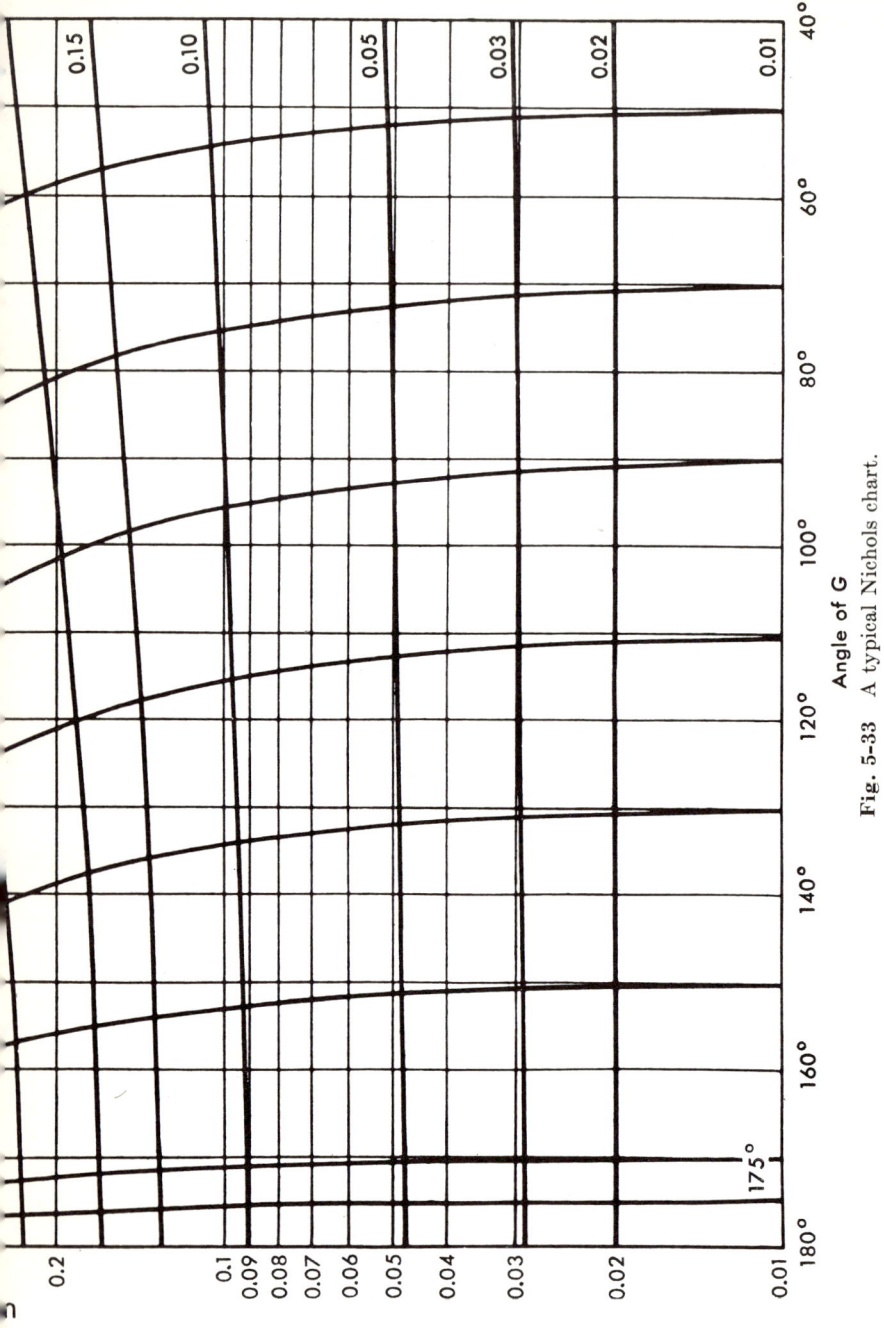

Fig. 5-33 A typical Nichols chart.

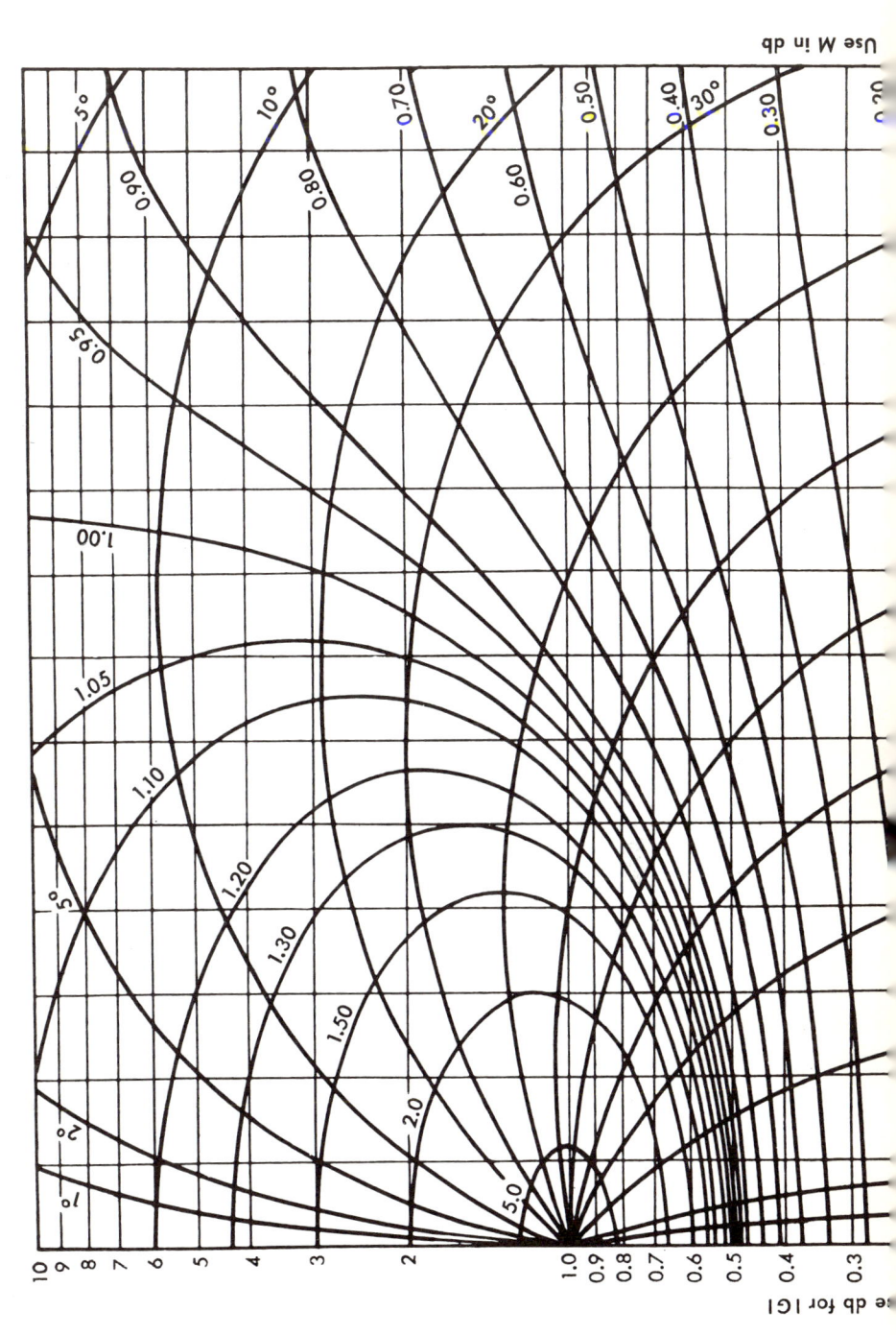

transfer function. For unity feedback systems ($H = 1$), the phase shift between input and output is a constant if the $G(j\omega)$ locus runs along one of the N circles.

5-11 Nichols Charts. A great number of extensions of Nyquist theory exist in the field today. The inverse G locus, the magnitude and phase plots for inverse G, M, and N contours, and Nichols charts are only a few. Reference to many of the texts cited in the bibliography[5,6,25,37] will yield a considerable amount of information about these techniques.

In Sec. 5-3, the Bode diagram is constructed by using decibel gain and the phase angle in degrees as vertical coordinates and the logarithm of frequency as the abscissa. The two plots are combined into one if the gain in decibels is plotted against the phase angle of $G(j\omega)$. Used in this fashion, the point $-1 + j0$ of the polar diagram maps into the 0-db and $-180°$ point on this rectangular diagram. Encirclements of this critical point are determined in the same manner as for the standard polar plot.

Passing from the polar system to the rectangular is essentially a transformation. M and N circles on the polar plot transform into M and N contours in the rectangular system. The plots of these contours for various constant values of M and N compose a "Nichols chart." One form of this chart is shown in Fig. 5-33. The use of the Nichols chart is similar to that as discussed in connection with the M and N circles using polar Nyquist diagrams except that a different set of coordinates is used. The rectangular system of coordinates is generally used in conjunction with decibel gain and phase vs. log frequency diagrams (Bode plots), since the same coordinates are plotted.

Nyquist diagrams, however plotted, are all equivalent diagrams, and no information can be inferred from one which cannot be similarly obtained from the other. The principal advantage of the Nichols chart lies in the ease with which system modifications are translated into changes in the diagram.

As an example of the use of the Nichols chart solution consider a system that is represented by an open-loop transfer function of the form

$$\frac{K}{s(s + 1)(0.2s + 1)} \qquad (5\text{-}67)$$

Suppose it is desired to find K for three values of M_p: 1.3, 2.0, 5.0. The Bode plot is first made for $K = 1$ as shown in Fig. 5-34 and the gain vs. phase plot is drawn from the Bode plot, as shown in Fig. 5-35, to the

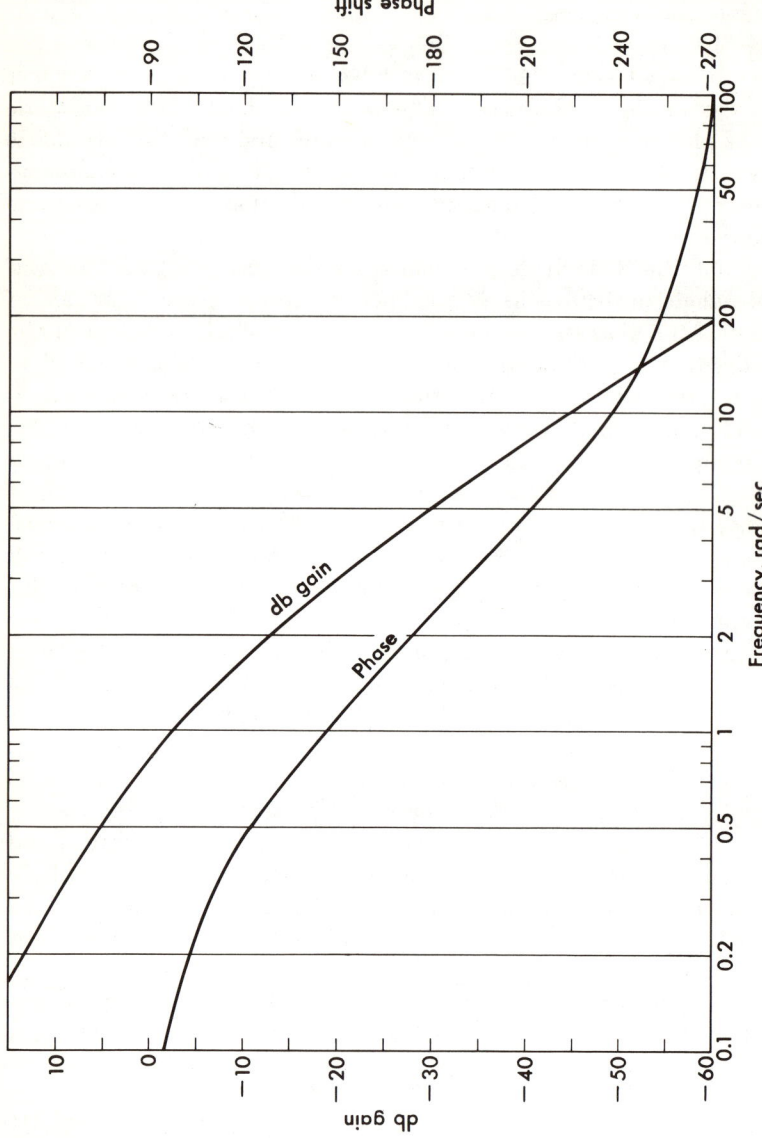

Fig. 5-34 Bode plot for $\dfrac{K}{j\omega(j\omega + 1)(0.2j\omega + 1)}$.

same scale as the Nichols chart of Fig. 5-33. The gain vs. phase plot is made on transparent paper so that it can be placed over the Nichols chart. The system gain is changed by moving the gain vs. phase plot vertically with respect to the Nichols chart. To find the system gain

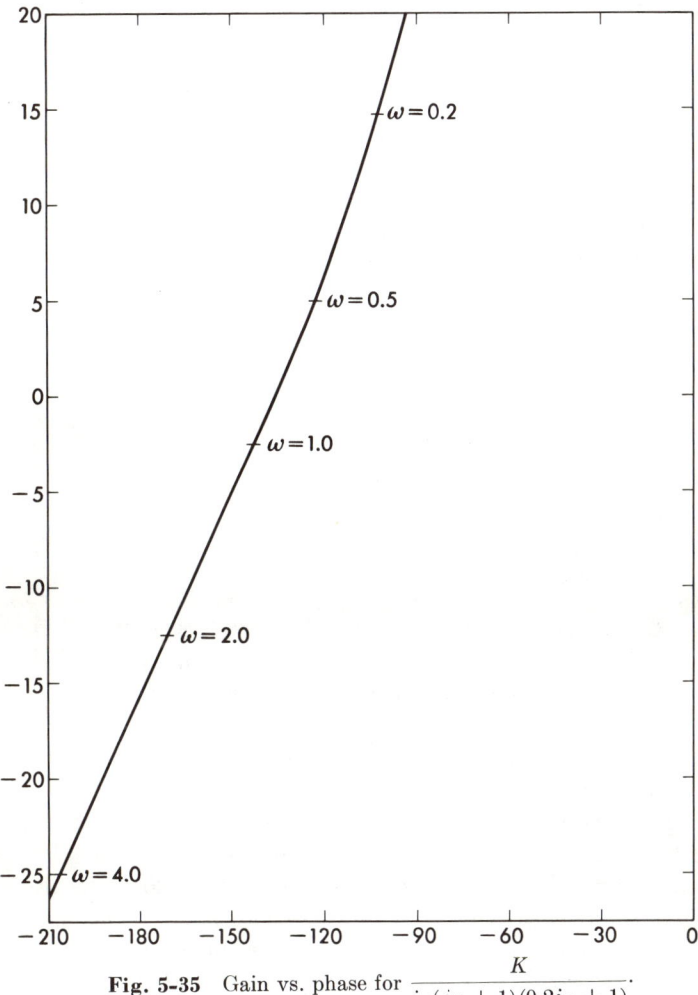

Fig. 5-35 Gain vs. phase for $\dfrac{K}{j\omega(j\omega + 1)(0.2j\omega + 1)}$.

to produce a given M_p, for example, $M_p = 1.3$, move the gain vs. phase plot until it is tangent to the $M_p = 1.3$ contour on the Nichols chart. The vertical distance moved is the system gain required to give an $M_p = 1.3$. By making the curve tangent to $M_p = 2.0$ and 5.0, the corresponding system gains can be determined.

194   Control System Design

The results are summarized as follows:

M_p	1.3	2.0	5.0
System gain K	0.8	1.5	3.4

Included in Appendix VII is a new method for determination of system stability. This method combines the Nichols chart and the root-locus method and is especially useful for more complex systems.

5-12 Determining the Degree of Stability of a System by Use of the Frequency-analysis Method. Besides answering the question as

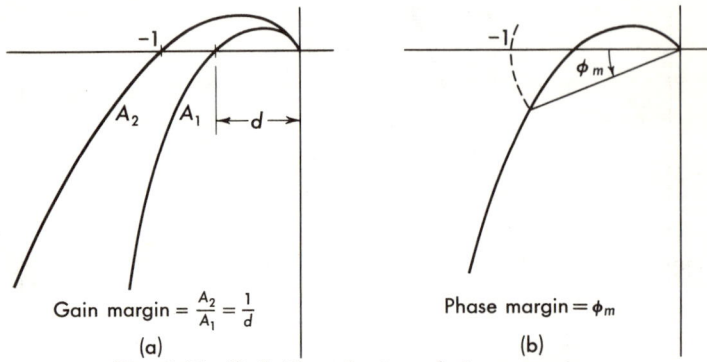

Fig. 5-36   Definition of gain and phase margin.

to whether a system is or is not stable, the frequency-analysis method gives quantitative information with respect to how stable the system is; i.e., an approximation to the damping ratio ζ of the dominant pair of roots of the transient component for the closed-loop system. For small values of damping ratio the $GH(j\omega)$ locus passes closer to the point $-1 + j0$ without enclosing it, as contrasted to large values of ζ which yield $GH(j\omega)$ plots that pass at a further distance from the point -1. Two quantities which give damping-ratio information, known as the phase and gain margins, are defined as follows:

The *gain margin*, defined in Fig. 5-36a, is the factor by which the magnitude of $GH(j\omega)$ at the phase of $-180°$ must be multiplied to produce marginal stability. Hence, if the system is stable with a gain $|GH(j\omega)| = A_1$, the gain can be increased, thus multiplied, by any value up to the gain margin before marginal stability occurs. Marginal stability results for a loop gain $|GH(j\omega)| = A_2$ that causes the locus of $GH(j\omega)$ to pass through the $-1 + j0$ point. This critical value of gain A_2 is thus related to A_1 by

$$A_2 = \text{gain margin} \times A_1 = \frac{1}{d} A_1 \qquad (5\text{-}68)$$

where d is shown in Fig. 5-36a and is measured along the $-180°$ polar line. Hence when the phase angle of $GH(j\omega)$ is $-180°$, the gain margin is the factor by which the original gain constant can be multiplied to produce marginal stability.

The *phase margin* ϕ_m is 180° plus the phase angle—expressed negatively—of $GH(j\omega)$ for which $|GH(j\omega)| = 1$, as is graphically indicated in Fig. 5-36b. On a polar plot this angle is easily found by placing the sharp point of a compass at the origin with the opposite end at the $-1 + j0$ point and then rotating the compass counterclockwise (usually) until it intersects the $GH(j\omega)$ contour, as shown in Fig. 5-36b. The included angle is the value of the phase margin ϕ_m.

The closeness of approach to the point $-1 + j0$ can be interpreted in terms of the damping ratio of the dominant pair of roots. In particular, the relationship between the damping ratio ζ and the phase margin ϕ_m can be calculated analytically for a second-order system. The curve relating phase margin to the damping ratio is shown in Fig. 5-37. The equation for this curve is derived in Appendix V. It is not possible to obtain the curve relating the damping ratio to the gain margin for a second-order system, since only for infinite frequency does the system have a phase angle of $-180°$. Hence, the system is stable for all values of gain, as shown in Chap. 4.

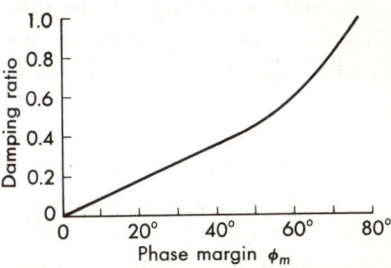

Fig. 5-37 Damping ratio ζ versus phase margin ϕ_m.

A good approximation to the ζ versus ϕ_m curve is provided by the linear portion of Fig. 5-37. Reference to Appendix V, where the curve is derived, indicates that in the region

$$0 < \phi_m < 40° \qquad (5\text{-}69)$$

the damping ratio is approximated by

$$\zeta \cong \frac{\pi}{360} \phi_m \qquad (5\text{-}70)$$

As an example of the use of the phase and gain margins, consider the example of Eq. (5-58). The phase and gain margins are read directly from Fig. 5-27. At the frequency where the amplitude characteristic crosses the 0-db line, the phase angle ϕ_0 is $-101°$. Thus the phase margin is 79°. The system of Fig. 5-22, however, has no gain margin, since the phase angle is $-180°$ only for the limiting case of $\omega = \infty$ and $GH(j\omega) = \infty$, whence multiplication of the gain produces the same infinite value. Often only one of the two margins (either the gain or phase margin) yields useful information; if so, as in this case, the meaningful one is used.

The *M criterion* is discussed in Sec. 5-9. The relative stability of a system can be ascertained from the nearness of approach of the $G(j\omega)$ locus to the $-1 + j0$ point. The estimation of the damping ratio from the Nyquist diagram gives only an approximate value, however. In many cases the nearness of approach to the point $-1 + j0$ is used to estimate the damping. More quantitatively, the smallest value of the ratio of the distance from the origin to a point on the $GH(j\omega)$ curve to the distance from the point $s = -1 + j0$ to the same point on the $GH(j\omega)$ curve may be taken as a measure of the nearness to the point $-1 + j0$. This definition is arbitrary and is of value only if its use gives satisfactory results. The usefulness of this definition can be evaluated by comparison of the results of its use with the results obtained by other methods, the root-locus technique of Chap. 4, for example.

The maximum value of $|M|$, denoted by the symbol M_p, can be related to the closed-loop damping ratio. Large values of M_p, which refer to a large resonant rise in the closed-loop frequency response, correspond to small values of damping ratio. A small resonant rise, small M_p, corresponds to a large value of damping ratio.

The exact relationship between the damping ratio and M_p can be determined for a second-order unity feedback system. Only for this system, for which

$$G(s) = \frac{K}{s(\tau s + 1)} \quad \text{and} \quad H(s) = 1$$

does a unique analytical relationship exist between M_p and ζ. This relationship, derived in Appendix V, is plotted in Fig. 5-38. The assumption is often made that this same relationship is approximately correct for higher-order systems. However, such use of Fig. 5-38 can be quite misleading, especially if the system has several pairs of complex conjugate roots near the imaginary axis. It is unfortunate, in fact, that this relationship is least correct to use for precisely those complex systems into which the engineer's experience and intuition afford least insight as how to proceed with design.

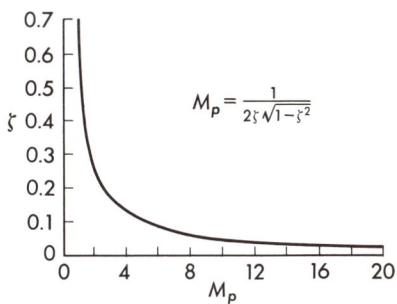

Fig. 5-38 Damping ratio ζ versus maximum M (M_p).

In summary, it may be remarked that the most important use of M_p, phase margin, and gain margin is in determining the "relative stability" of a system. Even though use of these quantities yields only approximate results in finding the absolute value of ζ, they are valuable for determining whether a parameter change in a particular system has resulted in a greater or lesser degree of stability.

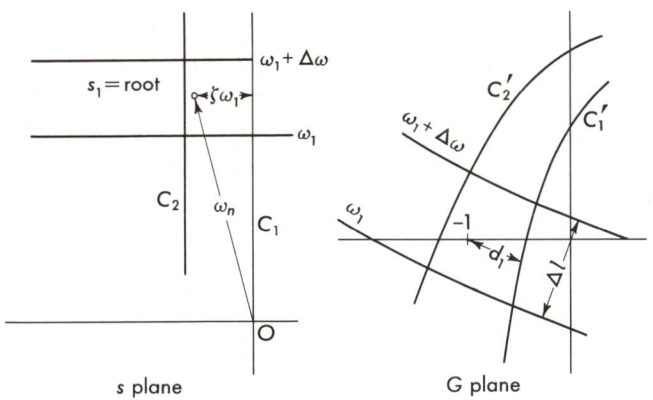

Fig. 5-39 Frequency-gradient method of approximating damping ratio.

The *frequency-gradient method* of estimating ζ is often more significant than the use of the phase or gain margin or the M criterion in estimating the damping of the least stable roots. An approximate expression for the damping ratio can be obtained from knowledge of the distance of nearest approach and of the frequency gradient along the $G(j\omega)$ locus. The equation is derived from a consideration of Fig. 5-39, in which all the terms are defined. The ratios of distances in the s plane are approximately equal, for small distances, to the ratios of distances in the $G(j\omega)$ plane. For small ζ, the distance $\zeta\omega_n$ is approximately $\zeta\omega_1$. The distance $\zeta\omega_1$ is related to d_1, the distance from the $-1 + j0$ point to the curve C_1, in the same manner in which the distance $\Delta\omega$ in the s plane is related to the distance Δl in the $G(j\omega)$ plane. Written in mathematical form,

$$\frac{\zeta\omega_1}{d_1} \approx \frac{\Delta\omega}{\Delta l} \tag{5-71}$$

Equation (5-71) can be rearranged and

$$\zeta \approx \frac{(\Delta\omega/\Delta l)d_1}{\omega_1} \tag{5-72}$$

This equation is most valid in the unusual cases where the damping is low. Since the M criterion often gives poorer results in these cases, this equation is a good complement to the M criterion. The degree of approximation becomes poorer, of course, as the distance of nearest approach becomes larger.

5-13 Comparison of Various Methods of Finding ζ. As an example of finding ζ for the least damped complex conjugate roots, the system of Fig. 5-40 is considered. The damping ratio ζ is computed by the following methods:

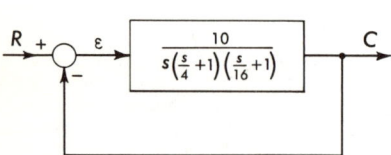

1. Root locus
2. M criterion
3. Phase margin
4. Frequency gradient

Fig. 5-40 Example system which is solved by several methods.

The root-locus plot for the system is shown in Fig. 5-41. The transfer function is written in the form appropriate for root-locus analysis as follows:

$$GH(s) = \frac{640}{s(s + 4)(s + 16)} \tag{5-73}$$

Three trials are sufficient to locate the point on the locus where $K = 640$. From Fig. 5-41 the damping ratio is found to be $\zeta = 0.139$.

Stability; the Frequency-analysis Method

The polar Nyquist plot for the system of Fig. 5-40, as plotted from the expression

$$G(j\omega) = \frac{10}{j\omega(j\omega/4 + 1)(j\omega/16 + 1)} \quad (5\text{-}74)$$

is shown in Fig. 5-42. Several M circles are drawn. The peak value M_p of M, which corresponds to the circle that is tangent to the $G(j\omega)$ plot, is 3.80. Figure 5-38 indicates that $M_p = 3.80$ corresponds to a damping ratio of $\zeta = 0.141$.

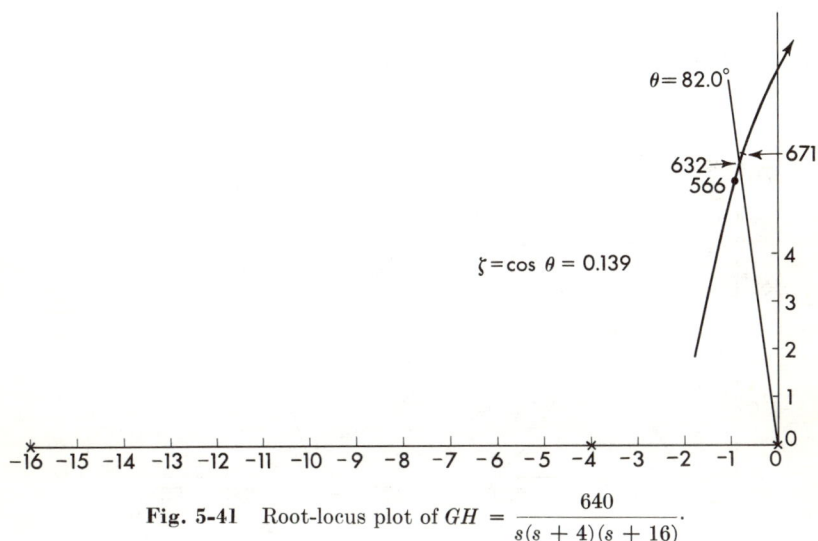

Fig. 5-41 Root-locus plot of $GH = \dfrac{640}{s(s+4)(s+16)}$.

The phase margin for this system, as found from Fig. 5-42, is

$$\phi_m = 16.7°$$

Figure 5-37 gives the damping ratio as $\zeta = 0.145$.

In the frequency-gradient method the damping ratio is estimated by measuring the pertinent lengths and noting the corresponding frequencies on Fig. 5-42. For this example,

$$d_1 = 0.60 \text{ in.} \qquad \omega_1 = 6 \text{ rad/sec}$$
$$\Delta l = 0.42 \text{ in.} \qquad \Delta\omega = 0.6 \text{ rad/sec}$$

Control System Design

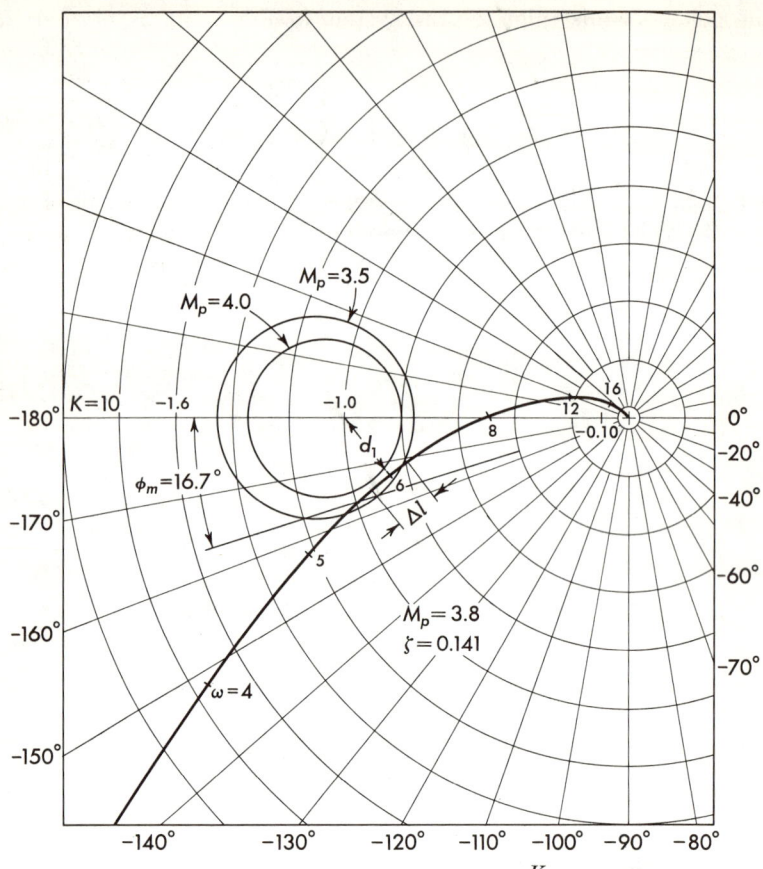

Fig. 5-42 Polar plot of $GH = \dfrac{K}{j\omega(1 + j\omega/4)(1 + j\omega/16)}$.

Substituting these values in Eq. (5-77) gives

$$\zeta \approx \frac{d_1}{\omega_1}\frac{\Delta\omega}{\Delta l} = \frac{(0.60 \text{ in.})(0.6 \text{ rad/sec})}{(6 \text{ rad/sec})(0.42 \text{ in.})} = 0.143 \quad (5\text{-}75)$$

The values of ζ found by the four methods are tabulated in Table 5-2.

Table 5-2 Value of ζ Found by Different Methods of Computation

Method of computation	Damping ratio ζ
Root locus	0.139
M criterion	0.141
Phase margin	0.145
Frequency gradient	0.143

Comparison of the various values of the damping ratio in this particular example indicates a greater degree of agreement than is normally to be expected. That the system is only of the third order and that the one real pole, $s = -6$, is so far to the left account for the close agreement.

In general, the more complex the system, the less agreement is to be expected for values similarly obtained. This is to be anticipated, because the performance of a complex system cannot be described by a single quantity ζ. Instead, it requires an understanding of the locations of the roots of the characteristic equations and how changes in system parameter affect these locations.

5-14 Closed-loop Frequency Response. In feedback-control-system design, the transfer function used in plotting the Nyquist diagram is not the overall closed-loop transfer function, nor is it the product of the transfer functions of the forward loop. It is, rather, the product of transfer functions of forward and feedback loops. However, the transfer function and, hence, the frequency response of the closed loop can usually be easily obtained. Thus, this transfer function for a single-loop system is

$$\frac{C(j\omega)}{R(j\omega)} = \frac{G(j\omega)}{1 + G(j\omega)H(j\omega)} = \frac{1}{H(j\omega)} \frac{G(j\omega)H(j\omega)}{1 + G(j\omega)H(j\omega)} \quad (5\text{-}76)$$

The frequency response of the entire system can be determined by manipulating the $GH(j\omega)$ function into the form given in the last part of Eq. (5-76). Frequently this may be required, for a closed-loop control system, to ensure that the specification is satisfied.

The M contours represent the magnitude of $G/(1 + G)$ [or, alternatively, $GH/(1 + GH)$]. When the values of M are read at points along the G plot, a closed-loop amplitude-frequency plot is formed. The closed-loop phase response is available by reading points from the N contours. The Nichols chart of Fig. 5-33 is useful in obtaining closed-loop frequency response. The $G(j\omega)$ locus is plotted on the Nichols chart in a fashion similar to that which gives Fig. 5-42. The closed-loop magnitude is read from the intersection of the $G(j\omega)$ locus with the M contour; the closed-loop phase response is read from the intersection of the $G(j\omega)$ locus with the N contour. By reading M and N at various values of frequency and drawing smooth curves through the points on an M versus N plot, the closed-loop frequency response is obtained.

If the system has a feedback function $H(j\omega) \neq 1$, the complex-number values read from the Nichols chart by use of $GH(j\omega)$ must be divided, usually graphically,* by $H(j\omega)$, as indicated by Eq. (5-76).

* Such a procedure is outlined in Ref. 5, pp. 266–274.

Effecting these plots, however, comprises one of the terminal design stages and comes only after questions of gain, stability, and steady-state errors have been satisfied. The plots are useful because they supply frequency-response specifications such as bandwidth.

5-15 Experimental Data. One of the most important uses of the frequency-analysis method of system analysis is in determining the transfer functions of certain components. Because of difficulty in formulating the differential equations of performance of pneumatic and hydrodynamic components, magnetic amplifiers, rotating components, and the like, analytic expressions for the transfer functions of such components may be difficult to obtain. Pneumatic valves, for example, do not readily lend themselves to analytic determination of their transfer functions. Very often, however, a frequency test can be run on these components and the decibel gain and phase shift vs. logarithmic frequency can be plotted. Then design can be facilitated by the use of approximate transfer functions determined from the experimentally obtained frequency plots. Although means of obtaining functional approximations to experimentally derived curves, as may be needed in network synthesis and other fields of electrical engineering, have been considered in great detail,[3] the asymptotic approximations afforded by the Bode plots are, perhaps, most directly helpful in practical control-system design.

Thus, illustratively, consider an example for which the experimental amplitude and phase characteristics of a system are as tabulated in

Table 5-3 Experimental Frequency Data

f	ω	Gain, db	Phase shift, deg
60	377	−7.75	−155
50	314	−4.3	−150
40	251	−0.2	−145
35	219	0.75	−140
25	157	5.16	−135
20	126	7.97	−120
16	100	10.53	−110
10	63	15.03	−100
7	44	16.90	−85
2.5	16	20.43	−45
1.3	8	21.6	−30
0.22	1.2	24.0	−5
0.16	1.0	24.1	0

Table 5-3. It is desired to obtain a transfer function which approximates these characteristics. If the characteristics are plotted, as in Fig. 5-43, against the logarithmic frequency, a series of straight-line asymptotes can be fitted to these data for both amplitude and phase. By use of the slopes and corresponding corner frequencies a transfer

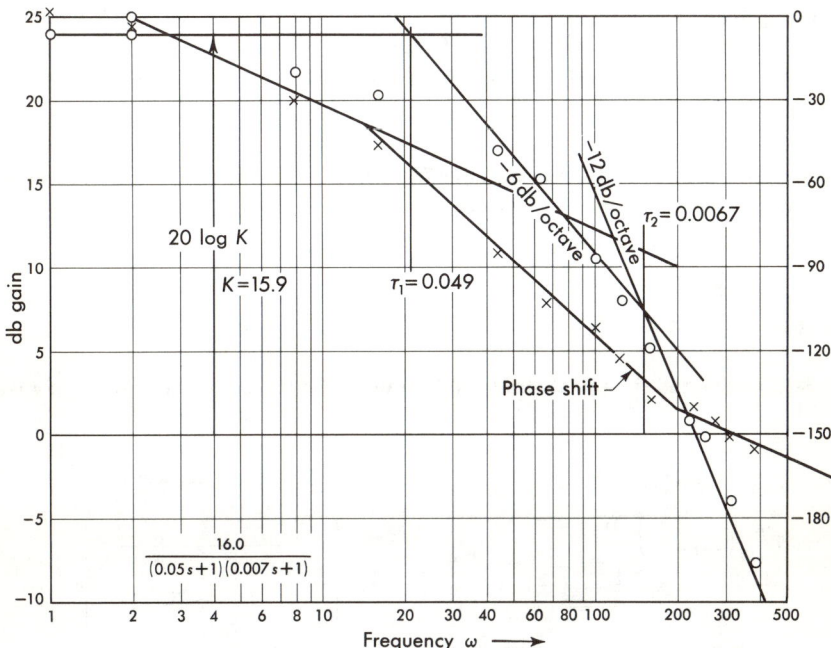

Fig. 5-43 Approximate transfer function for experimental data.

function is obtained. For the example given in Fig. 5-43 an approximate transfer function, as so calculated from the Bode plot, is

$$GH(s) = \frac{16.0}{(0.05s + 1)(0.007s + 1)} \quad (5\text{-}77)$$

Often the phase vs. logarithmic frequency, as calculated from the approximate transfer function, will not completely agree with the corresponding experimental curve. The problem of obtaining the best match for both amplitude and phase curves is simplified if linear asymptotes are used for both amplitude and phase curves and the two line-segment curves are utilized simultaneously. Here the phase-shift curve can be approximated by a straight line, as shown in Fig. 5-8b.

Use of this approximated transfer function, Eq. (5-77), in conjunction with the remaining analytically obtainable transfer functions permits the engineer to analyze the system. A more accurate study may require use of the actual element in conjunction with representation of the rest of the system on an analogue computer. In small firms, however, such a computer or simulator may be lacking and it may prove too expensive to rent time on one elsewhere. In such case, careful application of an experimentally determined transfer function can yield useful, although approximate, information by reasonable calculation effected in a short time.

Having once obtained a transfer function from this frequency analysis, the engineer can continue the synthesis of the control system by means of the root-locus techniques described in Chap. 4.

PROBLEMS

5-1. Derive the sinusoidal steady-state transfer functions for the systems of Fig. 5P-1.

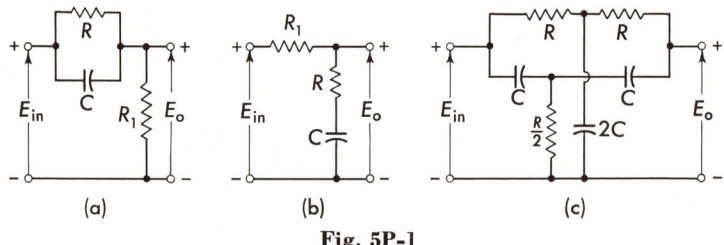

Fig. 5P-1

5-2. Plot the amplitude and phase response of the transfer function of Prob. 5-1 for $R_1 = 0.5R = 100$ kilohms and $C = 1$ μf.

5-3. Find the sinusoidal steady-state transfer functions for the systems of Prob. 2-3.

5-4. Show that the substitution $s = j\omega$ is applicable to finding the steady-state solution for any linear differential equation (system) with constant coefficients when the driving function is sinusoidal and the initial conditions are zero.

5-5. Plot the rise in the amplitude-frequency response vs. the damping ratio ζ for the function

$$G = \frac{E_o}{E_{\text{in}}} = \frac{1}{(1 - u^2) + j2\zeta u}$$

where $\quad u = \dfrac{\omega}{\omega_n}$

5-6. Use the asymptotic approximation to construct amplitude and phase diagrams vs. log frequency for the following transfer functions:

(a) $\dfrac{1}{s}$

(b) $\dfrac{1}{(s+5)(s+10)}$

(c) $\dfrac{1}{s(s+5)}$

(d) $\dfrac{5s+1}{20s+1}$

(e) $\dfrac{1}{s(4s^2+3s+2)}$

(f) $\dfrac{9(s^2+2s+1)}{3s^3+4s^2+7s+2}$

5-7. Sketch polar diagrams for the following transfer functions:

(a) $\dfrac{1}{\tau_1 s+1}$

(b) $\dfrac{\tau_1 s+1}{\tau_2 s+1}$

(c) $\dfrac{1}{(\tau_1 s+1)(\tau_2 s+1)}$

where $\tau_1 = 5$, $\tau_2 = 20$.

5-8. From the curves of Prob. 5-6, plot polar Nyquist diagrams for the given transfer functions. From the polar diagrams determine the stability of each system.

5-9. Construct amplitude and phase vs. log frequency diagrams for the transfer functions of Prob. 4-9. Estimate the gain that will result in a damping ratio ζ of 0.2.

5-10. Construct a Bode plot for $10K(s+1)/s^2(s+10)$ and discuss stability for various values of K.

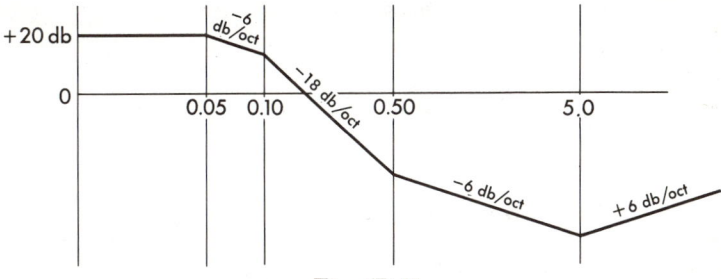

Fig. 5P-11

5-11. Obtain the transfer function for the amplitude curve shown in Fig. 5P-11.

5-12. Plot a polar Nyquist diagram for the function

$$GH = \dfrac{K}{s(s/4+1)(s/16+1)(4s+1)}$$

for $K = 1$ and $K = 4$. Which is stable?

5-13. For the system of Prob. 5-12 with $K = 1$, obtain an approximation to the damping ratio by each of the following four methods:
(a) M criterion
(b) Root locus
(c) Phase margin
(d) Frequency gradient
Compare results.

Ans. $\zeta = 0.16$

5-14. The open-loop response of a control system is given by

$$G = \frac{K}{s(s + 1)(s/5 + 1)}$$

Find K such that the closed-loop response has a damping ratio equal to 0.20, which corresponds to an $M_p = 2.54$. *Ans.* $K = 22$

5-15. Use the M criterion to determine the gain K that corresponds to a $\zeta = 0.2$ ($M_p = 2.54$) for the systems of Prob. 4-9.

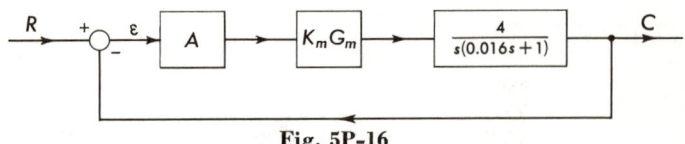

Fig. 5P-16

5-16. A position servo consists of an electronic amplifier (which has a transfer function of A), a magnetic power amplifier, and a motor. The motor transfer function is $4.0/s(0.016s + 1)$, and the transfer function of the magnetic amplifier is G_M. The block diagram for the system is shown in Fig. 5P-16. The experimental frequency data on the magnetic amplifier are shown in the table. Determine the gain A so that the system has a damping ratio of 0.20.

Frequency f, cps	Gain, db	Phase shift, deg
120	−7.8	−165
100	−4.3	−160
80	−0.2	−159
70	0.75	−151
50	5.16	−139
40	7.97	−128
32	10.53	−126
20	15.03	−90
14	16.90	−55
5	20.43	−37
2.5	21.6	−27
0.5	24.0	−10
0	24.0	0

Ans. $A = 1$

5-17. Use the Nichols chart to plot the closed-loop frequency response $G/(1 + G)$ for Prob. 5-14 with each of the following values of M_p: 1.3, 2.0, and 5.0.

5-18. Three identical positive-gain amplifiers with the individual characteristics shown in Fig. 5P-18 are ganged together. The 3-db points are as shown.

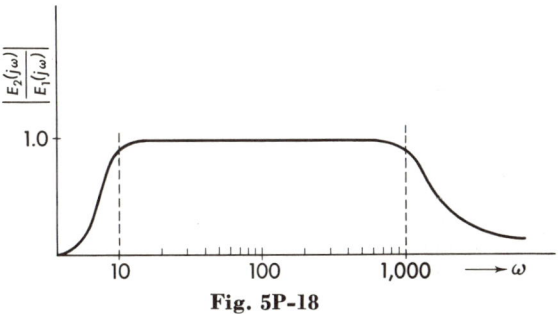

Fig. 5P-18

(a) What is the transfer function of the ganged amplifiers; i.e., what is the open-loop transfer function?

(b) Show in a polar plot the combined characteristic of the ganged amplifiers. Can the closed-loop system utilizing the ganged amplifiers in the open loop be unstable? Why?

5-19. Determine the transfer function from the following data:

ω	$G(j\omega)$	ω	$G(j\omega)$
0.1	$0 - j20$	2.0	$-0.4 - j0.6$
0.2	$-1.3 - j10$	4.0	$-0.2 - j0.2$
0.4	$-0.8 - j5$	8.0	$-0.05 - j0.008$
1.0	$-0.6 - j1.7$	16.0	$-0.008 + j0.0003$

5-20. In the speed-control system shown in Fig. 5P-20, set the gain A for a damping ratio of $\zeta = 0.2$. The experimental data are shown in the table.

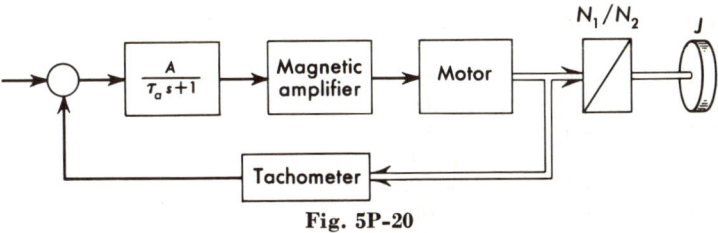

Fig. 5P-20

208 Control System Design

ω	φ, deg	Gain, db
1	−10	0
2	−30	0
5	−40	−2
10	−50	−6
25	−60	−11
50	−70	−20

Constants:
$J = 3 \times 10^{-5}$ oz-in.-sec^2
Motor $K_m = 20$ (volt-sec)$^{-1}$ (with load J reflected through the gear train)
               $\tau_m = 0.05$ sec
Gear train $N_1 = 100$
               $N_2 = 500$
Tachometer   0.05 volt/(rad/sec)
Amplifier $\tau_a = 0.1$ sec

5-21. The block diagram for an electromechanical integrator is shown in Fig. 5P-21. Use the servomotor in Eq. (7-47) with a 100:1 gear train and with the magnetic amplifier whose experimental data are shown.

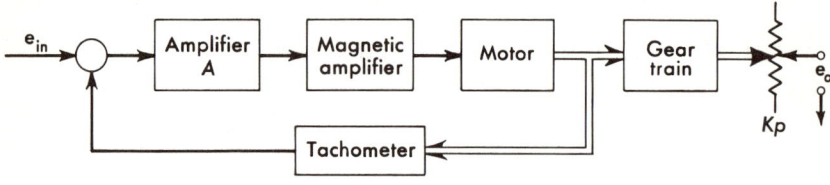

Fig. 5P-21

(a) So select K_p that the scale factor is unity, that is, $e_o = \int e_{in} \, dt$.
(b) So select A that $\zeta = 0.1$.
(c) Plot the closed-loop frequency response (amplitude and phase vs. frequency).

Frequency, cps	Gain, db	Phase angle, deg
0.6	0	−7
1.0	+1	−12
1.5	+2	−22
2.5	+4	−52
3.2	+4	−90
4.5	−5	−140
6.0	−10	−155
10.0	−18	−168
20.0	−32	−178

5-22. (a) Plot the Bode plot for the system

$$\frac{K}{s(s^2 + 10s + 100)}$$

and set the gain for a damping ratio of $\zeta = 0.05$. *Ans.* $K = 200$

(b) For the value of gain found in (a), sketch the *closed-loop* frequency response.

chapter 6

DESIGN OF FEEDBACK CONTROL SYSTEMS

6-1 Introduction. Feedback-control-system design involves compromise between the allowable steady-state error magnitude and the degree of stability desired. After the first system analysis it may be found that the system has too great a steady-state error (possibly K_v is too small). Alternatively, the steady-state error may be within specifications, but the system is too close to instability (i.e., the damping ratio is too small). In either case the system must be "equalized." In general terms, control-system equalization consists in changing the loop gain, adding zeros and poles, or changing time constants so that the steady-state error and the degree of stability are brought to within the desired specifications.

6-2 Equalization by Gain Adjustment. For example, consider the system of Fig. 6-1, which has an open-loop transfer function

$$G(s) = \frac{4.1A}{s(0.005s + 1)(0.016s + 1)} \tag{6-1}$$

Suppose that the steady-state error requirement is met by this system but the damping is too low. The reduction in gain necessary to increase the damping ratio from 0.1 to 0.2 is obtained from the root-locus plot, Fig. 6-2, as

$$K_{\zeta=0.2} = 0.64 K_{\zeta=0.1} \tag{6-2}$$

The error coefficient K_v, however, is reduced by the same factor as the gain. Although the stability is improved, the velocity error is increased. If the system is simple, it may be expedient to equalize by gain reduction. In general, an optimum system does not result, and in more complex cases gain variation alone is not sufficient.

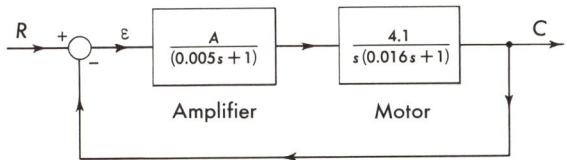

Fig. 6-1 Position servo system.

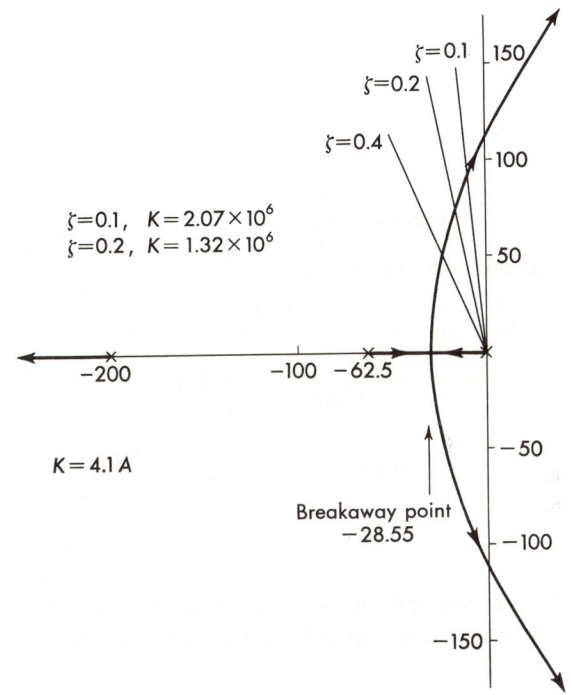

Fig. 6-2 Root-locus plot for the system with transfer function

$$\frac{4.1A}{(0.016)(0.005)} \frac{1}{s(s+200)(s+62.5)}$$

6-3 Equalization by Inserting a Network. When the system requires an improvement in the steady-state error or a degree of stability greater than that which can be obtained by gain variation, insertion of a network in the loop may be required. Consider again the system shown in Fig. 6-1. A network, of transfer function G_N, is inserted in the forward loop shown in Fig. 6-3. It is again desired to raise ζ to 0.2 but without sacrificing the gain magnitude.

The amplifier time constant is 0.005, and the amplifier gain is A. The network, shown in Fig. 6-4, is inserted in the loop at a point where

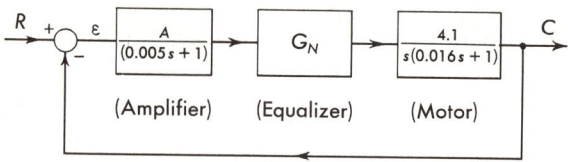

Fig. 6-3 Equalized position servo.

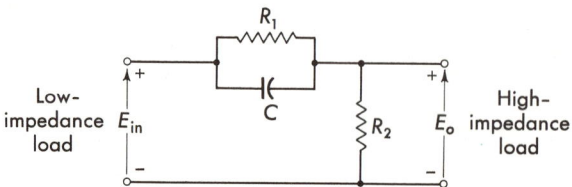

Fig. 6-4 A passive circuit equalizer.

the input impedance to the network is low (ideally zero) and the impedance which loads the network output is high (ideally infinite). The network transfer function chosen for this example is

$$G_N(s) = \frac{s + 1/\tau_1}{s + 1/\tau_2} \tag{6-3}$$

where $\tau_1 = R_1 C$ and $\tau_2 = \dfrac{R_1 R_2 C}{R_1 + R_2}$ (6-4)

As a first case, suppose $R_1 = R_2$ and it is required to determine the value of the RC product that will yield the maximum velocity-error coefficient K_v, corresponding to the minimum error for $\zeta = 0.2$. If RC is represented by τ, the open-loop transfer function is

$$G(s) = \frac{A(5.13 \times 10^4)(s + 1/\tau)}{s(s + 200)(s + 62.5)(s + 2/\tau)} \tag{6-5}$$

The transfer function of Eq. (6-5) is written in standard form for root-locus solution. The velocity coefficient is

$$K_v = \lim_{s \to 0} \frac{sA(5.13 \times 10^4)(s + 1/\tau)}{s(s + 200)(s + 62.5)(s + 2/\tau)}$$

$$K_v = \frac{A(5.13 \times 10^4)(1/\tau)}{(200)(62.5)(2/\tau)} \cong 2A \tag{6-6}$$

It is to be noticed that the network changes K_v by a factor of 2 because $R_1 = R_2 = R$ acts as a voltage divider. The method of optimizing this system is outlined as follows:

1. The poles and zeros are located on the s plane.
2. The $\zeta = 0.2$ line is drawn from the origin ($\theta = \cos^{-1} \zeta$, where θ is measured from the negative real axis).
3. The one variable pole and one variable zero are moved along the negative real axis. For different positions of pole and zero, the point is located on the $\zeta = 0.2$ line where the sum of the angles equals 180°.
4. The gain is determined at the point $G(s) = 1$. A closed-loop damping ratio of $\zeta = 0.20$ results for this specific value of τ.

A typical root-locus sketch is shown in Fig. 6-5 for one location of the equalizer zero and pole ($1/\tau = 120$). The zero and pole are moved

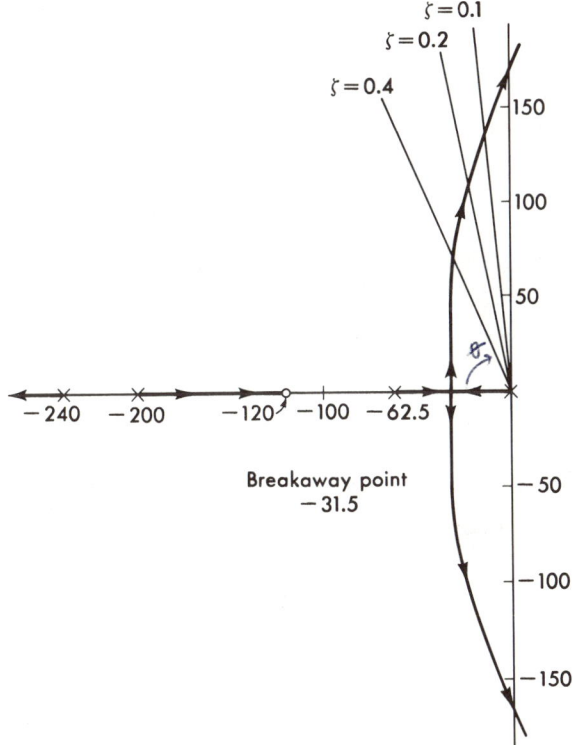

Fig. 6-5 A typical root-locus plot for

$$\frac{4.1A}{(0.016)(0.005)} \frac{s + 120}{s(s + 200)(s + 62.5)(s + 240)}$$

along the real axis until an optimum K_v is obtained for a specific damping ratio of $\zeta = 0.20$. Included in Fig. 6-6 is a plot of K_v versus the reciprocal time constant $1/RC = 1/\tau$. A maximum $K_v(=168)$ occurs for a $\tau = RC = 8.33$ msec. An improvement in K_v by approximately a factor of 3 results by inserting the network of Fig. 6-4 with $R_1 = R_2$, since the velocity constant is approximately 50 for $1/RC = 0$.

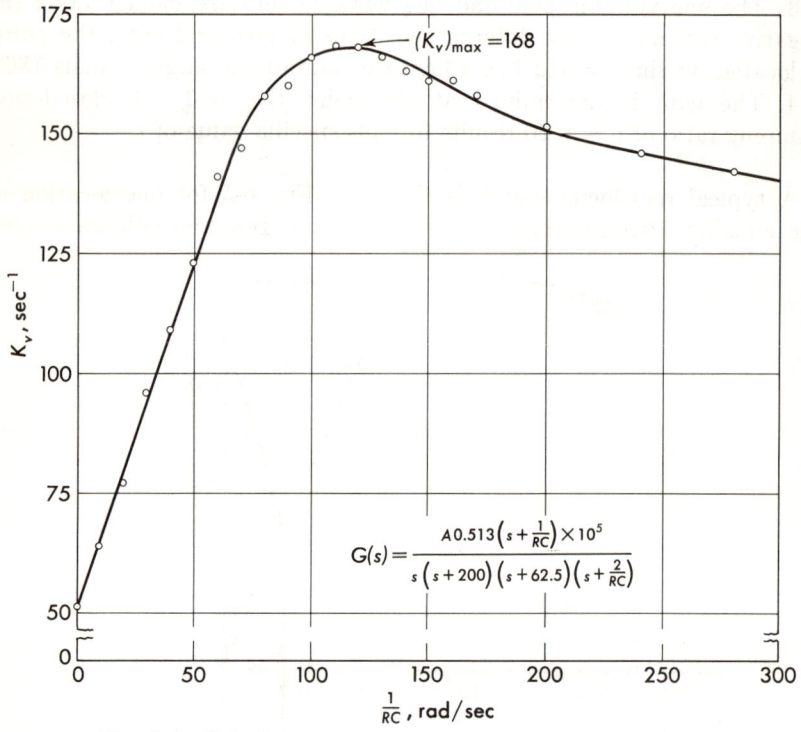

Fig. 6-6 Velocity-error constant K_v versus reciprocal of RC.

As the final step, suppose that R_1 is not held equal to R_2 but that in optimizing this system the amplifier gain is limited to 1,000. Since the zero and pole now bear no specific relation (τ_2/τ_1 is not necessarily equal to $\frac{1}{2}$), more freedom is available and a larger K_v is possible. The problem becomes somewhat complicated because an optimum must be found when two quantities are varying simultaneously. The problem is approached by holding the ratio of τ_2/τ_1 fixed and plotting a curve of K_v versus $1/\tau_1$. By taking several τ_2/τ_1 ratios, an optimum solution is achieved.

To determine the values of R_1, R_2, and C that give a maximum K_v, it is necessary to use a technique of successive approximations. The velocity constant K_v is given by

$$K_v = \lim_{s \to 0} sG(s) = \lim_{s \to 0} \frac{A 5.13 \times 10^4 (s + 1/\tau_1)}{(s + 200)(s + 62.5)(s + 1/\tau_2)}$$

$$K_v = A \frac{5.13 \times 10^4}{1.25 \times 10^4} \frac{\tau_2}{\tau_1} = 4.1 A \frac{\tau_2}{\tau_1}$$

(6-7)

where $\tau_2 = [R_1 R_2/(R_1 + R_2)]C$ and $\tau_1 = R_1 C$. It is solved by plotting a family of K_v versus $1/\tau_1$ curves for several specific ratios of τ_2/τ_1. Examination of these curves indicates the dependence of K_v upon the parameter τ_2/τ_1. Interpolation between the curves results in the maximum K_v possible with a fixed value of $A = 1,000$.

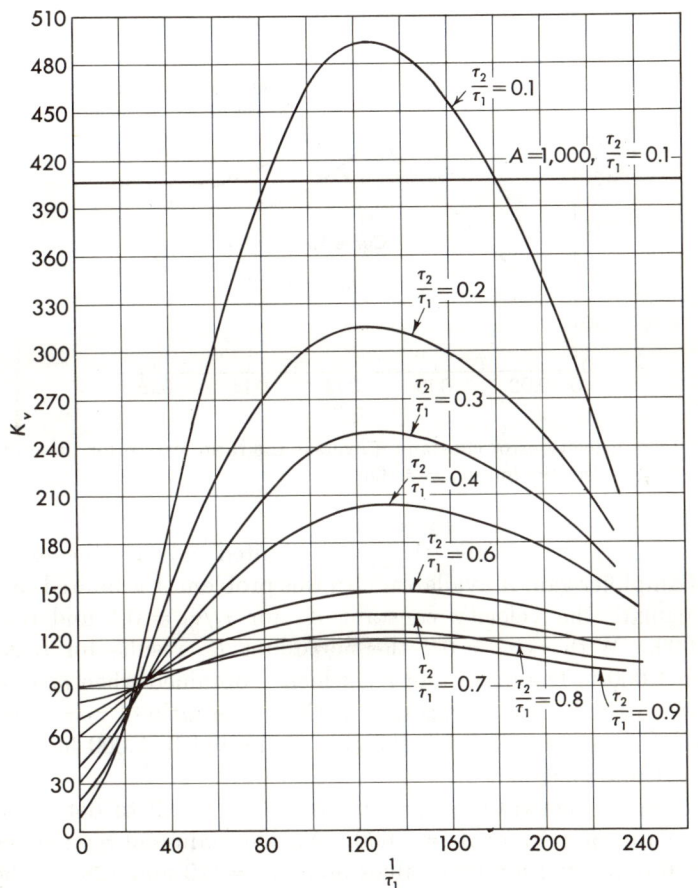

Fig. 6-7 Velocity-error constant K_v versus $1/\tau_1$ for different ratios of τ_2/τ_1.

The method of system optimization is outlined in the following step-by-step procedure:

1. Obtain curves of K_v versus $1/\tau_1$ for the specific parameter values of $\tau_2/\tau_1 = 0.1, 0.2, 0.3, 0.4, 0.6, 0.7, 0.8,$ and 0.9. The method used to obtain the curves is the same as in the preceding case. The curves for these ratios are shown in Fig. 6-7. The maximum value of K_v increases without limit as τ_2/τ_1 approaches zero. Each curve has only one maximum, which is located near the point $1/\tau_1 = 120$. As K_v increases,

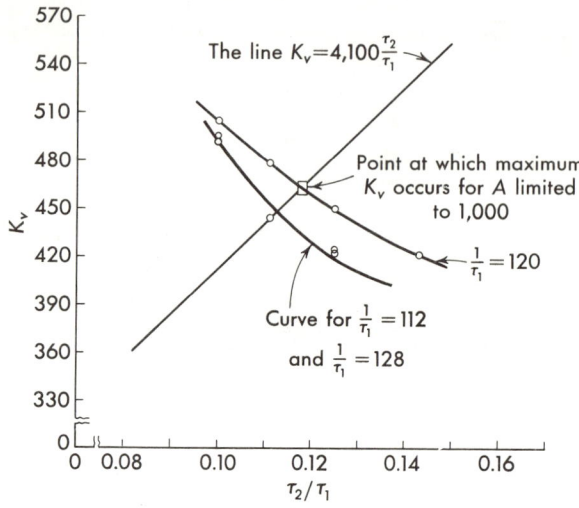

Fig. 6-8 The velocity-error constant K_v versus the ratio of τ_2/τ_1 for three different values of $1/\tau_1$. Also, the line $K_v = 4{,}100 \tau_2/\tau_1$.

A increases. This indicates that K_v is limited only by the maximum value of amplifier gain A available. In this problem A is limited to 1,000.

2. Calculate the velocity constant K_v for $\tau_2/\tau_1 = 0.1$ and 0.2 with $A = 1{,}000$. With $\tau_2/\tau_1 = 0.2$ the maximum K_v results for a gain of less than 1,000. The maximum K_v, which is obtained when $\tau_2/\tau_1 = 0.1$, occurs when the gain is in excess of 1,000. The ratio of τ_2/τ_1 necessary to obtain a maximum K_v with the gain A limited to 1,000 lies between $\tau_2/\tau_1 = 0.1$ and $\tau_2/\tau_1 = 0.2$:

3. Interpolate between $\tau_2/\tau_1 = 0.1$ and $\tau_2/\tau_1 = 0.2$ to determine the accurate value of τ_2/τ_1 for $A = 1{,}000$. Various values of τ_2/τ_1 are chosen between 0.1 and 0.2 for two values of $1/\tau_1$ ($=120$ and 128). The two corresponding curves, included in Fig. 6-8, are plotted for K_v versus

τ_2/τ_1. A line which satisfies the equation

$$K_v = 4{,}100 \frac{\tau_2}{\tau_1}$$

is drawn through these curves. Where this line intersects the curve, a ratio of τ_2/τ_1 that provides a maximum K_v for each value of $1/\tau_1$ is obtained.

4. Plot the maximum K_v values versus $1/\tau_1$, found in step 3, for $A = 1{,}000$. Figure 6-9 shows the plot of this example. The maximum occurs for $\tau_2/\tau_1 = 0.118$ and $1/\tau_1 = 120$. A maximum K_v of 463 is obtained. This is an improvement by a factor of nearly 10.

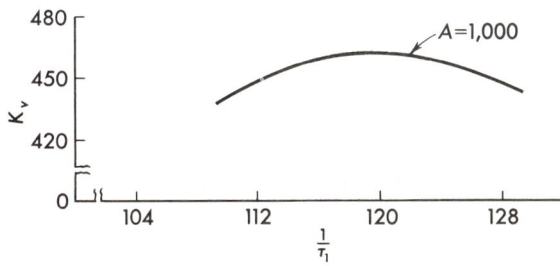

Fig. 6-9 The velocity-error constant K_v versus $1/\tau_1$ for $A = 1{,}000$.

5. Calculate the component values from the optimum conditions of step 4. From the values $1/\tau_1 = 120$ and $\tau_2/\tau_1 = 0.118$ the component values are specified as follows:

$$\frac{R_2}{R_1 + R_2} = 0.118 \quad \text{and} \quad C = \frac{1.112 \times 10^{-3}}{R_2} \qquad (6\text{-}8)$$

R_1, R_2, and C cannot be found in absolute value unless some other condition, such as input or output impedance, is specified. In a practical circuit the impedance level at the point where the network is to be inserted must be considered.

6-4 Method of Control-system Equalization. The example in the preceding section indicates the method of equalizing a system with a passive circuit. Although a procedure requiring a trial-and-error approach is presented, each trial is rapidly analyzed and a solution is obtained. The procedure can be generalized with the following outline:

1. Determine the constants and transfer functions of all the "unalterable" components. For example, a position-control system requires a motor to drive the output shaft. This is an unalterable component. Use frequency-analysis techniques to approximate the transfer functions of difficult components.

2. Determine the steady-state errors. Add any necessary poles at the origin to reduce this error. (In many systems this is not necessary, since the unalterable components contain poles at the origin.)

3. Determine the degree of stability of the basic system. If this is not satisfactory (that is, ζ too small), choose a network for insertion.

4. Move the zeros and poles of the network until a suitable degree of stability is obtained.

Since the first choice of a network requires some experience, the next sections consider some common networks that can be used to equalize simple control systems.

There is a myriad of electrical networks for compensation. These networks consist of zeros and poles which are introduced either to shift the unstable branches of the locus back into the left half plane or to alter the gain gradient (variation of a point on the locus with a change in gain) along the locus. It is often impossible to accomplish both of these effects with one network.

6-5 Passive Lead Network. The circuit of a passive lead network, sometimes referred to as a passive circuit differentiator, is shown in Fig. 6-10a. The transfer function for this network, with zero source and infinite load impedance, is

$$\frac{E_o}{E_{\text{in}}} = G(s) = \frac{s + 1/\alpha\tau}{s + 1/\tau} \tag{6-9}$$

where $\alpha = 1 + R_1/R_2$ and $\tau = R_1 R_2 C/(R_1 + R_2)$. The network is represented in the s plane of Fig. 6-10b by a pole at $s = -1/\tau$ and a zero at $s = -1/\alpha\tau$. The zero is closer to the origin than the pole, and the ratio of the two distances is α. The example in the preceding section indicates that an improvement by a factor of 10 in K_v is possible by appropriately choosing component values for the lead network. If τ is small, both the pole and the zero are remote and effectively cancel each other. As τ is made larger, the zero enters the region where the locus is affected. For very large τ, the pole also becomes significant.

The polar plot for the phase-lead network is obtained from the rearranged expression

$$\frac{E_o}{E_{\text{in}}} = G(s) = \frac{1}{\alpha}\frac{\alpha\tau s + 1}{\tau s + 1} \tag{6-10}$$

and is shown in Fig. 6-10c for several values of α. The maximum phase lead ψ_m which can be obtained from the network depends upon α, as shown in the figure. The maximum phase lead approaches 90° as α approaches infinity.

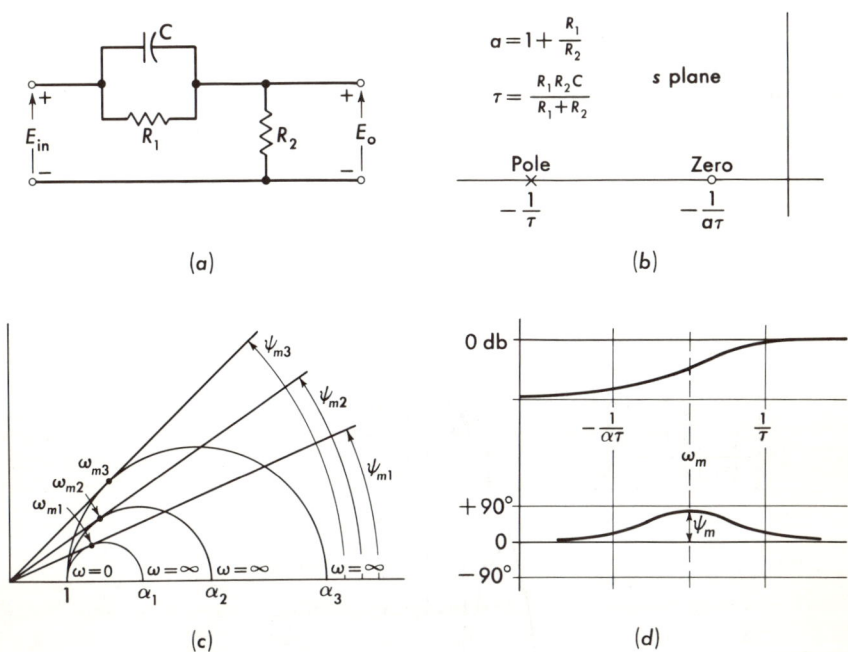

Fig. 6-10 (a) Passive circuit lead network; (b) zero-pole configuration for a lead network; (c) polar plot of lead network; (d) Bode plot of lead network.

The Bode plot for the lead network of Fig. 6-10a is sketched in Fig. 6-10d. The peak phase shift ψ_m and the frequency at which this maximum lead occurs can be related to the parameters α and τ as follows:

$$\omega_m = \frac{1}{\sqrt{\alpha\tau}} \quad \text{and} \quad \psi_m = \sin^{-1}\frac{\alpha - 1}{\alpha + 1} \tag{6-11}$$

As another example, consider the effect of inserting a passive circuit lead network into a third-order system. The open-loop transfer function for the system and equalizer is

$$GH = \underbrace{\frac{20K}{s(s+1)(s+5)}}_{\text{system}} \underbrace{\frac{s + 1/4\tau}{s + 1/\tau}}_{\text{equalizer}} \tag{6-12}$$

This is plotted in Fig. 6-11 for several values of τ. Points are connected on each locus for gains of 1, 2, 5, 10, and 15.

As can be seen from Fig. 6-11, the effect of the lead network depends upon both τ and K. Generally, the lead network produces the following results:

1. The damping ratio is increased and hence the overshoot is reduced.
2. The resonant frequency is increased and hence the bandwidth is usually increased and the rise time is faster.
3. For a given degree of stability, the velocity constant is usually increased.

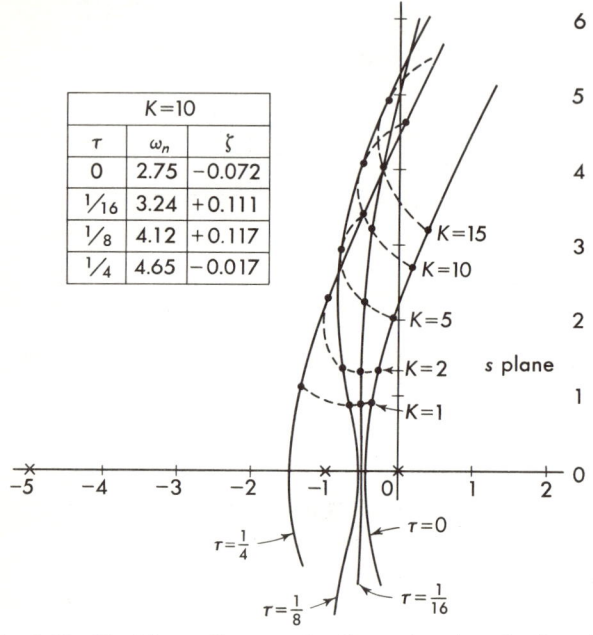

Fig. 6-11 Root-locus diagrams for the system transfer function

$$G(s)H(s) = \frac{20K}{s(s+1)(s+5)} \frac{s + 1/4\tau}{s + 1/\tau}$$

6-6 Passive Lag Network. A passive phase-lag network, often termed an integrator, is shown in Fig. 6-12a. For zero source and infinite load impedance the transfer function is

$$\frac{E_o}{E_{in}} = G(s) = \frac{1}{\alpha} \frac{s + 1/\tau}{s + 1/\alpha\tau} \tag{6-13}$$

where
$$\alpha = 1 + \frac{R_1}{R_2} \qquad \tau = R_2 C$$

Insertion of a lag network in a system produces a zero and pole in the s plane. The zero is located at $-1/\tau$, and the pole at $-1/\alpha\tau$. As can be seen from Fig. 6-12b, the pole is closer to the origin. If τ is large, both zero and pole are near the origin and show no stabilizing effect except to introduce another root which is near the zero (located at $1/\tau$) for all values of gain. Such a root gives rise to a response with a long time constant (approximately equal to τ). As τ is decreased, the zero

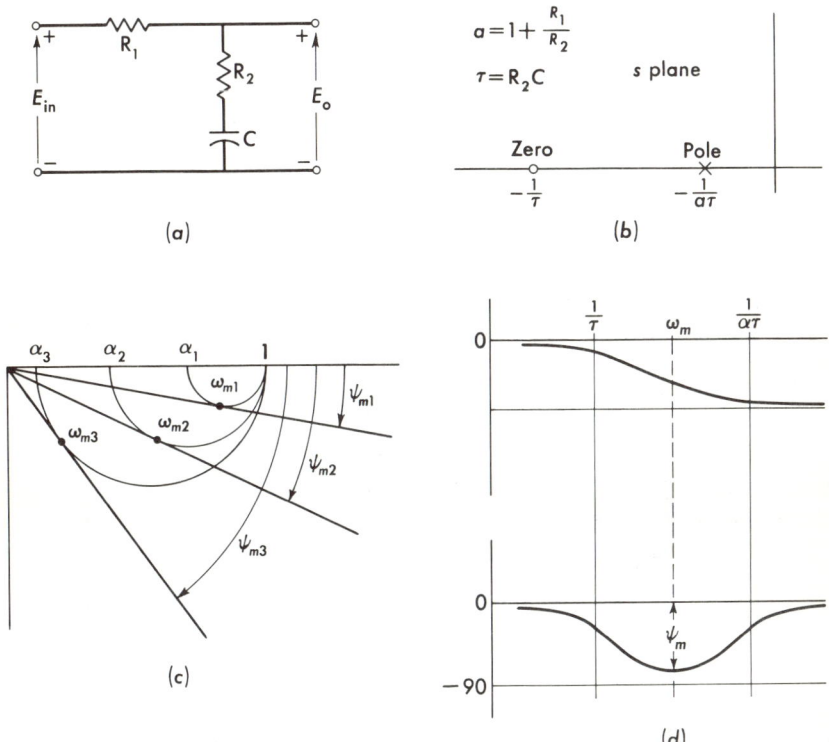

Fig. 6-12 (a) Passive circuit lag network; (b) zero-pole configuration for a lag network; (c) polar plot of phase-lag network; (d) Bode plot of phase-lag network.

approaches the other roots and has a stabilizing effect. For still smaller τ, the pole becomes effective and exerts a destabilizing effect.

The polar plot for the transfer function of Fig. 6-12a is depicted in Fig. 6-12c for three values of α. As α increases, approaching infinity, the maximum phase lag increases, and approaches 90°. The frequency ω_m at which the maximum occurs decreases with increasing α.

The Bode plot for the phase-lag network comprises two corner frequencies; it is shown in Fig. 6-12d.

222 Control System Design

Use of the phase-lag network in control-system design relies upon amplitude attenuation at high frequency. The phase-lead network utilizes the addition of phase lead to equalize the system.

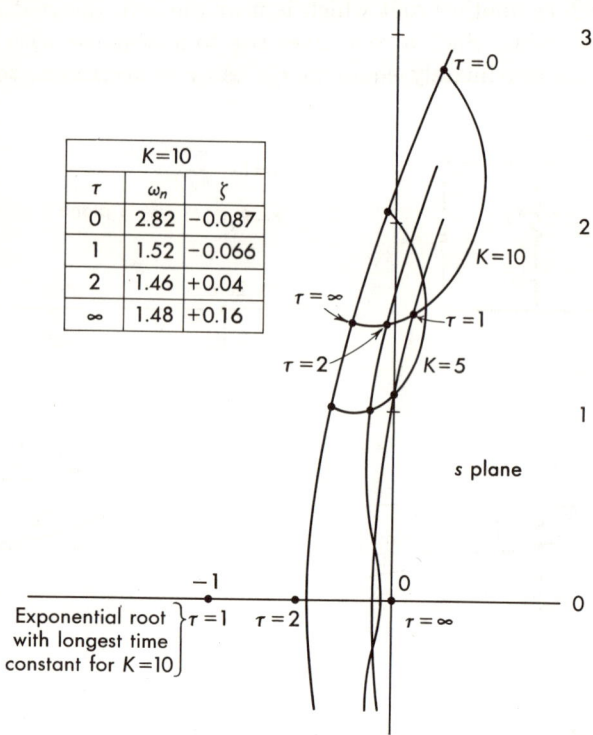

Fig. 6-13 Root-locus diagrams for the system transfer function

$$G(s)H(s) = \frac{5K}{s(s+1)(s+5)} \frac{s+1/\tau}{s+1/4\tau}$$

The system of Fig. 6-11 is again stabilized but with a lag network instead of a lead network. The transfer function for the system containing the lag network can be written

$$GH = \frac{5K}{s(s+1)(s+5)} \frac{s+1/\tau}{s+1/4\tau} \tag{6-14}$$

The root-locus plot for several values of τ is shown in Fig. 6-13. Points of constant K (5 and 10) are connected on each locus. Inspection of these loci indicates that with a careful choice of τ the system can be made more stable. It should be noticed, however, that with an improper choice of τ the system can be made less stable.

Lag networks possess certain disadvantages. If the time constant is too large, a long exponential root occurs in the system. The effect of this root can be minimized by keeping the zero close to the pole. The root-locus diagrams of Fig. 6-13 show the location of the root with the longest time constant $K = 10$. The choice of a value of τ which will provide effective stabilization with a satisfactory exponential time constant is often impossible.

A phase-lag network has the effect of reducing high-frequency gain. The results of inserting a phase-lag network into a control system can be summarized as follows:

1. For a fixed relative stability, the velocity constant is increased.
2. The system bandwidth is decreased, which results in a slower rise time.

6-7 Summary of Various Equalizer Networks. In an effort to present some of the possibilities available with passive circuit networks, Table 6-1 is included. The circuit for the network, the transfer function, and the zero-pole configuration are given for each case.

Only simpler networks are included in the table. Positions of the zeros and poles are given in the s plane or can be discerned from the transfer function. As an example, suppose it is desired to locate the poles of network d given in Table 6-1. The poles are found by factoring the denominator with the binomial expression

$$\frac{1}{\tau_1 \tau_2} \left[-\frac{\tau_1 + \alpha \tau_2}{2} \pm \sqrt{\frac{(\tau_1 + \alpha \tau_2)^2}{4} - \tau_1 \tau_2} \right] \qquad (6\text{-}15)$$

The locations of these poles are found by substituting numerical values into Eq. (6-15). The same denominator is common to several of the networks in Table 6-1.

Transfer functions for the networks of both Tables 6-1 and 6-2 are derived for zero source and infinite load impedance. When the network is used in an application where these impedance conditions are not fulfilled, the transfer functions deviate from the ideal.

6-8 Bridged-T Network. In many control systems, the open-loop transfer function contains at least one pair of complex conjugate poles. If the complex poles are near the $j\omega$ axis, one good way to equalize the

Table 6-1 Common Passive Circuit Equalizer Networks

	Circuit	Transfer Function	Pole-Zero Plot
(a)	R series, C shunt	$\dfrac{1}{T}\dfrac{1}{s+\frac{1}{T}}$ $T=RC$	pole at $-\frac{1}{T}$
(b)	C series, R shunt	$\dfrac{s}{s+\frac{1}{T}}$ $T=RC$	zero at origin, pole at $-\frac{1}{T}$
(c)	C_1, C_2 series; R_1, R_2 shunt	$\dfrac{T_1 T_2 s^2}{1+(T_1+aT_2)s+T_1 T_2 s^2}$ $a=1+\dfrac{R_1}{R_2}$ $T_1=R_1C_1$, $T_2=R_2C_2$	two zeros at origin, two poles
(d)	R_1, R_2 series; C_1, C_2 shunt	$\dfrac{1}{1+(T_1+aT_2)s+T_1 T_2 s^2}$ $a=1+\dfrac{R_1}{R_2}$ $T_1=R_1C_1$, $T_2=R_2C_2$	two poles
(e)	$R_1 \| C_1$ series; R_2 series with C_2 shunt	$\dfrac{(T_1 s+1)(T_2 s+1)}{T_1 T_2 s^2+(T_1+aT_2)s+1}$ $a=1+\dfrac{R_1}{R_2}$ $T_1=R_1C_1$, $T_2=R_2C_2$	two zeros, two poles
(f)	$R_1 \| C_1$ series; R_2, C_2 shunt	$\dfrac{1}{a}\dfrac{1+T_1 s}{1+\left(\frac{T_1}{a}+\frac{T_2}{b}\right)s}$ $\dfrac{1}{a}=\dfrac{R_2}{R_1+R_2}$ $\dfrac{1}{b}=\dfrac{R_1}{R_1+R_2}$ $T_1=R_1C_1$, $T_2=R_2C_2$	zero at $-\frac{1}{T_1}$, pole at $-\frac{1}{\frac{T_1}{a}+\frac{T_2}{b}}$
(g)	R series, L and R_L shunt	$\dfrac{R_L}{R+R_L}\dfrac{1+\frac{L}{R_L}s}{1+\frac{L}{R+R_L}s}$	zero at $-\frac{R_L}{L}$, pole at $-\frac{R+R_L}{L}$
(h)	L, R_L series; R shunt	$\dfrac{R}{R+R_L}\dfrac{1}{1+\frac{L}{R+R_L}s}$	pole at $-\frac{R+R_L}{L}$
(i)	$R \| C$ series; L, R_L shunt	$\dfrac{(R_L+Ls)(1+Ts)}{(R+R_L)+(R_L+L)s+LTs^2}$ $T=RC$	two poles, two zeros

system is by cancellation of the poles with a pair of complex zeros. None of the networks of Table 6-1 can be used for this purpose.

A bridged-T network* provides two complex zeros and two real-axis poles. As a result, the two complex zeros of this network can be used to cancel the two undesirable complex poles.

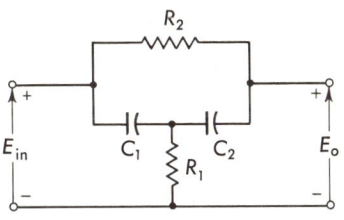

Fig. 6-14 Resistor-shunt bridged T.

A resistor-shunt bridged T is shown in Fig. 6-14. When the network is inserted at a point where the input impedance is low and the output impedance is high, the transfer function is

$$\frac{E_2(ju)}{E_1(ju)} = \frac{(1-u^2) + jurn}{(1-u^2) + jun} \qquad (6\text{-}16)$$

The following dimensionless quantities are defined:

$$u = \frac{\omega}{\omega_0} \qquad \omega_0 = \frac{1}{R_1C_1R_2C_2}$$
$$r = \frac{R_1(C_1 + C_2)}{R_1(C_1 + C_2) + C_2R_2} \qquad n = \omega_0[R_1(C_1 + C_2) + R_2C_2] \qquad (6\text{-}17)$$

The parameters can be so chosen that the two complex zeros of the bridged T cancel the two system complex poles.

6-9 Adjustable Networks. The ability to adjust a single potentiometer experimentally and hence alter the zero-pole configuration of an equalizer is often most valuable in control-system design. Besides reshaping the root-locus plots, an adjustable network is often necessary for setting the phase shift in an a-c system. Table 6-2 includes the circuits, transfer functions, and zero-pole configurations for several adjustable networks. The time constant τ for all networks is RC. The fraction of total rotation of the potentiometer is represented by γ.

Figure 6-15a presents an adjustable network which is valuable for phase-shifting a-c control systems. The transformer has a center-tapped secondary. The transfer function, for the conditions of small

---

* A design procedure for these bridged- and also parallel-T networks is included in Appendix VI. See also pages 117 to 123 of Ref. 20.

Table 6-2 Several Adjustable Networks

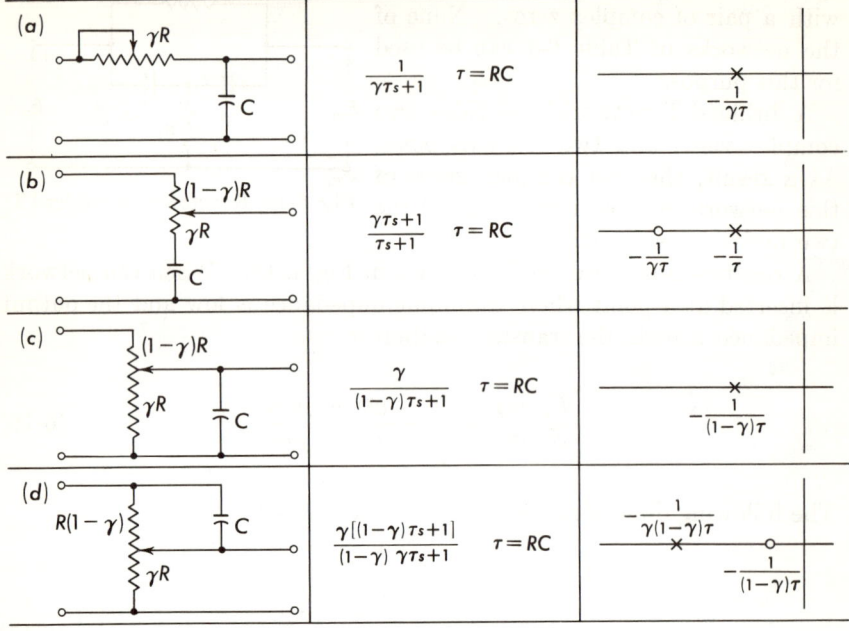

source and large load impedance, is

$$\frac{E_o}{E_{\text{in}}} = n\,\frac{1 - s\tau\gamma}{1 + s\tau\gamma} = n\,\frac{1 - j\omega\tau\gamma}{1 + j\omega\tau\gamma} = n\underline{/-2 \tan^{-1} \omega\tau\gamma} \qquad (6\text{-}18)$$

Since the voltages nE_{in} applied to this network are derived from a transformer, the transfer function of Eq. (6-18) is not applicable at zero frequency. If the voltages are derived at the output of a d-c phase inverter as shown in Fig. 6-15b, Eq. (6-18) is applicable at zero frequency. Equation (6-18) represents a so-called "all-pass network"; i.e., as γ is varied there results essentially no amplitude change, while the phase lag is variable from 0 to -180 for a fixed frequency. The zero-pole configuration for this phase shifter is shown in Fig. 6-16 for two specific values of γ.

6-10 Design of a Position Servo with Lead Network Equalization.
Three methods of equalizing a third-order control system are considered in this chapter. In this section the design is accomplished with a lead network. In Sec. 6-14 a rate generator is utilized, and in Sec. 6-23 a mechanical network is utilized. The same control system, shown in Fig. 6-17, is used for each case. The system, used to position an inertia

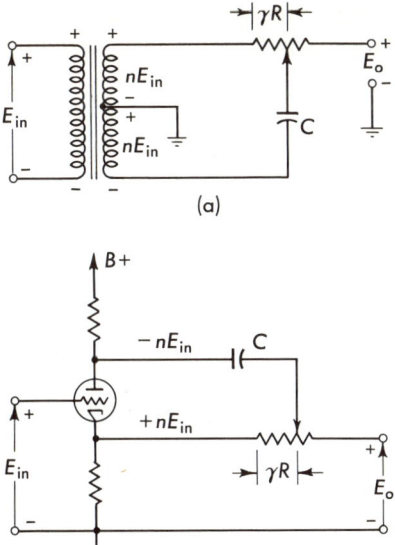

Fig. 6-15 (a) Adjustable phase-shift network (passive); (b) adjustable phase-shift network (active).

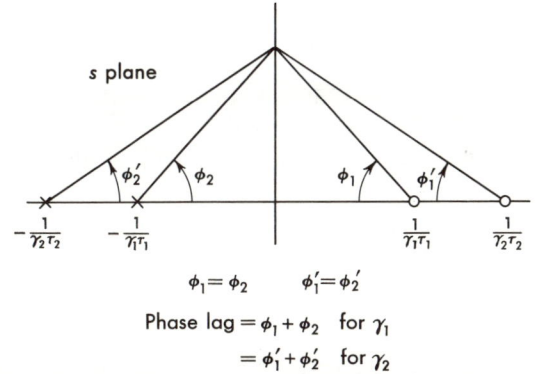

$$\phi_1 = \phi_2 \qquad \phi_1' = \phi_2'$$
$$\text{Phase lag} = \phi_1 + \phi_2 \text{ for } \gamma_1$$
$$= \phi_1' + \phi_2' \text{ for } \gamma_2$$

Fig. 6-16 Zero-pole configuration for adjustable phase-shift network.

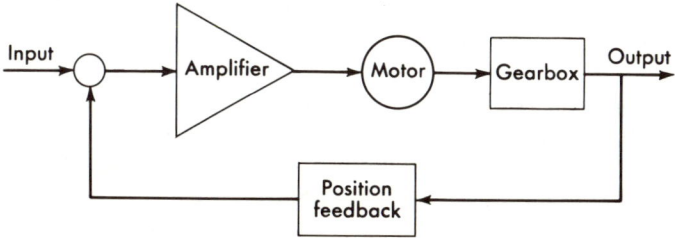

Fig. 6-17 Schematic of a control system.

(e.g., a tracking radar antenna drive or a gun directional system), comprises a motor and gearbox, an amplifier, and an output-error sensor.

The component transfer functions are as follows:

Amplifier.................... $\dfrac{A}{\tau_a s + 1}$

Motor....................... $\dfrac{K_m}{s(\tau_m s + 1)}$

Gearbox..................... 0.01 step-down

Position feedback............ K_p

Two values are taken for the amplifier time constant τ_a and the equivalent motor time constant τ_m as follows:

1. $\tau_a = 0.025$ and $\tau_m = 0.07$
2. $\tau_a = 0.005$ and $\tau_m = 0.02$

τ_m includes the effect of the motor inertia plus the reflected load moment of inertia. The transfer function of the unequalized system, for $\tau_a = 0.025$ and $\tau_m = 0.07$, is

$$C(s) = \dfrac{\dfrac{AK_m}{0.175s(s + 14.3)(s + 40)}}{1 + \dfrac{AK_m K_p}{0.175s(s + 14.3)(s + 40)}} \tag{6-19}$$

The root locus is plotted in Fig. 6-18. The system, with gain

$$K = 5.72 A K_m K_p$$

becomes unstable for gains above 32,300. The velocity-error constant K_v equals $0.01 A K_m K_p$. As the gain is increased, the velocity-error constant is increased proportionately and the steady-state error (reciprocal of the error constant) is decreased. In Fig. 6-19 the damping ratio ζ and the resonant frequency ω_n are plotted as a function of K_v.

The choice of α is not completely arbitrary. It is desirable theoretically to select α as large as possible. Unfortunately, in a lead network a larger α results in a greater low-frequency attenuation. A choice of $\alpha = 10$ is about as large as practical. In this design α is chosen equal to 10. For $\tau_a = 0.025$ and $\tau_m = 0.07$, K_v is plotted as a function of $1/\tau$ for various values of ζ in Fig. 6-20 and ω_n is plotted versus $1/\tau$ in Fig. 6-21. In Figs. 6-22 and 6-23, ζ and ω_n are plotted versus $1/\tau$ with $K = AK_mK_p/N\tau_a\tau_m$ for $\tau_a = 0.005$ and 0.02. These charts, which indi-

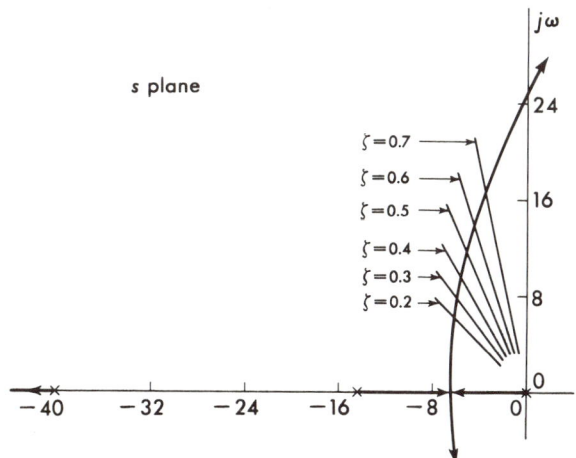

Fig. 6-18 Root locus of uncompensated network.

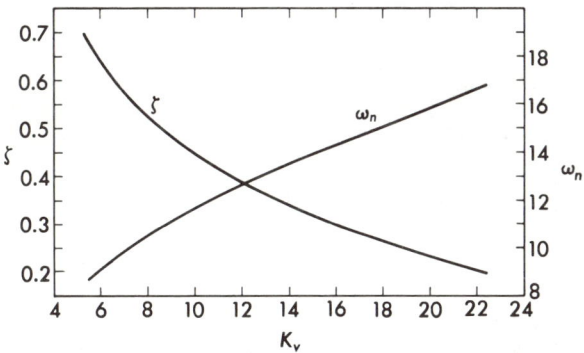

Fig. 6-19 ζ and ω_n versus the error parameter K_v.

cate the optimum values of $1/\tau$ for each value of ζ, can be used as a guide in the design of systems using a lead network.

6-11 Elementary Lattice and Ladder Synthesis. In more complicated systems it may be necessary to use a more elaborate network than those described in the preceding sections. A lattice network can be the solution. Any realizable RC transfer function can be realized

230   Control System Design

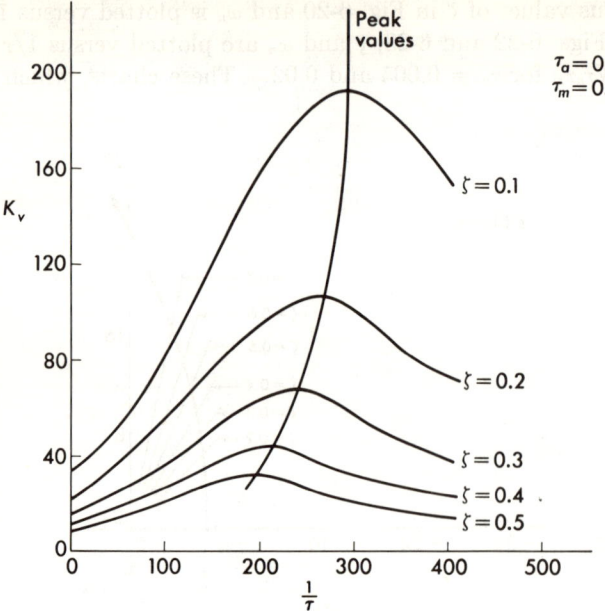

Fig. 6-20 K_v versus $1/\tau$.

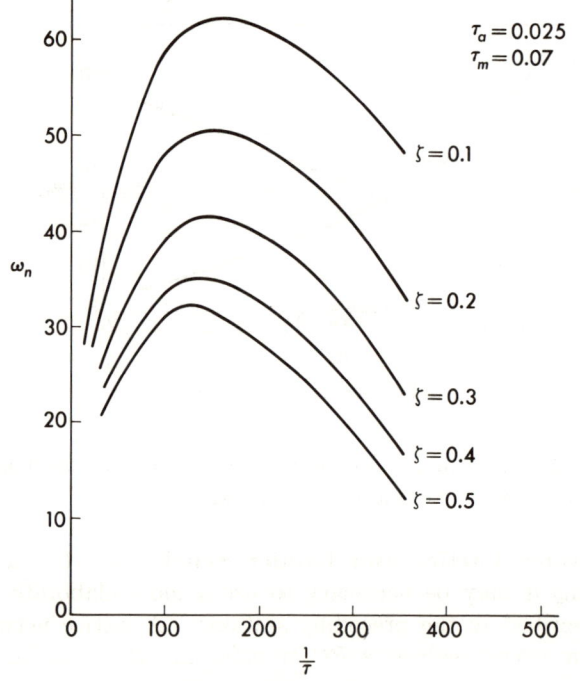

Fig. 6-21 ω_n versus $1/\tau$.

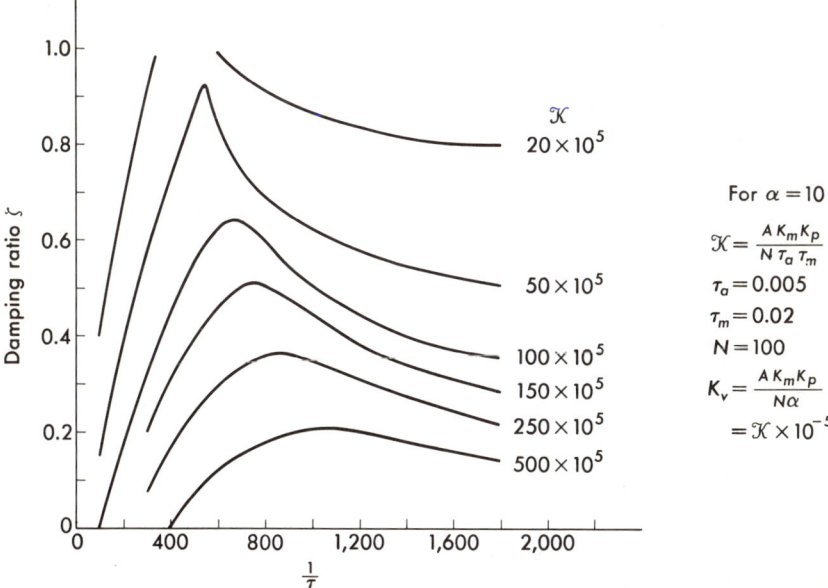

Fig. 6-22 ζ versus $1/\tau$ for the system

$$GH = \frac{K(s + 1/\alpha\tau)}{s(s + 50)(s + 200)(s + 1/\tau)}$$

by a lattice network, a typical one of which is shown in Fig. 6-24. The transfer function for this network is found for the condition of zero source and infinite load impedance. The equations of the two nodes E_2 and E_3 with the lower input node taken as ground are written

$$E_2\left(\frac{1}{Z_a} + \frac{1}{Z_b}\right) = E_{in}\frac{1}{Z_a}$$
$$E_3\left(\frac{1}{Z_a} + \frac{1}{Z_b}\right) = E_{in}\frac{1}{Z_b} \qquad (6\text{-}20)$$

Solution of these equations yields

$$E_o = E_2 - E_3 = \frac{E_{in}(1/Z_a - 1/Z_b)}{1/Z_a + 1/Z_b} = E_{in}\frac{Z_b - Z_a}{Z_b + Z_a} \qquad (6\text{-}21)$$

The two-terminal impedance functions Z_a and Z_b are found by identifying numerator and denominator with the desired transfer function.

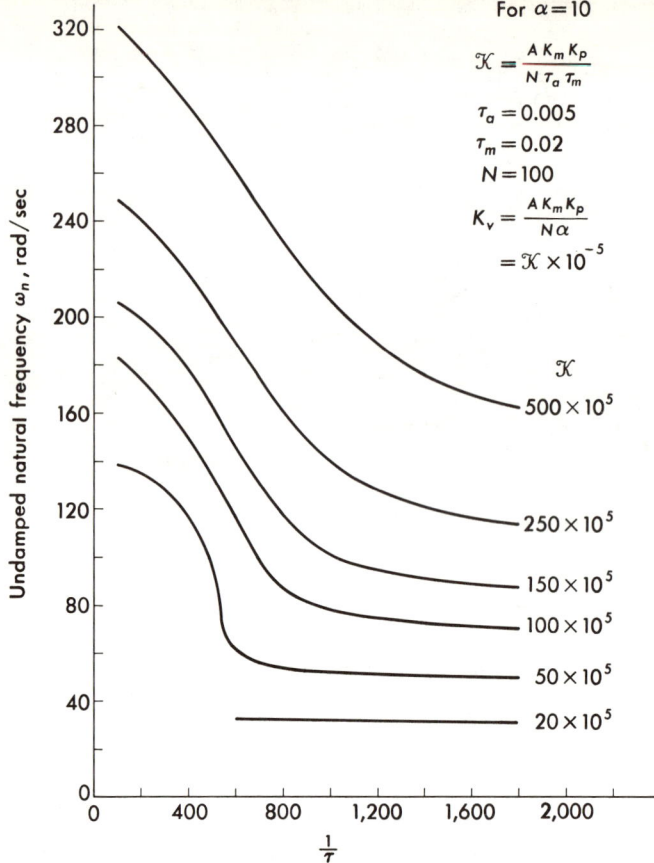

Fig. 6-23 ω_n versus $1/\tau$ for the system

$$GH = \frac{K(s + 1/\alpha\tau)}{s(s + 50)(s + 200)(s + 1/\tau)}$$

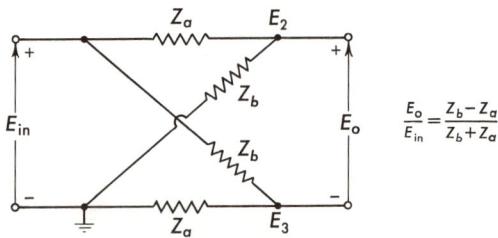

Fig. 6-24 Lattice network.

If the transfer function is denoted as $E_o/E_{in} = Z_{12}(s)/Z_{11}(s)$, the two-terminal impedances are found as follows:

$$Z_b - Z_a = Z_{12}(s)$$
$$Z_b + Z_a = Z_{11}(s) \qquad (6\text{-}22)$$

Adding and subtracting Eqs. (6-22) yields

$$Z_b = \tfrac{1}{2}[Z_{11}(s) + Z_{12}(s)] \qquad Z_a = \tfrac{1}{2}[Z_{11}(s) - Z_{12}(s)] \qquad (6\text{-}23)$$

Each of the functions Z_a and Z_b can be divided by $Kq(s)$, where $q(s)$ is any desired polynomial in the operator s. Since a ratio of impedance is taken in Eq. (6-21), the $Kq(s)$ cancels out of the fraction. $q(s)$ is inserted to aid in realizing the two-terminal impedance functions that result. Division of Eqs. (6-23) by $Kq(s)$ yields the equations

$$Z_b = \frac{1}{2}\frac{Z_{11}(s) + Z_{12}(s)}{Kq(s)} \qquad Z_a = \frac{1}{2}\frac{Z_{11}(s) - Z_{12}(s)}{Kq(s)} \qquad (6\text{-}24)$$

Consider, as an example, the following transfer function:

$$\frac{E_o}{E_{in}} = \frac{(s + 30)^2}{(s + 50)(s + 100)} = \frac{s^2 + 60s + 900}{s^2 + 150s + 5{,}000} \qquad (6\text{-}25)$$

The quantities Z_a and Z_b are found from Eqs. (6-23) as follows:

$$Z_b = \tfrac{1}{2}(s^2 + 150s + 5{,}000 + s^2 + 60s + 900)$$
$$Z_a = \tfrac{1}{2}(s^2 + 150s + 5{,}000 - s^2 - 60s - 900)$$

These quantities are divided by $Kq(s)$:

$$Z_a = \frac{90s + 4{,}100}{Kq(s)} = \frac{90}{K} + \frac{4{,}100}{Ks}$$

$$Z_b = \frac{2s^2 + 210s + 5{,}900}{Kq(s)} = \frac{2s}{K} + \frac{210}{K} + \frac{5{,}900}{Ks}$$

In this case it is convenient to take $Kq(s) = Ks$. The two-terminal impedances Z_a and Z_b are shown in Fig. 6-25. The numerical values become reasonable in size when K is taken equal to 10^{-2}.

In general, Z_a and Z_b can be built with passive elements if the subtraction necessary in Eq. (6-24) results in all positive components.

Practically, the lattice network has the disadvantage of possessing no common ground between input and output. To overcome this problem, it is necessary to drive the input with a balanced signal from, for example, a transformer or difference amplifier. Alternatively, the load may be driven in push-pull so that the common-ground problem can be

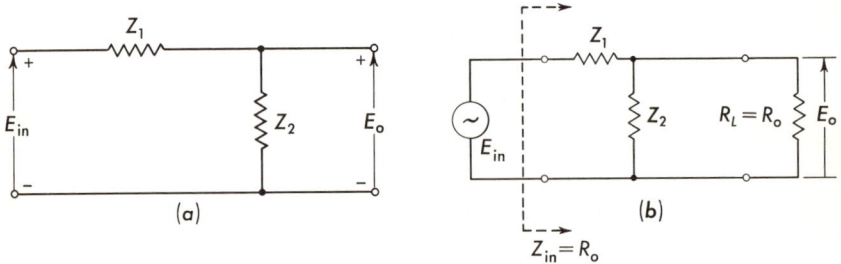

$$R_a = \frac{90}{K} = 9{,}000 \, \Omega \qquad L_b = \frac{2}{K} = 200 \, h$$

$$C_a = \frac{K}{4{,}100} = 2.44 \, \mu f \qquad R_b = \frac{210}{K} = 24{,}000 \, \Omega$$

$$Z_a = \frac{90}{K} + \frac{4{,}100}{Ks} \qquad C_b = \frac{K}{5{,}900} = 1.69 \, \mu f$$

$$Z_b = \frac{2s}{K} + \frac{210}{K} + \frac{5{,}900}{Ks}$$

Fig. 6-25 Lattice elements.

obviated. Another difficulty with the lattice network is the number of elements necessary. For the example given here, ten elements are necessary to achieve the desired transfer function.

Synthesis of the necessary two-terminal impedances Z_a and Z_b is accomplished by reference to Table 6-3. It is important to maintain low source and high load impedance.

Fig. 6-26 (a) Simpler ladder network; (b) loaded ladder network.

Owing to the disadvantage of an ungrounded network, considerable emphasis has been placed on three-terminal networks. One terminal is common to the input and the output and hence can be grounded. A ladder network is an example of such a grounded network. Since many synthesis procedures have been developed* (and are being developed), only a very simple and limited example is considered in this text. The ladder synthesis is based upon the circuit of Fig. 6-26a. When this

* See Ref. 38, pp. 190–220.

Design of Feedback Control Systems

Table 6-3 Passive Circuit Impedance Forms

	Circuit	Impedance		Circuit	Impedance
(a)	L, C in series; R in parallel	$\dfrac{R\left(s^2+\frac{1}{LC}\right)}{s^2+\frac{R}{L}s+\frac{1}{LC}}$	(j)	L, C, R	$\dfrac{s^2+\frac{1}{RC}s+\frac{1}{LC}}{s^2+\frac{1}{LC}}$
(b)	R, C, L	$\dfrac{R_s\left(s+\frac{1}{RC}\right)}{s^2+\frac{R}{L}s+\frac{1}{LC}}$	(k)	R, L, C	$\dfrac{s^2+\frac{1}{RC}s+\frac{1}{LC}}{s\left(s+\frac{R}{L}\right)}$
(c)	R, L, C	$\dfrac{s+\frac{R}{L}}{C\left(s^2+\frac{R}{L}s+\frac{1}{LC}\right)}$	(l)	C, R, L	$\dfrac{L\left(s^2+\frac{1}{RC}s+\frac{1}{LC}\right)}{s+\frac{1}{RC}}$
(d)	C, L, C_1	$\dfrac{s^2+\frac{1}{LC}}{C_1 s\left(s^2+\frac{1}{L}\frac{C_1+C}{C_1 C}\right)}$	(m)	R, C, C_1	$\dfrac{(C+C_1)\left[s+\frac{1}{R(C+C_1)}\right]}{CC_1 s\left(s+\frac{1}{RC}\right)}$
(e)	C, R, C_1	$\dfrac{s+\frac{1}{RC}}{C_1 s\left(s+\frac{1}{R}\frac{C_1+C}{C_1 C}\right)}$	(n)	L, C, L_1	$\dfrac{s\left(s^2+\frac{1}{C}\frac{L+L_1}{LL_1}\right)}{L\left(s^2+\frac{1}{LC}\right)}$
(f)	C, L, L_1	$\dfrac{LL_1 s\left(s^2+\frac{1}{LC}\right)}{(L+L_1)s^2+\frac{1}{C(L+L_1)}}$	(o)	L, R, L_1	$\dfrac{s\left(s+R\frac{L+L_1}{LL_1}\right)}{L_1\left(s+\frac{R}{L}\right)}$
(g)	R, L, L_1	$\dfrac{LL_1 s\left(s+\frac{R}{L}\right)}{(L+L_1)s+\frac{R}{L+L_1}}$	(p)	L, C, C_1	$\dfrac{(C+C_1)\left[s^2+\frac{1}{L(C+C_1)}\right]}{CC_1 s\left(s^2+\frac{1}{LC}\right)}$
(h)	R, L, R_1	$\dfrac{R_1\left(s+\frac{R}{L}\right)}{s+\frac{R+R_1}{L}}$	(q)	L, R, R_1	$\dfrac{(R+R_1)\left(s+\frac{1}{L}\frac{RR_1}{R+R_1}\right)}{s+\frac{R}{L}}$
(i)	R, C, R_1	$\dfrac{RR_1\left(s+\frac{1}{RC}\right)}{(R+R_1)\left[s+\frac{1}{C(R+R_1)}\right]}$	(r)	C, R, R_1	$\dfrac{R_1\left(s+\frac{1}{C}\frac{R+R_1}{RR_1}\right)}{s+\frac{1}{RC}}$

network is operated from zero or low source impedance and infinite or large load impedance, the transfer function can be written

$$\frac{E_o}{E_{in}} = \frac{Z_2}{Z_1 + Z_2} = \frac{1}{1 + Z_1/Z_2} \qquad (6\text{-}26)$$

When Eq. (6-26) is rearranged and the open-circuit impedances of Eq. (6-21) are utilized, the following equation results:

$$\frac{E_{in}}{E_o} - 1 = \frac{Z_{11}}{Z_{12}} - 1 = \frac{Z_{11} - Z_{12}}{Z_{12}} = \frac{Z_1}{Z_2} \qquad (6\text{-}27)$$

As an example of this procedure, consider the following problem: After the design of a control system, it is found that an equalizer which has the following transfer function is required:

$$\frac{E_o}{E_{in}} = \frac{(s + 30)^2}{(s + 50)(s + 100)} = \frac{Z_{12}}{Z_{11}} \qquad (6\text{-}28)$$

Find a network that yields this transfer function.

Substituting Eq. (6-28) into Eq. (6-27), the ratio of Z_1/Z_2 is found:

$$\frac{Z_1}{Z_2} = \frac{s^2 + 150s + 5{,}000}{s^2 + 60s + 900} - 1 = \frac{90s + 4{,}100}{s^2 + 60s + 900}$$

To find Z_1 and Z_2, identify numerators and denominators as follows:

$$Z_1 = \frac{90s + 4{,}100}{Kq(s)} \qquad Z_2 = \frac{s^2 + 60s + 900}{Kq(s)}$$

where, as in the lattice synthesis, $Kq(s)$ is any desired polynomial in the operator s which is inserted to aid in realizing the two-terminal impedance functions. Since the ratio of Z_1 to Z_2 is taken, $Kq(s)$ cancels out of the fraction. In this example take $Kq(s) = Ks$ and

$$Z_1 = \frac{90}{K} + \frac{4{,}100}{Ks} \qquad Z_2 = \frac{s}{K} + \frac{60}{K} + \frac{900}{Ks}$$

In this case the two-terminal networks are easily identified by inspection. The networks, which are shown in Fig. 6-27, consist of a resistor and capacitor in series for Z_1 and a series R, L, C combination for Z_2. Reasonable component sizes are obtained when K is taken equal to 10^{-2}.

If the subtraction $E_{in}/E_o - 1$ performed in Eq. (6-27) results in all positive signs, the Z_1 and Z_2 impedance can be found. A series of two-terminal impedance networks and their corresponding impedance functions are included in Table 6-3.

Practically, it is not possible to obtain zero source and infinite load impedance. A 10:1 impedance ratio is usually sufficient to yield the desired transfer function. That is, the source impedance should be one-tenth the magnitude of the network input impedance at the operating frequencies; the load impedance should be 10 times the magnitude of the network output impedance at the operating frequencies. When a

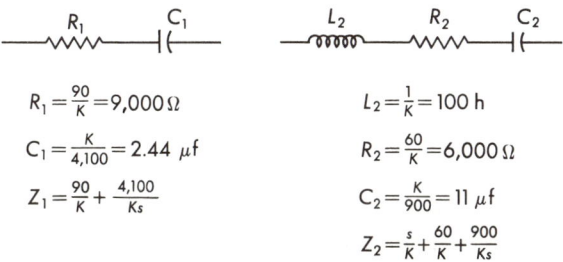

Fig. 6-27 Elements for a ladder network synthesis.

10:1 impedance change is not possible, a constant-resistance ladder network may prove useful. The equations for such a network are obtained from the circuit shown in Fig. 6-26b. The transfer function is written

$$\frac{E_o}{E_{in}} = \frac{R_o Z_2}{R_o Z_1 + R_o Z_2 + Z_1 Z_2} = \frac{Z_2/R_o}{Z_1/R_o + Z_2/R_o + (Z_1/R_o)(Z_2/R_o)} \quad (6\text{-}29)$$

and the input impedance is given by

$$Z_{in} = \frac{Z_1 R_o + Z_2 R_o + Z_1 Z_2}{R_o + Z_2}$$

$$Z_{in} = \frac{Z_1/R_o + Z_2/R_o + (Z_1/R_o)(Z_2/R_o)}{1 + Z_2/R_o} R_o \quad (6\text{-}30)$$

For the condition that the input impedance is equal to the load impedance, that is, $Z_{in} = R_o$, Eq. (6-30) is equaled to R_o with the result

$$\frac{Z_1}{R_o} + \frac{Z_2}{R_o} + \frac{Z_1 Z_2}{R_o R_o} = 1 + \frac{Z_2}{R_o} \quad (6\text{-}31)$$

When Eq. (6-31) is substituted into Eq. (6-29), the following design expression results:

$$\frac{E_o}{E_{in}} = \frac{Z_2/R_o}{1 + Z_2/R_o} \quad \text{or} \quad \frac{Z_2}{R_o} = \frac{1}{E_{in}/E_o - 1} \quad (6\text{-}32)$$

Similarly, the design equation for Z_1 is written from Eq. (6-31) as

$$\frac{Z_1}{R_o} = \frac{1}{1 + Z_2/R_o} = 1 - \frac{E_o}{E_{in}} \quad (6\text{-}33)$$

Application of Eqs. (6-32) and (6-33) for the constant-resistance ladder is similar to that for the simple ladder network.

A full treatment of network synthesis would require a volume in itself. The references[17,38] included at the conclusion of this book provide considerable additional material on network synthesis.

6-12 Active Network Synthesis. Much is known about the characteristics and synthesis of two-terminal networks. The synthesis of four-terminal active networks can be based upon passive two-terminal

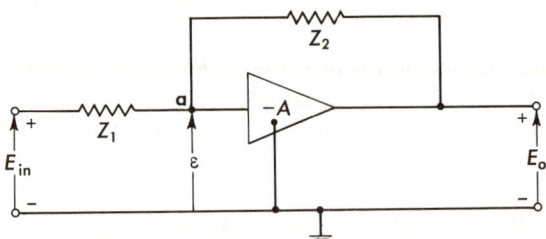

Fig. 6-28 Operational amplifier used for active network synthesis.

networks. The use of an amplifier, however, does not result in a transfer function differing from a ratio of two polynomials. The constant multiplier may be changed, but the form of the transfer function remains the same.

One of the most direct methods of active network synthesis lies in the operational amplifier approach. Consider the network shown in Fig. 6-28. The transfer function is found by summing the currents flowing into node a as follows:

$$\frac{E_{in} - \epsilon}{Z_1} + \frac{E_o - \epsilon}{Z_2} = 0 \quad (6\text{-}34)$$

The sum is equal to zero, since no current flows into the amplifier input. If the gain of the amplifier $-A$ is large, the error ϵ is negligible and the transfer function takes the form

$$\frac{E_o}{E_{in}} = -\frac{Z_2}{Z_1} \qquad (6\text{-}35)$$

The transfer function is simply the ratio of two-terminal impedance functions the synthesis of which is given in preceding sections.

As an example, if $Z_1 = R_1$ and $Z_2 = 1/sC_2$, the transfer function becomes

$$\frac{E_o}{E_{in}} = -\frac{1}{R_1 C_2 s} \qquad (6\text{-}36)$$

which is an integrator (pole at the origin). If $Z_1 = 1/sC_1$ and $Z_2 = R_2$, the transfer function is

$$\frac{E_o}{E_{in}} = -R_2 C_1 s \qquad (6\text{-}37)$$

which is a differentiator (zero at the origin).

Consider again the example of Eq. (6-28) with a negative sign inserted:

$$\frac{E_o}{E_{in}} = \frac{-(s+30)^2}{(s+50)(s+100)} = \frac{-(s+30)}{s+50} \frac{s+30}{s+100} \qquad (6\text{-}38)$$

Equating Eqs. (6-35) and (6-38) and solving for the two-terminal impedance functions,

$$Z_2 = \frac{s+30}{s+50} \quad \text{and} \quad Z_1 = \frac{s+100}{s+30} \qquad (6\text{-}39)$$

Reference to Table 6-3 [see (g), (h), (i), (q), (r)] indicates several two-terminal networks that will satisfy Eq. (6-39).

This active network synthesis does not require the subtraction that both the ladder and lattice synthesis procedures require. The two-terminal impedances are positive and can be synthesized easily. If the network is to be used at zero frequency, however, the high-gain amplifier must operate at zero frequency. A d-c amplifier is difficult to design and operate. If the low-frequency response is not required because of the form of the network (a differentiator, for example, has zero output for direct current), an a-c amplifier can be used. However, when the

240 Control System Design

network requires d-c response, the need for a d-c amplifier limits the use of this method of equalization.

6-13 Parallel Equalization. Preceding methods of equalization are termed "series equalization" because networks with the appropriate zero-pole configuration are inserted in the forward (or series) loop. In "parallel equalization" the equalizing network is inserted in the feedback path of the control system. Use of this form of equalization usually leads to multiple-loop* systems. Since there are a large number of a-c transducers, this method provides an especially good means for stabilizing control systems where the signal is suppressed-carrier (alternating current).†

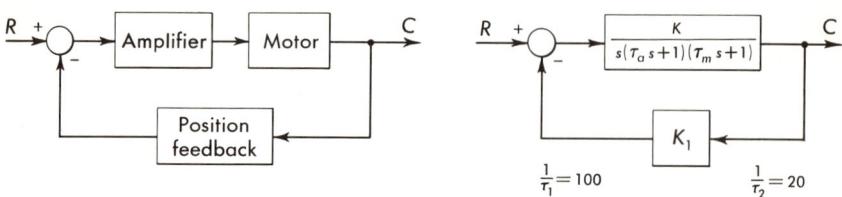

Fig. 6-29 Block diagram of position servo.

Consider the example of Fig. 6-29, where the simple position servo of Chap. 1 is shown. The amplifier time constant is τ_a, and τ_m is the motor time constant. For a gain K of 24×10^4 the system roots, shown in the root-locus sketch of Fig. 6-30, cross the j axis and the system becomes unstable. Suppose that an output-shaft rate signal is added to the position signal and fed back to the input, as in Fig. 6-31. In this figure α is the ratio of the position feedback voltage to the rate feedback voltage. The same numerical values as used in Fig. 6-30 are inserted into the system of Fig. 6-31. When α is taken equal to 10, the following open-loop transfer function results:

$$GH = \frac{K_1 K_2 (s + 10)}{s(s + 20)(s + 100)} \qquad (6\text{-}40)$$

The root-locus sketch for this function is included in Fig. 6-30. The insertion of the equalizer in the feedback loop adds a zero in the s plane. The system is now stable with the value of gain ($K = 24 \times 10^4$) at which it was previously unstable.

---

* See Sec. 4-14 for a discussion of the root-locus analysis of multiple-loop servos. Section 6-15 presents an example of a multiple-loop system.

† Alternating-current systems are discussed in Sec. 6-17.

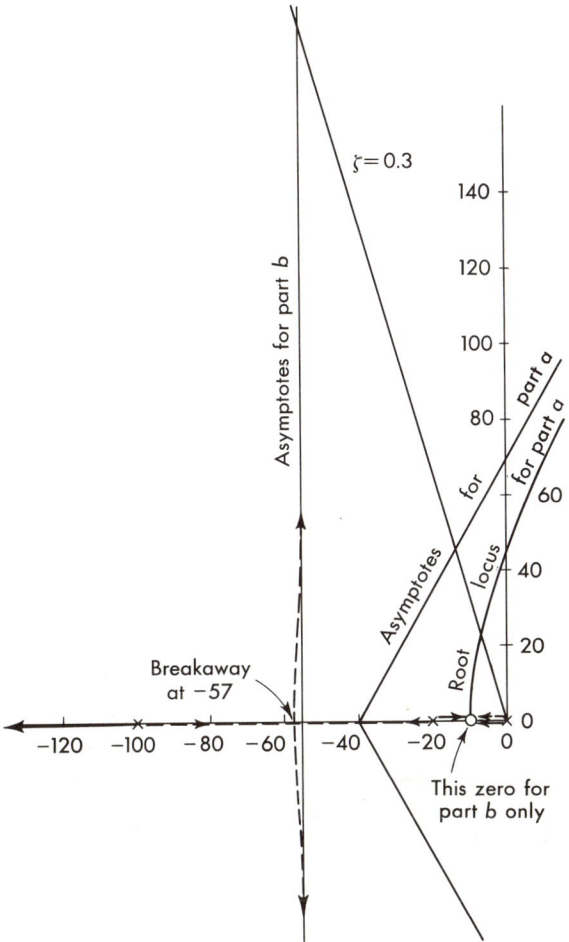

Fig. 6-30 (a) Root locus for

$$\frac{K}{s(s+20)(s+100)}$$

(b) Root locus for

$$\frac{K(s+10)}{s(s+20)(s+.100)}$$

In general, rate generators provide signals proportional to shaft velocity. These components are discussed in Chap. 7. When rate generator signals are added to the position signals, a zero of the following form is inserted into the s plane:

$$K(s + \alpha) \tag{6-41}$$

where α is the ratio of position to rate signal. When a rigid reference frame is not available, as, for example, in airborne or shipboard applications, a gyroscope* is customarily used to measure simultaneously the position and rate of the object being controlled. Rate feedback provides an effective stabilization method when output motion is available. In many cases—an all-electric system is an example—the output is not mechanical, and hence a rate generator cannot be used. In these cases, however, some type of rate signal can usually be obtained.

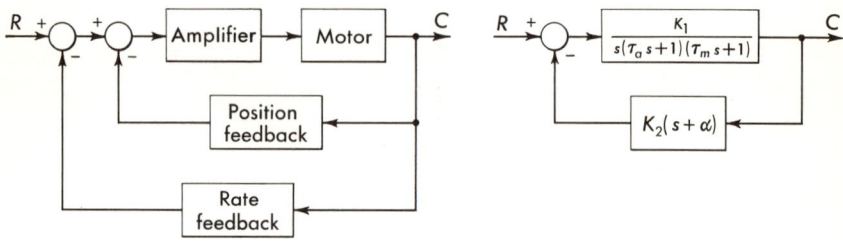

Fig. 6-31 Position servo with rate feedback.

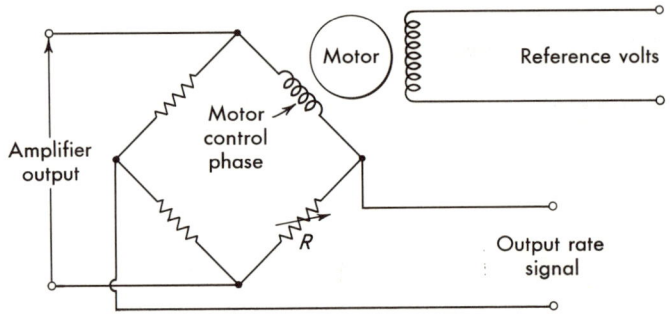

Fig. 6-32 Rate signal taken from the back emf of a-c motor.

Although many motor manufacturers build a-c and d-c rate generators, they are often expensive. Use of motor back emf, developed by both a-c and d-c motors when rotating, may be a convenient expedient to obtain the rate signal. The back emf is proportional to velocity and can be used to stabilize feedback control systems in the same manner that a rate-generator signal can be used. The output of the power device (possibly an amplifier) is applied across a bridge circuit as in Fig. 6-32.

The bridge is balanced to yield zero output with full-load voltage across the motor windings and with the motor stalled. The resistance

---

* See Chap. 7 for a discussion of both position and rate gyroscopes.

R is varied to balance the bridge when the motor is stalled. As the motor is allowed to rotate, the back emf unbalances the bridge and an output rate signal results. This signal is fed back to the input in the same manner as the signal from a rate generator would be utilized.

6-14 Design of a Position Servo with Rate Feedback Equalization.
The system equalized in Sec. 6-10 with a lead network is optimized in

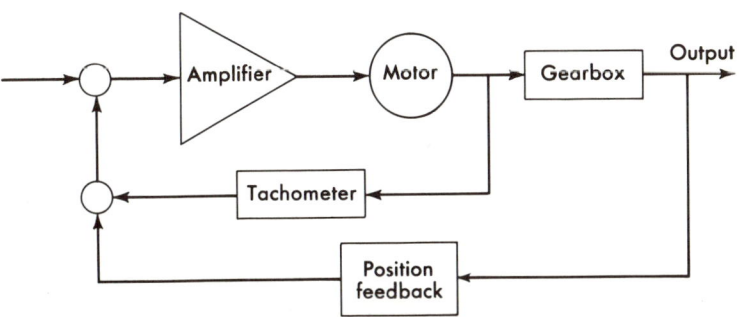

Fig. 6-33 System block diagram. Position plus tachometer feedback.

this section with a rate generator as shown in Fig. 6-33. The third-order system is optimized with several root-locus plots, as shown on Fig. 6-34. The open-loop transfer function for this system is

$$GH = \frac{AK_mK_R}{\tau_a\tau_m} \frac{s + K_p/NK_R}{s(s + 1/\tau_a)(s + 1/\tau_m)} \qquad (6\text{-}42)$$

$$GH = \frac{(AK_mK_R/\tau_a\tau_m)(s + \alpha)}{s(s + 1/\tau_a)(s + 1/\tau_m)} \qquad (6\text{-}43)$$

where $\alpha = K_p/NK_R$ is the ratio of position to rate feedback. τ_m is the equivalent time constant of the motor and includes the moment of inertia of the motor plus the reflected moment of inertia of the load. As α is varied, the zero moves and the system performance varies. Since

$$K_v = \lim_{s \to 0} sGH = \alpha AK_mK_R$$

the velocity constant varies with α. Variation of K_v with α for various values of ζ is shown in Fig. 6-35. Amplifier and effective motor time constants are taken as follows:

$$\tau_a = 0.025 \qquad \tau_m = 0.07$$

In Fig. 6-36 the damping ratio and natural frequency are plotted as functions of K_v. In Figs. 6-37 and 6-38 the same plots are shown for $\tau_a = 0.005$ and $\tau_m = 0.02$. These charts are useful as a guide to the design of systems with rate feedback.

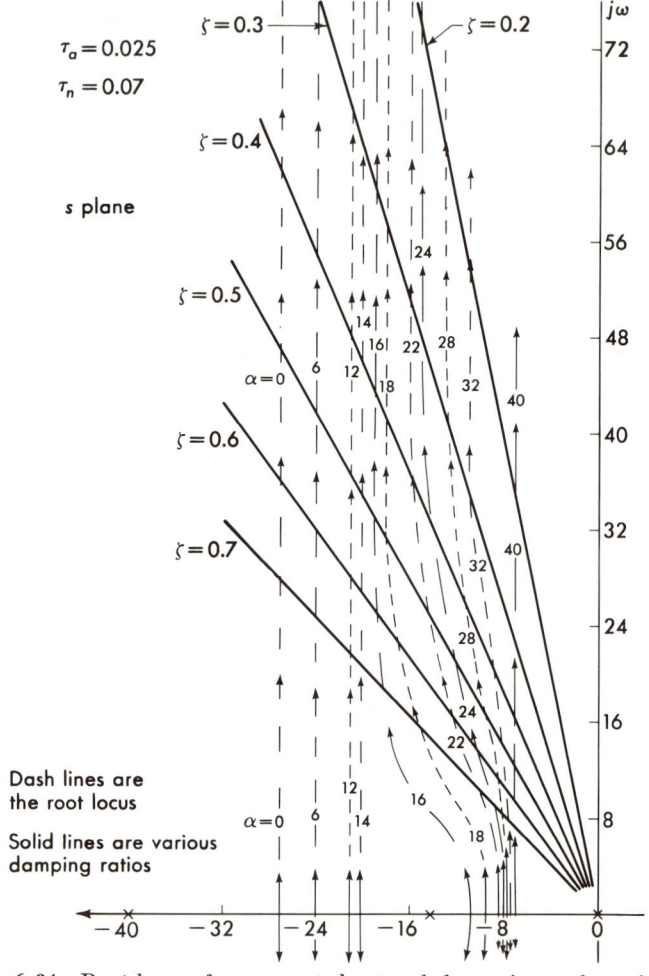

Fig. 6-34 Root locus of compensated network for various values of α.

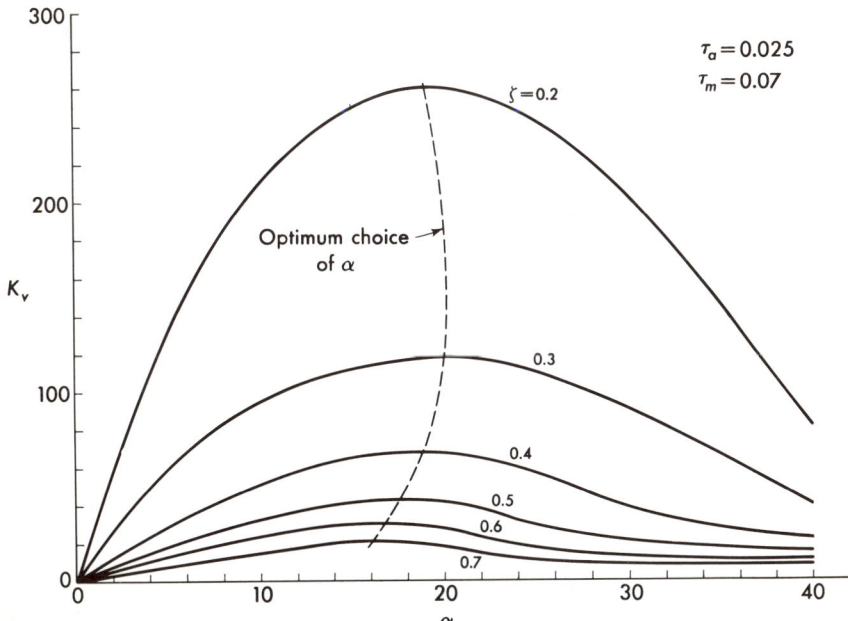

Fig. 6-35 Velocity-error parameter K_v versus α.

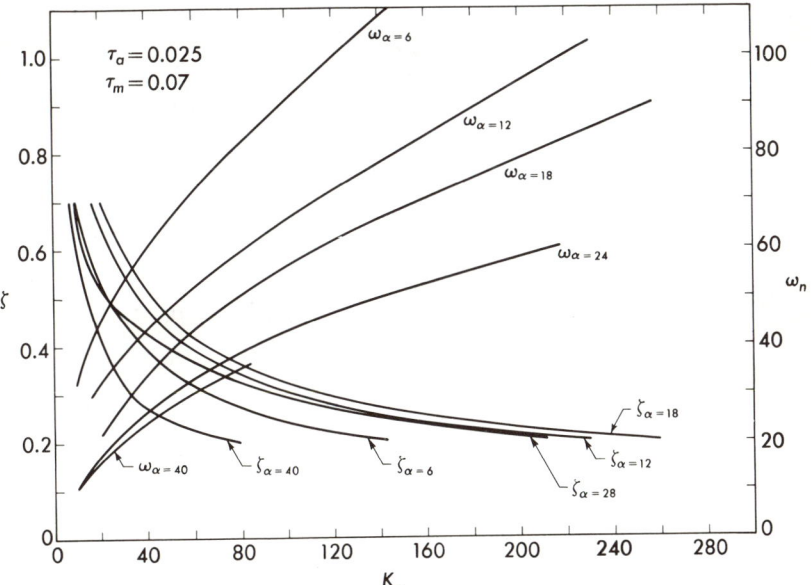

Fig. 6-36 Damping ratio and natural frequency vs. velocity-error parameter.

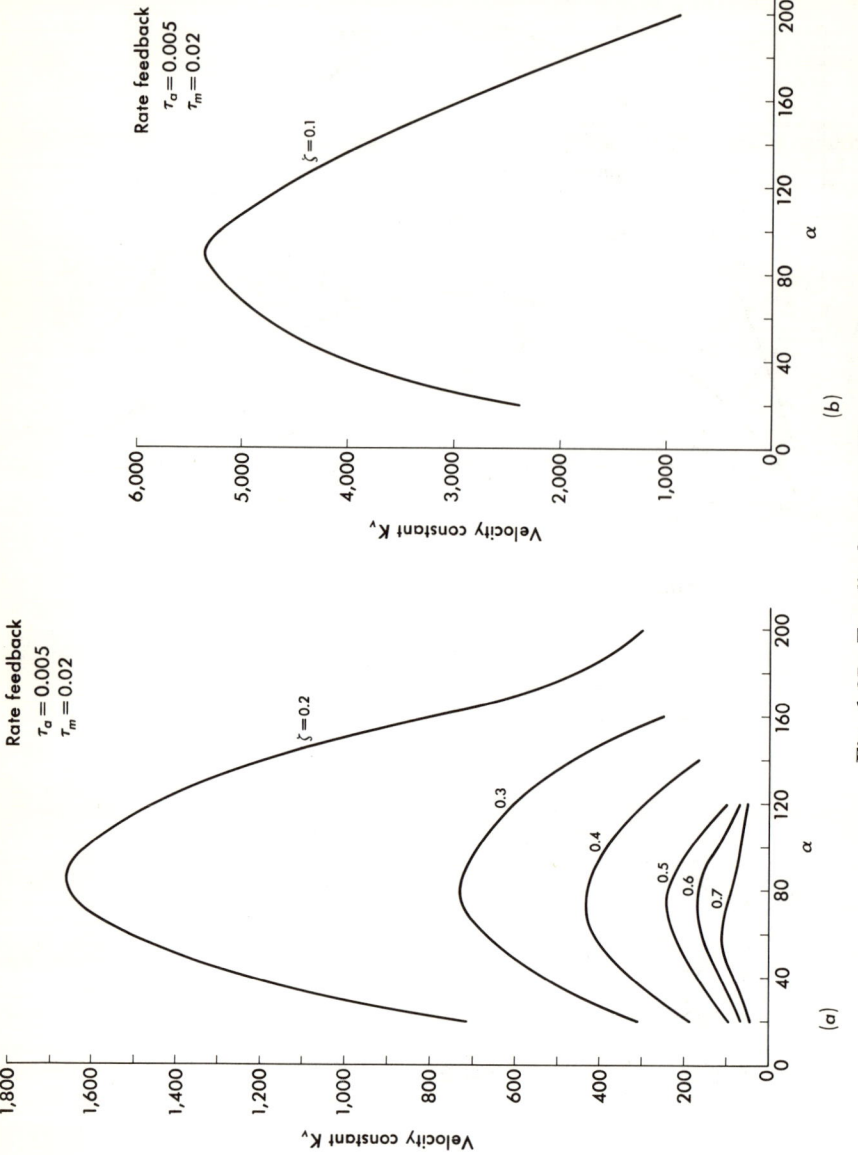

Fig. 6-37 Equalized system (K_v versus α).

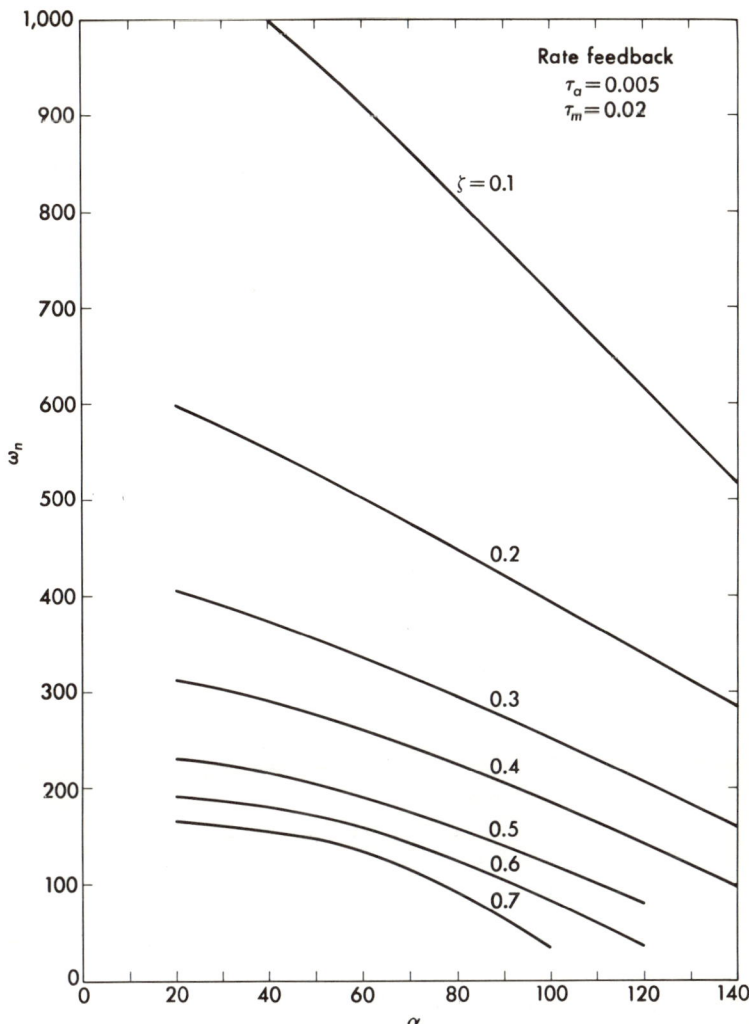

Fig. 6-38 Equalized system (ω_n versus α).

6-15 Multiple-loop Control Systems. Parallel equalization often leads to systems that comprise more than one feedback loop. In some cases multiple loops are inherent in the system; in others they are intentionally introduced to improve the transfer function of certain components.

Block diagrams of some common types of multiple-loop systems are shown in Fig. 6-39. Figure 6-39a is typical of systems in which a minor loop is position feedback around an electric or pneumatic control motor to convert it from an integrating device to a positioning device. Figure 6-39b is typical of the cross-coupling problems often encountered in missile and space control systems.

Use of the Chap. 1 block-diagram identities usually permits the reduction of the block diagrams and transfer functions. The root-locus

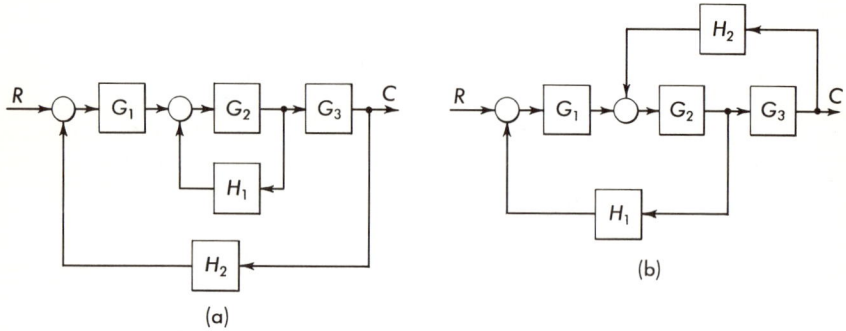

Fig. 6-39 Typical multiple-loop control systems.

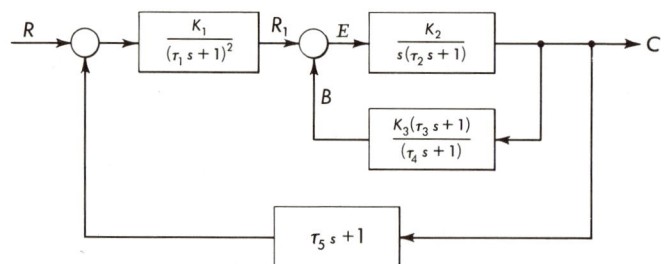

Fig. 6-40 An example of a multiple-loop control system.

method, as discussed in Sec. 4-14, is a convenient method of analyzing multiple-loop systems.

To illustrate the use of a second loop to stabilize a system, consider Fig. 6-40. Transfer functions of the individual blocks are given in the figure. Numerical values for the constants are taken as follows:

$$K_2 = 300 \text{ sec}^{-1} \qquad \tau_2 = 1.0 \text{ sec}$$
$$K_3 = 0.1 \qquad \tau_3 = 0.1 \text{ sec} \qquad (6\text{-}44)$$
$$\tau_1 = 0.1 \text{ sec} \qquad \tau_4 = 0.3 \text{ sec}$$

The problem is to determine whether rate feedback (represented by τ_5) can stabilize the system when K_1 is adjusted to give a natural frequency of 1.5 cps for the closed-loop system.

Design of Feedback Control Systems 249

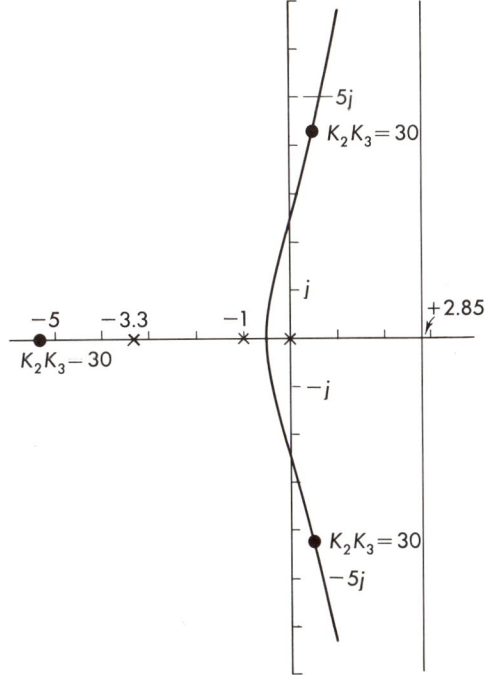

Fig. 6-41 The root-locus diagram for the minor loop.

The root-locus diagram, shown in Fig. 6-41 for the minor loop, is first plotted. For $K_2K_3 = 30$, the roots of the system are located as follows:

$$s = -5.35 \qquad s = +0.5 + j4.25 \qquad s = +0.5 - j4.25 \quad (6\text{-}45)$$

The inner loop can now be replaced by a single transfer function in which the roots of Eq. (6-45) become the poles of the new transfer function. Care must be exercised, however, in writing this transfer function. The roots of the characteristic equation are found from a solution of

$$\frac{K_2K_3(\tau_3 s + 1)}{s(\tau_2 s + 1)(\tau_4 s + 1)} = -1 \quad (6\text{-}46)$$

The roots s_1, s_2, and s_3 are found from this equation without making a distinction between the forward and feedback transfer functions. In writing the overall transfer function C/R_1, however, a distinction does arise. The correct form is found from an examination of the overall transfer function:

$$\frac{C}{R_1} = \frac{\dfrac{K_2}{s(\tau_2 s + 1)}}{1 + \dfrac{K_2 K_3(\tau_3 s + 1)}{s(\tau_2 s + 1)(\tau_4 s + 1)}} \tag{6-47}$$

$$\frac{C}{R_1} = \frac{K_2(\tau_4 s + 1)}{K_2 K_3 \left[\dfrac{s(\tau_2 s + 1)(\tau_4 s + 1)}{K_2 K_3} + (\tau_3 s + 1)\right]}$$

$K_2 K_3$ is factored from the denominator of Eq. (6-47) so that the dimensions are correct when substituting the roots

$$\left(1 + \frac{s}{s_1}\right)\left(1 + \frac{s}{s_2}\right)\left(1 + \frac{s}{s_3}\right) = \frac{s(\tau_2 s + 1)(\tau_4 s + 1)}{K_2 K_3} + \tau_3 s + 1 \tag{6-48}$$

As a check on the constants $K_2 K_3$, set $s = 0$ in Eq. (6-48), with the result that both sides of the equation reduce to unity. Hence the inner-loop transfer functions must be written as follows:

$$\frac{C}{R_1} = \frac{(1/K_3)(\tau_4 s + 1)}{(1 + s/s_1)(1 + s/s_2)(1 + s/s_3)} \tag{6-49}$$

The roots of the inner loop become the poles in the outer-loop root locus. Generally, the zeros of the overall transfer function are the zeros of the forward transfer function and the poles of the feedback function. The constant is readily found by setting $s = 0$. For this example, the poles are located at the points indicated in Eq. (6-45).

In addition two remaining poles exist at $s = -10$. Zeros are located at $s = -1/\tau_5$ and at $s = -3.33$ for the minor loop function.

Root-locus diagrams for the complete system are shown in Fig. 6-42 for $\tau_5 = 0.5$, 1.0, and 2.0 sec as well as the limiting curve for $\tau_5 \to \infty$. The loci of this series show that for $\omega_n = 9.4$ a value of $\tau_5 = 1.0$ yields marginal stability. As τ_5 increases, the degree of stability improves. The limiting value of $\tau_5 = \infty$ gives a $\zeta = +0.044$. Although the system is made stable, further equalization is required, since this system cannot be considered satisfactory.

6-16 Comparison of Various Equalizers. In an attempt to indicate a comparison of several types of equalizers, the following system is studied:

$$G(s) = \frac{K}{s(1 + 0.5s)(1 + 0.2s)} \tag{6-50}$$

Design of Feedback Control Systems 251

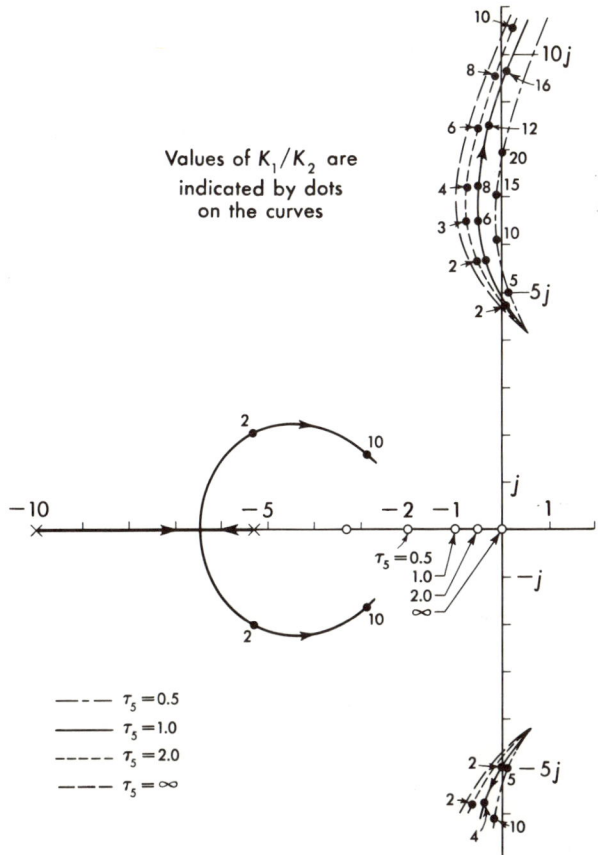

Fig. 6-42 Root-locus diagram of the multiple-loop system of Fig. 6-39.

Although this system is only third order, the wide variety of results indicate the degree of freedom afforded the control-system engineer in synthesizing a particular system. The results apply only to this particular system, which is equalized by the following methods:

1. No compensation
2. Series lead network of the form

$$G_1(s) = \frac{0.25(1 + 0.3s)}{1 + 0.075s} \quad (6\text{-}51)$$

3. Tachometer plus unity feedback of the form

$$H(s) = 1 + 0.3s \quad (6\text{-}52)$$

252 Control System Design

4. Series-parallel lead network of the form

$$G_1 = \frac{1 + s/6}{1 + s/60} \qquad H = \frac{1 + s/6}{1 + s/60} \qquad (6\text{-}53)$$

For each equalizer, the system is analyzed by both Bode and root-locus methods. A typical root-locus diagram is included in Fig. 6-43.

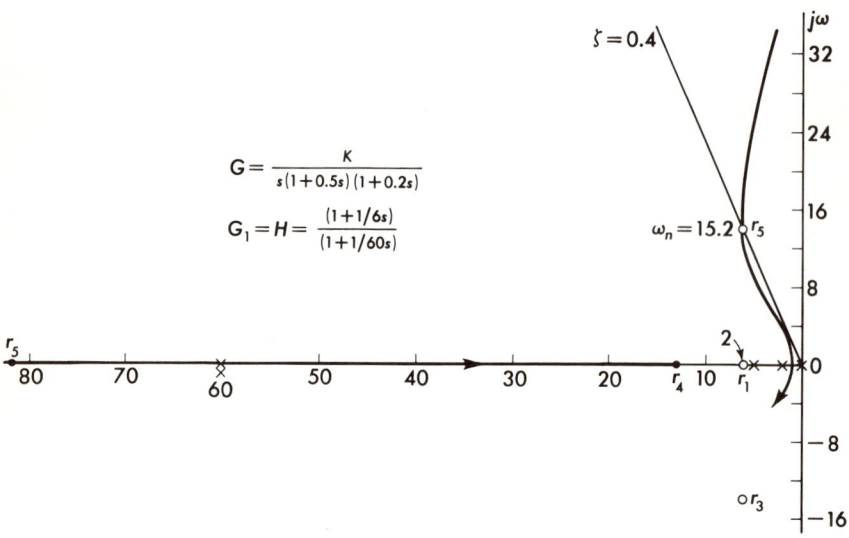

Fig. 6-43 Root-locus diagram for a system with series-parallel compensation.

All the intermediate Bode and root-locus diagrams are not included; however, the results are summarized in Table 6-4. All systems are designed for $\zeta = 0.4$.

The results tabulated in Table 6-4 are further compared in Fig. 6-44, where the response to a step function is plotted for each system. The time base for all cases is the same. In this particular comparison the last equalizer yields the most desirable solution.

6-17 A-C Control Systems. For the following reasons, it may be advantageous to use an a-c system:

1. Drift (random fluctuations of d-c level) in d-c amplifiers. Alternating-current amplifiers can be coupled by using transformers or capacitors with no problem of drift.

Table 6-4 Comparison of Various Equalizer Networks

Summary of results	No compensation $G = \dfrac{K}{s(1+0.5s)(1+0.2s)}$		Series lead network $\dfrac{0.25(1+0.3s)}{1+0.075s}$		Tachometer plus unity feedback $H(s) = 1 + 0.3s$ $K_t = 0.3$		Series-parallel lead networks $G_1 = \dfrac{1+s/6}{1+s/60}$ $H = \dfrac{1+s/6}{1+s/60}$
	Bode	Root-locus	Bode	Root-locus	Bode	Root-locus	Root-locus
Performance characteristics:							
Velocity constant K_v	1.59	1.3	3.16	2.9	2.14	2.1	5.4
ω_R, rad/sec	1.26	1.25	2.4	2.4	2.55	2.43	10
Transient rise time	.....	1.6	.....	1.2	.....	0.91	0.24
Resonant peak, db	+3	+3	+3	+2.7	+0.5	+0.35	+0.2

2. Simplicity and versatility of a-c components.
3. Higher accuracy of a-c components.

Analysis of a-c systems is similar to that of d-c systems. It is not necessary to take notice of the carrier or a-c signal presence during the initial

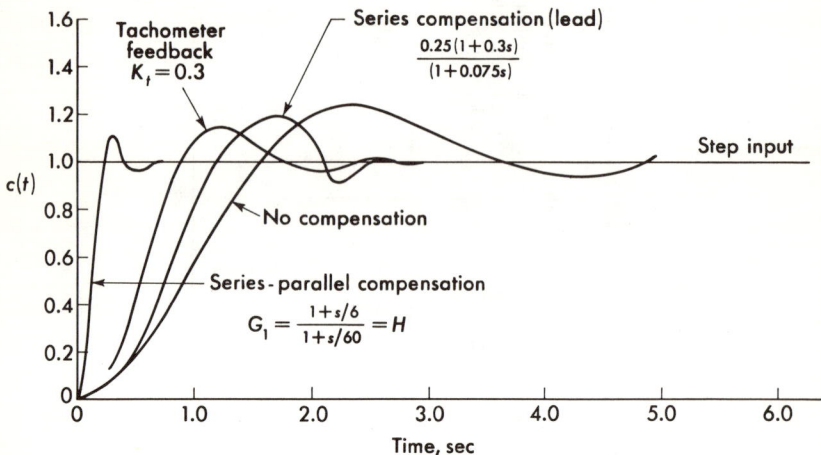

Fig. 6-44 Transient response for various equalizer networks.

phases. For example, to a first approximation transfer functions of a-c and d-c servomotors are the same. The practical method of mechanizing the required zero-pole configuration, however, is different from the d-c case. The next few sections consider the special techniques that can be used for the design of a-c equalizers.

6-18 Suppressed-carrier Modulation. The form of the signal that is present in a-c servos is shown in Fig. 6-45 for a sinusoidal modulating

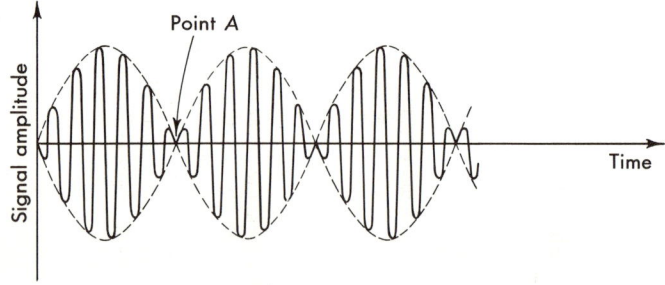

Fig. 6-45 Suppressed-carrier signal for a sinusoidal modulating voltage.

signal. This signal is termed "double-sideband suppressed-carrier." The equation for this signal is given as follows:

$$e = \sin \omega_s t \sin \omega_c t = \tfrac{1}{2}[\cos (\omega_c - \omega_s)t - \cos (\omega_c + \omega_s)t] \quad (6\text{-}54)$$

where ω_s is the signal frequency and ω_c is the carrier frequency. Typical carrier frequencies are 60, 400, and higher. Signal frequencies are usually small: 0 to 10 cps. The name for this type of modulation arises from the second part of Eq. (6-54). Notice that only the frequencies

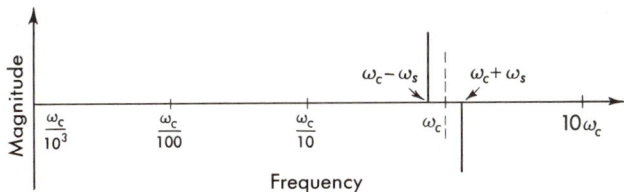

Fig. 6-46 Frequency spectrum for suppressed-carrier signal.

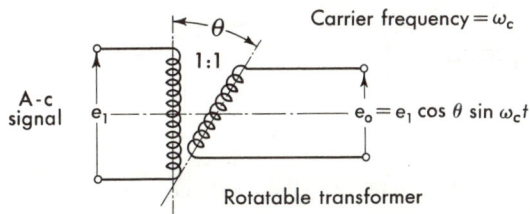

Fig. 6-47 Rotatable transformer produces a suppressed-carrier signal.

$\omega_c + \omega_s$ and $\omega_c - \omega_s$ exist in this expression. Since there is no component at the carrier frequency, this type of modulation is termed "suppressed-carrier." A frequency spectrum* (i.e., a plot of harmonic magnitudes plotted against frequency) is plotted in Fig. 6-46 on a logarithmic scale.

Of more importance than its name is how the signal arises in feedback control systems. The special transformer, shown in Fig. 6-47, has a secondary that can be rotated mechanically with respect to the primary. When the mechanical angle θ between the windings is 0°, full voltage e_1 is developed across the secondary. As θ is varied, a voltage is developed across the output. For θ near 90°, this voltage can be written as

$$e_o = e_1 \sin \omega_c t \cos (90 - \theta) = e_1 \sin \theta \approx (e \sin \omega_c t)\theta \quad (6\text{-}55)$$

* See Ref. 14, p. 143.

It is this type of signal that is developed at the output of a-c position pickoffs. (Chapter 7 considers several types of a-c transducer.) A pickoff, which is described by Eq. (6-55), produces a linear voltage with respect to shaft position for small angular rotations.

The sign of the signal is contained in the phase of the carrier signal. In the region where the signal amplitude approaches the origin (point A in Fig. 6-45), the carrier signal shifts phase by 180°. This is also shown in Eq. (6-55). The carrier signal changes sign as θ changes sign. If an all-a-c system is employed, the torque output from the motor is proportional to the magnitude of the signal. The direction of the torque applied depends upon the phase of the carrier. If a low-frequency signal is required, a phase-sensitive demodulator* must be used to restore correct polarity information to the output.

6-19 A-C System Equalization. Many types of a-c systems exist. In some, a-c signals exist throughout the loop; in others, at least a part

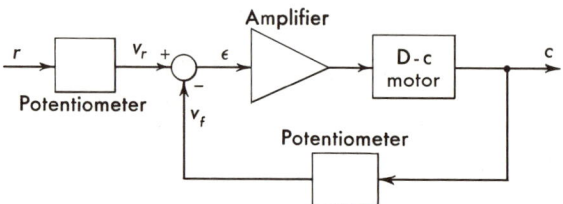

Fig. 6-48 An all-d-c control system.

of the control system employs d-c components. This latter variety normally requires a modulator and demodulator to transform from direct to alternating current and vice versa. A modulated carrier system usually employs envelope feedback. (That is, the feedback signal is also suppressed carrier. An a-c rate generator would provide such a signal.) Although various types of a-c control systems are in use, the method of analysis of such systems is essentially the same as in the d-c case.

Consider the simple d-c positioner which is shown in Fig. 6-48. In this system a potentiometer, excited with a d-c voltage, is used to transduce the input and output shaft position into voltage. A d-c amplifier and d-c motor complete the system.

A hybrid a-c–d-c system is shown in Fig. 6-49. The potentiometers are excited by a d-c voltage. The error voltage is amplified in a d-c preamplifier. The d-c signal is now suppressed-carrier-modulated.

---

* In theory such a demodulator adds in a carrier or reference signal. The sum is then rectified and filtered to yield the low-frequency signal. See Ref. 1, chaps. 9 and 5, for demodulator circuits.

Design of Feedback Control Systems

This can be accomplished with a solid-state chopper or with electronic circuits. The a-c signal is amplified and used to drive the a-c motor.

An all-a-c system is shown in Fig. 6-50. In this system all components operate with alternating current; the amplifier is transformer or resistance-capacitance coupled, the subtraction is performed with transformers, and the position pickoffs are induction type.

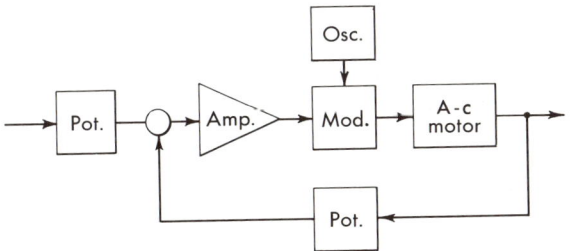

Fig. 6-49   Hybrid a-c–d-c control system.

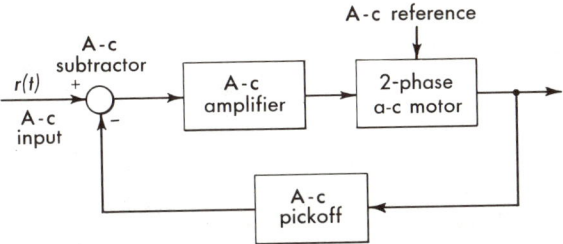

Fig. 6-50   An all-a-c control system.

If equalization is necessary in the control system, special techniques must be employed so that the equalizer operates only on the signal and not on the carrier. Any of the d-c equalizers of the preceding section operate only on a low-frequency signal. If the signal of Eq. (6-54) were inserted into one of these "d-c equalizers," the output would not have the correct form. For example, if the signal of Eq. (6-54) were inserted into the lead network of Fig. 6-12a, the only result, with $\omega_c \gg 1/RC$, would be attenuation of the output by a constant amount

$$\frac{R_2}{R_1 + R_2} \tag{6-56}$$

The network has no equalizing effect upon the suppressed-carrier signal. Since the zeros and poles of the network are located in the low frequencies, 0 to 10 cps in most applications, the normal d-c network has no effect in the region of the carrier frequency, ω_c, 400 cps.

The block diagram of Fig. 6-51 represents a common method of equalizing a-c systems: demodulation-modulation technique. All the advantages of a-c pickoffs and a-c amplifiers are maintained. The a-c signal is reduced to direct current (low frequency) with the phase-sensitive demodulator. The equalization is accomplished with the networks, discussed previously, and the signal is again returned to alternating current with a modulator. Practically, the system of Fig. 6-51 is mechanized in a number of ways. For example, the equalized d-c signal can be used to drive a magnetic amplifier* whose output is alternating current. Alternatively, the equalized d-c signal might be used directly to drive a d-c actuator through a d-c power stage.

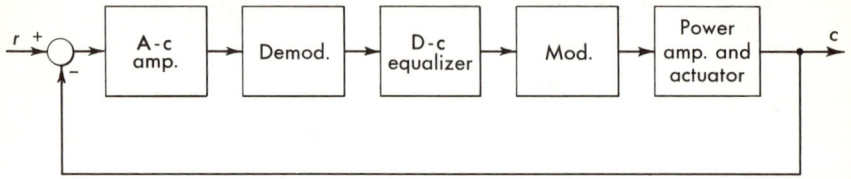

Fig. 6-51 Demodulation-modulation method of equalizing a-c systems.

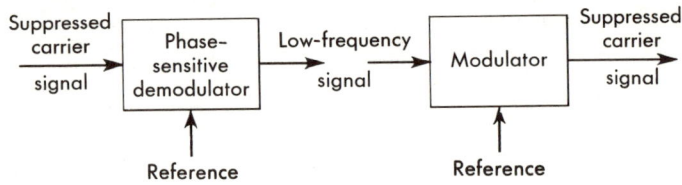

Fig. 6-52 Block diagram for a demodulator and a modulator.

Phase-sensitive demodulators and modulators necessary for the equalization of a-c systems act as switches which are controlled by a reference signal of the same frequency as the carrier frequency. Circuits for both demodulators and modulators are discussed in Chap. 7. It is sufficient to consider a demodulator as a unit that reduces the suppressed-carrier a-c signal to direct current (or low frequency) and a modulator as a unit that modulates the signal on a carrier. This is shown in the block diagram of Fig. 6-52. A hybrid a-c–d-c system, in which the suppressed carrier is demodulated to a low-frequency signal, can be equalized with the d-c networks of this chapter. The method of design is, of course, identical with that for an all-d-c system.

6-20 Carrier Networks. It is possible to stabilize a-c systems with passive circuits in the forward loop without reducing the a-c signal to

---

* Magnetic amplifiers are discussed in Chap. 7.

direct current. The networks that accomplish this are termed "carrier networks." Bridged- and parallel-T networks are two such carrier networks. The circuit diagram and equations for a bridged-T network are shown in Sec. 6-8. The amplitude (in decibels) and phase vs. frequency curves (on a logarithmic scale) are shown in Fig. 6-53.

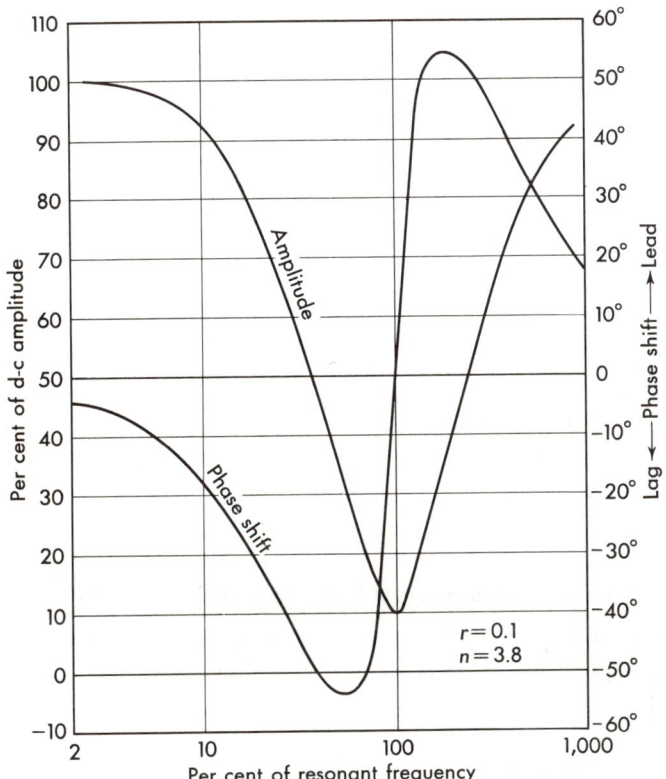

Fig. 6-53 Typical amplitude and phase response of a bridged-T notch network.

The ability of a bridged-T network to stabilize an all-a-c system can be understood by reference to Fig. 6-54. Equation (6-54) and Fig. 6-46 show that a suppressed-carrier signal contains two components—one at $\omega_c - \omega_s$ and the other at $\omega_c + \omega_s$. The signal varies about the carrier at ω_c exactly as the d-c signal varies about zero frequency.

In Fig. 6-54 are shown the amplitude and phase of a lead network. In the same figure are shown the amplitude and phase of a bridged-T network with the carrier frequency reduced to zero ($\omega_c = 0$). The similarity between these two networks is seen from this comparison. When the suppressed-carrier signal is passed through a bridged-T network, the sidebands (at $\omega_c \pm \omega_s$) respond in an analogous fashion to direct current through a lead network. Hence, a bridged-T network equalizes a-c suppressed-carrier systems in manner similar to that in which a passive circuit differentiator stabilizes a d-c system. Except

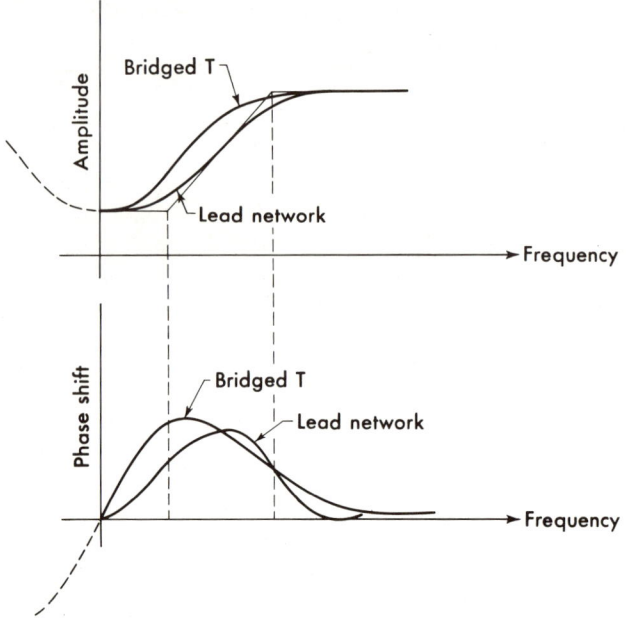

Fig. 6-54 Comparison of bridged-T and lead networks.

for low-frequency (60-cps) systems, carrier networks usually find little use. Reasons for this are summarized as follows:

1. Low-Q networks attenuate the sidebands at $\omega_c \pm \omega_s$ so that the signal-to-noise ratio is increased. This effect is less at the lower frequencies (60 cps).

2. Shift of the reference frequency causes the sidebands to receive nonsymmetrical amplitude and phase response, and hence distortion results.

The general procedure for synthesizing an a-c equalizer is summarized as follows:

1. Determine the correct d-c network from the stability analysis.
2. Find the network with approximately the same gain and phase characteristics about ω_c (carrier frequency) as the d-c network has about zero frequency.

Design of Feedback Control Systems

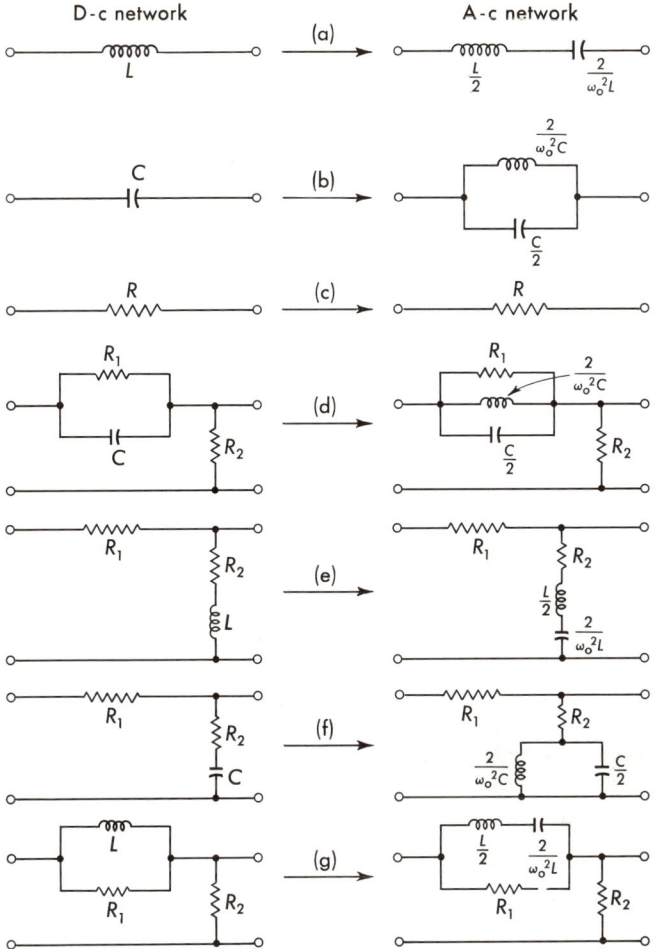

Fig. 6-55 Alternating-current network equivalents for d-c networks.

This latter problem is solved, when $\omega_c \gg \omega_s$, by a transformation of frequency.* The $j\omega$ in the transfer function of the network,

$$\frac{E_o}{E_{in}} = G(j\omega) \tag{6-57}$$

is replaced by a function of frequency $h(j\omega)$. Provided the maximum signal frequency is less than the carrier frequency, it can be shown* that

---

* See Ref. 38, pp. 401–409, for a more extensive treatment of this subject.

a frequency transformation of the form

$$h(j\omega) = j\frac{\omega}{2}\left(1 - \frac{\omega_c^2}{2\omega^2}\right) \tag{6-58}$$

transforms the d-c network into the desired a-c equivalent. An inductance L in the original d-c network is replaced by an impedance

$$j\omega\frac{L}{2} + \frac{\omega_0^2 L}{j2\omega} \tag{6-59}$$

which is an inductance in series with a capacitance, as shown in Fig. 6-55a. A capacitance C is replaced by a network with an admittance

$$j\frac{\omega C}{2} + \frac{\omega_0^2 C}{j2\omega} \tag{6-60}$$

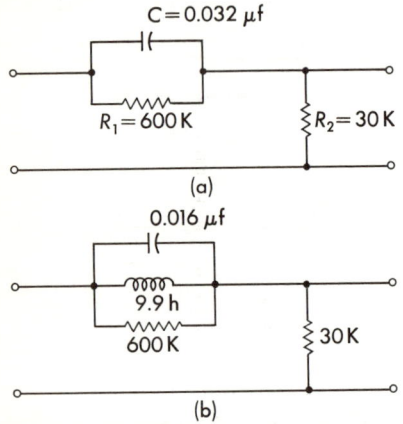

Fig. 6-56 (a) Direct-current network equalizer; (b) alternating-current equivalent network for d-c equalizer.

which is a capacitance in parallel with an inductance, as shown in Fig. 6-55b. Resistances remain unchanged. Figure 6-55 includes the a-c equivalents for several d-c lead and lag networks.

As an example, suppose the network shown in Fig. 6-56a was required to equalize the d-c system. The transformation of this d-c network to operate with a 400-cps carrier system results in the network shown in Fig. 6-56b.

Practical application of these networks requires:

1. High-Q inductors to prevent sideband attenuation
2. Closely controlled carrier frequency to prevent frequency shift

The value of either L or C requires adjustment so that the resonant circuit can be accurately tuned to the carrier frequency.

6-21 Electromechanical Networks. Because of the analogy (Chap. 2) between electrical and mechanical networks, it is reasonable to expect that mechanical components can be utilized for stabilizing feedback control systems. Since a two-phase motor* converts the suppressed-carrier-voltage input into a low-frequency shaft position, the motor functions as a demodulator. This section treats the electromechanical integrator and differentiator.

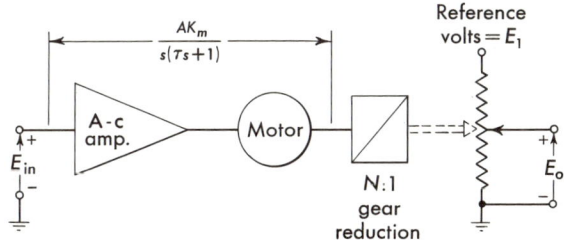

Fig. 6-57 Block diagram of an electromechanical integrator.

An electromechanical integrator is shown in the block diagram of Fig. 6-57. The transfer function E_o/E_{in} is given by

$$\frac{E_o}{E_{in}} = \frac{E_1 A K_m/N}{s(\tau s + 1)} \qquad (6\text{-}61)$$

where the quantities are as defined in Fig. 6-57. For low frequencies, the motor time constant is neglected and the system is described by the equation

$$\frac{E_o(j\omega)}{E_{in}(j\omega)} \approx \frac{E_1 A K_m}{N} \frac{1}{j\omega} \qquad (6\text{-}62)$$

In the low-frequency range, this equation represents an integration:

$$e_o = \frac{E_1 A K_m}{N} \int_0^t e_{in} \, dt \qquad (6\text{-}63)$$

An integrator is used in a system to reduce the steady-state error to zero. For a small constant input the motor will continue to turn at constant velocity. The integrator output is taken from a potentiometer which can be excited with either an alternating or unidirectional voltage.

Figure 6-58 shows the block diagram of an electromechanical lead network. The transfer function of the closed loop is

$$\frac{E_o}{E_{in}} = \frac{A}{1 + \dfrac{A K_m/N}{s(\tau s + 1)}} = \frac{A s(\tau s + 1)}{s(\tau s + 1) + A \dfrac{K_m}{N}} \qquad (6\text{-}64)$$

* See Chap. 7 for a more detailed discussion of two-phase motors.

For large forward loop gain A and for small values of motor time constant τ, the transfer function is

$$\frac{E_o}{E_{in}} \approx \frac{N}{K_m} s \qquad (6\text{-}65)$$

This equation represents a differentiation:

$$e_o = \frac{N}{K_m} \frac{de_{in}}{dt} \qquad (6\text{-}66)$$

Hence the network of Fig. 6-58 is a suppressed-carrier differentiator. In Figs. 6-57 and 6-58 the amplifiers are so tuned that a high gain is obtained with a narrow bandwidth and noise is kept to a minimum.

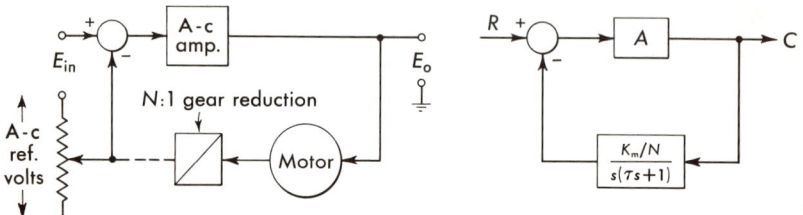

Fig. 6-58 Block diagram of electromechanical differentiator.

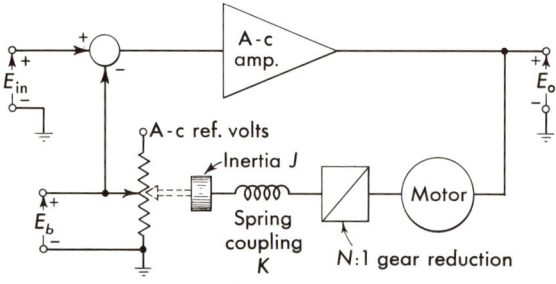

Fig. 6-59 An electromechanical network equalizer.

Various transfer functions are obtained if other mechanical components, such as damping and spring rate, are added to the motor shaft. For example, consider the transfer function for the electromechanical system of Fig. 6-59. The following equation can be written:

$$NL = K_1 E_a - K_2 s \frac{\bar{\theta}_1}{N} \qquad (6\text{-}67)$$

where K_1 and K_2 are as defined in Sec. 1-3.

$$NL = N^2 J_m s^2 \frac{\bar{\theta}_1}{N} + K\left(\frac{\bar{\theta}_1}{N} - \theta_2\right) \tag{6-68}$$

and

$$Js^2 \bar{\theta}_2 + K\left(\bar{\theta}_2 - \frac{\bar{\theta}_1}{N}\right) = 0 \tag{6-69}$$

When these equations are solved, the feedback transfer function H becomes

$$H = \frac{E_b}{E_o} = \frac{K_s}{N^2(JJ_m/K)s^3 + (JK_2/K)s^2 + (N^2 J_m + J)s + K_2} \tag{6-70}$$

With a high-gain amplifier in the forward loop the overall transfer function is well approximated as follows:

$$\frac{E_o(s)}{E_{\text{in}}(s)} = \frac{A}{1 + AH(s)}$$

Hence
$$\frac{E_o(j\omega)}{E_{\text{in}}(j\omega)} = \frac{A}{1 + AH(j\omega)} = \frac{1}{1/A + H(j\omega)}$$

Now, for the range ω for which $|H(j\omega)| \ll A$,

$$\frac{E_o(j\omega)}{E_{\text{in}}(j\omega)} \cong \frac{1}{H(j\omega)} \tag{6-71}$$

The overall transfer function is approximated as follows:

$$\frac{E_o(j\omega)}{E_{\text{in}}(j\omega)} = \frac{1}{K_s}\left[N^2 \frac{JJ_m(j\omega)^3}{K} + \frac{JK_2(j\omega)^2}{K} + (N^2 J_m + J)j\omega + K_2\right] \tag{6-72}$$

This system produces three zeros whose locations depend upon the mechanical constants.

6-22 Various Methods of Stabilizing A-C Systems. Because of the widespread use of a-c systems, especially 400-cycle, many schemes have been advanced for stabilizing them. The methods primarily deal with changes in the load that result in output-shaft damping. The application of these devices is generally limited to small-instrument-type a-c servos whose output is mechanical motion. Although these methods are fairly inexpensive, the damping is usually improved at the expense of the steady-state error. Although the units shown in the figures are for rotational-motion systems, the extension to linear-motion outputs is clear.

a. Coulomb Friction. Friction surfaces rubbing on the output shaft, as shown in Fig. 6-60, produce output-shaft damping. For systems where accuracy requirements are not severe, the friction opposes shaft motion and creates an inexpensive and simple damper. Since this damper is nonlinear,* care must be exercised when using it. The disadvantages of coulomb-friction damping are summarized:

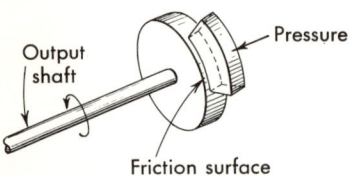

Fig. 6-60 A coulomb-friction damper.

1. Life is limited by wear.
2. As the unit wears, the degree of damping changes.
3. Output power is wasted.
4. Static and velocity errors in shaft position are introduced.

b. Viscous Friction. This damping can be obtained, as shown schematically in Fig. 6-61, by mounting paddles on a shaft. The paddles are caused to move through grease or oil, depending upon the motion and the amount of damping required. The rotating paddle disk exerts a restraining torque proportional to velocity. This damping unit is simple

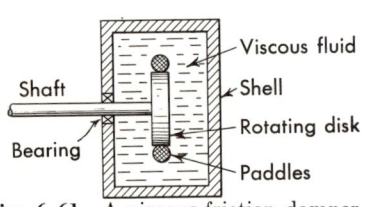

Fig. 6-61 A viscous-friction damper.

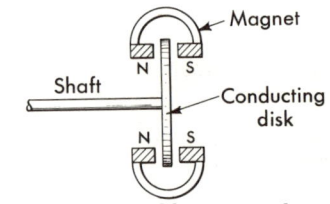

Fig. 6-62 An eddy-current damper.

and rugged and has long life. The disadvantages of viscous damping are summarized:

1. The viscosity and hence the degree of damping are temperature sensitive.
2. Some inertia is added to output shaft.

c. Eddy-current or Foucault Damping. This type of damping, shown in Fig. 6-62, produces an approximate viscous damping by means of a

---

* See Sec. 8-3 for a discussion of coulomb friction. Chapter 8 considers the analysis of systems containing such nonlinearities.

magnetic field. A conducting (copper or aluminum) cup is attached to the output shaft and is rotated through a magnetic field. The eddy currents induced in the cup set up a magnetic field which reacts with the original field in such a manner that a retarding torque is produced on the output shaft. This unit is easily constructed, is highly reliable, and has no fluid to leak out. Disadvantages of eddy-current damping are summarized:

1. The maximum damping available is small.
2. The resistance of the disk and hence the degree of damping are temperature sensitive.

d. Direct Current through Motor Windings. When the torque output of instrument servos is sufficient, a certain degree of damping can be obtained simply by running a small amount of direct current through the a-c windings. An example of this technique is shown in Fig. 6-63. A diode in a parallel-resistance network permits a controllable amount of direct current through the motor reference phase. The disadvantages of this type of damping are:

1. Only a small amount of damping is possible.
2. Motor torque is reduced.

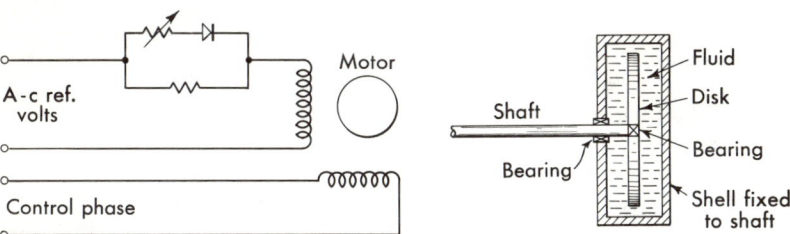

Fig. 6-63 Direct current through a-c motor results in motor damping.

Fig. 6-64 Viscous acceleration damping.

e. Damping Proportional to Acceleration. Both viscous and magnetic acceleration dampers are utilized in position systems. The viscous acceleration damper is shown in Fig. 6-64. The unit is similar to the viscous velocity damper of Fig. 6-61, except that the shell is fixed to the shaft and the disk is free to rotate. When the velocity suddenly changes, the moment of inertia of the disk produces a restraining torque on the shaft. The unit dissipates energy only when the velocity changes; it is an acceleration damper. The unit produces no additional velocity error. The disadvantages are summarized:

1. Shaft moment of inertia is greatly increased.
2. The mechanical construction is more complex; the unit requires a special bearing.

6-23 Design of a Position Servo with an Acceleration Damper. The equivalent circuit for a motor driving an inertia load J and an

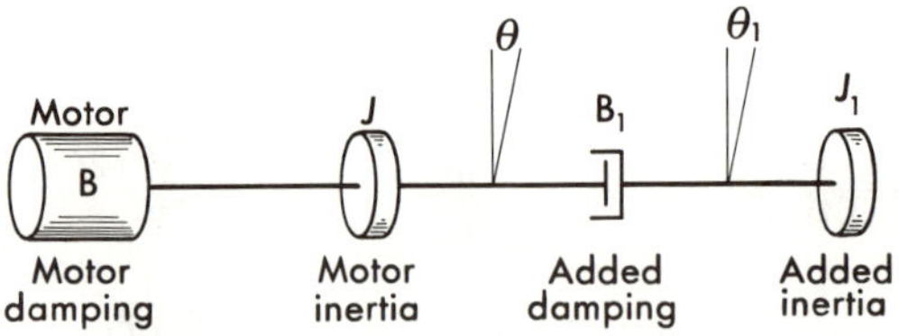

Fig. 6-65 Acceleration damper.

acceleration damper is shown in Fig. 6-65. Equations for this system are written as follows:

$$KE_{\text{in}} - Bs\bar{\theta} = Js^2\bar{\theta} + B_1 s(\bar{\theta} - \bar{\theta}_1) \tag{6-73}$$

$$0 = J_1 s^2\bar{\theta}_1 + B_1 s(\bar{\theta} - \bar{\theta}) \tag{6-74}$$

Solving for $\bar{\theta}$,

$$\bar{\theta} = \frac{\begin{vmatrix} KE_{\text{in}} & -B_1 s \\ 0 & J_1 s^2 + B_1 s \end{vmatrix}}{\begin{vmatrix} Js^2 + (B + B)s & -B_1 s \\ -B_1 s & J_1 s^2 + B_1 s \end{vmatrix}} \tag{6-75}$$

The determinant is solved for $\bar{\theta}$:

$$\frac{\bar{\theta}}{E_{\text{in}}} = \frac{(K/J)(s + B_1/J_1)}{s\left(s^2 + \dfrac{J_1 B + J_1 B_1 + B_1 J}{J_1 J}s + \dfrac{B_1 B}{J_1 J}\right)} = \frac{(K/J)(s + \omega_1)}{s(s + \omega_2)(s + \omega_3)} \tag{6-76}$$

This network is applied to the position controller of Fig. 6-17. Because the damper cannot be separated from the motor, the single transfer function of Eq. (6-76) replaces the motor load plus damper. The complete transfer function is

$$GH = \frac{200A}{s + 200} \frac{(K/J)(s + \omega_1)}{s(s + \omega_2)(s + \omega_3)} \frac{1}{100}$$

The numbers correspond to $\tau_m = 0.02$ sec and $\tau_a = 0.005$.

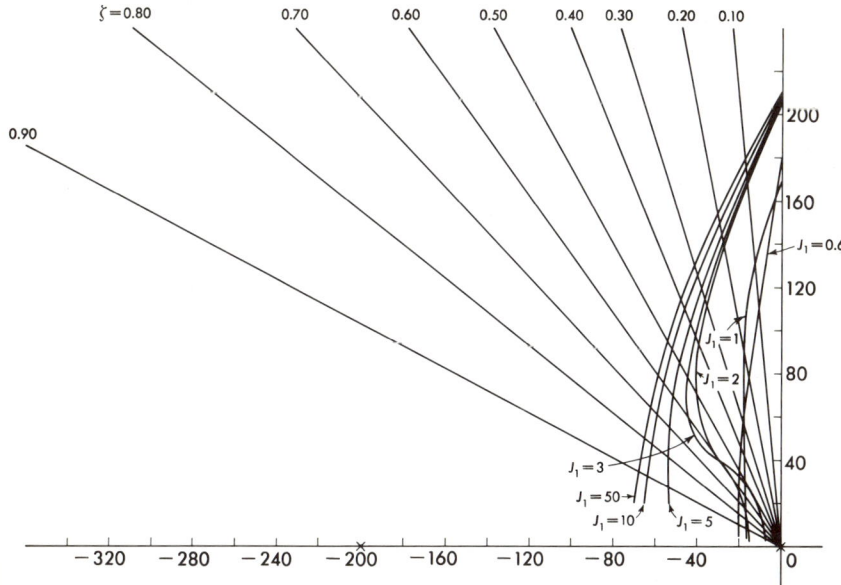

Fig. 6-66 Root-locus plots, inertially damped system.

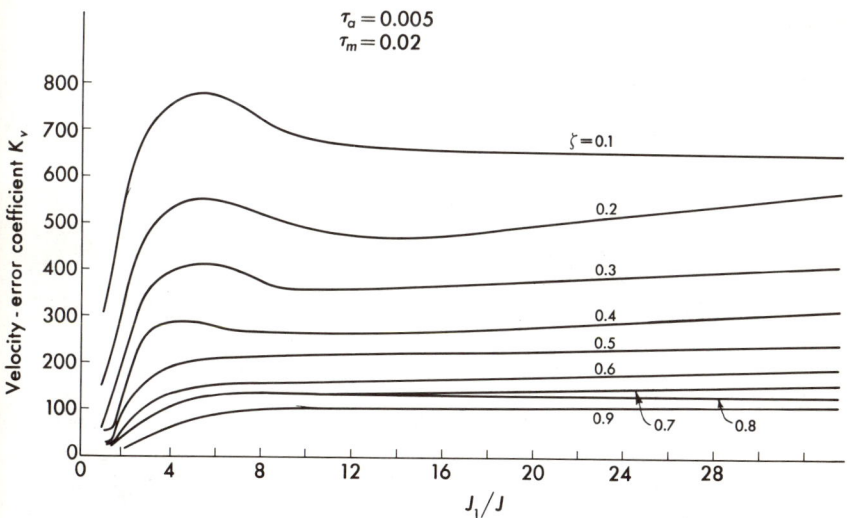

Fig. 6-67 Velocity error coefficient versus J_1/J.

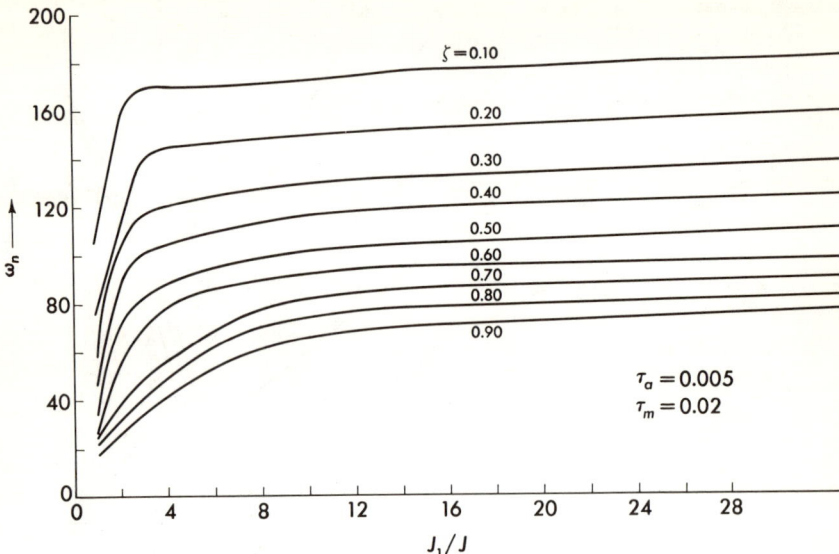

Fig. 6-68 System natural resonant frequency versus J_1/J.

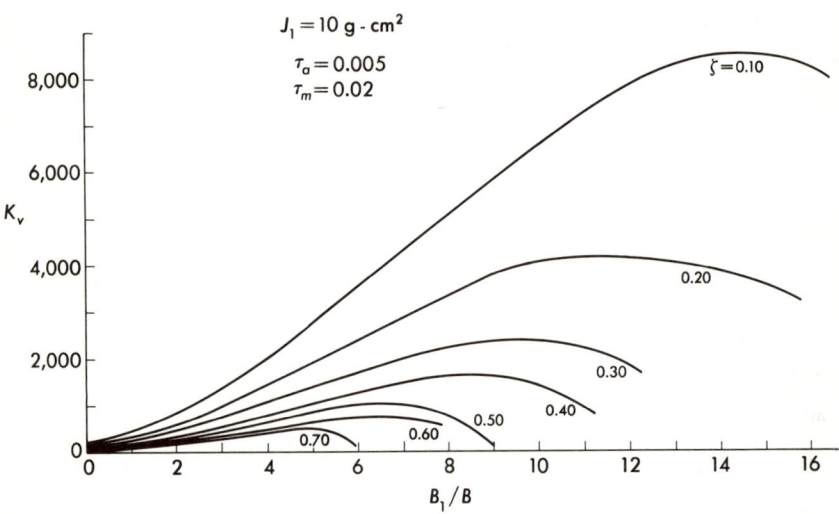

Fig. 6-69 Velocity error coefficient versus B_1/B.

A root-locus study (an example is shown in Fig. 6-66) provides design charts for use with an inertially damped motor. The curves of Figs. 6-67 and 6-68 show the variation of K_v and ω_n with J_1/J. The damping ratio ζ is the parameter in these curves. Figures 6-69 and 6-70 show the variation of K_v and ω_n versus B_1/B with ζ as a parameter.

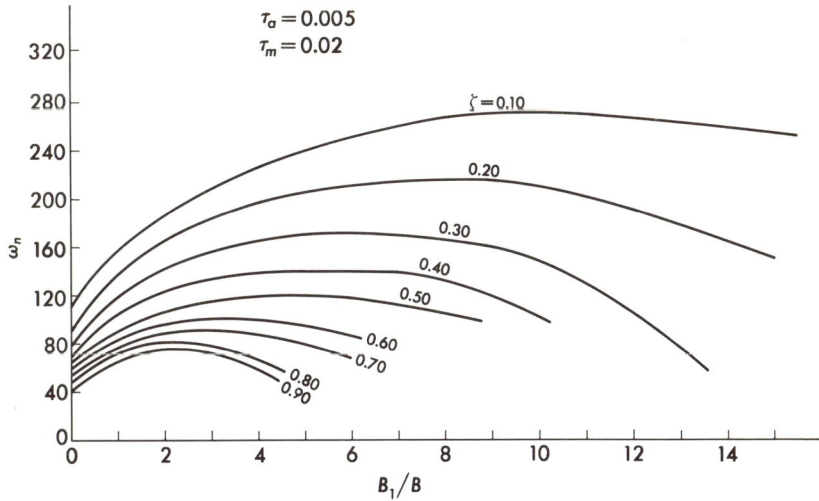

Fig. 6-70 System natural resonant frequency versus B_1/B.

6-24 Practical Considerations in A-C Control-system Design. In Sec. 6-16 several methods of control-system equalization are compared. The comparison does not take into account that the system may utilize alternating current. This problem should be considered after the equalizer transfer function is found. The series lead network can be inserted into the a-c system in several ways:

1. Use a demodulator and modulator (see Fig. 6-51).
2. Frequency-transform the network as shown in Sec. 6-20.

An a-c rate generator geared to the output shaft easily produces the $H(s) = 1 + 0.3s$.

The last system is least suitable for an a-c system, since two d-c networks, which would require two sets of demodulators and modulators, would be required.

In the design of an a-c system, the same methods of analysis (cf. Chaps. 3 to 5) are used as in the design of d-c systems. The main difference between the d-c and a-c control-system design problem lies in the practical area of mechanizing the network.

Of great importance in the design is the carrier phase shift. For full torque on the two-phase control motor, the control and reference phases must be 90° apart. If the phase is 60°, the power is 70 per cent of maximum, since the torque is a sine function of the phase shift. Because of unaccounted-for phase shifts through the system, it is usually necessary

272   Control System Design

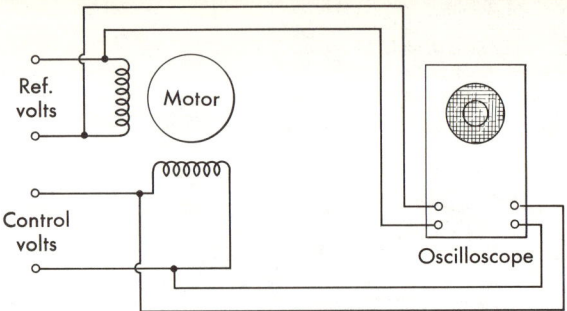

Fig. 6-71   Experimental checking of the phase shift in two-phase a-c motor.

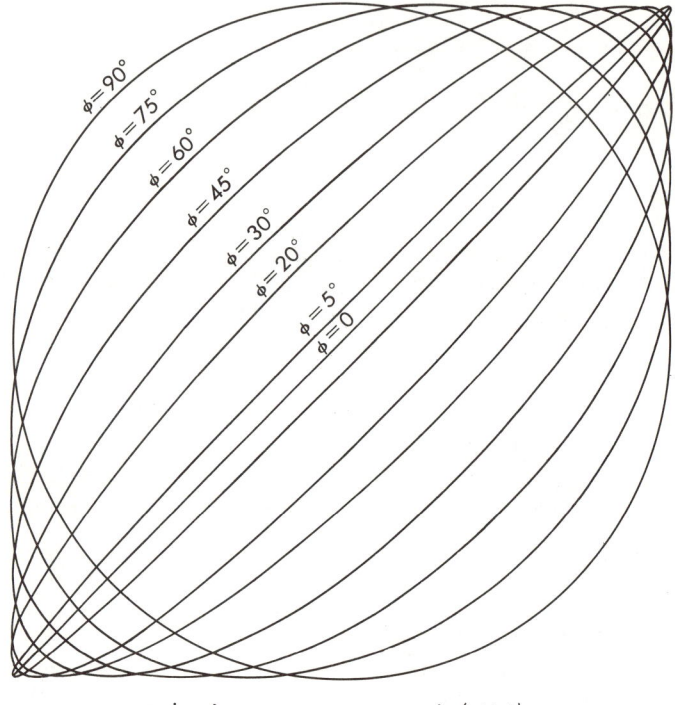

$x = \sin \omega t$ $y = \sin (\omega t + \phi)$
Fig. 6-72   Oscilloscope ellipses for various phase shifts between x and y axes.

to provide additional phase-shifting networks.   With the reference and control phases of the control motor on the X and Y plates of an oscilloscope as in Fig. 6-71, the pattern is a circle when the oscilloscope channels have the same gain.   The chart of Fig. 6-72 shows ellipses that are seen on the oscilloscope for several phase differences between the x and y axes. Often an estimated phase shift is suitable for approximating purposes (especially in the laboratory). The student should familiarize himself with Fig. 6-72. Poor performance in a-c servos is frequently attributable to phase errors between the reference and control phases.

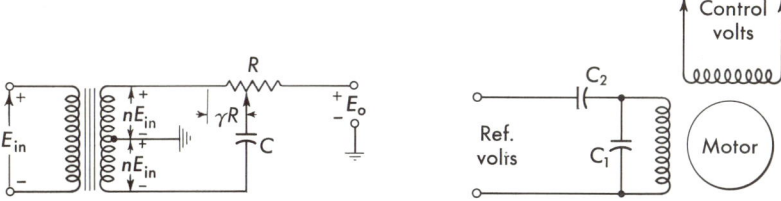

Fig. 6-73 Phase-shift network for control phase of a-c control system.

Fig. 6-74 Phase-shift network for reference phase of a-c motor.

Phase-shift correction can be accomplished in either the control or reference phases. The adjustable network of Fig. 6-73, discussed in Sec. 6-9, provides a simple method of inserting an adjustable amount

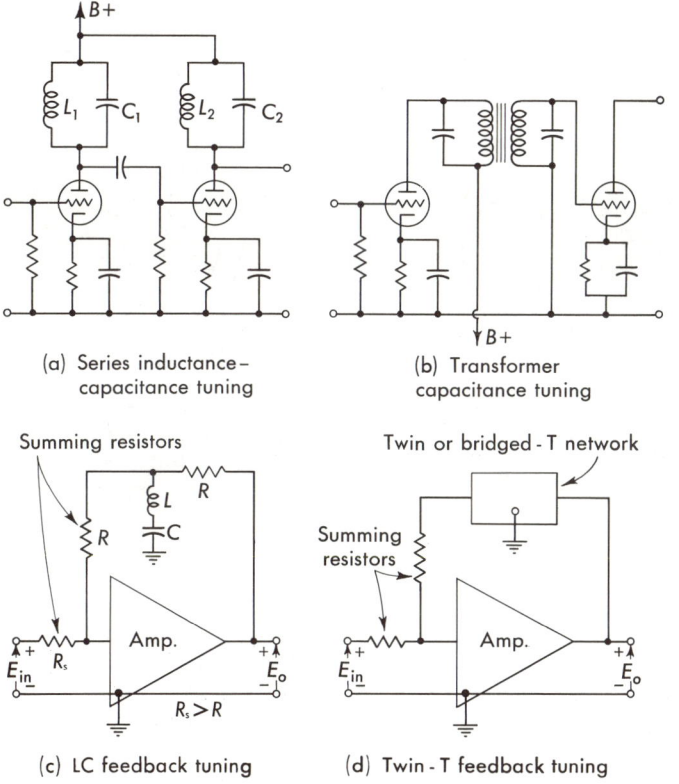

(a) Series inductance–capacitance tuning

(b) Transformer capacitance tuning

(c) LC feedback tuning

(d) Twin-T feedback tuning

Fig. 6-75 Tuned amplifiers used to eliminate undesirable frequencies.

274 Control System Design

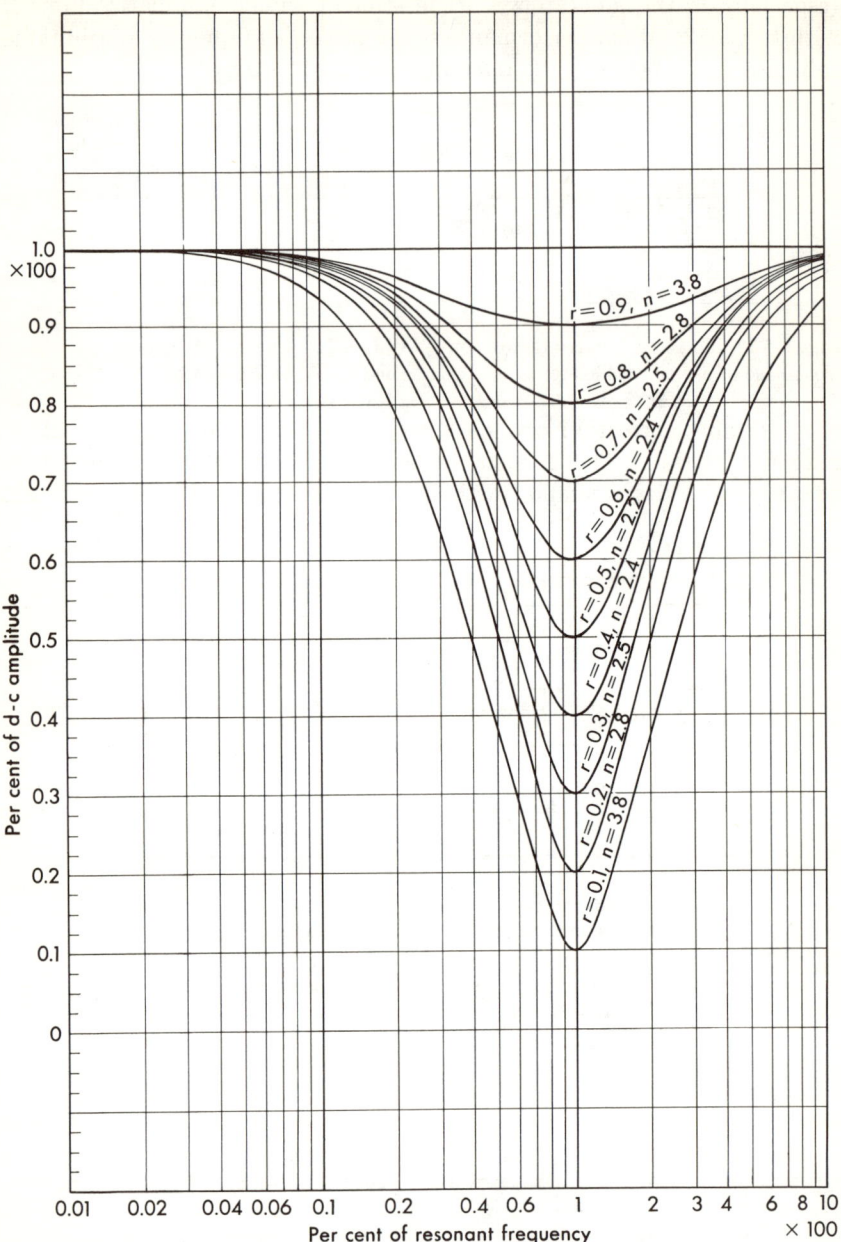

Fig. 6-76a Amplitude response for the bridged-T networks.

Design of Feedback Control Systems 275

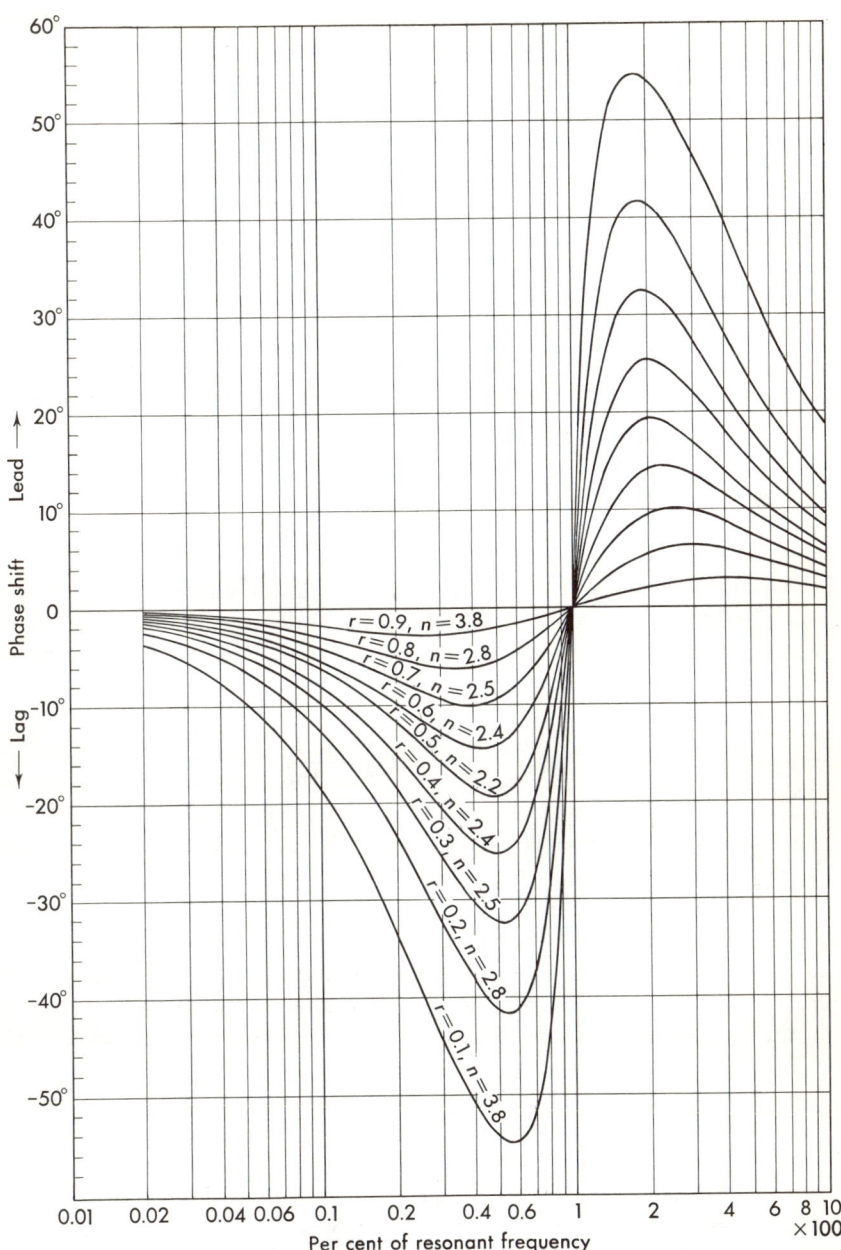

Fig. 6-76b Phase-angle response for the bridged-T networks.

of phase shift in the control phase. Since this network can be placed ahead of the power amplifier, it is required to carry little power.

Shifting the reference phase is conveniently done by a pair of capacitors. One method is shown in Fig. 6-74. Since these capacitors are inserted at a high power level, it is usually not feasible to make them continuously variable. Many manufacturers suggest the size of capacitor required for suitable operation of their motors. In any case the final selection of these components is best made in the laboratory.

In the design of a-c control systems, the signal-to-noise ratio is considerably improved by using tuned amplifiers. The design is based upon the concept of eliminating from the response characteristics of the system all frequencies except in the narrow region about the carrier frequency. Careful design of the tuned circuit is necessary, since sharp tuning adds phase lag to the signal. Parallel-tuned LC circuits, as shown in Fig. 6-75a, are inserted in the forward path and produce a narrow-band system. Tuned transformer coupling, as shown in Fig. 6-75b, produces essentially the same results as in Fig. 6-75a, since both methods are applied to the forward loop of the amplifier.

An alternative method of tuning the amplifier is shown in Fig. 6-75c. A series LC coil is used in the feedback path of the amplifier. For large values of gain A, the transfer function of the amplifier with a network $H(s)$ in the feedback is

$$\frac{E_o}{E_{\text{in}}} = \frac{A}{1 - AH} \approx \frac{1}{-H} \qquad (6\text{-}77)$$

The gain of the closed-loop amplifier is large only when H is small. In particular, at a frequency when $H = 0$, the overall gain is A. Outside this frequency band the gain drops off.

These networks suffer from temperature sensitivity of iron-core inductors. Especially for transistor amplifiers, interstage coupling transformers have an input magnetizing inductance comparable with the output impedance of the transistor. As the temperature varies, the magnetizing impedance varies and the gain of the stage may vary by a factor of 2. To avoid such problems created by the use of iron-core inductors, RC networks, such as bridged- and parallel-T networks, are used, as shown in Fig. 6-75d. Figure 6-53 shows the frequency response for one bridged-T network. Figures 6-76a and b show the amplitude and phase response for a bridged-T network for various notch depths. The network is driven from a low source impedance and into a large output impedance. A design procedure for bridged- and parallel-T networks is included in Appendix VI.

PROBLEMS

6-1. For each of the following forward loop transfer functions choose the gain necessary to obtain $\zeta = 0.2$ and $\zeta = 0.4$. Assume $H = 1$.

(a) $\dfrac{K}{s(0.005s + 1)(0.016s + 1)}$

(b) $\dfrac{K(s + 10)}{s(0.005s + 1)(0.016s + 1)}$

(c) $\dfrac{K}{s(0.005s + 1)^2(0.016s + 1)}$

6-2. Determine the transfer function G_e of a network that will stabilize the system of Fig. 6P-2. The input and output impedances on the network are respectively zero and infinite.

$$Ans.\ G_e = \frac{(s + \alpha)(s + \beta)}{(s + \gamma)(s + \delta)}$$

Fig. 6P-2

6-3. It is desired to support a body electrostatically. The block diagram of a control system to do so is shown in Fig. 6P-3.

$\dfrac{1/M}{s^2 - K_2/M}$ = transfer function of body supported with an electrostatic field

F_s = force supplied by control system

K_b = transfer function of position pickoff

$AG(s)$ = amplifier and equalizer transfer function

K_e = transfer function of force field

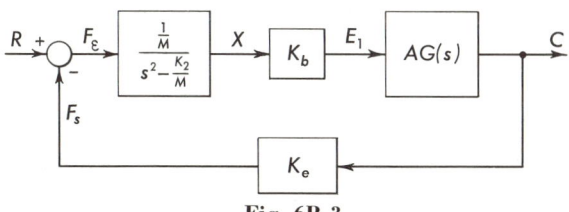

Fig. 6P-3

The values are

$$K_b = 1.1 \times 10^2 \text{ volts/in.}$$
$$K_e = 2.5 \times 10^{-5} \text{ lb/volt}$$
$$M = 10^{-3} \text{ lb-sec}^2/\text{ft}$$
$$\sqrt{\frac{K_2}{M}} = 315 \text{ rad/sec}$$
$$A = \text{amplifier gain}$$

278 Control System Design

(a) Take $G(s) = \alpha s + 1$ and choose α and A for the optimum system. Compute the steady-state error to a step function input.

(b) Take $G(s) = \alpha s + 1 + \gamma/s$ and choose α, γ, and A for an optimum system.

6-4. The system of Fig. 6P-4 represents the basic form of strip-chart recorder. An unknown d-c voltage $e_x(t)$ is to be dynamically recorded. A reference voltage E_R is accurately maintained across the slide-wire potentiometer. In most commercial systems E_R is automatically compared with the voltage of a *standard cell* and the same control system that is shown in Fig. 6P-4 serves as the self-correcting feedback system.

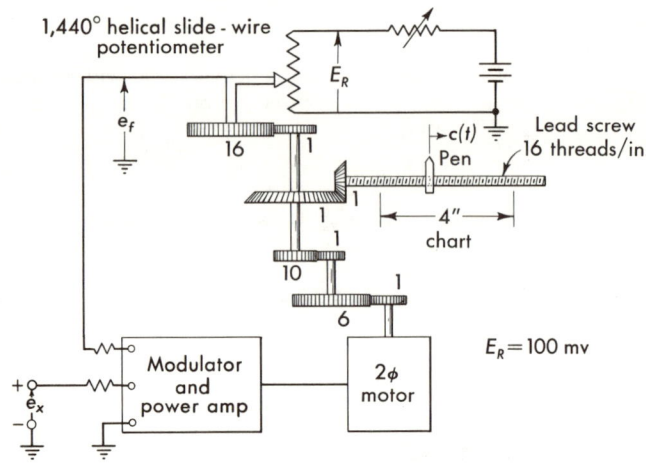

Fig. 6P-4

The operation entails a comparison of $e_x(t)$ and $e_f(t)$. The difference voltage $e_x(t) - e_f(t)$ is converted into an a-c error voltage through the use of a chopper modulator. The a-c error voltage drives a power amplifier which excites the control phase of a two-phase control motor. The motor jointly drives the lead screw mechanism and the helical potentiometer in the proper direction to reduce the error signal.

Motor Data

Stall torque............. 3.84 in.-oz with 74 volts applied to the control winding
No-load speed............ 3,600 rpm
Rotor inertia............ 0.96×10^{-3} oz-in.-sec^2

The gear train moment of inertia referred to the motor shaft is found to be 1.63×10^{-3} oz-in.-sec^2. Neglect friction and backlash in this study. The transfer function of the modulator-amplifier combination is found to be A. Let $c(t)$ be the linear displacement, in inches, of the pen on the strip chart.

(a) Formulate a block diagram for the system using $e_x(t)$ as the input and $c(t)$ as the controlled output.

(b) What is the necessary value of gain A for a damping ratio of 0.7, and what is the corresponding undamped angular frequency? *Ans.* $A = 2,400$

Note: This information can be obtained directly from the equations.

(c) With the gain set as in (b), what is the steady-state error, in inches, for a step input of 100 mv?

(d) If you were to use this recorder, to what value would you set the damping and why?

6-5. The system of Fig. 6P-5 is a speed regulator in which the speed depends upon a potentiometer setting e_r. Tachometric feedback is employed to compare a voltage e_b, which is proportional to the angular velocity of the output shaft, with the reference voltage e_r. The error voltage $e_r - e_b$ is amplified and used to excite the

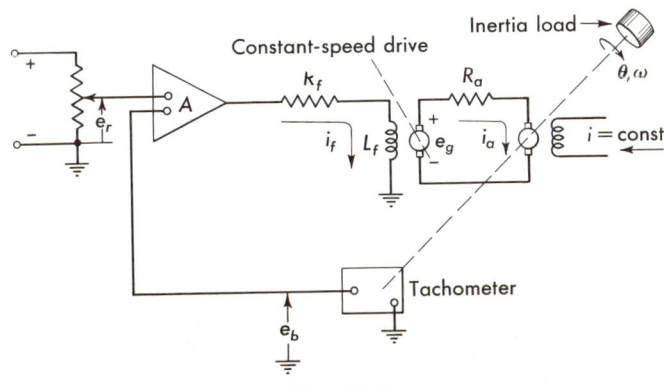

Fig. 6P-5

generator field. The generator then drives the motor and load. Assume no back emf induced in the field of the alternator and assume zero output impedance for the amplifier. The back emf of the motor is $K\omega$.

$$e_g = K_f i_f = R_a i_a + K\omega \qquad K_b = 0.1 \text{ volt}/(\text{rad}/\text{sec})$$

Motor torque $L_m = K_a i_a \qquad K_a = 0.5 \text{ lb-ft}/\text{amp}$

$$e_b = K_b \omega \qquad J = 0.25 \text{ slug-ft}^2$$

$$K_f = 50 \text{ volts}/\text{amp} \qquad K = 0.3 \text{ volt}/(\text{rad}/\text{sec})$$

$$A = 50 \qquad R_f = 25 \text{ ohms}$$

$$L_f = 1.2 \text{ henrys} \qquad R_a = 1.5 \text{ ohms}$$

(a) Form the block diagram for the system. Express individual transfer functions in terms of the system parameters, that is, K_a, K_b, J, A, etc.

(b) Write the closed-loop transfer function $\bar{\omega}/E_r$.

(c) Determine the stability of the system.

6-6. A feedback control system is described by the transfer function

$$G(s) = \frac{K}{s^2(s+1)}$$

(a) For what values of K is the system stable? *Ans.* Unstable for all K

(b) Insert a series equalizer $G_e(s) = (1/\alpha)[(\alpha\tau s + 1)/(\tau s + 1)]$ and choose the best values of α and τ.

(c) Is there any other equalizer that may be superior?

6-7. For the multiple-loop control system shown in Fig. 6P-7, $G_1 = 8/s^3$ and $G_2 = s^n$. For what values of n is the system stable? *Ans.* No value of n

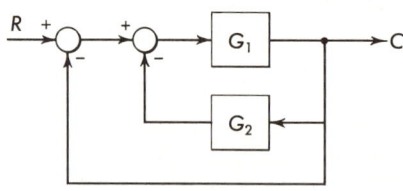

Fig. 6P-7

6-8. It is desired to select a lead network of the form

$$G_N = \frac{1 + 4\tau s}{1 + \tau s}$$

and to insert this in the control system of Fig. 6P-8. Obtain a plot of the damping ratio ζ and the undamped natural resonant frequency ω_n as a function of τ for $0 \leq \tau \leq 0.75$. What is the optimum setting for τ?

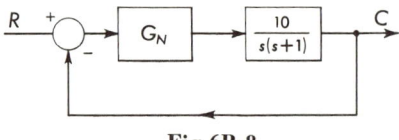

Fig 6P-8

6-9. Find lattice networks for the following transfer voltage ratios:

(a) $\dfrac{s^2}{(s + 6)(s + 50)}$

(b) $\dfrac{s(s + 1)}{s^2 + s + 10}$

(c) $\dfrac{s^2 + 9}{(s + 12)(s + 20)}$

(d) $\dfrac{s^2 + s + 10}{(s + 10)^2}$

6-10. Find grounded networks for the transfer voltage ratios of Prob. 6-9.

6-11. Synthesize the negative of the transfer voltage ratios of Prob. 6-9 with active networks (as in Sec. 6-12).

6-12. Determine the transfer function for the electromechanical equalizer shown in Fig. 6P-12. Would this system have any practical use?

6-13. Design a resistor-shunt and a capacitor-shunt bridged-T network with a resonant frequency of 400 cps, a notch depth of 0.2, and a d-c impedance level of 50 kilohms.

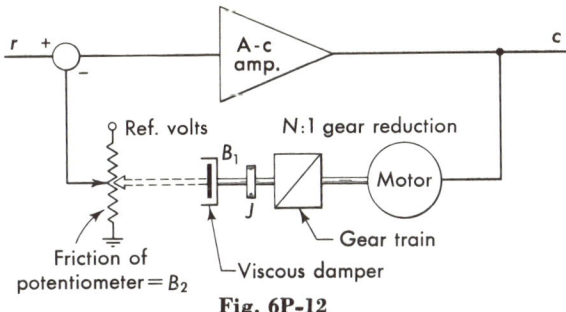

Fig. 6P-12

6-14. Derive the equation for the phase shifter of Fig. 6-73 and show how this network is useful for a-c control systems.

6-15. Determine the amplitude and phase response for the LC tuned amplifier of Fig. 6-75c. Assume that the high-gain amplifier has infinite input and zero output impedance. Take

$$L = 10 \text{ henrys} \quad C = 0.015 \text{ }\mu\text{f} \quad R = 10 \text{ kilohms}$$

6-16. A light-seeker control system consists of two photocells mounted on a bracket attached to a shaft which in turn is driven by a motor. The system is shown schematically in Fig. 6P-16. Equalize this system by means of passive-element networks. The specification requires an overshoot of 5 per cent and minimum response time (i.e., minimize the rise time).

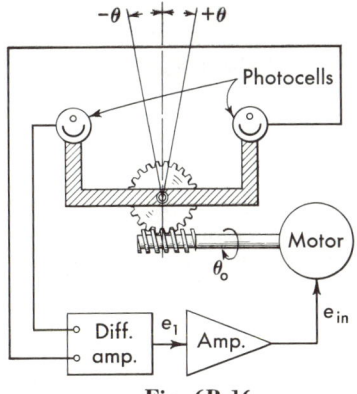

Fig. 6P-16

The output of the differential amplifier is a voltage which is proportional to θ and is positive for positive θ and negative for negative θ. The transfer function of the motor is

$$\frac{\bar{\theta}_o}{E_{\text{in}}} = \frac{100}{s(s + 10)}$$

Let $E_1/\theta = 10$ volts/deg and θ_o turn through 20° to cause θ to turn through 1°. The system is to use suppressed-carrier alternating current.

6-17. In a certain industrial application it is desired to keep the flow of liquid from a tank constant and independent of the amount of liquid in the tank. A feedback control system which performs this operation consists of a motor-driven gate valve, an amplifier, and a flow meter which produces a voltage output proportional to the quantity of liquid flowing [60 volts/(ft^3/sec)]. A schematic diagram of the

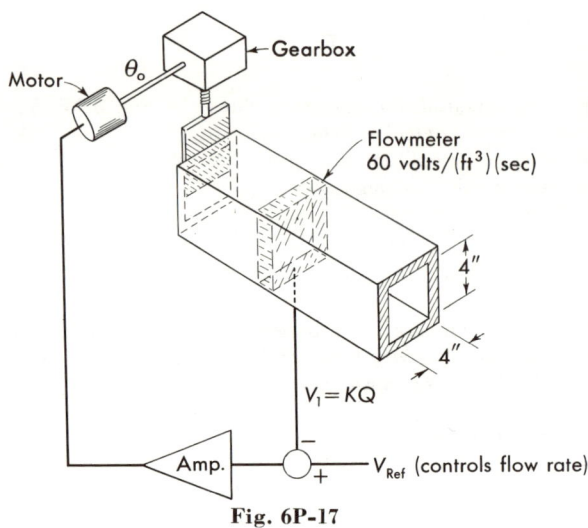

Fig. 6P-17

system is shown in Fig. 6P-17. For simplicity, the pipe is rectangular. Assume the flow rate is proportional to the area of the pipe at the gate. The gate valve travels 1 in./100 rotations of θ_o. The maximum flow rate with the valve completely open is 0.1 ft^3/sec. The transfer function of the motor is $150/s(s+5)$. What value of gain would give a ζ of 0.3 for this system? How would the system be built if an a-c motor and amplifier were to be used? *Ans. A* = 20

6-18. The system of Fig. 6P-18 utilizes rate feedback for stability. Set the gain K so that the overshoot produced by a step function input does not exceed 1.20 times the steady-state value. *Ans. K* = 110

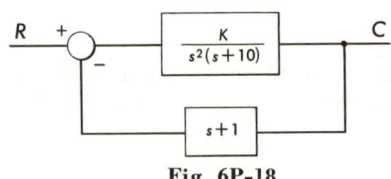

Fig. 6P-18

6-19. Design a position control system to drive a large antenna. The system, shown in Fig. 6P-19, uses position plus rate feedback.
(a) Set the scale factor to be $10°$/volt
(b) Select R_1, R_2, and R_3 in the system for a damping ratio of $\zeta = 0.4$
(c) Calculate the error due to a load torque of 1 ft-lb

Component values:
$$J_{\text{load}} = 1.5 \text{ lb-in.-sec}^2$$
$$L_0 = 40 \text{ in.-lb}$$
$$\omega_0 = 4,000 \text{ rpm}$$
$$J_m = 0.002 \text{ lb-in.-sec}^2$$
$$\text{Rated control voltage} = 115 \text{ volts}$$
$$A_1 = -10,000$$
$$A_2 = 10$$
$$N_1 = 800 = N_3$$
$$N_2 = 40$$
$$\text{Tachometer constant} = 30 \text{ volts}/1,000 \text{ rpm}$$
$$\text{Potentiometer constant} = 10 \text{ volts/rev}$$

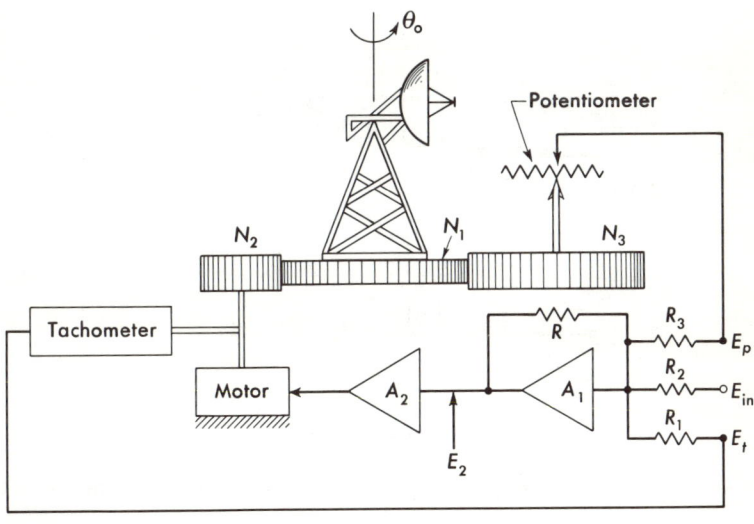

Note: $\dfrac{E_2}{R} = -\dfrac{E_p}{R_3} - \dfrac{E_t}{R_1} - \dfrac{E_{in}}{R_2}$

Fig. 6P-19

284   Control System Design

6-20. Design a centrifuge which drives an inertia J at a precise angular velocity. The system is as shown in Fig. 6P-20. Set the scale factor to be 1,000 rpm/volt. Discuss the stability and steady-state errors and suggest an equalizer.

Component values:

$$J_{\text{load}} = 1.5 \text{ lb-in.-sec}^2$$

$$L_0 = 40 \text{ in.-lb}$$

$$\omega_0 = 4{,}000 \text{ rpm}$$

$$J_m = 0.002 \text{ lb-in.-sec}^2$$

Rated control voltage = 115 volts d-c

$$N_1 = 800$$

$$N_2 = 40$$

$$N_3 = 20$$

Tachometer constant = 30 volts/1,000 rpm

Ans. $R_1/R_2 = 120$

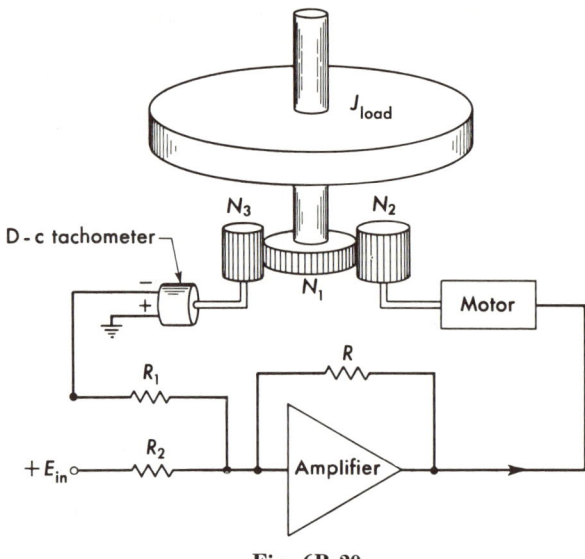

Fig. 6P-20

6-21. Use a lag network $\dfrac{1}{\alpha} \dfrac{s + 1/\tau_n}{s + 1/\alpha\tau_n} = G_n$ to equalize the third-order system

$$G = \frac{K}{s(\tau s + 1)[(\tau/10)s + 1]}$$

Select $\alpha = 10$ and determine the optimum value of τ_n.

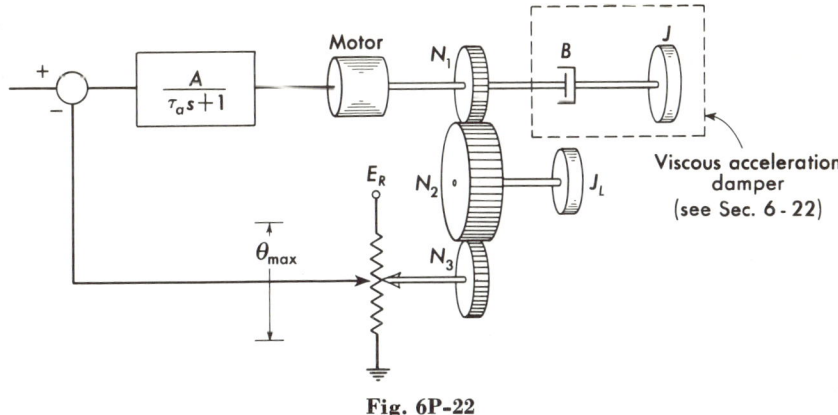

Fig. 6P-22

6-22. Figure 6P-22 is the schematic diagram for a position servo with acceleration damping. The transfer function of the unloaded motor, i.e., when driving only its own inertia J_m, is

$$\frac{K_m}{s(\tau_m s + 1)}$$

where

$J_m = 1.2 \times 10^{-5}$ oz-in.-sec

$K_m = 10$ (volt-sec)$^{-1}$

$\tau_m = 0.02$ sec

$\tau_a = 0.002$ sec

and the remaining constants are $N_1 = 20$, $N_2 = 200 = N_3$ (N_i = number of teeth on gear i), $E_R = 10$ volts, $\theta_{max} = 3{,}600°$, $J_L = 0.12$ oz-in.-sec. Assume zero moment of inertia in the gear train.

(a) With no damper, select A for $\zeta = 0.2$ and determine K_v and ω_n.

(b) With the acceleration damper attached, determine the values of B and J which will yield a maximum K_v for a fixed $\zeta = 0.2$.

chapter 7

SERVOMECHANISM COMPONENTS

7-1 Introduction. A transducer is an instrument that converts an input signal such as mechanical shaft rotation, pressure, or angular velocity into an output signal of another form, usually electrical. Although the approximate mathematical transfer functions and block diagrams are easily derived, the manufacture of these components is often difficult because of accuracy and reliability requirements.

Characteristics of a transducing element can be summarized as follows:

1. Accuracy. Since the transducer is often placed outside the closed loop or in the feedback loop, its accuracy must be the same as that of the overall closed-loop system.
2. Low power. Transducers are required to handle only low amounts of power (maximum of 2 watts).
3. High quality. The construction is usually of the highest possible quality. Low noise, freedom from harmonics and quadrature, high linearity, and low friction are just a few of the requirements. Harmonics and quadrature which are generated by the transducer must be within limits which prevent saturation of the follow-up amplifiers.
4. Reliability. Because a wide variety of quantities (acceleration, pressure, angle of attack, Mach number, etc.) must be transduced into an electrical signal, many "gadgets" have been invented. The importance of reliability in control equipment has caused considerable emphasis to be placed on transducer reliability.

This chapter considers some of the important servo components and transducers, such as gyros, motors, and amplifiers, used in the design of feedback control systems. For convenience, various types of transducers are grouped according to the quantity measured, e.g., position, velocity, and acceleration.

7-2 Potentiometers.* The system engineer has a wide class of instruments available for the measurement of position. The most important

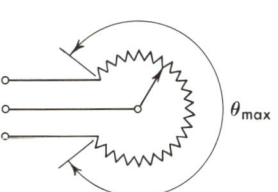

Fig. 7-1 Single-turn potentiometer.

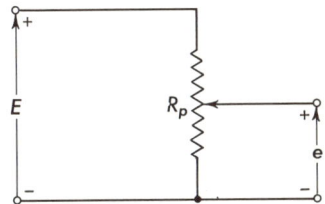

Fig. 7-2 Schematic diagram of a potentiometer.

types that are considered in this chapter are conductive instruments (potentiometers) and inductive instruments (synchros, resolvers, E pickoffs).

A potentiometer, comprising a slider which moves along a resistance element, is shown in Fig. 7-1. A more useful schematic, with excitation E and output voltage e included, is shown in Fig. 7-2. The potentiometer transfer function is determined from the voltage gradient (volts per radian or volts per inch) along the potentiometer. The potentiometer of Fig. 7-2 has a voltage E applied across a total angle of θ_{max}, in degrees. The transfer function is

$$\frac{e}{\theta} = \frac{E}{\theta_{max}} \qquad (7\text{-}1)$$

Fig. 7-3 A single-turn precision potentiometer. (*Courtesy of Electro-Mec Instrument Corp., Long Island City.*)

The most common potentiometer is one utilizing a linear resistance element; it finds wide use in computers and for position comparison in feedback control systems. Single-turn potentiometers generally have a usable rotation of less than 360°. The remaining angle is available for conducting overtravels (10 to 15° arcs at each end of the resistance element). A

---

* Much of the material of this section has been taken from Ref. 31.

typical unit is shown in Fig. 7-3 and in the outline drawing of Fig. 7-4. The accuracy is proportional to the diameter of the potentiometer and to the mechanical precision.

Since the potentiometer slider does not move along a continuous wire, the transfer function is not a continuous curve, but a succession of steps. As the slider moves from turn to turn, the voltage increases in a stepwise fashion. The stepwise output voltage ΔE is the total voltage E divided by the number of turns n of resistance wire. For a wiper which touches

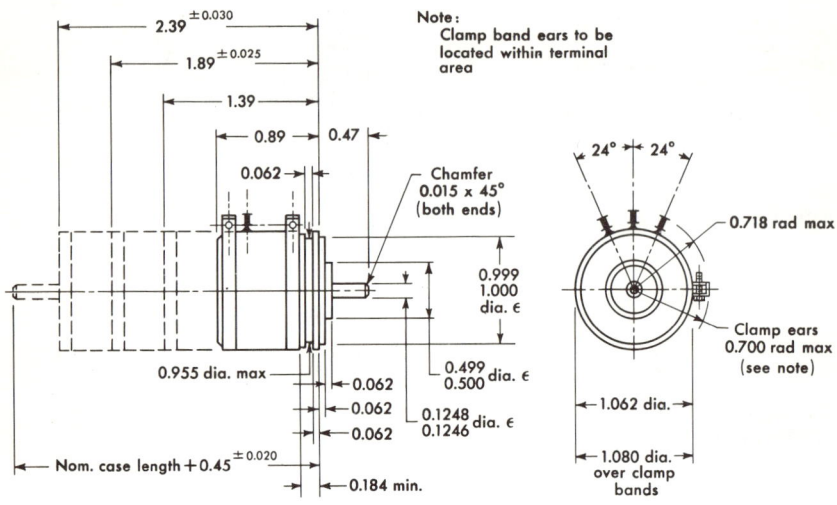

Fig. 7-4 Outline dimensions of a single-turn precision potentiometer. (*Courtesy of Electro-Mec Instrument Corp., Long Island City.*)

only one wire at a time, the resolution, or accuracy of the shaft position readout, is given by

$$\text{Resolution} = \frac{\Delta E}{E} = \frac{E/n}{E} = \frac{1}{n} = \frac{1}{\text{no. of turns on potentiometer}} \quad (7\text{-}2)$$

Since the potentiometer accuracy is limited by the size of the voltage steps, and since a system may tend to oscillate between two adjacent wires, multirevolution or multiturn potentiometers have been developed. This type of potentiometer has a helical resistance element. As the slider rotates, it travels along the helix. The greater length of wire results in improved resolution. A typical unit is shown in the cutaway view of Fig. 7-5.

Multiturn potentiometers can be more accurate and have finer resolution, but they suffer from large size and additional friction. Resolution of

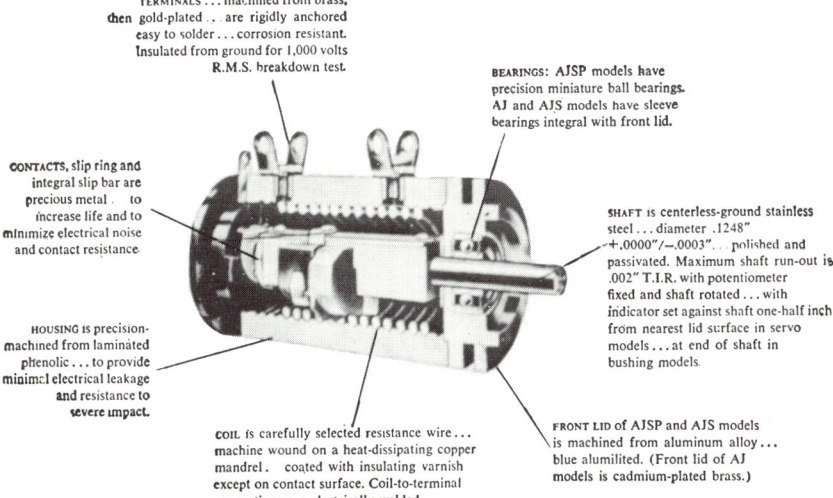

Fig. 7-5 Cutaway view of a typical multiturn precision potentiometer. (*Courtesy of Helipot Division, Beckman Instruments, Fullerton, Calif.*)

the "pot" indicates the accuracy to which any shaft setting can be made. The absolute, theoretical accuracy of a potentiometer is one-half the resolution. This is shown by the curve of Fig. 7-6.

The transfer function of Eq. (7-1) represents a straight-line input-output relationship. The deviation of the potentiometer from this straight line is a measure of the linearity accuracy. Two definitions of linearity are in common use.

Independent linearity (used in connection with precision potentiometers) is the maximum allowable deviation from the best straight line that can be drawn through a plot of the actual points of voltage on the voltage vs. rotation curve. Tolerance is expressed as a percentage of the maximum voltage output. This is shown in Fig. 7-7a, where the straight line has been oriented to fit the actual output curve best.

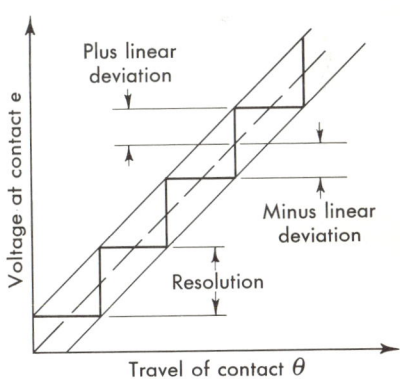

Fig. 7-6 Potentiometer resolution.

Zero-based linearity (used in connection with a rheostat or variable resistance) is the maximum allowable deviation from the best straight line that can be drawn through the voltage points and also pass through zero voltage at zero shaft displacement. This is shown in Fig. 7-7b.

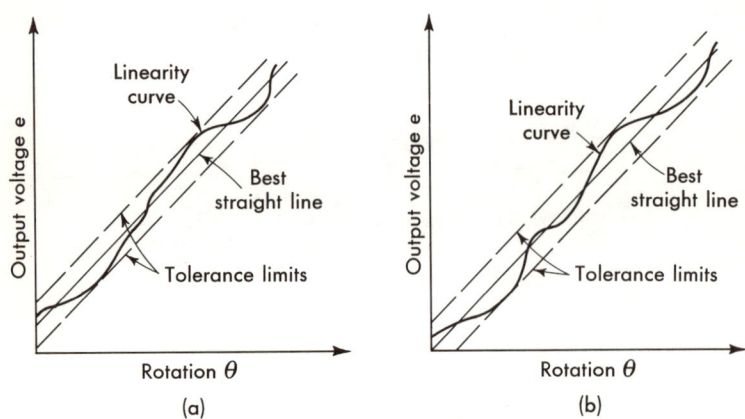

Fig. 7-7 Independent and zero-based potentiometer linearity.

Use of a potentiometer requires that circuit resistances be considered. With a constant voltage impressed across the coil of a linear potentiometer, the voltage may not follow the linear transfer function of Eq. (7-1). As current is drawn through the slider, a so-called "loading error" results. The loading error varies with slider position. The output voltage is given by the expression

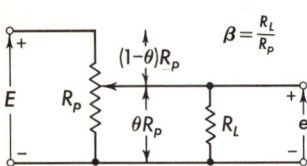

Fig. 7-8 Equivalent circuit for a loaded potentiometer.

$$\frac{e}{E} = \frac{\theta}{1 + (\theta/\beta)(1 - \theta)} \qquad (7\text{-}3)$$

where β is the ratio of load resistance R_L to potentiometer resistance R_p. θ is the setting of the potentiometer and is the fraction of the total resistance. An equivalent circuit for a potentiometer working into a load resistance is shown in Fig. 7-8.

The deviation or error from the straight-line curve is given by

$$\frac{\epsilon}{E} = \frac{(\theta^2/\beta)(1 - \theta)}{1 + (\theta/\beta)(1 - \theta)} \qquad (7\text{-}4)$$

The loading error, plotted in Fig. 7-9, varies with slider position. The error is zero at both ends of the coil and has a maximum value at approximately two-thirds rotation. The error varies with load resistance. A load resistance 10 times that of the potentiometer resistance produces a maximum error of 1.5 per cent of the applied voltage. An error of 0.15

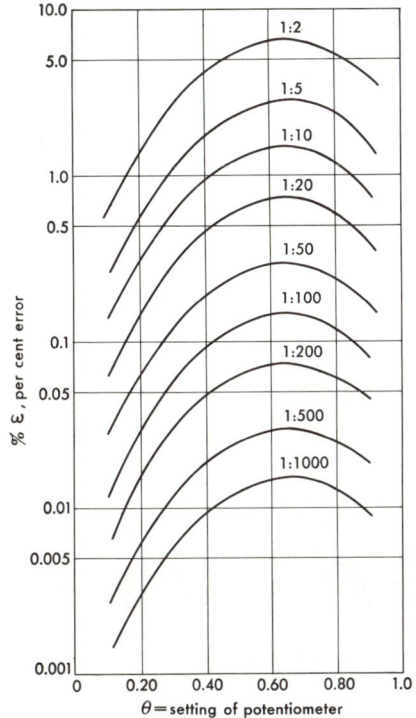

Fig. 7-9 Potentiometer loading error.

per cent results when the load resistance is 100 times the potentiometer resistance. The effect of loading can be reduced by several means:

1. Wind on the potentiometer a slight nonlinear function which compensates for the loading.
2. Locate a tap at the $\frac{2}{3}$ point on the potentiometer. An appropriate resistor connected between the tap and the 100 per cent point compensates for the loading.

The loading error of potentiometers, as shown in Fig. 7-9, is normally considered detrimental. When a potentiometer is loaded appropriately, many nonlinear functions can be obtained with the linear, nontapped

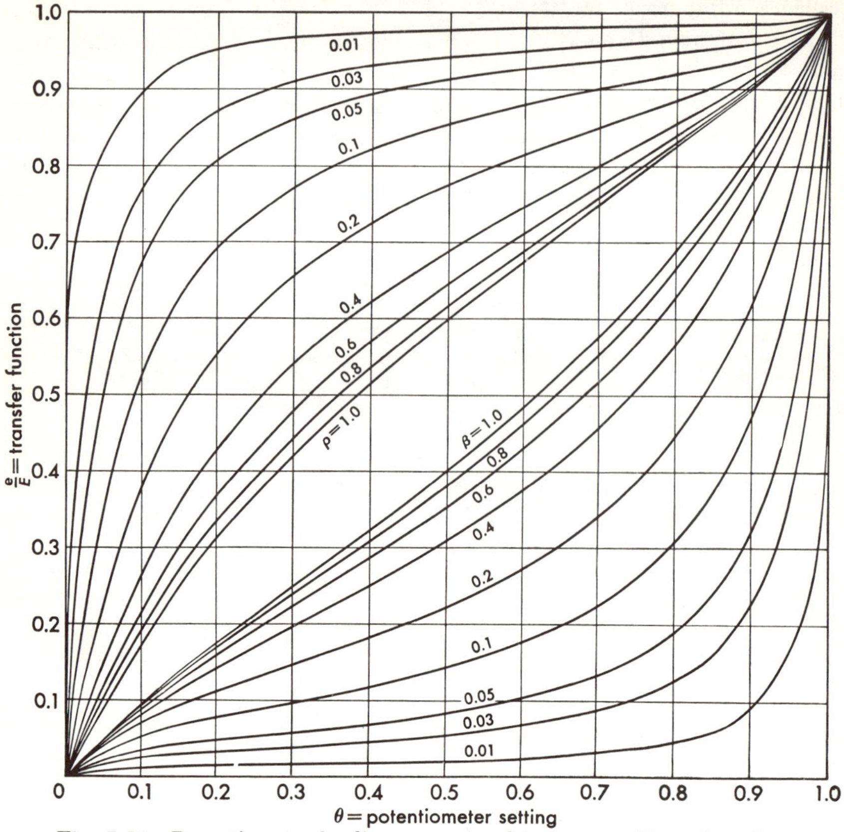

292 Control System Design

Fig. 7-10 Potentiometer loading curves used to form nonlinear functions.

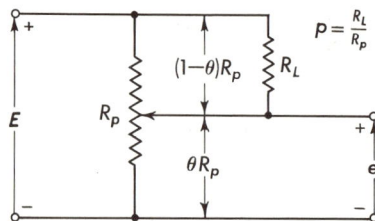

Fig. 7-11 Equivalent circuit for a loaded potentiometer.

potentiometer. In this application, the loading curves become useful. Equation (7-3) produces the nonlinear curves of Fig. 7-10 for β in the range 0.01 to 1.0. When the potentiometer is loaded on the top, as shown in Fig. 7-11, the potentiometer transfer function is given by

$$\frac{e}{E} = \frac{1 - \theta + \rho}{1 - \theta + \rho/\theta} \tag{7-5}$$

Equation (7-5) plots into the curves of Fig. 7-10 when ρ is in the range 0.01 to 1.0.

Other types of nonlinear functions can be obtained by either of the following methods:

1. A standard linear potentiometer with voltage taps and resistance loading
2. A nonlinear winding which incorporates a variation of resistance as θ varies

In general, the same characteristics are inherent in the precision nonlinear potentiometer as in precision linear units (i.e., resolution, loading error, etc.). Deviation of the output voltage from the prescribed function is termed "conformity" rather than linearity.

The potentiometer offers many advantages in terms of versatility, ease of forming functions, and ability to use either a-c or d-c excitation. There are two problems in the use of potentiometers: electrical noise generation and reliability.

The physical form of a potentiometer varies with the manufacturer. Figure 7-3 shows a typical single-turn precision potentiometer, and Fig. 7-5 is a cutaway view of a multiturn unit. The latter unit is of $\frac{7}{8}$ in. diameter and 1.587 in. body length and is available with a 0.25 per cent linearity tolerance. The construction details of a small ($\frac{7}{8}$ in. diameter by $\frac{1}{2}$ in. length), single-turn potentiometer are shown in Fig. 7-12.

Fig. 7-12 A $\frac{7}{8}$-in.-diameter subminiature precision potentiometer. (*Courtesy of DeJur-Ansco Corp., Long Island City.*)

In computer applications it is necessary to "gang," or assemble several potentiometers on a single shaft. The unit shown in Fig. 7-13 permits the ganging of up to 20 units on a single shaft. The individual units are held together with circular clamps. When circuit elements require change, the potentiometers are easily removed and replaced.

The problem of potentiometer resolution (Fig. 7-6) has been approached in several ways. Smooth and stepless operation, low noise, long life, and low starting torque are some of the characteristics of a slide-wire potentiometer. These potentiometers may have a resistance of 2,500 ohms in a multiturn unit. Another approach to the problem of resolution is through the use of a deposited-film potentiometer. A resistance material is evaporated and deposited on a ceramic disk. A slider passes over this deposited material to form the potentiometer. Although these units do not have "infinite" resolution, they do exhibit a resolution which is 20 to 70 times better than that of a comparable wire-wound unit.

When translational motion is to be measured, a linear-motion potentiometer can be used instead of an angular potentiometer with a rack and pinion. Good resolution is possible, and strokes of 1 to 15 in. are available.

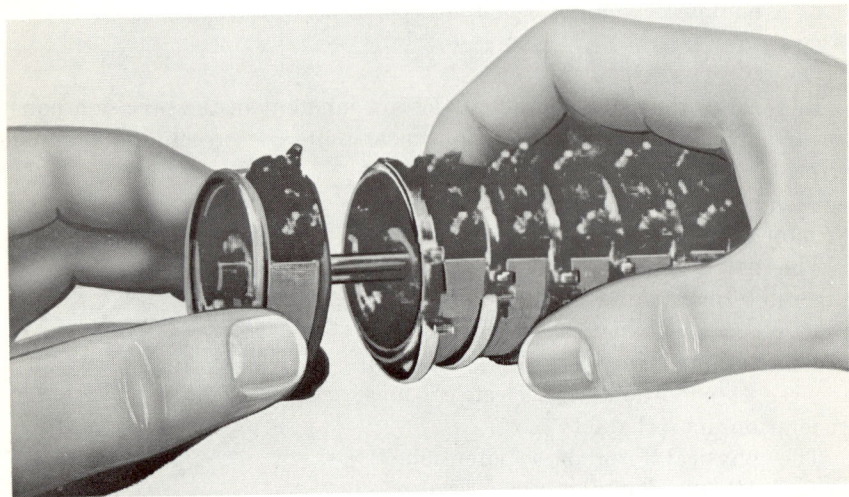

Fig. 7-13 Ganging several potentiometers on a single shaft. (*Courtesy of Fairchild Controls Corp., Hicksville, N.Y.*)

A large variety of taps, resistance, shaft extensions, stops, and tapped continuous-rotation units gives a wide latitude to potentiometer applications. The factors to be considered in potentiometer selection can be summarized as follows:

Electrical suitability
1. Linearity as required by the application
2. Tolerance on total resistance and shaft angle
3. Resolution error and its effect on system stability and wear
4. Sufficient electrical insulation strength
5. Freedom from stray capacitive and inductive effects
6. Generation of undesirable radio interference

Mechanical suitability
1. Adequate life to meet anticipated performance cycle
2. Adaptability to environmental conditions: temperature, humidity

3. Ability to withstand vibration without excessive contact bounce
4. Precision of mounting surface and shaft extension
5. Provision for easy zeroing and calibration
6. Acceptable starting friction

7-3 Induction-type Position Indicators.* Induction-type transducers are available for position measurement. Although these units are limited to a-c operation and are more costly, they exhibit stepless output, are small in size, and have a high degree of reliability.

The schematic of an induction component which is essentially a transformer with a rotatable secondary is shown in Fig. 7-14. The rotor winding is excited with an a-c voltage, and the output voltage, which is induced in the stator winding, depends on the shaft angle. The output voltage is proportional to the fraction of the flux cut by the secondary coil. When the primary and secondary coils are lined up with the flux, the entire primary voltage is transformed. As the secondary is rotated through an angle θ, the output voltage depends upon this angle:

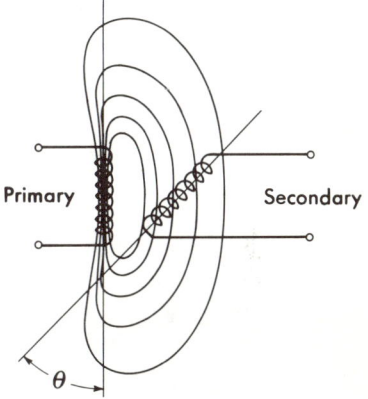

Fig. 7-14 Principle of operation of induction components.

$$E_{\text{out}} = K \cos \theta \qquad (7\text{-}6)$$

The advantages of induction units can be listed as follows:

1. High reliability (no contact moving along a sawlike resistance element, only sliding contacts: slip rings).
2. Long life (less wear).
3. Smooth, continuous output voltage.
4. Low noise generated in the system.
5. No radio interference generated.
6. Isolation of input and output circuits because of transformer action.

---

* The author is indebted to Mr. Sidney Davis for much of the material in these sections. The reader is referred to his excellent paper "Rotating Components for Automatic Control" (Ref. 10).

296 Control System Design

The disadvantages of these units are summarized:

1. Applicable to a-c signals only.
2. Difficult to obtain good linearity.
3. Phase shift varies with temperature.
4. Less versatile for forming functions.
5. Generate harmonics and quadrature.

Physically, the induction components have the appearance of small motors. Typical units are shown in Fig. 7-15, and a cutaway view is included in Fig. 7-16. The unit contains a wound rotor mounted on

Fig. 7-15 Photograph of a typical induction-type transducer. (*Courtesy of Ford Instrument Co., Long Island City.*)

bearings. The connections to the rotor are carried through slip rings and are brought out to connectors mounted at one end of the instrument.

7-4 Resolvers. Figure 7-17 shows a cutaway view of a typical resolver. The input winding is the stator, which is fixed to the housing of the instrument. The lead-in wires for the rotor, which is the output winding, are connected through slip rings, since the rotor revolves on precision bearings.

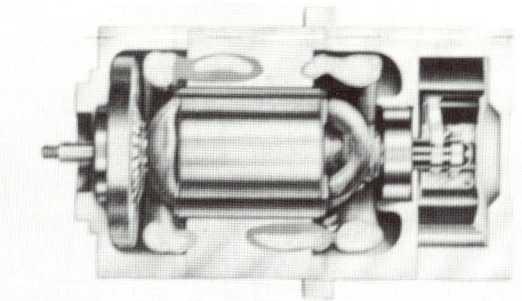

Fig. 7-16 Cutaway view of an induction-type transducer. *(Courtesy of Ford Instrument Co., Long Island City.)*

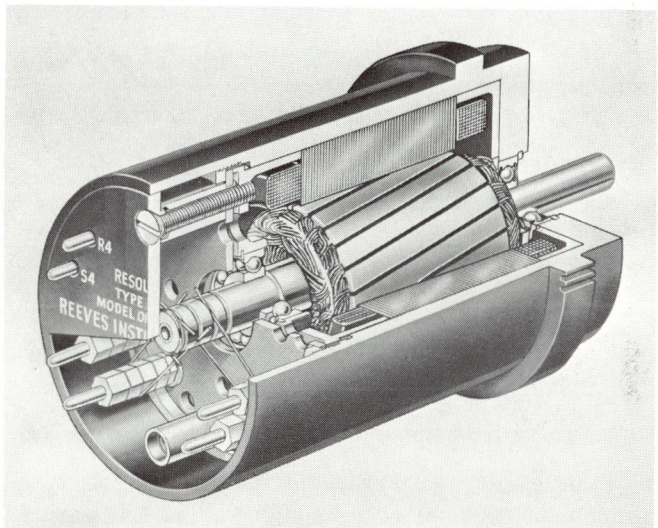

Fig. 7-17 Resolver cutaway. *(Courtesy of Reeves Instrument Corp., Garden City, L.I., N.Y.)*

The schematic for a simple two-winding resolver is shown in Fig. 7-18. The relationship between input voltage E_{s42}, output voltage E_{r13}, and shaft angle θ is

$$E_{r13} = K_1 E_{s42} \cos \theta \tag{7-7}$$

where K_1 is the coupling factor of the winding, or the transformation ratio of the resolver.

A slightly more complex unit, a three-winding resolver, is shown in Fig. 7-19. Here the two rotor windings are spaced 90 mechanical degrees apart to provide both sine and cosine functions of the input voltage according to the defining equations

$$E_{r13} = K_1 E_{s42} \cos \theta$$
$$E_{r24} = K_3 E_{s42} \sin \theta \tag{7-8}$$

This type may also be built with two perpendicular stator windings and a single rotor winding if required for some specific application.

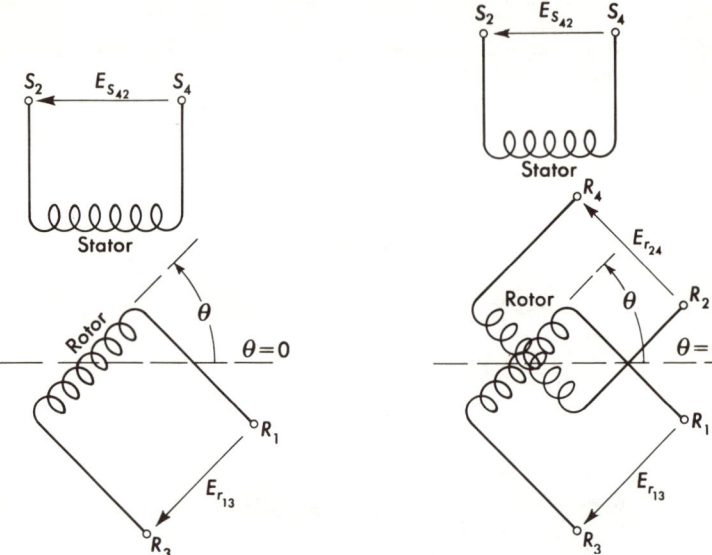

Fig. 7-18   Two-winding resolver.   Fig. 7-19   Three-winding resolver.

The most commonly used resolver is the four-winding type which is utilized for complex trigonometric functions. The schematic for a four-winding resolver which comprises two stator windings and two rotor windings with conventional terminal designations and voltages is shown in Fig. 7-20. The equations are

$$E_{r13} = K_1 E_{s42} \cos \theta + K_2 E_{s31} \sin \theta$$
$$E_{r24} = K_3 E_{s42} \sin \theta - K_4 E_{s31} \cos \theta \tag{7-9}$$

This equation is valid for counterclockwise shaft rotation as viewed from the shaft-extension end; that is, θ is defined as positive for counterclockwise shaft rotation.

Precision resolvers find wide use in many types of control systems, computers, and automatic control systems. Representative examples are discussed in this section.

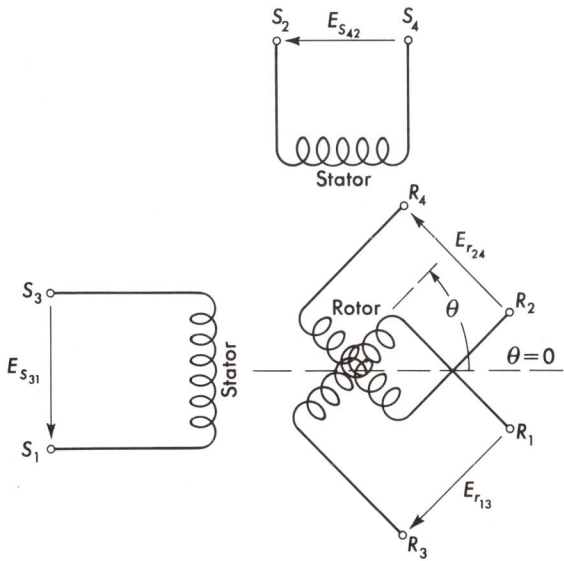

Fig. 7-20 Four-winding resolver with polarities specified.

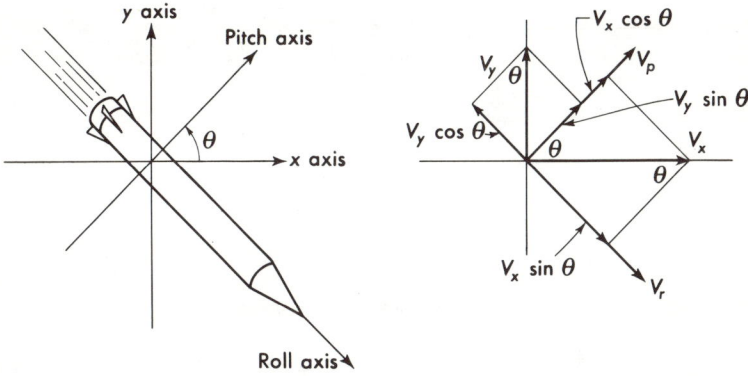

Fig. 7-21 Transformation of coordinates.

Transformation of Coordinates. Many aircraft and space vehicles require information generated in one coordinate system to be transformed into another coordinate system. For example, components of velocity measured with an inertial navigator must be transformed into aircraft pitch and roll velocities. This transformation is accomplished, as shown in Fig. 7-21, with a precision resolver with the following transformation equations:

$$V_{\text{pitch}} = V_y \sin\theta + V_x \cos\theta$$
$$V_{\text{roll}} = V_y \cos\theta - V_x \sin\theta \qquad (7\text{-}10)$$

The resolver system shown in Fig. 7-22 permits conversion from cartesian (x, y) to polar (R, θ) coordinates. In this system, voltages corresponding

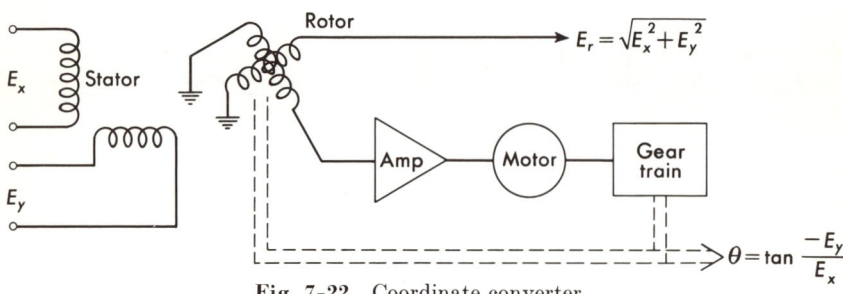

Fig. 7-22 Coordinate converter.

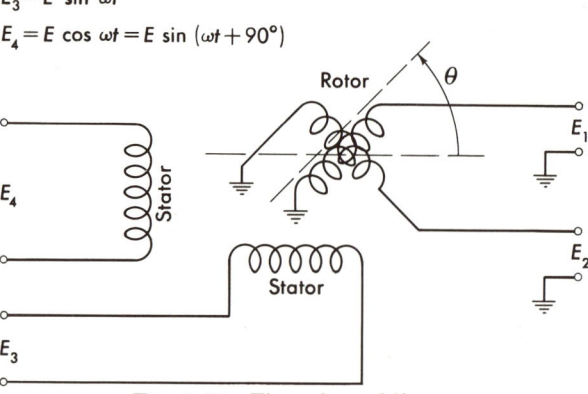

Fig. 7-23 Time-phase shifter.

to the x and y components (E_x and E_y) are applied to the two stator windings. One of the rotor windings is used as the error detector for a follow-up servo, which continually drives the resolver shaft to a null. The opposite resolver winding is kept aligned with the vector sum E_r of E_x and E_y; $E_r = \sqrt{E_x^2 + E_y^2}$. The angle θ through which the resolver shaft is rotated defines the other polar coordinate $\theta = \tan^{-1}(E_y/E_x)$.

Time-phase Shifter. A resolver can be connected in several ways to provide a variable amount of time-phase shift. Figure 7-23 provides variable time-phase shift when two voltages separated 90° in time phase are available. The resolver output is

$$E_1 = E_4 \sin \theta + E_3 \cos \theta$$
$$E_2 = E_4 \cos \theta - E_3 \sin \theta \tag{7-11}$$

with $E_3 = E \sin \omega t$ and $E_4 = E \cos \omega t = E \sin (\omega t + 90°)$.
These equations become

$$\frac{E_1}{E} = \cos \omega t \sin \theta + \sin \omega t \cos \theta = \sin (\omega t + \theta)$$
$$\frac{E_2}{E} = \cos \omega t \cos \theta - \sin \omega t \sin \theta = \cos (\omega t + \theta) \tag{7-12}$$

and hence E_1 and E_2 can be shifted in time phase simply by rotating the resolver. An interesting use for this phase shifter is found in the testing

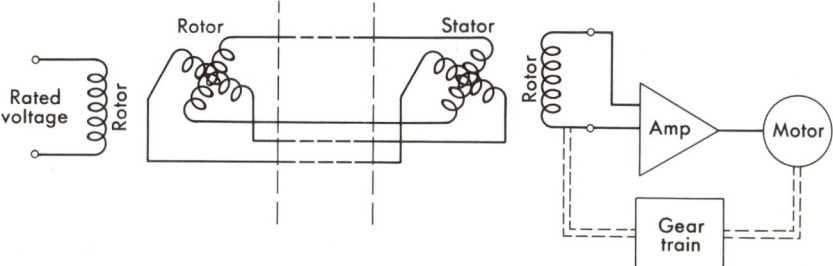

Fig. 7-24 Data transmission.

of feedback control systems for frequency response. The phase shift of a device under test can be measured directly by comparing on an oscilloscope the sinusoidal output from the system with the phase-shifted output from a resolver. For each reading the phase shifts of the two signals are made equal by adjusting the resolver shaft. The in-phase condition is noted on an oscilloscope, and the system phase shift can be read directly from a calibrated dial on the resolver shaft.

Data Transmission. For the transmission of angular position data, a four-wire resolver system, as shown in Fig. 7-24, provides the highest possible accuracy. Although it is possible to transmit data with only three wires (one lead common to both rotor windings), the four-wire system is considerably more accurate.

A voltage is applied to one stator winding of the resolver transmitter. The position of the rotor is transmitted as the sine and cosine of the angle, and these voltages are applied to the two stator windings of the resolver receiver. The rotor of the resolver is used as the input to a follow-up control system which maintains the second resolver shaft aligned with the shaft of the resolver transmitter.

7-5 Synchros. Depending upon the manufacturer, this unit may be known by the name selsyn, autosyn, or telesyn. A unit usually consists of a Y-connected stator winding and a single-phase rotor winding. Depending on its function, a unit can be used as either a torque transmitter or a position indicator.

Torque units are connected as shown in Fig. 7-25. The excitation voltage produces a magnetic field in the synchro transmitter which causes current to flow in the interconnecting wires and hence establishes a

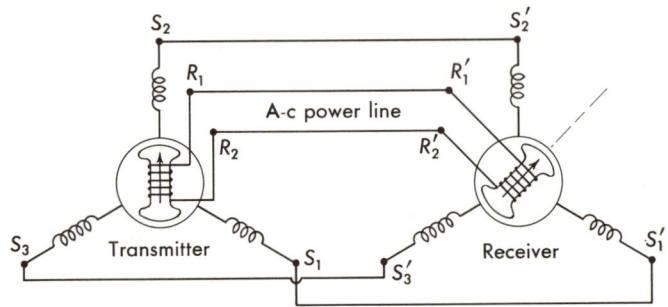

Fig. 7-25 Schematic of two synchros used to transmit torque.

magnetic field in the synchro receiver. This receiver field interacts with the field established by the rotor current and aligns the rotor and stator fields. Because of losses in the system, the power at the receiver is less than that at the input. The torque per angle of misalignment is called the synchro torque gradient. When the transmitter shaft is turned, the shaft of the receiver rotates through the same angle to an accuracy which depends largely upon the load but also upon electrical accuracy. Table 7-1 indicates the performance specifications for typical units.

A synchro, which is similar in appearance to other induction components, is shown in Fig. 7-26. The transmitters and receivers are similar with the exception of a damping device used on the receiver. An input inertia and a dissipation element permit the receiver to follow with a minimum of oscillation. Even with a damper on the receiver, large inertias should not be coupled to the shaft, since they tend to produce oscillations. Large coulomb-friction loads (potentiometers, etc.) are also objectionable, since this type of load causes a positional error.

Table 7-1 Performance Characteristics of Various Types of Synchros*
(All units weigh 4 oz)

Type	Voltage gradient, mv/deg	Torque gradient, mg-mm/deg	Error (maximum), min
Transmitter	204	3,750	10
Control transformer	403	.....	10
Repeater	...	2,050	30
Differential	196	.....	10
Resolver	439	.....	10

* Courtesy of Kearfott Division, General Precision Aerospace, Little Falls, N.J.

Another useful synchro for torque or signal transmission is the differential synchro. This unit is similar in objective to a mechanical differential, as is found in an automobile. The differential synchro has a Y-connected stator and rotor and is identified by the three slip rings. The unit is

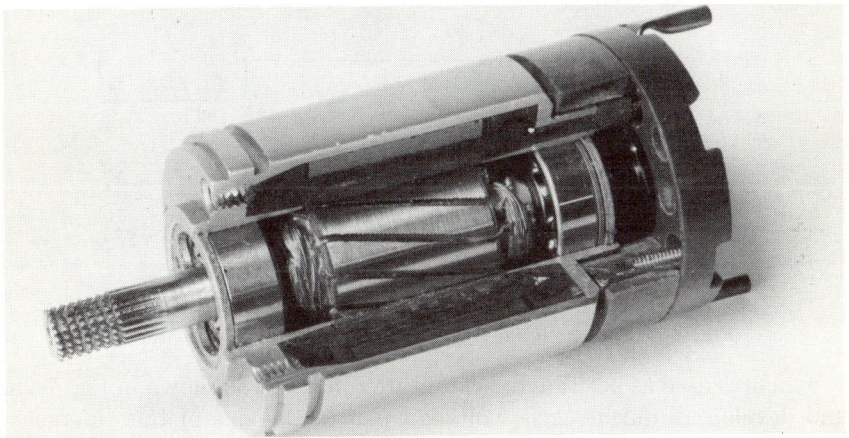

Fig. 7-26 Cutaway view of a military size 11 synchro. (*Courtesy of Kearfott Div., General Precision Aerospace, Little Falls, N.J.*)

connected between a synchro transmitter and receiver, as shown in Fig. 7-27. The synchro receiver angle θ_R is given by $\theta_R = \theta_G + \theta_{DG}$, where θ_G is the synchro generator angle and θ_{DG} is the differential generator angle.

Standard torque units have certain disadvantages that make them inadequate for most control design:

304 Control System Design

1. Poor accuracy when driving a friction load (e.g., potentiometer stack)
2. Poor damping when driving an inertia load
3. Less versatility than a control system in setting the damping (e.g., by electrical means)

Torque synchros are mainly used to present display information such as dial readings.

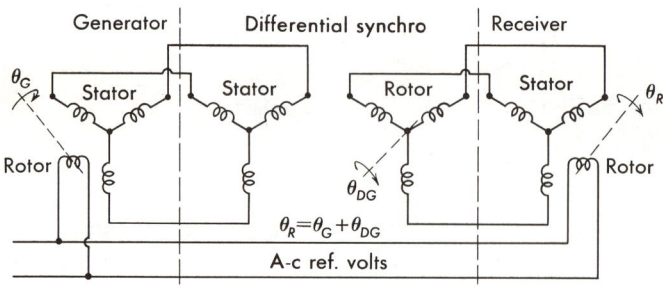

Fig. 7-27 Schematic diagram of a differential synchro used in a torque synchro system.

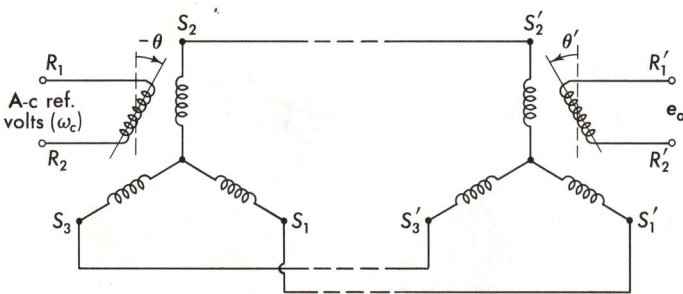

Fig. 7-28 Schematic of synchros used as position indicators.

Synchros used as position indicators are connected as shown in Fig. 7-28 and develop an output-error voltage equal to the sine of the difference between the input and the output shaft angles. This error signal is amplified and used to drive a motor that restores input-output correspondence by properly positioning the output shaft. Output torque can be as high as desired. Since the torque is supplied by a servomotor driven by an amplifier, mechanical requirements of low bearing friction and built-in damping are not as important as in torque units.

Synchros used in this application, termed control synchros, differ from the torque units in the impedance level of the windings. Input impedance is high, so the units do not draw much power. Torque gradient is kept to a minimum to eliminate torque transmission of the ordinary torque units. The torque synchro rotor aligns itself with the field. The position synchro is rotated by the servo until the secondary rotor is aligned perpendicular to the magnetic field, since at this position the input to the servoamplifier is zero. Hence the rest position for a torque synchro differs by 90 electrical degrees with respect to the null position for a position synchro.

With voltages impressed across corresponding stator leads of the synchro transmitter shown in Fig. 7-28, proportional fluxes are produced. These fluxes, which add vectorially to produce a resultant flux, have the same angular position with respect to the synchro receiver stator coils as the transmitter rotor has to its stator coils. Neglecting source impedance, the voltage appearing across $R_1' - R_2'$ (Fig. 7-28) may be expressed as

$$E_{R_1'R_2'} = K_s \sin (\theta - \theta') \tag{7-13}$$

For small errors, $\theta - \theta'$ can be substituted for $\sin (\theta - \theta')$ with the result

$$E_{R_1'R_2'} = K_s(\theta - \theta') \tag{7-14}$$

where K_s is the voltage sensitivity of the synchro (volts per radian misalignment) and ω_c is the carrier frequency.

7-6 Reluctance Pickoffs. Position instruments in great variety are built on the variable-reluctance principle. These units are used to measure both angular and linear motion. They are often built into other instruments. For example, pressure transducers, accelerometers, and gyroscopes often utilize a variable-reluctance pickoff. A linear-motion unit, the so-called "E pickoff," is analyzed in this section. The name comes from the mechanical construction, shown in Fig. 7-29.

An a-c flux, shown by the dashed lines, is established in the magnetic circuit. The flux at one instant of time is in the direction of the arrows.

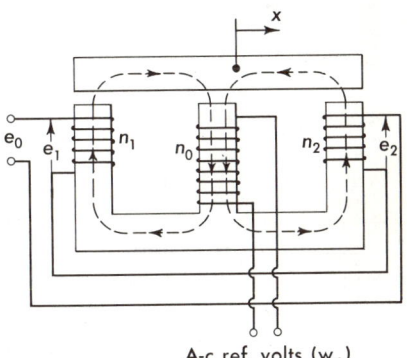

Fig. 7-29 E-type variable-reluctance position pickoff.

If the movable portion of the unit (termed an "armature") is located symmetrically with respect to the E-shaped unit (termed a "stator"), the flux in the outer legs is equal. When coils n_1 and n_2 are equal but wound in opposition, the voltage induced across the coils is equal. Hence, with the unit in the symmetrical position the output voltage $e_o = e_1 - e_2 = 0$. If the armature is displaced an amount x, the flux through n_1 decreases and the flux through n_2 increases. The voltage e_2 becomes greater than e_1, and $e_o = e_1 - e_2$ assumes a value. For small displacements and with careful shaping of the magnetic circuit, the transfer function of the unit is

$$e_o = K_s x \qquad (7\text{-}15)$$

where K_s is the pickoff sensitivity expressed in volts per unit length. ω_c is the reference carrier frequency.

When the E pickoff is properly designed, there is small reaction torque on the I member over a wide range of motion. Sensitivity of these variable-reluctance pickoffs is extremely high. It is not uncommon for these units to sense changes of several microinches.

7-7 Measurement of Velocity. Velocity pickoffs or tachometers develop an output voltage proportional to the shaft angular velocity. The tachometer output voltage is used for damping in position servos or as an output comparison unit for velocity servos. Although there are numerous types of velocity pickoffs, the induction or a-c generator and the d-c tachometer have the greatest application.

An ideal tachometer should have the following characteristics:

1. The output voltage is linear with respect to shaft speed.
2. The output is relatively free from harmonics and quadrature.
3. At any one speed the output voltage is proportional to input voltage.
4. High sensitivity, i.e., appreciable output voltage, with small shaft speed.
5. Low output voltage when the shaft is not moving.

7-8 Induction Tachometers. The a-c tachometer, shown schematically in Fig. 7-30, resembles a two-phase a-c motor. An a-c reference voltage is applied to one phase of the tach generator, and a voltage of reference frequency and amplitude proportional to shaft speed is generated on the other phase. The sinusoidal reference voltage which is applied to the primary winding sets up in the generator a sinusoidally varying flux of constant amplitude. The secondary winding is located mechanically 90° with respect to this flux, so that with zero velocity no voltage is induced in the secondary. As the shaft is rotated, currents that generate a flux are induced in the rotor. The rotor flux adds to the reference flux and shifts the direction of the total flux. Total flux now has a component in the direction of the 90° winding, and a voltage is generated in the winding. With proper design the flux component in the 90° winding is made proportional to velocity; hence, the output voltage is proportional to velocity.

Characteristics of an a-c generator are best summarized in the equation

$$e_o = K_s \omega \qquad (7\text{-}16)$$

where K_s is the voltage gradient, in volts per radian per second, and ω is the shaft speed. A good instrument has a constant K_s invariant with shaft speed ω, voltage level, and carrier frequency. Phase shift of the output voltage is either 0 or 180° with respect to the reference voltage, depending on whether the shaft direction is positive or negative.

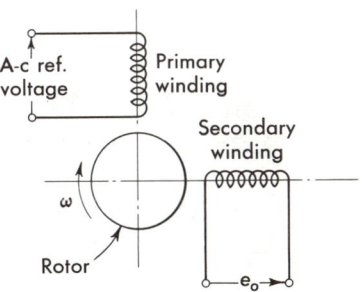

Fig. 7-30 Schematic diagram of an induction tachometer.

The voltage gradient K_s is found from the manufacturer's literature. Often the manufacturer gives the K_s information as follows: Output volts at 1,000 rpm = 30 volts with rated reference voltage applied. With data in this form the gradient is assumed linear and K_s is computed as follows:

$$K_s = \frac{3}{1{,}000(2\pi/60)} = 2.87 \times 10^{-2} \text{ volts/(rad/sec)} \qquad (7\text{-}17)$$

Equation (7-16) is written alternatively as

$$\frac{E_o}{\theta} = K_s s \qquad (7\text{-}18)$$

where s is the Laplace-transform operator.

In physical size and appearance, any of these rate generators (either alternating- or direct-current) are similar to the synchros. One type of unit is shown in Fig. 7-31. This rate generator is known as a "drag-cup tachometer." It incorporates a cylindrical drag cup which serves as the rotating element. This cup, which is made of conducting material, rotates in the air gap between a cylindrical stator and a core, both of precision-machined laminated magnetic steel (cf. Fig. 7-32). In small units (size 8 and 11) the rotor is magnetic material with a thin copper sleeve over the cylindrical rotor.

The instrument functions as a two-phase induction generator. When one stator phase is excited by a constant-frequency, constant-amplitude voltage source, the output voltage from the other stator phase has an

Fig. 7-31 Alternating-current rate generator. (*Courtesy of Ford Instrument Co., Long Island City.*)

Fig. 7-32 Sketch of drag-cup a-c tachometer.

amplitude proportional to the speed of the shaft and a frequency identical with that of the excitation voltage. Typical characteristics for an a-c rate generator are summarized in Table 7-2.

Table 7-2 Characteristics of a Temperature-compensated 400-cps Rate Generator*

Input voltage	115
Maximum total residual rms voltage (null), mv	100
Output voltage per 1,000 rpm	$3.6 \pm 0.4 / \pm 90°$
Output impedance at 1,800 rpm, ohms	26,000
Moment of inertia, oz-in.^{2}	0.075

* Courtesy of Ford Instrument Co., Long Island City.

Besides their application to damp closed-loop systems, rate generators are used for integrating and computing. As an example of the latter application consider the block diagram shown in Fig. 7-33. The overall transfer function is

$$\frac{C}{R} = \frac{AK_m}{s[(\tau_m s + 1) + AK_s K_m]} \qquad (7\text{-}19)$$

For large amplifier gain the transfer function reduces to

$$\frac{C}{R} \approx \frac{1}{K_s s} \qquad (7\text{-}20)$$

Hence the circuit of Fig. 7-33, which uses a rate generator in the feedback path, comprises an approximate integrator. The accuracy of such an

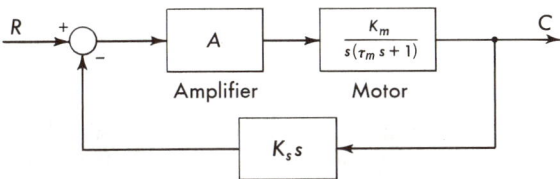

Fig. 7-33 Block diagram of an electromechanical integrator which uses an a-c rate generator.

integrator is limited by the approximation made (that is, $A \gg 1$). The accuracy is also limited by the following characteristics of the tachometer:

1. Threshold (zero output voltage even with some finite shaft velocity)
2. Variation of K_s with velocity, applied voltage, or environment

7-9 D-C Tachometers. Although a-c rate generators are most common, d-c generators are often required for d-c control applications which are similar to the above-mentioned a-c applications: d-c control-system damping, d-c velocity-control-system comparison, etc. Owing to the necessity for brushes and commutation in d-c machines, d-c rate generators are less accurate, exhibit greater friction, produce both electrical and radio noise, are subject to drift, and quite often have a different K_s in one direction than the other.

The d-c generator is similar in appearance and operation to the d-c motor. A fixed field is established with a direct current through a field coil. As the rotor windings cut the constant magnetic field, a voltage proportional to velocity is generated in the rotor windings. The transfer function can be written

$$\frac{E_o}{\theta} = K_s s \tag{7-21}$$

where K_s is the d-c generator gradient in volts per radian per second.

7-10 Permanent-magnet Tachometers. Alnico permanent magnets are often employed as the fixed field in d-c tachometers. These generators are smaller in size, are more reliable, and have a larger gradient K_s than the wound-field units.

Permanent magnets are stabilized against demagnetizing effects such as mechanical shock and short circuits by cycling an alternating field through the magnetic circuit after initial charging. The instrument is made insensitive to temperature changes by using a "carpenter metal" shunt in the magnetic path. This shunt has a permeability which changes with temperature in an opposite sense to the permeability changes of Alnico. This technique maintains an essentially uniform flux in the magnetic core. Adjustable magnetic cores (or slugs) permit an adjustment of K_s.

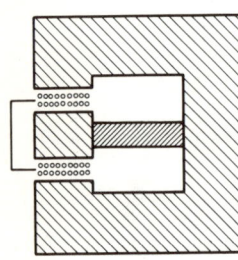

Fig. 7-34 Sketch of d-c-linear-motion rate pickoff.

Permanent magnets are often used for measuring linear-motion velocities. This type of pickoff consists of a coil mounted on the movable member and a magnet mounted on the fixed member. A sketch of a linear-motion d-c rate pickoff is shown in Fig. 7-34. As the magnet moves in and out of the coil, a voltage proportional to linear velocity is generated at the output terminals of the coil. The proportionality constant K_s is the voltage gradient in volts per inch per second.

7-11 Measurement of Acceleration. Although many types of acceleration transducers are manufactured, most of these instruments are based upon a common principle of operation. The instrument measures the motion of a restrained mass when it is subjected to an acceleration. Variations of linear and angular accelerations, a fluid or a solid mass, fluid or pneumatic damping, a-c or d-c output, etc., do not change the fundamental theory underlying the operation of the accelerometer. Two types of accelerometer are considered. One is basically a mechanical unit in which the acceleration measurement depends upon the mechanical system, and the other is a force-balance unit in which a feedback control system corrects some of the basic mechanical system inaccuracies.

7-12 Mechanical Accelerometer. A "seismic," or mechanical, accelerometer schematic is depicted in Fig. 7-35. The instrument consists of a mass suspended from a frame by a spring. Damping is provided either mechanically or electrically, and a pickoff measures the mass position y with respect to the frame. In Fig. 7-35 x is the displacement of the frame with respect to the body whose acceleration is to be measured. K is the spring constant of the suspension, and B is the viscous damping constant. Since y is measured with respect to the frame, the force on the mass due to the spring is $-Ky$ and the force due to the damper is $-B\,dy/dt$. The motion of the mass is $y - x$. The Laplace-transformed equation for the sum of the forces on the mass is

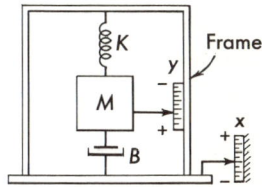

Fig. 7-35 Schematic diagram of a seismic accelerometer.

$$Ms^2(Y - X) + BsY + KY = 0 \qquad (7\text{-}22)$$

where zero initial conditions are assumed. Equation (7-22) is rearranged:

$$(Ms^2 + Bs + K)Y = Ms^2 X = MA \qquad (7\text{-}23)$$

where $s^2 X = A$ is the acceleration to be measured.

The transfer function of the accelerometer is

$$Y = \frac{MA}{Ms^2 + Bs + K} = \frac{A}{s^2 + (B/M)s + K/M} \qquad (7\text{-}24)$$

This equation is simplified to

$$\frac{Y}{A} = \frac{1}{s^2 + 2\zeta_n \omega s + \omega_n^2} \qquad (7\text{-}25)$$

where $\qquad \omega_n = \sqrt{\frac{K}{M}} \qquad$ and $\qquad \zeta = \frac{B}{2\sqrt{KM}} \qquad (7\text{-}26)$

The output of the accelerometer is measured with any of the position instruments discussed in Sec. 7-2. Commonly, either a linear-motion potentiometer or an E pickoff is used to measure y. If the sensitivity (transfer function) of the position instrument is K_s, the accelerometer transfer function is

$$\frac{Y}{A} = \frac{K_s}{s^2 + 2\zeta\omega_n s + \omega_n^2} \tag{7-27}$$

The block diagram is shown in Fig. 7-36.

Accelerometers of this type fit in the category of spring-mass instruments. In these instruments the acceleration to be measured appears as a proportional force (or torque) which acts upon a small mass (or moment of inertia). This force is balanced against a spring, and the deflection of this spring becomes a measure of the force and therefore the acceleration.

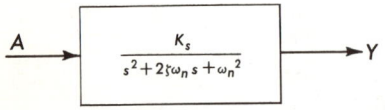

Fig. 7-36 Block diagram of a mechanical accelerometer.

For static measurements, only a spring and a mass or inertia are necessary. Under dynamic conditions damping is required to dissipate vibrational energy. Accelerometers employ various forms of damping. They have an accuracy which depends directly upon the linearity of the spring. Hysteresis, nonlinearities, or nonsymmetrical properties of this spring

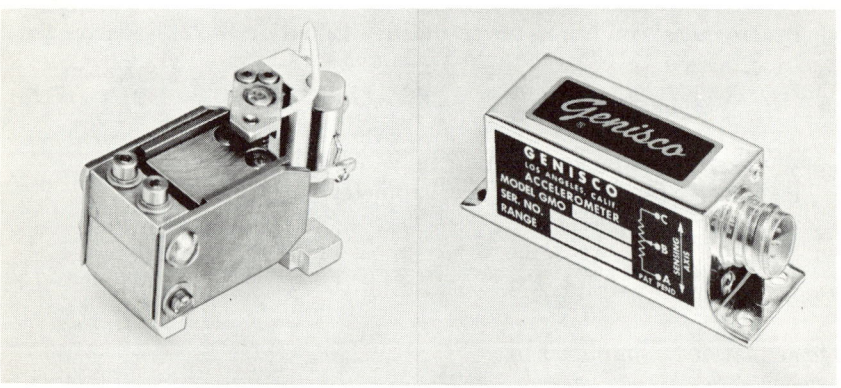

Fig. 7-37 A small and rugged seismic accelerometer with potentiometer output. (*Courtesy of Genisco, Inc., Los Angeles, Calif.*)

will result in errors in the instrument. For this and other reasons which are related to open-loop computations, units of this type are generally less accurate and have a smaller range than those of the force-balance or "servo" type, which are described in the next section. Seismic accelerometers are small and rugged, however, and can provide a high output voltage from a potentiometer output.

Figure 7-37 shows a typical unit that weighs 5 oz and is $1\frac{3}{32}$ by $1\frac{1}{32}$ by $1\frac{7}{8}$ in. in overall size. The $\pm 2g$ to $\pm 120g$ output is linear within 1 per cent of full scale. Silicone oil shear damping is provided. Use of bimetallic damping plates which vary dimensionally with temperature provides relatively constant output from -50 to $+200°$F. Output is measured with a potentiometer or a switch.

Small seismic accelerometers are designed with strain gauges used to transduce the motion of the seismic mass. A $\pm 100g$ unit is shown in Fig. 7-38. The frequency response is flat from 0 to 2,100 cps. Owing to the stepless character of a strain gauge, the ultimate resolution of these

Fig. 7-38 Seismic accelerometer with strain-gauge output. (*Courtesy of Statham Laboratories, Inc., Los Angeles, Calif.*)

accelerometers is infinite. Use of "gas damping" provides constant-frequency response from -65 to $+250°$F.

The frequency response of seismic instruments [Eq. (7-27) for $s = j\omega$] is shown in Fig. 5-3a (amplitude) and b (phase shift). So that the acceleration measurements will be without waveform change, the widest possible frequency range (bandwidth) is desired. The least requirements are a flat amplitude and a linear phase shift. Note that in Fig. 5-3a the $\zeta = 0.707$ curve gives a flat amplitude response within 90 per cent of the natural frequency. The phase-shift curve, Fig. 5-3b, is nearly linear. For these reasons, accelerometers (and many other instruments) are commonly damped at 0.7 critical damping. As is to be expected, increased damping will decrease the magnitude of the response peak in the neighborhood of the natural frequency, and thus transient oscillations due to high-frequency excitation are minimized.

7-13 Force-balance Accelerometers.

A force-balance accelerometer, shown in the schematic of Fig. 7-39, overcomes some of the disadvantages of the seismic instruments. In this system a mass is allowed to move along the acceleration-sensitive axis. The position of this mass is measured with a position pickoff, perhaps of the E type. The output voltage from this position transducer is amplified with a high-gain

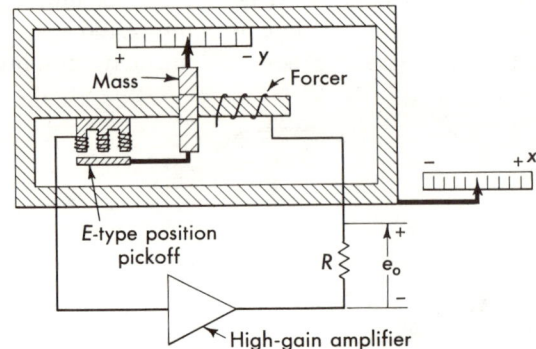

Fig. 7-39 Schematic diagram of force-balance accelerometer.

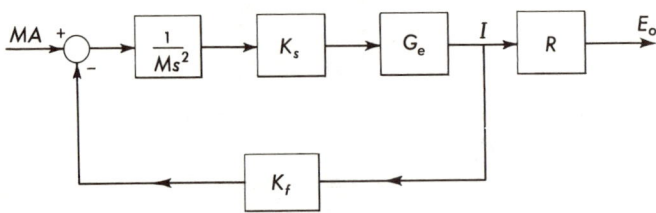

Fig. 7-40 Block diagram of force-balance accelerometer.

amplifier whose output is a current. The current flows through the windings of a forcer* which forces the mass back to its original null position. The forcer current, necessary to zero the position of this mass, is proportional to acceleration, and it is measured as a voltage across a resistor in series with the forcer coil. In this system high damping, which is independent of temperature, and good accuracy are obtained by means of appropriate equalization.

Analysis of the force-balance accelerometer of Fig. 7-39 is based upon the block diagram of Fig. 7-40. If the mass is free to move along the rod with essentially zero damping, the force on the mass is related to the

---

* See Sec. 7-18 for a discussion of a forcer.

displacement as follows:

$$F = Ms^2(Y - X) \tag{7-28}$$

or, rewritten,

$$Y = \frac{1}{Ms^2}(F + Ms^2 X) = \frac{1}{Ms^2}(F + MA) \tag{7-29}$$

s is the Laplace-transform operator, and zero initial conditions are assumed. Equation (7-29) is represented by the first block whose output is Y, which is converted to a voltage with a position pickoff (transfer function K_s) and amplified and equalized (KG_e). Current from the amplifier passes through a forcer, with transfer function K_f in force units per milliampere. This current forces the mass back to a null position. The output voltage is taken at the output of a final block with transfer function R. The force fed back cancels the input acceleration.

With an equalizer of the form

$$G_e(s) = \alpha s + 1 \tag{7-30}$$

that is, position plus rate compensation, the system can be made stable. The damping is electrically controllable (as α is varied) and so is independent of temperature effects. The closed-loop system has the equation

$$\frac{E_o}{MA} = \frac{\dfrac{RK_s K(\alpha s + 1)}{Ms^2}}{\dfrac{K_s K(\alpha s + 1)K_f}{Ms^2} + 1} \tag{7-31}$$

which reduces to

$$\frac{E_o}{A} = \frac{RK_s K(\alpha s + 1)}{s^2 + 2\zeta\omega_n s + \omega_n^2} \tag{7-32}$$

where

$$\omega_n^2 = \frac{K_s KK_f}{M} \quad \text{and} \quad \zeta = \frac{\alpha}{2}\sqrt{\frac{K_s K_f K}{M}} \tag{7-33}$$

In addition to an $\alpha s + 1$ equalizer it is necessary to add an integrator, K/s. Without the integrator the accelerometer has a steady-state error.

Force-balance instruments, which are applicable to many types of measurement, have an advantage in that the mechanical system inaccuracies, such as spring hysteresis, are reduced. Accuracy of the force-balance system depends, to a large extent, on the accuracies with which one can build a forcer.

If temperature compensation, such as a Curie shunt* across the magnetic path of the forcer, and appropriate equalization to account for variations of the copper with temperature are used, an accelerometer of high accuracy and fine resolution can be obtained. Instruments of this type are capable of measuring accelerations of the order of $0.001g$. Since the unit is always operating about a null, the linearity can be improved, by factors of 10, over completely mechanical accelerometers.

7-14 Pressure Transducers. Static and differential pressure transducers are built in a variety of forms, but most are based upon a similar principle of operation. Mechanical motion is produced in the pressure

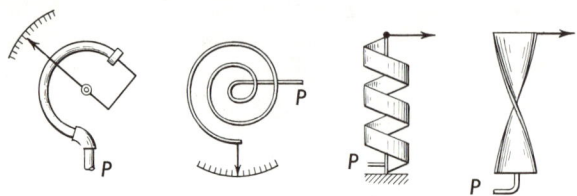

(a) "C" Bourdon   (b) Spiral   (c) Helical   (d) Twisted
Fig. 7-41 Sketch of four different Bourdon-tube arrangements.

transducer by the expansion of a diaphragm or a Bourdon tube.† Pressure transducers used for control-system applications produce an output voltage proportional to input pressure.

For ranges from 0 to 30 up to 0 to 10,000 psi,‡ Bourdon tubes are usually used. For pressure ranges of 0 to 1 in. of water up to 30 psi, diaphragms are commonly used.

Four general types of Bourdon tubes are used for measuring pressure (Fig. 7-41):

1. *C type.* Tube is bent in the form of a circular arc.
2. *Spiral type.* Tube is wound on itself in the form of a spiral.
3. *Helical type.* Tube is wound in the form of a helix.
4. *Twisted type.* Tube is twisted in a helix.

The principle of operation of all types is identical. Practical considerations dictate the type used by the gauge manufacturers.

* A Curie shunt is the name usually given carpenter metal whose reluctance varies with temperature in the opposite direction as does that of iron. See Sec. 7-18 for more information on forcers.
† This tube takes its name from Eugene Bourdon, a French engineer who, in 1849, patented a metallic tube designed to deflect proportionally to a change in pressure.
‡ Pounds per square inch.

A twisted Bourdon tube, shown in Fig. 7-42, used in conjunction with a variable-reluctance a-c pickoff, produces a voltage output proportional to

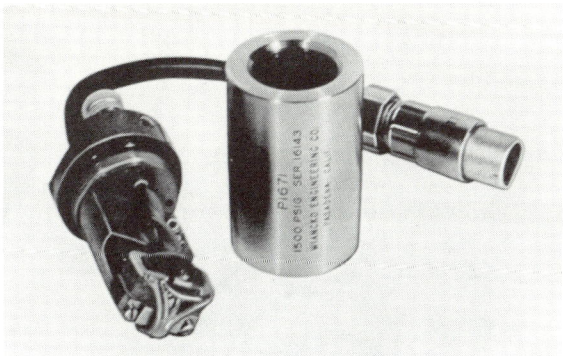

Fig. 7-42 Aircraft-type pressure transducer using a twisted Bourdon tube. (*Courtesy of Wiancko Engineering Co., Pasadena, Calif.*)

pressure. Both absolute- and differential-pressure instruments are available. One unit has the following characteristics:

Scale factor.................. 100 mv/mm Hg
Pressure range.............. 0 to 1,500 psig
Size........................ $2\frac{1}{4}$ by $2\frac{1}{4}$ by $4\frac{1}{2}$ in.
Excitation volts............ 30 volts at 400 cps
Linearity................... ±2%

7-15 Control Motors.* The output torque required for many control systems is supplied by control motors. These motors are commonly driven by an electronic, transistor, or magnetic amplifiers. For control-system applications motor performance is specified not only in terms of load power requirements but also in terms of the special control-system requirements. Both a-c and d-c control motors are manufactured. Speed-torque characteristics of both types, to a first approximation, are represented by the ideal curves of Fig. 7-43. A typical control motor is shown in the cutaway view of Fig. 7-44. This unit operates at 400 cps and is a two-phase induction motor. Load characteristics for a motor of this type are shown in Fig. 7-45. When the speed-torque curve of Fig. 7-45 is linearized, an approximate transfer function can be obtained.

---

* See Ref. 1, chap. 9, for additional discussion of motors.

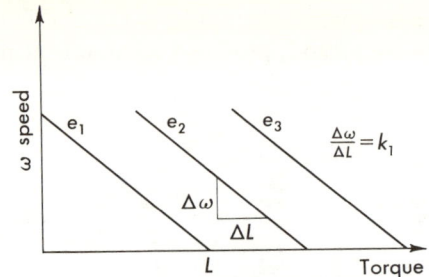

Fig. 7-43 Linearized speed-torque characteristics for a control motor.

Fig. 7-44 Cutaway of a miniature control motor. *(Courtesy of Kearfott Div., General Precision Aerospace, Little Falls, N.J.)*

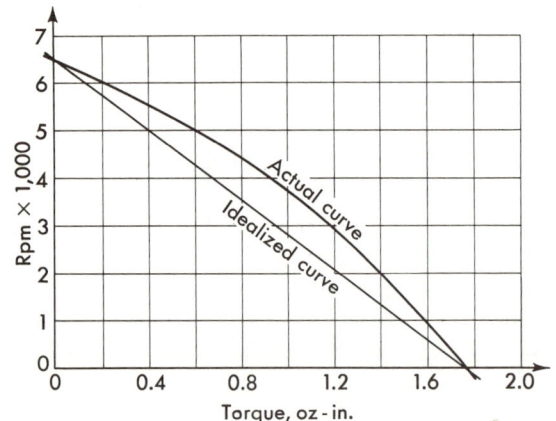

Fig. 7-45 Typical speed-torque characteristics for a control motor.

The two phases of an a-c motor are termed "excitation," or fixed, and "control," or variable. Normally, a reference voltage is applied to the fixed phase and the amplified error voltage is applied to the control phase.

If the fixed field is excited and the control field is shorted, the output torque is zero. If, while in this state, the shaft is turned with the fingers, a distinct dragging tendency is felt. This is "internal" electrical damping; the motor shaft feels as if someone had dipped the motor into molasses. As the shaft is turned, currents are induced in the rotor. The magnitude of these currents is proportional to the velocity in the same manner as in the a-c rate generator. This internal damping is reduced unless the control winding is driven from a low-impedance source.

Suppose full reference voltage is applied to the fixed phase while a variable voltage is applied to the control phase. As the control voltage is increased, the stall torque increases uniformly. The greater the control voltage, the greater the torque.

Hence the motor produces a damping torque proportional to velocity and another torque proportional to control voltage. This physical reasoning is borne out in the following analysis based on the speed-torque curves of Fig. 7-43. These curves, together with a knowledge of load conditions and motor inertia, are sufficient for determining the motor transfer function. An inertia load J is postulated, and a transfer function is found from the linearized speed-torque curves of Fig. 7-43.

Since the equation for a straight line is $y = mx + b$, the equations for each of the straight lines is written

$$L = a\omega + b \qquad (7\text{-}34)$$

where L is the torque delivered by the motor, ω is the angular velocity of the motor shaft, and m and b are constants. Equation (7-34) is the general equation of a straight line. Depending upon the type of control motor (d-c or a-c), different constants a and b are obtained. The slope of all lines is negative and is represented by the same constant

$$a = -m \qquad (7\text{-}35)$$

The quantity b depends upon the control voltage. For $L = 0$, the intersections of the straight lines with the vertical axis yield the no-load-speed vs. voltage curve shown in Fig. 1-10. Slope of the linear portion of the curve (near the origin) is k/m. The equation for this curve is

$$\omega_c = \frac{ke}{m} \qquad (7\text{-}36)$$

where e is the voltage applied to the motor. Substituting for ω from Eq. (7-36) into Eq. (7-34) for $L = 0$,

$$b = ke \qquad (7\text{-}37)$$

The equation for a motor is rearranged and written

$$L + m\omega = ke \qquad (7\text{-}38)$$

where L = torque delivered by motor
e = applied voltage
ω = motor velocity

Suppose the motor drives an inertia load $\bar{L} = Js^2\bar{\theta}$, where $\bar{\theta}$ is the angle of the motor shaft. The shaft velocity $\bar{\omega}$ is

$$\bar{\omega} = s\bar{\theta} \qquad (7\text{-}39)$$

The equation is obtained by combining the above expressions to

$$Js^2\bar{\theta} = kE_{\text{in}} - ms\bar{\theta} \qquad (7\text{-}40)$$

which is rearranged as follows:

$$\bar{\theta} = \frac{kE_{\text{in}}}{s(Js + m)} = \frac{K_m}{s(\tau s + 1)} \qquad (7\text{-}41)$$

where $K_m = k/m$ and $\tau = J/m$.

Both a-c and d-c motors can be approximated by this expression for the transfer function and by the block diagram of Fig. 7-46. With d-c motors, however, brush friction near $\omega = 0$ causes a larger error. It must be emphasized that the block diagram of Fig. 7-46 and the transfer function of Eq. (7-41) are applicable only for a motor driving an inertia load.

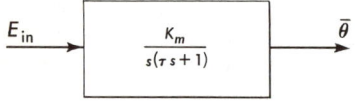

Fig. 7-46 Block diagram for a motor driving an inertia load.

In the actual design of a control system it is necessary to obtain numerical values for the transfer function of Eq. (7-41) from the manufacturer's literature. A speed-torque curve similar to Fig. 7-45 together with a table of performance data similar to Table 7-3 is usually supplied by the manufacturer. The slope m is found from

$$m = \frac{\text{stall torque (rated voltage)}}{\text{no-load speed (rated voltage)}} \qquad (7\text{-}42)$$

Table 7-3 Performance Data on a Typical 400-cps Control Motor

Rotor inertia, g-cm². 3.3
Torque at stall, oz-in. 1.53
No-load speed, rpm. 5,300
Torque at maximum power output, oz-in. 0.85
Speed at maximum power output, rpm. 2,800
Theoretical acceleration at stall, rad/sec². 32,600
Operating temperature range, °C. . . . . −54 to +125
Weight, oz. 7.3

	Fixed phase	Control phase
Voltage, volts.	115	115
Maximum power output, watts.	1.8	1.8
R (stall), ohms.	490	490
X (stall), ohms.	890	890
Z (stall), ohms.	1,030	1,030

The other constant is

$$k = \frac{\text{stall torque}}{\text{rated control voltage}} \quad (7\text{-}43)$$

The stall torque is read from the speed-torque curve at the point where the speed is zero, and the no-load speed is read at the point where the torque is zero. For the motor of Fig. 7-45

Stall torque = L_0 = 1.75 oz-in.

No-load speed = ω_c = 6,500 rpm = $\dfrac{(6,500)(2\pi)}{60}$ = 680 rad/sec

Rated control voltage = 115 volts

The constants are

$$k = \frac{1.75}{115} = 1.52 \times 10^{-2} \text{ oz-in./volt}$$

$$m = \frac{1.75}{680} = 2.57 \times 10^{-3} \text{ oz-in.-sec} \quad (7\text{-}44)$$

For an inertia load which is equal to the inertia of the motor

$$J = (2)(0.9) = 1.8 \text{ g-cm}^2$$

$$= \frac{1.8}{980} \text{ g-cm-sec}^2 = \frac{1.8}{980}(0.0137) \text{ oz-in.-sec}^2$$

$$J = 2.51 \times 10^{-5} \text{ oz-in.-sec}^2$$

The transfer function quantities are

$$\text{Time constant} = \tau = \frac{J}{m} = \frac{2.51 \times 10^{-5} \text{ oz-in.-sec}^2}{2.57 \times 10^{-3} \text{ oz-in.-sec}} = 0.00976 \text{ sec} \tag{7-45}$$

$$\text{Motor constant} = K_m = \frac{k}{m} = \frac{1.52 \times 10^{-2} \text{ oz-in./volt}}{2.57 \times 10^{-3} \text{ oz-in.-sec}}$$

$$= 5.91 \text{ (volt-sec)}^{-1} \tag{7-46}$$

The approximate transfer function is

$$\frac{\bar{\theta}}{E_{\text{in}}} = \frac{5.91}{s(0.0096s + 1)} \tag{7-47}$$

Figure 7-47 shows a 60-cps servomotor.

If a family of speed-torque curves is available or can be run, a more accurate transfer function can be obtained. Speed-torque curves for

Fig. 7-47 Low-inertia 60-cycle control motor. (*Courtesy of Diehl Mfg. Co., Somerville, N.J.*)

various control voltages of a typical servomotor were run on a dynamometer* and are plotted in Fig. 7-48. The intersections with the $\omega = 0$ axis of the various curves are taken from Fig. 7-48 and plotted in Fig. 7-49.

* The dynamometer consists of a d-c motor coupled to the test motor through a calibrated torsion wire. The angular deviation between the two motors (which is converted into torque units) is read by the apparent stopping of the motion with a stroboscope, from which the speed is also read. The d-c motor acts as a variable load. This makes an excellent servolab experiment.

Servomechanism Components 323

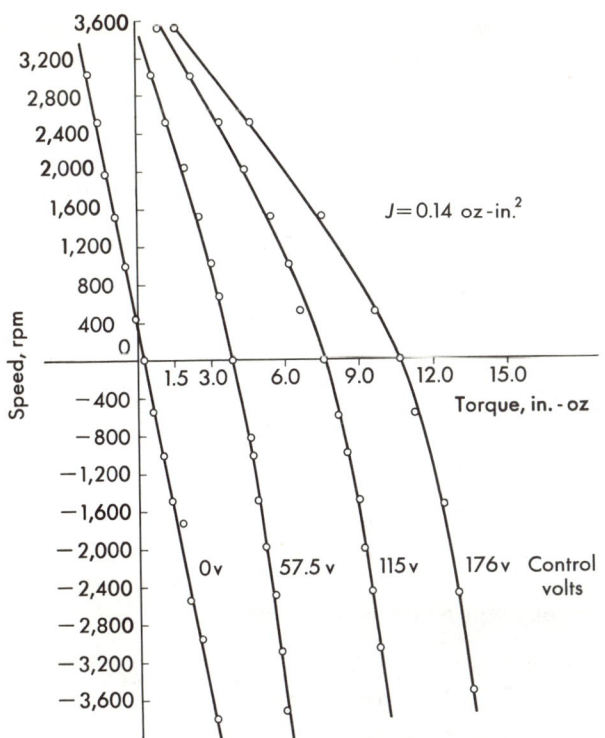

Fig. 7-48 Speed-torque curve for 5-watt 115-volt 60-cps two-phase control motor.

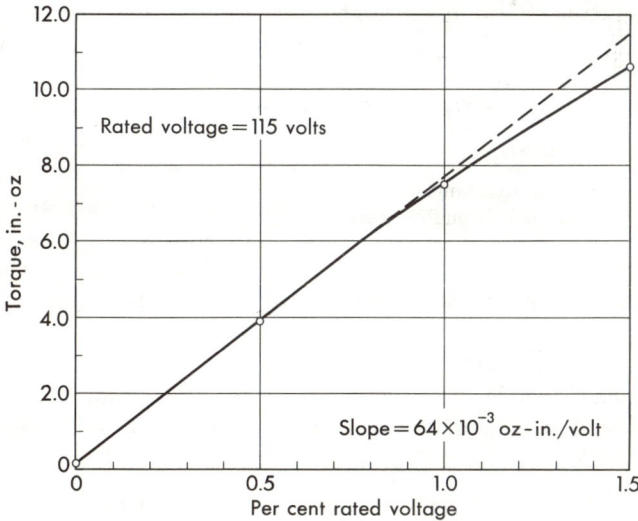

Fig. 7-49 Torque-voltage curve taken from speed-torque curve.

The slope of the speed-torque curve for zero control voltage is used for m, since the servo operates near the null (zero control volts):

$$m = \frac{+1.5 - (-1.5)}{(2\pi/60)[-1{,}700 - (2{,}350)]} = -\frac{3}{(4{,}050)(6.28/60)}$$
$$= 7.09 \times 10^{-3} \text{ oz-in.-sec} \quad (7\text{-}48)$$

The slope of the linear portion of Fig. 7-49 is used to determine the second constant:

$$k = \frac{3.9 - 0.2}{(0.5)(115)} = 64 \times 10^{-3} \text{ oz-in./volt} \quad (7\text{-}49)$$

The transfer function quantities are found from these values and the motor inertia ($J_m = 0.14$ oz-in.^{2}). For an inertia load which is equal to the motor inertia, multiply J by 2 and convert units:

$$J = \frac{(2)(0.14)}{(32)(12)} \frac{\text{oz-in.}^2}{\text{in./sec}^2} = 0.73 \times 10^{-3} \text{ oz-in.-sec}^2 \quad (7\text{-}50)$$

The time constant and motor constant are

$$\tau = \frac{J}{m} = \frac{0.73 \times 10^{-3}}{7.09 \times 10^{-3}} = 0.103 \text{ sec}$$

$$K_m = \frac{k}{m} = \frac{64 \times 10^{-3}}{2.09 \times 10^{-3}} = 9.04 \text{ (volt-sec)}^{-1} \quad (7\text{-}51)$$

Most control motors operate at a high rpm with a low torque. The load, however, is usually driven at a slower speed with far greater torque required. This state of affairs requires that a gearbox be used between the motor and the load. It is necessary to determine the transfer function of a control motor driving an inertia load through a gear train, as shown in Fig. 7-50. The effect of the gear train* is twofold:

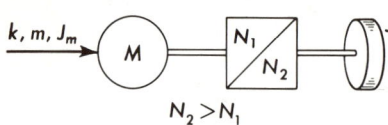

Fig. 7-50 Schematic diagram of motor driving an inertia load through a gear train.

1. The load inertia is reflected across the gear train as $(N_1/N_2)^2 J_L$. This is added to the motor inertia J_m to form an equivalent inertia $J_e = J_m + (N_1/N_2)^2 J_L$.

* See Secs. 2-12 and 7-19 for a more detailed discussion of gear-train performance.

2. The motor constant K_m, in rad/(sec/volt), is reduced by the gear ratio $K_m N_1/N_2$.

The transfer function is written

$$\theta = \frac{(N_1/N_2)K_m}{s\left[\dfrac{J_m + (N_1/N_2)^2 J_L}{m} s + 1\right]} \tag{7-52}$$

Again it must be emphasized that this equation is for an inertia load only.

7-16 A-C Control Motors. The two-phase a-c servomotor, shown in the schematic diagram of Fig. 7-51, satisfies the requirements for the output unit of an a-c system. The main, or "fixed," field is continuously excited from the reference line. The voltage is adjusted to give a maximum torque per control field watt. The control field is separated 90° in space from the fixed field. The control field is driven with a suppressed-carrier signal, which must be 90° phase-shifted in time with respect to the reference voltage. The control field and the reference field are usually similar in power rating. Often, however, the control field is of a higher impedance level, so that a good impedance match to the amplifier is achieved.

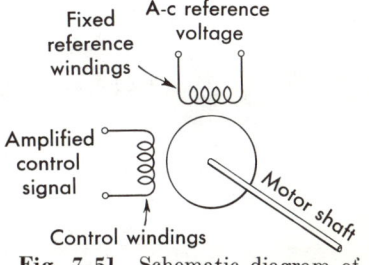

Fig. 7-51 Schematic diagram of a two-phase control motor.

A variety of special motor designs is available to the control engineer. A partial list of a-c motor types follows.

a. Squirrel Cage. This is an induction motor which has a series of shorted conducting bars located around the periphery of the rotor. This type of motor has excellent performance characteristics and a high ratio of output to input power. A typical size 15 motor unit is shown in Fig. 7-44.

b. Drag Cup. This type of motor is similar to the squirrel-cage motor except that the rotor is a drag cup of conducting material. The mechanical construction is similar to that of the rate generator shown in the schematic of Fig. 7-32. All the heavy iron laminations which provide the low-reluctance flux path are stationary, and only a light cup is rotating. The advantages of this type of motor are (1) constant devel-

326   Control System Design

oped torque independent of rotor position, (2) high torque-to-inertia ratio, and (3) low bearing friction.

  c. Motor-Tachs. An integral motor and a-c tachometer provide a power source with an output-velocity voltage available to be fed back for damping. Since these units utilize one housing and one shaft, a saving in size and cost results. The resulting motor has a greater adjustability of damping. A size 5 combination unit, which is shown in

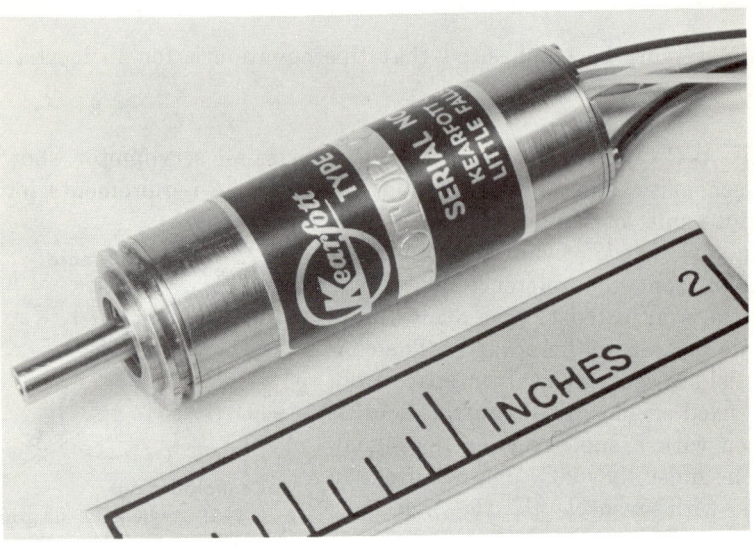

Fig. 7-52 Combination servomotor-generator. (*Courtesy of Kearfott Div., General Precision Aerospace, Little Falls, N.J.*)

Fig. 7-52, is only $\frac{1}{2}$ in. OD and $1\frac{1}{2}$ in. long. Direct-current motors and tachometers on one shaft are also available for use in d-c systems.

7-17 D-C Control Motors. To a first approximation, the transfer function of Eq. (7-41) is applicable to d-c control motors. A more accurate but more complicated transfer function would include the effects of winding time lags, armature inductance, etc. Speed-torque characteristics are greatly affected by the type of excitation, i.e., series, shunt, or fixed excitation. Direct-current machines develop large power output in a small size. Extreme environmental conditions are withstood by sealing the units.

  The major application of d-c motors in aircraft control systems is in power actuators, where weight and space limitations demand a unit with a large power-to-volume ratio.

The disadvantages of d-c motors are:

1. Brush and commutator maintenance
2. Radio interference generated
3. D-c amplifier drift (develop an output with no input)
4. Difficult to match amplifier to motor, since it is impossible to use a transformer

Permanent-magnet excitation finds wide application in d-c control motors, since it overcomes some of the above-cited disadvantages.

7-18 Forcers. In many applications, such as the force-balance accelerometer, a force (or a torque) with no appreciable motion involved is required. The term "forcer" or "torquer" is applied to such a unit. Figure 7-53 shows a simplified diagram of a forcer. A coil is free to move a small distance through the field produced by a permanent magnet. A current through the field coil produces a flux which interacts with the fixed flux of the permanent magnet and hence develops a force whose sign depends upon the polarity of the magnet and direction of current through the field winding. Frequently the coil is fixed and the magnet moves.

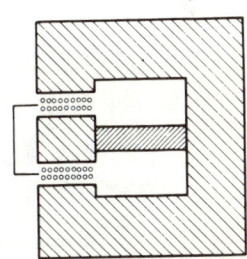

Fig. 7-53 Simplified diagram of a forcer.

Careful stabilization of the magnet and compensation of the magnetic path for temperature changes can result in a precision instrument whose equation can be written

$$f = K_s i \tag{7-53}$$

where K_s is the sensitivity of the instrument in force units per ampere. The unit resembles the d-c rate pickoff described in Sec. 7-10. The methods of stabilization are similar.

7-19 Gear Trains. Feedback control systems make wide use of high-quality gear trains for the following reasons:

1. Mechanical matching of the motor to the load. Since a servomotor often operates at high speed but low torque, a gear train is required to drive the load with a greater torque but less speed.

2. Gain adjustment in the closed loop.
3. Adding and subtracting mechanical signals
4. Scale changing as in a two-speed system.
5. Reversing direction of rotation.

For many systems which use a potentiometer for transducing the shaft position, the motor must run at a reasonable velocity and the potentiometer can move through only 1 to 10 turns. A precision, size 11 speed reducer is shown in Fig. 7-54. The unit is 1.062 in. in diameter

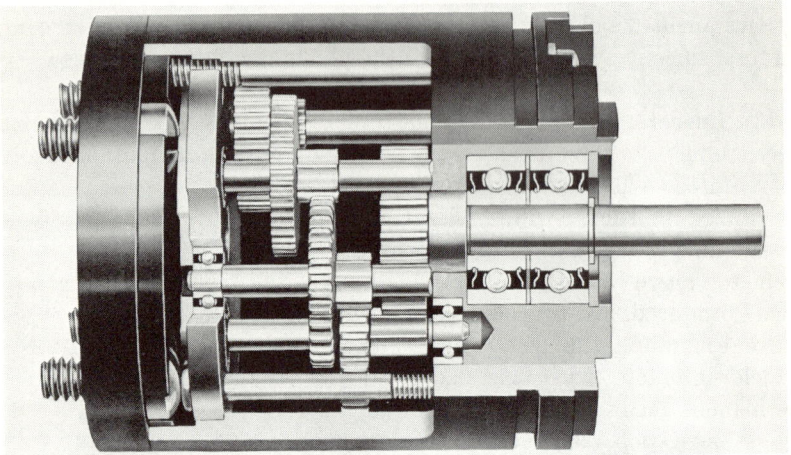

Fig. 7-54 Cutaway of a precision miniature speed reducer. (*Courtesy of Bowmar Instrument Corp., Fort Wayne, Ind.*)

and 1.656 in. long. The unit drives a load torque of 25 in.-oz, and gear ratios up to 14,000:1 are possible.

7-20 Choice of Optimum Gear Ratio. Perhaps the simplest method of selecting gear ratios for feedback control systems is to arrange for the maximum load power demand to match the maximum motor output capabilities. Suppose the maximum load demand is 1.6 watts at 100 rpm. The speed-torque curve of a typical two-phase a-c induction servomotor, such as shown in Fig. 7-55, is essentially a straight line between zero torque–maximum speed and zero speed–maximum torque. Somewhere on this curve the product of torque and speed will be maximum, and this is actually the maximum power output point of the motor. If the speed-torque characteristic is assumed a straight line, the maximum power point is halfway up that line. If, however, it is not a straight line and has some curved shape, the maximum power point will have to be calculated. In the case of Fig. 7-55 the maximum power point is 2,600 rpm. Therefore, knowing that the maximum load power demand occurs at 100 rpm and the maximum motor power delivery is 2,600 rpm, the quotient of motor speed divided by load speed yields the optimum gear ratio for this match, or 26:1.

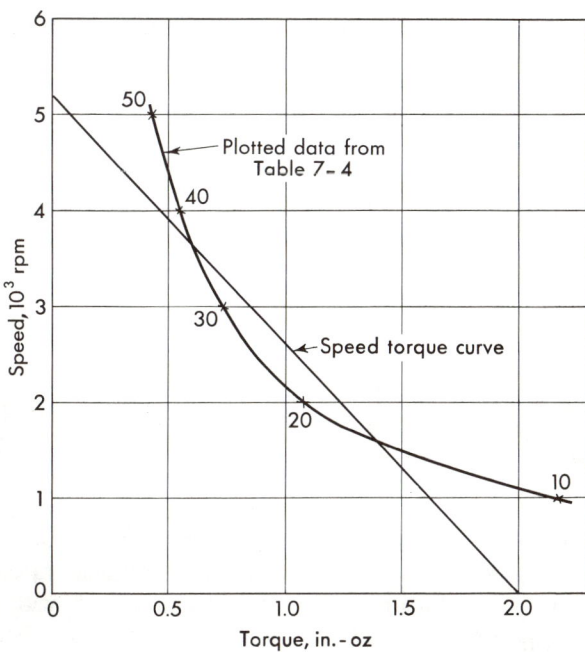

Fig. 7-55 Choice of optimum gear ratio.

Although the maximum-power method is adequate for many design problems, it is often wise to check it by a method based on a mechanical impedance match. The load-speed and torque conditions for maximum power demand are known. The speed is 100 rpm, and the torque is given by

$$L = \frac{1.356 P \times 10^3}{\omega} \tag{7-54}$$

where L = torque, in.-oz
P = mechanical power, watts
ω = speed, rpm

For this application,

$$L = \frac{(1.356)(10^3)(1.6)}{100} = 21.7 \text{ in.-oz} \tag{7-55}$$

In this method various possible gear ratios between motor and load are arbitrarily assumed. Based on each assumption; the motor speed and motor torque are calculated and arranged in tabular form. As an example, Table 7-4 shows the calculation of a series of assumed gear ratios and motor conditions. Each gear ratio results in motor-speed and motor-torque points which are plotted directly on the motor-speed-torque curve. Each assumed gear ratio yields a point on the curve. These are connected to form a curve that has a relation to the motor speed-torque curve such as shown in Fig. 7-55.

Table 7-4 Optimum Gear Ratio

Assumed gear ratio	Motor speed, rpm	Motor torque, in.-oz
10	1,000	2.17
20	2,000	1.08
30	3,000	0.724
40	4,000	0.543
50	5,000	0.434

All portions of the gear-ratio curve that lie beneath the motor speed-torque curve represent satisfactory gear ratios. Any gear ratios above the normal motor speed-torque curve are unusable and will not give the required load performance. If the whole gear-ratio curve lies outside the motor speed-torque curve, the motor has insufficient power and will not do the job. It can readily be seen from this curve that a gear-ratio value of 26 is adequate for the job and will work satisfactorily. If the gear-ratio curve barely penetrates the motor speed-torque characteristics, the motor is marginal.

7-21 Gyroscopes. A gyroscope consists of a wheel mounted on a shaft and arranged to be spun at high angular velocity. Frequently, the wheel is mounted in a system of gimbals that permits complete freedom of movement on all three axes.

The most useful characteristic of the gyroscope is its tendency to maintain its spin axis in a fixed direction in space. This phenomenon is best explained by a consideration of rotational dynamics.

A simple model is used to gain insight into the operation of the gyroscope. In this model the effects of moments of inertia of the wheel and gimbal system about axes other than the spin axis are neglected. The

Servomechanism Components

resulting equation will fail to show certain characteristics. For many purposes, however, the results are adequate. For this derivation the following nomenclature is needed:

M_x, M_y, M_z = components of angular momentum about x, y, and z axis
I_s = moment of inertia of wheel about spin axis
ω_s = angular velocity of wheel
$H = I_s\omega_s$ = angular momentum of wheel

These quantities are demonstrated in Fig. 7-56.

At time $t = 0$, suppose the torque Q_x is applied about the x axis by pressing down on the gyro housing at point p. Initially, $M_z = I_s\omega_s = H$, $M_x = M_y = 0$, and the angular momentum of spin lies along OZ and has a magnitude H. Since the rate of change of angular momentum of a system is equal to the applied torque,* the following expression can be written:

$$Q_x = \frac{dM_x}{dt} \quad (7\text{-}56)$$

This is expressed in different form,

$$dM_x = Q_x\,dt \quad (7\text{-}57)$$

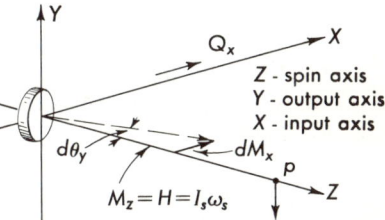

Fig. 7-56 Vector diagram of a simplified gyro system.

If this term is added vectorially to the initial angular momentum (see Fig. 7-56), a new value is obtained. $M_z + dM_x$ is separated from the initial value by an angle $d\theta_y$ which, from Fig. 7-57, is given by

$$d\theta_y = \frac{dM_x}{H} = \frac{Q_x\,dt}{H} \quad (7\text{-}58)$$

which is rewritten as

$$\omega_y = \frac{d\theta_y}{dt} = \frac{Q_x}{H} \quad (7\text{-}59)$$

The gyro is thus rotating about the OY axis with a velocity ω_y.

Equation (7-59) demonstrates the fundamental gyroscopic law: A torque about any axis other than the spin axis produces a velocity about the axis which is orthogonal to the applied-torque axis. Because of

---

* In linear mechanics, the rate of change of linear momentum equals the applied force $F = d(mv)/dt$.

this property the gyro is an important instrument for measuring torques ($Q_x = H\omega_y$) by measuring this orthogonal or precession velocity.

Alternatively, a large gyroscope can be used to obtain a stabilizing countertorque. For applied torques about the x or y axis, the gyro supplies an equal countertorque which prevents motion in the direction of the applied torque as long as the gyro can precess. Once the precession angle θ has reached 90°, the gyroscope is in a state of "gimbal lock" and ceases to function as described. In the state of gimbal lock the OZ axis has precessed into the OX axis, about which the torque is being applied. With a torque applied about the OX, or spin, axis, the gyro ceases to produce a countertorque. Thus the gyro tends to rotate to align its spin axis in the direction of applied torque.

7-22 More Complete Mathematical Treatment of a Gyroscope. In this section the more general gyroscopic equations are derived. Such quantities as spring constant, damping, and moment of inertia about the two axes of interest are considered, and their effects are discussed. The equations of this section are concerned with the single-degree-of-freedom gyroscope. In the subsequent derivation, the following assumptions are made:

1. Small-angle assumption (that is, $\sin \theta \approx \theta$ and $\cos \theta \approx 1$).
2. Moments of inertia about the x and y axes are considered constant during the period studied.
3. The gimbal system is sufficiently symmetric that products of inertia are negligible.

If the gyroscope is mounted in a gimbal system and the position of the spin axis is displaced from the OZ axis by small angles, θ in azimuth and ϕ in elevation (Fig. 7-57), the following additional nomenclature is needed:

J = moment of inertia of wheel, case, and gimbal system about x axis; this includes inertia of platform or other mounting of gyroscope
I = moment of inertia of wheel, case, and gimbal system about y axis
L = torque applied about x axis
U = disturbing torque applied about y axis

The remaining symbols are identical with those given earlier. The inertial torques which act in the two directions x and y are first obtained. The remaining torques are summed about appropriate axes, and application of Newton's law yields the desired equations: the precession angle θ and the gimbal angle ϕ as functions of L and U.

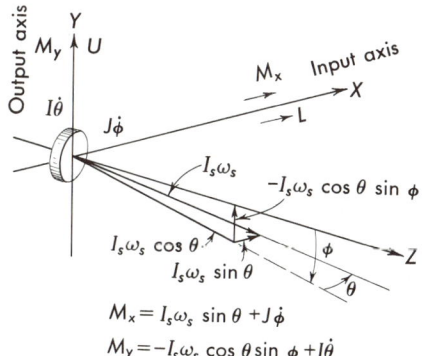

Fig. 7-57 Coordinate system for more complete gyro equations.

Because a single-degree-of-freedom gyro is considered here, the only axis capable of moving is the Y or output axis of Fig. 7-57. The gyro input, which is a torque, appears about the input axis. The applied torque L is the desired input, and the precession velocity θ the desired output.* Undesired torques appearing about the output axis are termed "disturbing torques," since they produce drift of the gyroscope. The greatest limitation in the use of gyroscopes lies in the random disturbing torques U which cause undesired velocities ϕ. Such torques as caused by spring constant K and damping D about this axis are especially important. A gyro gimbal system is shown in Fig. 7-58.

Consider the angular momentum of the spin axis $H = I_s \omega_s$ as shown in Fig. 7-57. For small angular displacements θ and ϕ about the y and x axes the projections of H in the directions of the x and y axes are

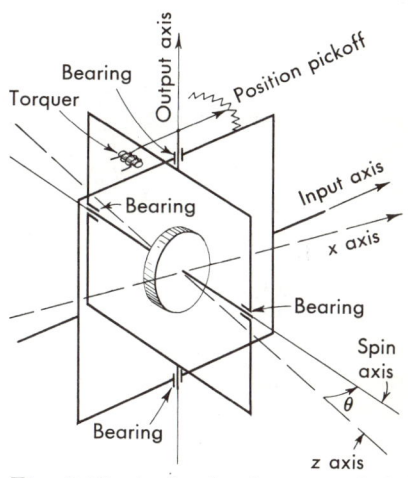

Fig. 7-58 A sketch of a gyro gimbal system.

---

* A dot over the variable means differentiation with respect to time; $\dot{\phi} = d\phi/dt$.

respectively

$$H \sin \theta \quad \text{and} \quad -H \cos \theta \sin \phi \qquad (7\text{-}60)$$

When combined with the angular momentum due to the precession velocity,

$$M_x = H \sin \theta + J \frac{d\phi}{dt} \qquad (7\text{-}61)$$

$$M_y = -H \cos \theta \sin \phi + I \frac{d\theta}{dt} \qquad (7\text{-}62)$$

Since the angles are small, $\sin \theta$ can be replaced by θ and $\cos \theta$ by 1. In addition, Eqs. (7-61) and (7-62) are Laplace-transformed:

$$\bar{M}_x = H\Theta + Js\Phi \qquad (7\text{-}63)$$

$$\bar{M}_y = -H\Phi + Is\Theta \qquad (7\text{-}64)$$

The bars over the variables indicate that the equations have been transformed where s is the Laplace-transform operator. Since the time rate of change of momentum is equal to the applied torque,

$$\bar{Q}_x = \mathcal{L}\frac{dM_x}{dt} = s\bar{M}_x = Hs\Theta + Js^2\Phi \qquad (7\text{-}65)$$

$$\bar{Q}_y = \mathcal{L}\frac{dM_y}{dt} = s\bar{M}_y = -Hs\Phi + Is^2\Theta \qquad (7\text{-}66)$$

The damping and spring torque about the input axes are negligibly small compared with the applied torque, so they are neglected with the result $\bar{Q}_x = L$. The damping and spring restoring torques about the output axis are included:

$$\bar{Q}_y = U - Ds\Theta - K\Phi \qquad (7\text{-}67)$$

where D is the damping coefficient and K the restoring torque coefficient, both about the output axis. Upon substituting into Eqs. (7-65) and (7-66),

$$L = (Js^2)\Phi + (Hs)\Theta \qquad (7\text{-}68)$$

$$U = (-Hs)\Phi + (Is^2 + Ds + K)\Theta \qquad (7\text{-}69)$$

where s denotes the Laplace-transform operator (zero initial conditions). Solution yields the following two equations:

$$\Phi = \frac{(L/J)(Is^2 + Ds + K) - U(H/J)s}{s^2(Is^2 + Ds + K + H^2/J)} \tag{7-70}$$

$$\Theta = \frac{L(H/J) + Us}{s(Is^2 + Ds + K + H^2/J)} \tag{7-71}$$

Equations (7-70) and (7-71) are the gyro equations. Depending upon the gimbal system and the particular use of the gyro, various of the terms can be neglected.

Since U in Eqs. (7-70) and (7-71) is the transform of the disturbing torque, the transfer function from torque input to precession angle output is given by

$$\frac{\Theta}{L} = \frac{H/J}{s(Is^2 + Ds + K + H^2/J)} \tag{7-72}$$

when U is set equal to zero.

7-23 The Free Gyro. If there are no restraining gimbals in any direction on the gyro, the gyro wheel remains fixed in space and the gyro measures the angular position of the aircraft with respect to the gyro as a reference. Gyros used in this manner establish an inertial reference of relatively low accuracy. In order to obtain the approximate transfer functions for a gyro in this mode of operation, set L, I, D, and K equal to zero in Eq. (7-70) and obtain

$$\mathcal{L}^{-1} s\phi = \frac{d\phi}{dt} = \frac{u}{H} \tag{7-73}$$

Similarly, if U, I, D, and K are set equal to zero in Eq. (7-71), a simplified equation results for the other axis:

$$\mathcal{L}^{-1} s\theta = \frac{d\theta}{dt} = \frac{l}{H} \tag{7-74}$$

When the gyro is used in this form, various mechanical members must remain constant with respect to time, temperature, and hysteresis. Hence, used in this mode, gyro drift rates of fractions of a degree per minute are common.

Figure 7-59 shows a cutaway view of a two-degree-of-freedom gyro measuring displacement about two axes by means of a pickoff placed on each gimbal axis. The complete symmetry of motor and inner gimbals enables this unit to sustain high vibrational requirements up to $9g$ at 1,000 cps with a drift of less than $2°/\text{min}$. The drift rate is less than $0.2°/\text{min}$ under normal operation. The unit has $360°$ continuous rotation on the outer gimbal and $\pm 85°$ on the inner gimbal. A thin, pancake-like synchro is used on the outer gimbal, and a potentiometer on the

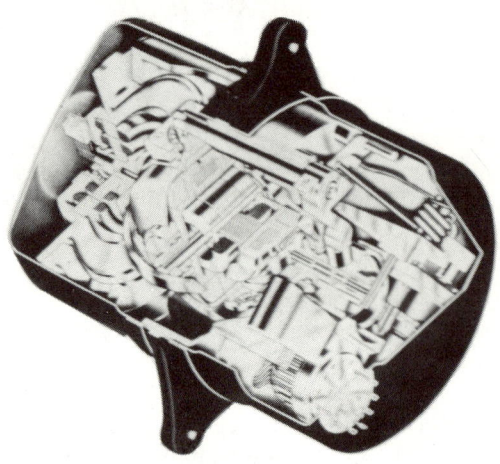

Fig. 7-59 Cutaway view of a free gyro. (*Courtesy of Summers Gyroscope Co., Santa Monica, Calif.*)

inner gimbal. The gyro fits into a hermetically sealed case $3\frac{5}{8}$ in. in diameter and $5\frac{1}{4}$ in. long.

7-24 The Rate Gyro. It is frequently necessary to measure the rate of roll, pitch, or yaw for the purpose of damping undesirable oscillations of the aircraft about these axes. A low-accuracy gyro, called a rate gyro, is used for this purpose. In order to understand the operation of a rate gyro, set U equal to zero and equate (7-70) and (7-71), eliminating L. The following expression results:

$$s\Phi = \frac{1}{H}(Is^2 + Ds + K)\Theta \qquad (7\text{-}75)$$

For small I and D or also with slowly varying time functions Eq. (7-75) reduces to

$$\theta = \frac{H}{K}\frac{d\phi}{dt} \qquad (7\text{-}76)$$

Equation (7-76) indicates that a signal which is proportional to the rate $d\phi/dt$ is obtained. Practically, the large value of K is obtained by supporting the output axis with a torsional spring.

A rate gyro produces velocity signals; i.e., it yields a voltage proportional to the time derivative of the input-axis angle. Any low rate drift of this angle is differentiated and hence of little importance. Also, since

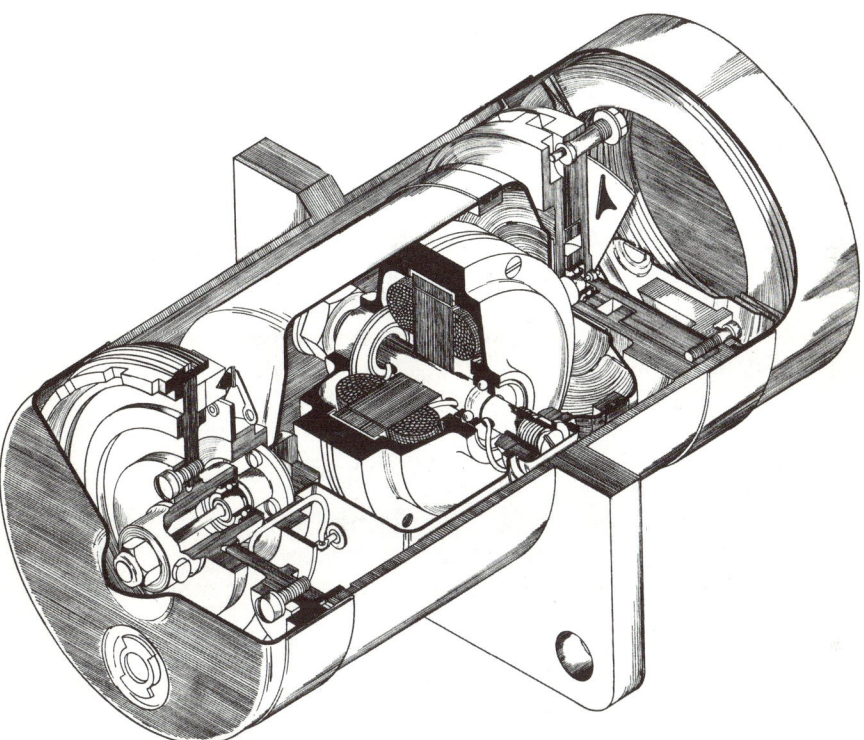

Fig. 7-60 Cutaway view of a type R-160 rate gyro. (*Courtesy of Whittaker Gyro, Division of Telecomputing Corp., Van Nuys, Calif.*)

gyros of this type are used primarily for damping in systems, high accuracy and low drift are not essential.

A cutaway view of a rate gyro is shown in the sketch of Fig. 7-60. This unit utilizes an induction-type pickoff with a linearity of ±1 per cent. The gyro resolution is less than 0.1°/sec. The unit is 3.3 in. long and is 2.0 in. in diameter. In Table 7-5 are included other technical data for this gyroscope.

Table 7-5 Rate-gyroscope Data

Motor.................. Hysteresis synchronous
Speed.................. 12,000 rpm
Start time.............. 30 sec max
Angular momentum..... 200,000 dyne-cm-sec
Gimbal deflection....... 3.2° full scale
Rate range............. 100°/sec
Gimbal restraint........ Torsion bar
Natural frequency....... 20 cps
Damping............... 40-centistoke silicon fluid, paddle wheel, and variable-orifice mechanism. Temperature controlled by a bimetallic element over range of −40 to 185°F. Damping coefficient between 0.4 and 0.7
Life.................... 1,000 hr
Hermetically sealed

7-25 The Restrained Gyro. If only one axis is free to move with respect to the case, the gyro becomes a restrained, or single-degree-of-freedom, gyro. If K, D, and U are set equal to zero in Eq. (7-71) (note that H^2/JI is much larger than 1), the approximate transfer function for a gyro in this mode is given by

$$\Theta \approx \frac{L}{sH} \qquad (7\text{-}77)$$

Because of the form of this transfer function, i.e., the $1/s$ multiplying the torque, this gyro is often termed an "integrating gyro." A more complete transfer function can be found by setting U and K equal to zero as follows:

$$\frac{\Theta}{L} = \frac{H/J}{s(Is^2 + Ds + H^2/J)} \qquad (7\text{-}78)$$

Gyroscopes of this type are quite accurate and are used to measure angular accelerations. Drift rates of 0.1°/hr are possible. Because of the high accuracy, considerable care must be taken in the design of the output axis. Any spring or damping torques that appear about the output axis of the gyro produce a precession velocity ϕ about the input axis [cf. Eq. (7-70)].

The wheel and gyro rotor of many precision gyroscopes are encased in a fluid-tight sphere which is floated in fluid. The gas within the sphere may be nitrogen or some other inert gas at atmospheric pressure. The case should be made of magnetic material so that any stray fields produced by the motor are shielded from the sensitive pickoff on the gyro.

Servomechanism Components 339

Single-degree-of-freedom gyros can be supported in a hydraulic or pneumatic bearing. The gyro weight is often buoyed up with a flotation fluid which has a viscosity that is stable with temperature and is large enough to produce a high damping. The flotation fluid reduces the gimbal bearing loading, which in turn lowers the gimbal bearing friction. It also provides good shock cushioning for the gimbal during vibration and acceleration conditions. An expansion device is incorporated on the side of the gyro to provide for expansion of the fluid with temperature and altitude. The fluid must be inert so it does not react with any of the exposed wiring or metal within the unit.

Typical values for an accurate gyroscope are

$$\omega_s = 24{,}000 \text{ rpm} = 2{,}515 \text{ rad/sec}$$

(in actual practice the frequency used to drive the wheel should be slightly different from 400 cps to prevent interaction with other a-c quantities) and

$$I_s = 1.2 \times 10^3 \text{ g-cm}^2 = 1.2 \times 10^3 \text{ dyne-cm-sec}^2$$

$$H = I_s \omega_s = 3 \times 10^6 \text{ dyne-cm-sec}$$

$$I = 10^4 \text{ g-cm}^2 = 10^4 \text{ dyne-cm-sec}^2$$

$$D = 10^5 \text{ dyne-cm-sec}$$

$$J = 10^6 \text{ g-cm}^2 = 10^6 \text{ dyne-cm-sec}^2$$

$$K = 4 \times 10^4 \text{ dyne-cm/rad}$$

Some of the typical terms are

$$\frac{H^2}{J} = 9 \times 10^6 \text{ dyne-cm/rad} \qquad (7\text{-}79)$$

(notice that this is considerably larger than the output axis spring rate $H^2/J = 9 \times 10^6 \gg K = 4 \times 10^4$) and

$$\sqrt{\frac{H^2}{JI}} = \text{nutation frequency} = 30 \text{ rad/sec} \qquad (7\text{-}80)$$

The nutation frequency is the undamped natural resonant frequency of the characteristic equation. With $D = 0$ in the denominator of either

Eq. (7-70) or (7-71), the characteristic equation is

$$Is^2 + \frac{H^2}{J} = 0 \qquad (7\text{-}81)$$

The roots of this equation are

$$s_i = \pm_j\omega_n = \pm j\,\frac{H}{\sqrt{IJ}} \qquad (7\text{-}82)$$

ω_n, the nutation frequency, is the frequency at which the undamped gyro would oscillate and is a figure of merit of the gyroscope.

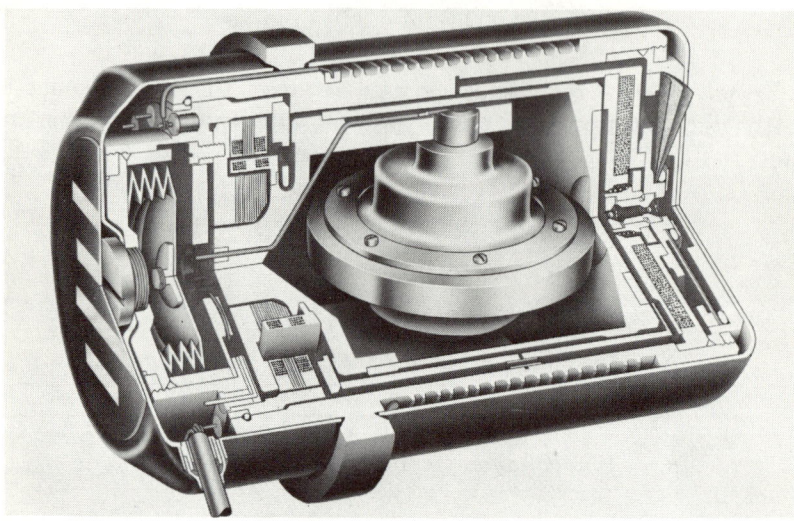

Fig. 7-61 Cutaway view of a single-degree-of-freedom gyro, the HIG gyro. (*Courtesy of the Aeronautical Division, Minneapolis-Honeywell Regulator Co., Minneapolis, Minn.*)

Figure 7-61 shows a cutaway sketch of a single-degree-of-freedom gyro with a floated gimbal construction. The gyro element, a symmetrical wheel, is mounted in a sealed container. A precise frequency source is required to keep H $(=I\omega_s)$ constant and hence induce little error into the torque equations (7-70) and (7-71).

The position of the gyro gimbal is indicated by an electromagnetic pickoff. Torque is applied to the gimbal by an electromagnetic (permanent-magnet type) torque generator (torquer). The sensitivity of the gimbal pickoff varies with the primary excitation current and frequency. The drift rate is less than 0.1°/hr.

The gyro is maintained at an elevated temperature by an electric heater. A built-in temperature-sensitive resistance element measures and controls the gyro temperature. Operating temperature is maintained by cycling the heater on and off by means of an external temperature control.

7-26 The Vertical Gyro. The vertical gyro, which is a special class of free gyro, is used to indicate the direction of the vertical. A pendulous

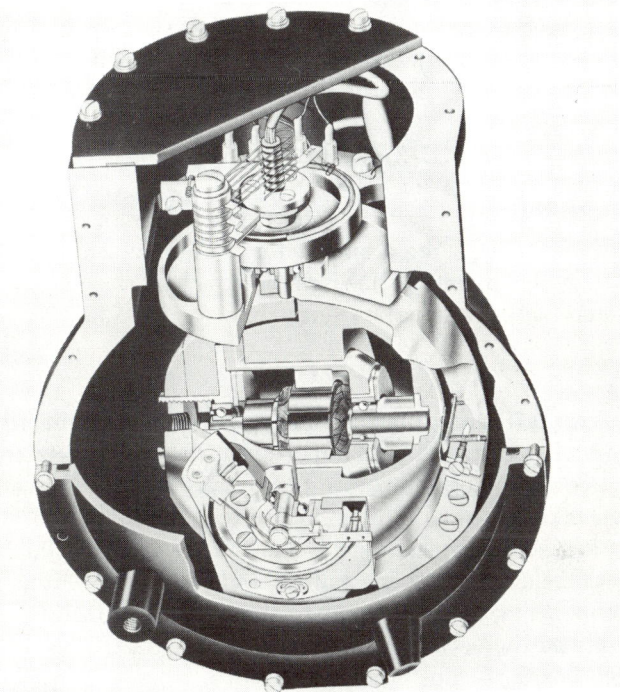

Fig. 7-62 Cutaway view of a vertical gyro. (*Courtesy of Summers Gyroscope Co., Santa Monica, Calif.*)

body applies torque electrically, hydraulically, or pneumatically to the appropriate free gyro axis with the result that the gyro is caused to precess so that one of its axes is aligned with respect to local gravity. The process of aligning the axis with respect to gravity is termed "erection." In order that a gyro will respond in a negligible fashion to accelerations (other than the earth's gravitation) of the aircraft, various schemes such as erection cutouts, limiters, and filters are employed.

Erection rates may vary from 10 to 20 min, and drift in these gyros may be from $\frac{1}{2}$ to 1°/min. Vertical gyros are commonly used to detect or to control changes in attitude about the pitch and roll axes of an airborne vehicle. Sensitivity is obtained about these two axes by mounting the gyro with its spin axis vertical.

A cutaway view of a roll-and-pitch vertical gyro is shown in Fig. 7-62. This unit is approximately 4 in. in diameter by $5\frac{1}{2}$ in. long and weighs $3\frac{1}{2}$ lb. Potentiometers on the inner and outer gimbals yield pitch and roll angles with an accuracy of $1\frac{1}{2}°$.

From a practical point of view the precession direction of a gyro can be determined as follows: When a torque is applied about a particular axis of the gyro (any axis other than the spin axis), the gyro will rotate in such a fashion that the spin axis will precess into the axis about which the torque is applied.

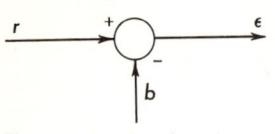

Fig. 7-63 Symbol for a subtractor.

7-27 Subtractors. The symbol shown in Fig. 7-63, which is always seen in control-system applications, is defined in Chap. 1 to be a subtractor. This element symbolizes the property that the signal to the right is equal to the input minus the fed-back quantity. In equation form this is written as

$$\epsilon = r - b \tag{7-83}$$

Since the error introduced by the comparison unit goes directly into the system error, the subtractor is an important item in control-system design.

In practice the subtractor can take many forms. Sometimes the subtraction takes place as a difference of torques applied to an axis (as in the gyro-stabilized platform). The subtraction can take place in the magnetic field (synchro transmitter and receiver). Often, however, special components must be used to perform this function. Some of them, described here, are differential gearboxes, transformers, difference amplifiers, and resistance networks.

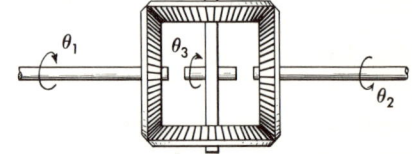

Fig. 7-64 Mechanical subtractor, a differential gearbox.

7-28 The Differential Gearbox. Figure 7-64 indicates, in schematic form, a differential gear system which yields an output shaft position equal to the difference between θ_1 and θ_2. If $\omega_1 = \omega_2$, then the inner shaft velocity will be zero. If $\omega_1 \neq \omega_2$, then θ_ϵ assumes a value proportional to the difference of position as follows:

$$\theta_\epsilon = \tfrac{1}{2}(\theta_1 - \theta_3) \tag{7-84}$$

where θ_1 and θ_2 are the angles of rotation of the end gears and θ is the angular rotation of the spider gears or differential housing.

7-29 Transformers. For a-c systems, transformers are used in a variety of ways to obtain a subtraction. In Fig. 7-65 three methods of transformer subtraction are indicated. The transfer function for all these connections is identical; i.e.,

$$e_\epsilon = N(e_{\text{in}} - e_f) \tag{7-85}$$

The choice of one particular circuit of Fig. 7-65 is a compromise between size (that of Fig. 7-65c is the smallest) and the null condition (that of Fig. 7-65a is most desirable). In Fig. 7-65a the difference is taken in the voltages at the secondary of the transformer. Even in the

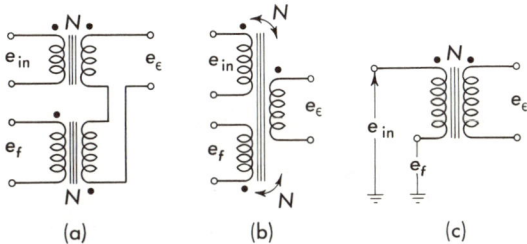

Fig. 7-65 A-c subtraction using transformers.

null condition ($e_\epsilon = 0$), the flux in each transformer core is maintained at a large value because the e_{in} and e_f may be large. In Fig. 7-65b the difference is taken in the flux within the inner core. Since the fluxes from the primary windings cancel, the excitation of the transformer is so small that near the null condition the output e_ϵ is often distorted. The configuration of Fig. 7-65c is the smallest, but it has less desirable characteristics. The subtraction takes place in the voltages at the primary to the transformer. Near the null condition, the transformer excitation is small and the output voltage e_ϵ is distorted from sinusoidal. Also, for appropriate operation, the internal impedance of the sources e_{in} and e_o must be nearly equal.

The use of transformers in computer and control-system applications has changed the design and appearance of these units. Low phase shift, constant transfer function (turns ratio) with changing load and changing temperature, and small size are characteristic of these units. Both laminated and toroidal cores are used. Phase shift of 10 milliradians and transfer function (turns ratio) accuracy of 1 part in 10,000 are available with high-quality toroids.

7-30 Difference Amplifier. The schematic of a difference amplifier for either a-c or d-c signals is shown in Fig. 7-66. The gain expressions for the difference amplifier can be obtained by reference to a text* on

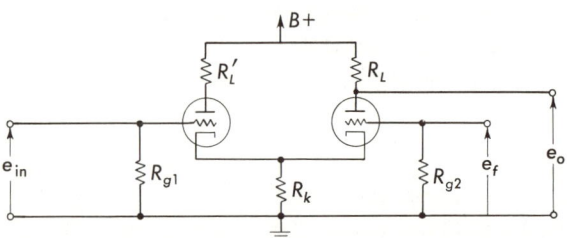

Fig. 7-66 Schematic diagram of a difference amplifier.

vacuum-tube or transistor circuits. For $(R_L + r_s)/(1 + \mu) \ll R_K$ the transfer function is given by

$$e_o = \frac{\mu R_L}{R_L + 2r_s} (e_{in} - e_f) \qquad (7\text{-}86)$$

where μ is the plate resistance and r_s is the amplification factor of both tubes. Careful adjustment of circuit values (especially R'_L) results in a good subtractor. It is important that the common cathode resistor R_k be large.

7-31 Resistance Subtraction. Common in feedback amplifier design is resistance differencing, shown schematically in Fig. 7-67. The amplifier, which can be either d-c or a-c, has high negative gain $-A$. If ϵ is the voltage of the node at the amplifier input, the sum of the currents into the node can be written as

$$\frac{e_{in} - \epsilon}{R_1} + \frac{e_f - \epsilon}{R_2} + \frac{e_o - \epsilon}{R_3} = 0 \qquad (7\text{-}87)$$

Since the amplifier has large gain, ϵ is small and the output is

$$e_o = -\frac{R_3}{R} (e_{in} + e_f) \qquad (7\text{-}88)$$

where $R_1 = R_2 = R$. If e_{in} and e_f are fed into the summing point with opposite signs, a difference is obtained. If $R_1 \neq R_2$, the gain of the

* See Ref. 32, pp. 111–117.

feedback signal can be changed with respect to the gain of the input signal:

$$e_o = -\frac{R_3}{R_1}e_{\text{in}} - \frac{R_3}{R_2}e_f \qquad (7\text{-}89)$$

Any number of signals can be added or subtracted at the input simply by bringing more resistors into the summing point, as shown in Fig. 7-68. An amplifier of this type finds wide application in analog computers.[35]

When the subtraction is performed on d-c signals (for d-c systems), a d-c amplifier must be used. For suppressed-carrier summing, an a-c

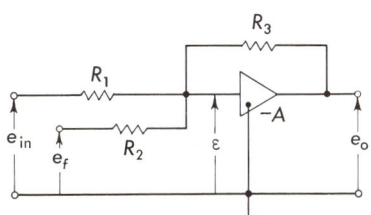

Fig. 7-67 Resistance subtraction using a high-gain amplifier.

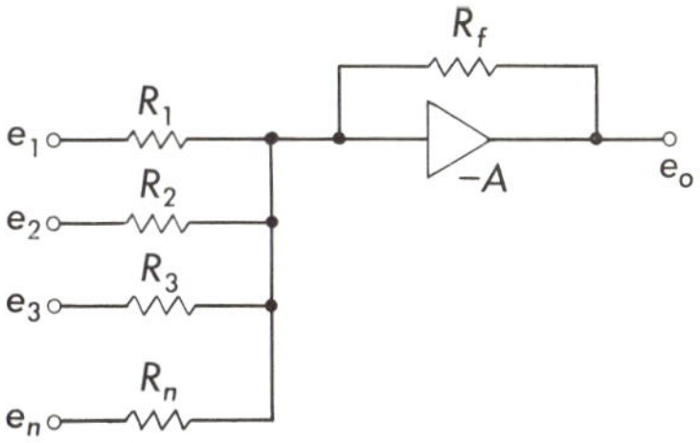

Fig. 7-68 Addition of many inputs.

amplifier can be used. Just as the other control-system components—motors, gear trains, synchros, potentiometers—are available from manufacturers, so also can package amplifiers be obtained.

7-32 Demodulators and Modulators. Suppressed-carrier signals, which are found in many a-c systems, are converted to direct current with a phase-sensitive demodulator. A modulator converts the d-c signal to alternating current. Application of both phase-sensitive demodulators and modulators is indicated in Chap. 6. The output d-c signal from a "phase-sensitive" demodulator is proportional to the amplitude of the a-c input. The polarity, however, depends upon the phase of the signal with respect to the reference. If the suppressed-carrier signal is in phase with the reference, the output is positive; if the signal is 180° out of phase with the reference, the output is negative.

The operation of a phase-sensitive demodulator is explained with the aid of Fig. 7-69. An a-c signal which is suppressed-carrier-modulated with a sinusoid is shown in Fig. 7-69a. When a reference signal, which is shown in Fig. 7-69b, is added, the waveform of Fig. 7-69c is obtained. Rectification of the signal shown in Fig. 7-69c yields a signal, presented in Fig. 7-69d, which is filtered to produce the low-frequency sinusoid of Fig. 7-69e.

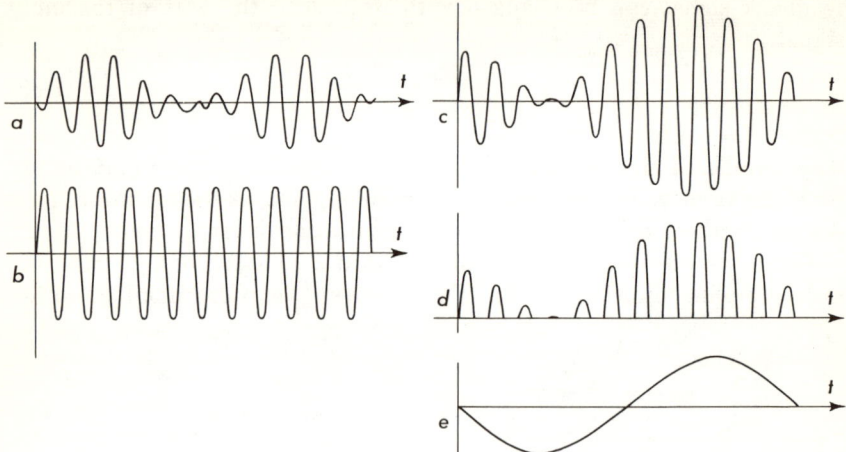

Fig. 7-69 Waveforms demonstrating phase-sensitive demodulation.

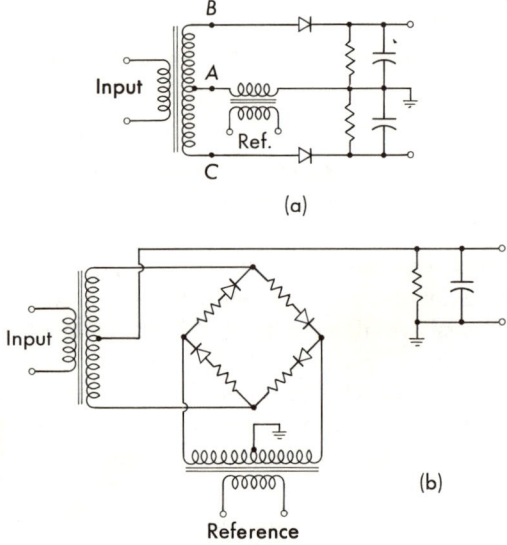

Fig. 7-70 Demodulator circuits.

There are many practical circuits* that accomplish the demodulation shown in Fig. 7-70. Two typical circuits are shown in Fig. 7-70. A half-wave diode demodulator is shown in Fig. 7-70a. The reference voltage is applied symmetrically to both diodes. With the input signal zero, the diodes conduct when points B and C are positive. During

* See Ref. 1, pp. 68–86.

conduction, equal currents flow through the load resistors, but in opposite directions. Hence the output voltage is zero. During the negative half cycle the diodes do not conduct and the output voltage is zero.

As an a c signal is applied, the voltage on the secondary of the transformer adds to the reference voltage on one side of the center tap and subtracts from the reference voltage on the other side of the center tap. During the positive (conducting) cycle one diode conducts more than the other. The voltages dropped across the load resistors are no longer equal, and the net voltage across the output assumes a value. This voltage is proportional to the input provided the reference voltage is larger than the amplitude of the signal voltage. The output voltage is smoothed with an RC filter.

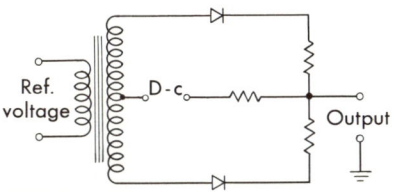

Fig. 7-71   Schematic of a modulator.

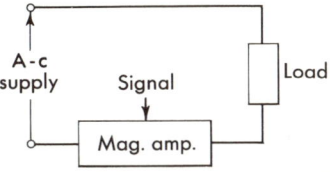

Fig. 7-72   Magnetic amplifier inserted between source and load.

The "ring demodulator" of Fig. 7-70b provides a full-wave rectified voltage with respect to ground. The analysis of the circuit is similar to that of Fig. 7-70a. The upper diodes conduct during the first half cycle, and the lower diodes during the second half cycle. An RC filter is again used to smooth the output.

Most phase-sensitive demodulator circuits can be reversed and used as modulators. A phase-sensitive modulator produces an a-c voltage which is proportional to the input d-c voltage magnitude and has a phase either 0 or 180° with respect to the reference, depending upon the polarity of the d-c input. Figure 7-71 shows a circuit for a modulator.

7-33 Magnetic Amplifiers. This type of amplifier is often superior to a vacuum-tube or transistor amplifier because of its greater dependability and reliability, longer life, and, in some cases, smaller size and lighter weight. Magnetic amplifiers are essentially variable-impedance devices. Their operation depends in general on their insertion between a source of a-c power and a load in a circuit such as that shown in Fig. 7-72. By virtue of the controllable impedance offered by the magnetic amplifier, the output circuit current and, therefore, the load power and voltage can be controlled. The impedance of the amplifier may be varied by an independent source of control power. Amplification occurs

348   Control System Design

because the power requirements of the control source may be many times less than the controlled power supplied to the load.

PROBLEMS

7-1. For the circuit of Fig. 7P-1 determine the loading error as a function of potentiometer setting for the following values of θ: 0.2, 0.4, 0.6, 0.8, and 1.00. Let $R_p = R_L$ and assume that the potentiometer is center-tapped.

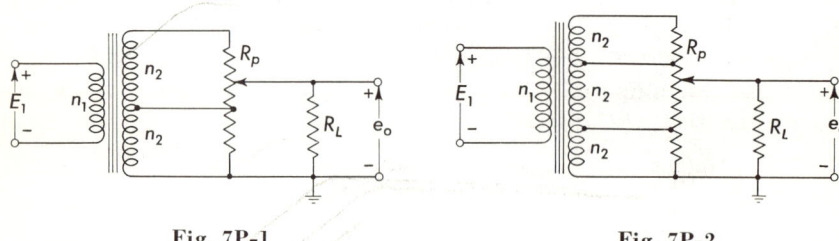

Fig. 7P-1   Fig. 7P-2

7-2. Repeat Prob. 7-1 for two equally spaced potentiometer taps, as shown in Fig. 7P-2, with $R_p = R_L$. Generalize the result to i taps and determine the effect of taps on potentiometer loading.

7-3. For the loading curve of the circuit of Fig. 7-10, with $\beta = 0.2$, fit a power series equation of the form

$$\frac{e}{E} = A\theta^n$$

Determine A and n for the best fit and find the greatest error. n is not necessarily an integer. *Ans.* $A = 1$, $n = 0.675$

7-4. Design an electromechanical integrator utilizing an a-c rate generator with a $K_s = 2.87 \times 10^{-2}$ volt/(rad/sec). Lay out a system block diagram such that an output rate of 0.1 rad/sec results when a step of 1.0 volt is applied to the input.

7-5. Find a passive circuit equalizer to stabilize the accelerometer which is described by Fig. 7-40.

7-6. Lay out a block diagram for a force-balance pressure transducer. Discuss the advantages and disadvantages of such a unit.

7-7. Mach number is defined by the ratio of true air speed to the speed of sound measured in the same units and under the same conditions. Hence the equation

$$M = \frac{V_t}{C} = \frac{\text{true air speed}}{\text{speed of sound}}$$

If P_s is the static atmospheric pressure and P_p is the stagnation or pitot pressure, the Mach number is defined by the following equations:

$$\frac{P_p}{P_s} = (1 + 0.2M^2)^{3.5} \quad \text{for } M \leq 1.0$$

$$\frac{P_p}{P_s} = \frac{167M^2}{(7 - 1/M^2)^{2.5}} \quad \text{for } M \geq 1.0$$

Design a servo comprising two pressure transducers, amplifier, motor, and function potentiometers which will compute Mach number.

7-8. Obtain the transfer function for the motor and gear train whose characteristics are shown in Fig. 7P-8. The motor inertia is 2.28×10^{-4} oz-in.-sec², the load inertia is 3 oz-in.-sec, and the gear ratio coupling the motor to the load is 100:1.

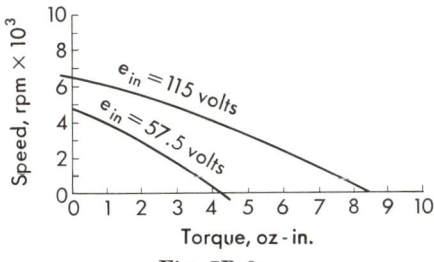

Fig. 7P-8

7-9. Derive the equations for the difference amplifier of Fig. 7-66.

7-10. The following numerical values represent a single-degree-of-freedom gyro:

$$H = 1 \times 10^6 \text{ dyne-cm-sec}$$
$$I = 10^4 \text{ g-cm}^2$$
$$D = 10^5 \text{ dyne-cm-sec}$$
$$J = 10^6 \text{ g-cm}^2$$
$$K = 5 \times 10^4 \text{ dyne-cm/rad}$$

(a) Find the gyro nutation frequency.
(b) Obtain the transfer function θ/L and locate the poles in the s plane.

7-11. Use the gyro of Prob. 7-10 in the block diagram of Fig. 7P-11.
(a) Let $G_2(s) = $ const and determine stability and steady-state errors.
(b) Let $G_2(s) = 1 + \alpha s$ and choose α and A for the optimum system.
(c) Let $G_2(s) = 1 + \alpha s + \gamma/s$ and choose α, γ, and A for the optimum system.
Note: An optimum system has zero steady-state error in θ.

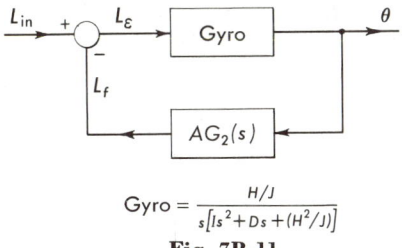

$$\text{Gyro} = \frac{H/J}{s[Is^2 + Ds + (H^2/J)]}$$

Fig. 7P-11

7-12. Find the system damping ratio ζ and the undamped resonant frequency ω_n for the system of Fig. 7P-12 with the following values:

$$J_m = 1.2 \text{ g-cm}^2$$
$$J_L = 2,500 \text{ g-cm}^2$$
$$A = 1,000$$
$$N_1 = 20$$
$$N_2 = 1,000$$
$$\theta_{max} = 360°$$
$$\text{Motor stall torque} = L_0 = 0.30 \text{ oz-in.}$$
$$\text{No-load speed} = \omega_0 = 800 \text{ rad/sec}$$
$$\text{Rated control volts} = 115 \text{ volts}$$

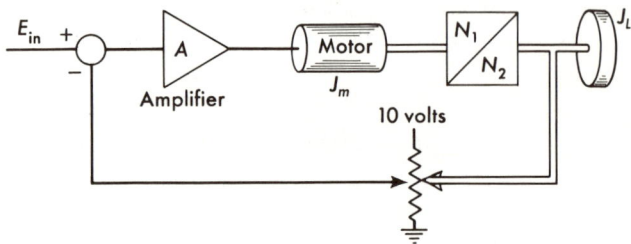

Fig. 7P-12

7-13. Design a torque-balance accelerometer using the gyro as shown in Fig. 7P-13. Use the gyro component values given in Prob. 7-10 and use the servomotor in Eq. (7-47). Other values are $N = 100$, $K_p = 1$ volt/deg, $ml = 0.001$ lb-sec^2.
 (a) Draw the block diagram.
 (b) Calculate the scale factor, in volts/(ft/sec^2).
 (c) Equalize the system and obtain a maximum gain for $\zeta = 0.2$.

Notice that the unbalance mass m is so placed that the instrument is sensitive only to accelerations in the direction of the H vector.

7-14. The system in Fig. 7P-14 is a force-balance accelerometer. As an acceleration a is applied in the direction shown, the inertia disk J rotates and causes a signal to be applied to the amplifier through the position pickoff. The signal is equalized and used to drive the motor (with constants K_m, τ_m, J_m). The shaft is coupled through a viscous coupler to the inertia J. The resulting torque, which is proportional to the velocity of the motor, nulls the torque applied by the input acceleration. Assume that $\theta_1 \ll \theta_0$.
 (a) Draw the block diagram.
 (b) Show that this system will measure linear acceleration.
 (c) Calculate the scale factor E/a in volts/(ft/sec^2).
 (d) Suggest an equalizer.

Servomechanism Components 351

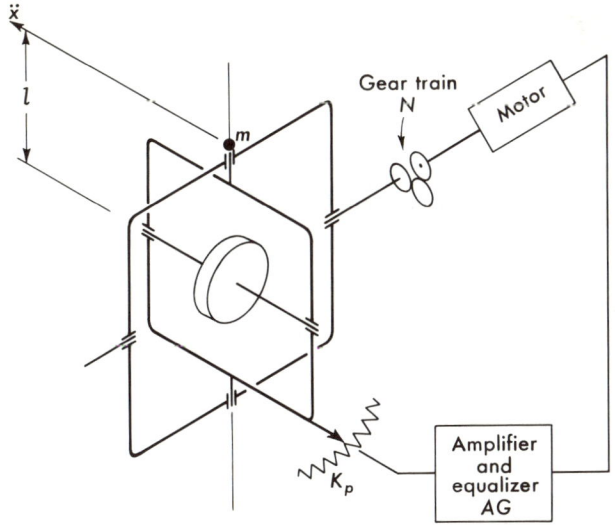

Fig. 7P-13

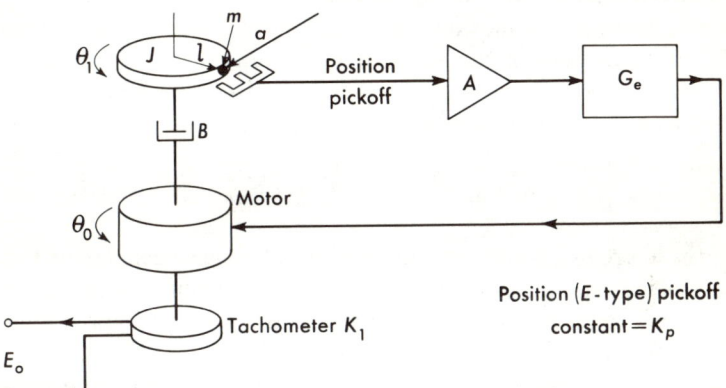

Fig. 7P-14

chapter 8

NONLINEARITIES IN CONTROL-SYSTEM DESIGN

8-1 Classification of Nonlinearities. In preceding chapters, only linear control-system equations have been considered. In these chapters, amplifiers have been assumed to operate only in their linear ranges; motor speed-torque curves have been linearized (cf. Chap. 1). These systems have been described by a linear differential equation of the form

$$a_n \frac{d^n y}{dt^n} + a_{n-1} \frac{d^{n-1} y}{dt^{n-1}} + \cdots + a_1 \frac{dy}{dt} + a_0 y = f(t) \qquad (8\text{-}1)$$

where the a_i are real constants independent of time or dependent variable. These systems have been assumed linear to simplify design considerations. Since system performance may deviate from that predicted from linear theory, this assumption is further evaluated in this chapter.

Actually, the springs and viscous dampers (Fig. 1-6) and the equations developed in Chap. 2 are linear only within restricted regions. For values near the origin, or null, and for large excursions, these units deviate from linearity. A spring actually may have the characteristics shown in Fig. 8-1. This spring has a preload force producing a discontinuity near the origin. For large displacements the spring force rate $\Delta f/\Delta x$ increases owing to interference of the spring coils.

Familiar stress-strain diagrams, gas-expansion laws, and even the simple pendulum cannot be described in all regions with linear equations. The electrical engineer is familiar with nonlinearities due to saturation of iron-core inductors and rotating components using iron magnetic circuits. Nonlinear vacuum-tube and transistor characteristics are fur-

Nonlinearities in Control-system Design 353

ther examples. These are just a few of the common nonlinear components with which control-system engineers grapple in the design of feedback control systems.

Why does the solution of nonlinear differential equations require special treatment? Basically the answer is that the principle of superposition is invalid. The solution of differential equations by Laplace-transform or other methods depends upon superposition.

Nonlinearities in control systems are conveniently divided into two classes: (1) those which are inherent in the system and (2) those which are purposely inserted into the system to improve the design. The first class includes all the unavoidable phenomena such as saturation, threshold, and backlash. Component manufacturers spend considerable time and money to reduce these nonlinearities in their equipment.

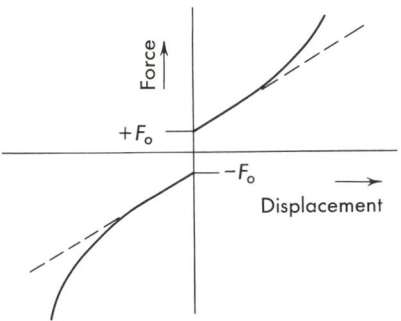

Fig. 8-1 Force vs. displacement of a preloaded spring.

Nonlinearities of the second class—those intentionally inserted—are finding increased application in extending the control possibilities of all-linear systems. Automatic gain control, which changes the gain of the system with signal level, and the aircraft acceleration limiter, which limits the maximum fin displacement depending upon the acceleration loads, are two examples of the second type of nonlinearity.

Control-system nonlinearities are also classed according to their magnitude, small or large. Unavoidable nonlinearities are often small, since every attempt is usually made to reduce them. The response of systems containing small nonlinearities usually deviates little from the linear design. The nonlinearities which are purposely inserted into the system are usually of larger magnitude. The differences between these two are summarized:

1. Analysis of systems with large nonlinearities is usually more difficult than for small nonlinearities.
2. New phenomena such as sudden jump in frequency response and subharmonic resonance often result when the nonlinearity is large.

No *general* method of analysis and design has been developed for solving nonlinear systems. Methods have been developed to treat certain nonlinearities, and for this reason nonlinearities have been divided into the two basic types, small and large. With few reservations, small nonlinearities are generally treated in an indirect manner. An analysis or design is made upon the linear system with constant coefficients. The effect of the nonlinearity is considered as a deviation from this linear result.

For large nonlinearities not only is there no general analysis of these systems but even an analysis of a given system does not reveal all details of its operation. For instance, the mode of attack used to demonstrate subharmonic frequency response is entirely different from that used to demonstrate jump phenomena. Since nonlinear system behavior is a function of input, it is not sufficient to design for simple functions such as a step or ramp. In fact, deliberately introduced nonlinearities may improve a system for certain inputs and degrade its performance for others.

The field of nonlinear systems is quite broad and is continually expanding as a result of research activity in this field. The material presented in this chapter is intended as an outline, and the reader is referred to the bibliography included at the end of this book for additional details.[2,22,24,36]

8-2 Linearization of Small Nonlinearities. Because of the additional complexity in designing nonlinear feedback control systems, the first analysis given here is based upon a linearized, although approximate, equation. For example, the motor speed-torque curves are linearized (cf. Chaps. 1 and 7) by drawing straight lines in the region of use.

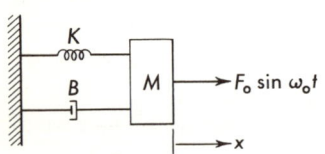

Fig. 8-2 Mechanical system used to demonstrate equivalent viscous damping.

8-3 Equivalent Damping. As an aid to linearizing nonlinear functions, energy techniques are often applied. Figure 8-2 shows a mechanical system containing a viscous damper. The response x is found from impedance techniques (cf. Chap. 2):

$$x = A \sin(\omega t + \theta) \qquad (8\text{-}2)$$

where A is the amplitude of the displacement and depends upon F_0, B, K, and M as follows:

$$A = \frac{F_0}{\sqrt{(K - M\omega^2)^2 + B^2\omega^2}} \qquad (8\text{-}3)$$

The energy dissipated per cycle, which, in this system, is dissipated in the damper B, is

$$\int_0^{2\pi/\omega} F_B \, dx = \int_0^{2\pi/\omega} F_B \frac{dx}{dt} \, dt \qquad (8\text{-}4)$$

where F_B is the damping force and dx/dt is the velocity. Since all quantities are sinusoidal, the energy dissipated is

$$W_B = \int_0^{2\pi/\omega} \left(B \frac{dx}{dt}\right) \frac{dx}{dt} \, dt = B \int_0^{2\pi/\omega} \left(\frac{dx}{dt}\right)^2 dt \qquad (8\text{-}5)$$

Since $dx/dt = A\omega \cos(\omega t + \theta)$, the integral of Eq. (8-5) becomes

$$W_B = B \int_0^{2\pi/\omega} A^2 \omega^2 \cos^2(\omega t + \theta) \, dt \qquad (8\text{-}6)$$

which upon integration becomes

$$W_B = BA^2 \omega \pi \qquad (8\text{-}7)$$

As an example of the use of Eq. (8-7), consider the problem of finding an equivalent viscous damper (force vs. velocity) to represent the coulomb- or dry-friction damper shown in Fig. 8-3. The energy dissipated per cycle with this type of damping is calculated and equated to Eq. (8-7). Solution for B yields the equivalent damping. The energy dissipated per quarter cycle is

$$W_B = \int_0^{\pi/2\omega} F \, dx = F_1 A \qquad (8\text{-}8)$$

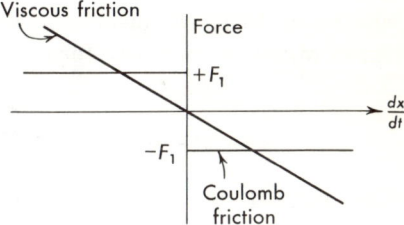

Fig. 8-3 Coulomb- or dry-friction force curve.

The energy dissipated in the complete cycle is $4F_1 A$, and the equivalent viscous damping constant is found by equating Eq. (8-7) to Eq. (8-8):

$$B_{eq} = \frac{4F_1 A}{A^2 \pi \omega} = \frac{4F_1}{A \pi \omega} \qquad (8\text{-}9)$$

This expression for B_{eq} depends upon the amplitude A of the motion. For sinusoidal, steady-state motion, the amplitude is easily found from

Eq. (8-3):

$$A = \frac{F_0}{\sqrt{(K - M\omega^2)^2 + [(4F_1/A\pi\omega)\omega]^2}} = \frac{F_0}{\sqrt{(K - M\omega^2)^2 + 16F_1^2/A^2\pi^2}} \quad (8\text{-}10)$$

Solving for A,

$$A = \frac{F_0(16/\pi^2)[1 - \tfrac{16}{2}(F_1^2/F_0^2)]^{\frac{1}{2}}}{K - \omega^2 M} \quad (8\text{-}11)$$

8-4 Equivalent Spring Constant. For small displacements from an operating point, the slope of the restoring force near the point can be used to predict performance. Consider, for example, the amplifier saturation curve of Fig. 8-4. Suppose, because of bias, noise, quadrature voltages, etc., the operating point occurs at 20 mv. For stability analysis the incremental slope $\Delta e_o/\Delta e_{in}$ at the operating point can be used. This gain is considerably less, as can be seen from Fig. 8-4, than the gain at null.

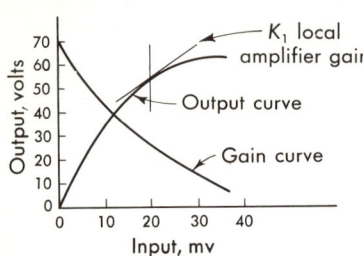

Fig. 8-4 Equivalent gain for a saturated amplifier.

8-5 The Describing-function Method.* If a sinusoidal function is the input to a nonlinear element, the output is periodic. A Fourier series of the output waveform can be made, as shown in the schematic

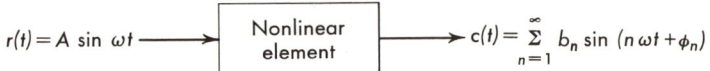

Fig. 8-5 Block diagram demonstrating the describing-function method.

of Fig. 8-5. The higher harmonics can be neglected for either of the following reasons:

1. The transmission of the higher harmonics in the system is less than the transmission of the fundamental. This is the behavior of a system whose frequency response resembles a low-pass filter, which is usually the case for a control system.

2. The signal amplitude is such that the effect of the nonlinearity is small and the higher harmonics present are negligible.

* The first presentation of this work in the United States is credited to R. J. Kochenburger, Ref. 21.

Nonlinearities in Control-system Design 357

If these characteristics are present and no subharmonics are present, the nonlinearity can be represented by an equivalent linear transfer

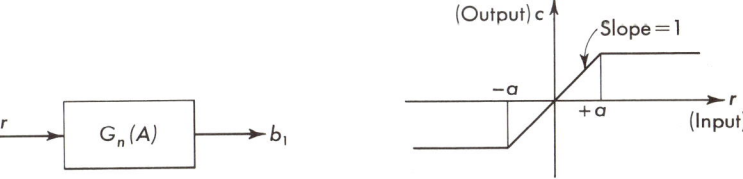

Fig. 8-6 Block diagram for describing function.

Fig. 8-7 Saturation, single-valued nonlinearity.

function. This transfer function, termed the "describing function," depends upon the signal amplitude:

$$G_n = \frac{\text{complex amplitude of output fundamental}}{\text{complex amplitude of input sinusoid}} \quad (8\text{-}12)$$

G_n, the describing function, may be real or complex. It is represented by the block diagram of Fig. 8-6.

If the nonlinearity is single-valued, as in the case of amplifier saturation shown in Fig. 8-7, G_n is a real quantity whose magnitude depends upon the amplitude of the input signal. If the nonlinearity is double-valued, as in the case of backlash shown in Fig. 8-8, G_n is a complex quantity whose magnitude and phase depend upon the amplitude of the input signal. If the nonlinearity occurs in one part of a differential equation, as does that due to an iron-core inductor, a nonlinear damper, the describing function depends upon both frequency and amplitude. Only real describing functions are considered in this text.*

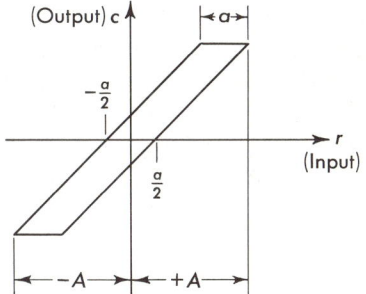

Fig. 8-8 Backlash, double-valued nonlinearity.

The nonlinear system is approximated by a linear system, and a block diagram for the system is drawn as in Fig. 8-9 with the limitations described previously. The frequency-analysis technique of Chap. 5 is applicable under these assumptions. As the amplitude of the sinusoidal input to the nonlinear element varies, the gain and phase of G_n change and hence produce a change in gain and phase in the closed-loop system.

* Additional information is found in Refs. 15, 21, and 38.

358   Control System Design

8-6 Describing Functions for Common Nonlinearities.* Some of the unavoidable nonlinearities such as saturation and threshold that are encountered in control-system design are considered in this section.

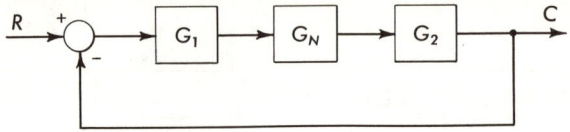

Fig. 8-9   Block diagram for describing-function approximation.

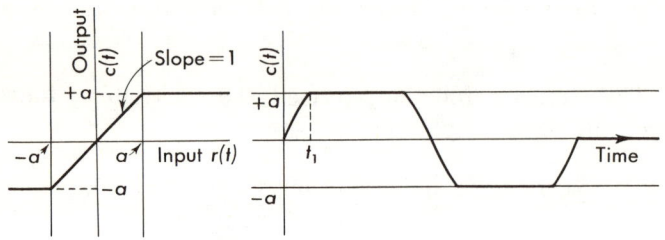

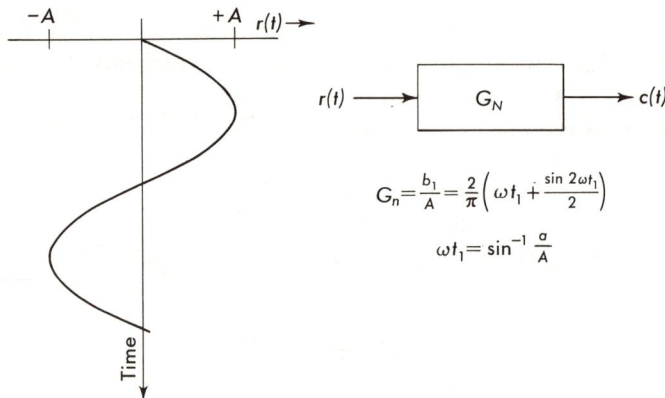

Fig. 8-10   Waveform and describing function for saturation.

8-7 Describing Function for Saturation. The phenomenon of saturation, or limiting, exists in many physical systems. Figure 8-7 depicts an idealized form of saturation. The output is proportional to the input, with unity proportionality constant, until the output reaches $\pm a$. For larger amplitudes the output remains at the value $\pm a$. If the amplifier should have a gain K rather than unity, the amplifier can be broken into two blocks placed in series:

* See also Ref. 15 on this same topic.

Nonlinearities in Control-system Design

1. A linear block with transfer function K (the amplifier gain in the unsaturated region)
2. A nonlinear block consisting of G_n, the describing function for an amplifier of unity gain and limited at $\pm a$

When a sinusoidal input (e.g., a voltage) $r(t) = A \sin \omega t$ is applied to the saturating element input (e.g., a transistor amplifier), the output has the form shown in Fig. 8-10. Peaks of the sinusoid, above amplitude $\pm a$, are clipped off. The output $c(t)$ has the form

$$c(t) = \sum_{n=1}^{\infty} b_n \sin n\omega t \qquad (8\text{-}13)$$

Since the output (shown in Fig. 8-10) is antisymmetrical with respect to the origin, only the $\sin n\omega t$ terms appear in the output Fourier series. Because the sinusoid is symmetrical about $\pi/2$ in the interval 0 to π, only the odd harmonics appear. The coefficients b_n are found from the Fourier series expansion

$$b_n = \frac{4}{\pi} \int_0^{\pi/2} c(t) \sin n\omega t \, d\omega t \qquad n = 1, 3, 5, \ldots \qquad (8\text{-}14)$$

where

$$c(t) = \begin{cases} A \sin \omega t & \text{for } 0 < t < t_1 \\ a & \text{for } \omega t_1 < \omega t < \dfrac{\pi}{2} \end{cases} \qquad (8\text{-}15)$$

When Eq. (8-15) is substituted into Eq. (8-14), the values of b_n are given by

$$\frac{b_n}{A} = \frac{2}{n\pi} \left[\frac{\sin (n-1)\omega t_1}{n-1} + \frac{\sin (n+1)\omega t_1}{n+1} \right] \qquad (8\text{-}16)$$

The describing function b_1/A is given by the expression

$$\frac{b_1}{A} = \frac{2}{\pi} \left(\omega t_1 + \frac{\sin 2\omega t_1}{2} \right) \qquad (8\text{-}17)$$

The equations for the two higher harmonics b_3/A and b_5/A are

$$\frac{b_3}{A} = \frac{2}{2\pi} \left(\frac{\sin 2\omega t_1}{2} + \frac{\sin 4\omega t_1}{4} \right) \qquad (8\text{-}18)$$

and

$$\frac{b_5}{A} = \frac{2}{5\pi} \left(\frac{\sin 4\omega t_1}{4} + \frac{\sin 6\omega t_1}{6} \right) \qquad (8\text{-}19)$$

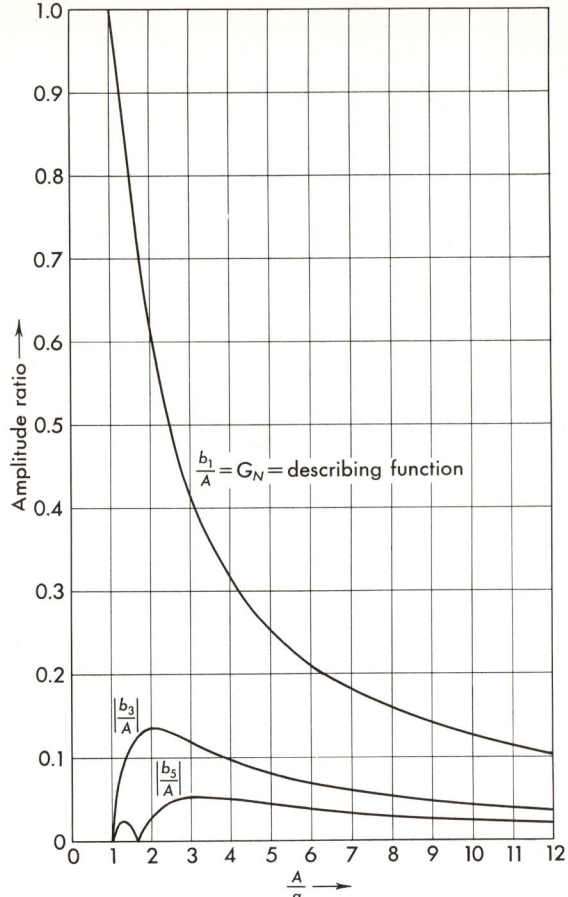

Fig. 8-11 Plot of the describing function for saturation, with third and fifth harmonics included.

Since the output fundamental is in phase with the input sinusoid, the describing function is a real quantity, with zero phase shift. It depends only upon the signal level a/A. The describing function b_1/A and the third and fifth harmonics are plotted in Fig. 8-11.

8-8 Describing Function for Threshold. In many control systems, the input to a component must exceed a certain minimum value before any output is realized. A motor, for example, requires a minimum finite voltage before any motion results. The minimum value is termed

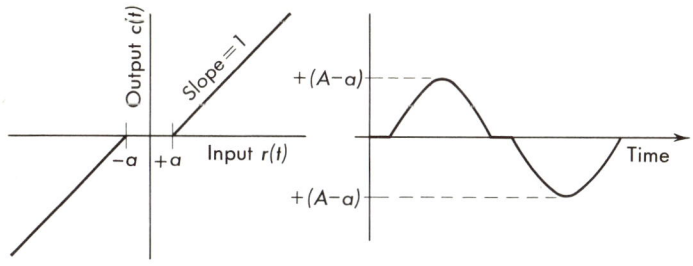

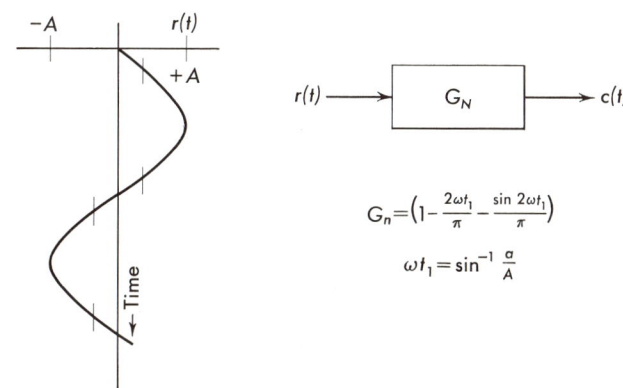

Fig. 8-12 Waveform and describing function for threshold.

"threshold," and inputs below the minimum value fall in a region termed "dead space." This "undesirable" effect is shown on Fig. 8-12, and the output is given by

$$c(t) = \sum_{n=1}^{\infty} b_n \sin n\omega t \tag{8-20}$$

The coefficients b_n are found from the Fourier expansion

$$b_n = \frac{4}{\pi} \int_0^{\omega/2} c(t) \sin n\omega t \, d\omega t \qquad n = 1, 3, 5, \ldots \tag{8-21}$$

where
$$c(t) = \begin{cases} 0 & \text{for } 0 < \omega t < \omega t_1 \\ A \sin \omega t - a & \text{for } \omega t_1 < \omega t < \frac{\pi}{2} \end{cases} \tag{8-22}$$

and
$$\omega t_1 = \sin^{-1} \frac{a}{A}$$

362   Control System Design

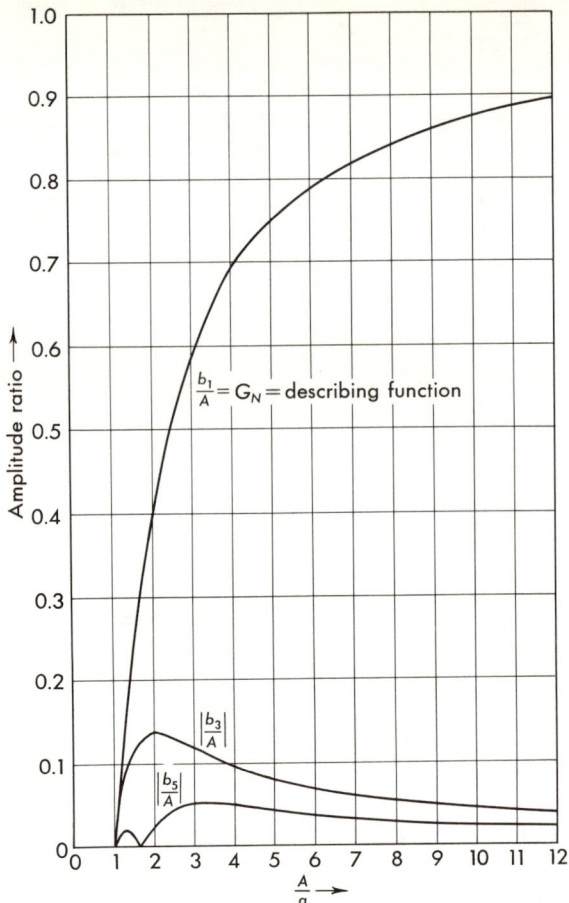

Fig. 8-13 Plot of describing function, with third and fifth harmonics, for threshold.

The output is represented by a sine series only, since $c(t)$ has odd symmetry with respect to the origin. Because the output is symmetrical about $\pi/2$ in the interval 0 to π, only odd harmonics occur. When Eq. (8-22) is substituted into Eq. (8-21), the following equation results:

$$\frac{b_n}{A} = \frac{+2}{n\pi}\left[\frac{\pi}{2} - \frac{\sin(n-1)\omega t_1}{n-1} + \frac{\sin(n+1)\omega t_1}{n+1}\right]$$

$$\text{for } n = 1, 3, 5, \ldots \quad (8\text{-}23)$$

where $\omega t_1 = \sin^{-1} a/A$. The describing function b_1/A, given by

$$\frac{b_1}{A} = \frac{2}{\pi}\left(\frac{\pi}{2} - \omega t_1 - \frac{\sin 2\omega t_1}{2}\right) \quad (8\text{-}24)$$

is plotted in Fig. 8-13 along with the third and fifth harmonics. Because the output fundamental is in phase with the input, the describing function is a real number independent of frequency. Notice that the third and fifth harmonics are the same in magnitude for threshold and saturation. Since the higher harmonics are used to indicate the error approximating the nonlinear function, only the magnitude is plotted.

8-9 Use of the Describing Function in Control-system Design. In order to understand the use of the describing-function method in control-system design, consider the effect of the two nonlinearities discussed

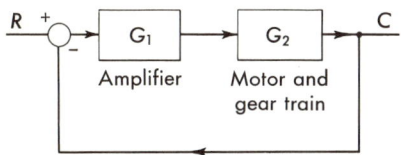

Fig. 8-14 Block diagram of a simple position servo.

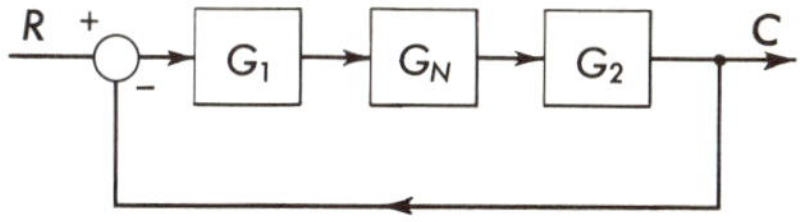

Fig. 8-15 Block diagram of a position servo.

in preceding sections on the simple position servo shown in Fig. 8-14. The system consists of an amplifier with a transfer function

$$G_1 = \frac{K}{1 + 0.2s} \tag{8-25}$$

and a motor gear train with a transfer function

$$G_2 = \frac{1}{s(1 + 0.5s)} \tag{8-26}$$

Each of the nonlinearities is included in the system whose open-loop transfer function is

$$G_1 G_2 = \frac{K}{s(1 + 0.2s)(1 + 0.5s)} \tag{8-27}$$

Since the describing function is derived for an element with a sinusoidal input, the use of this method is valid only near the $j\omega$ axis. Care must be used when extending the method to derive transient information. The frequency-response method is most applicable, since it is based upon sinusoidal driving functions. Both root-locus and frequency response methods are used in the example.

The gain K is set at 1.60, and the system is redrawn to include G_n in Fig. 8-15. The describing function G_n for amplifier saturation is simply a gain change (cf. Fig. 8-11). Use of the describing-function method with frequency analysis is shown in Fig. 8-16. Since the describing function is simply a gain change, the response is found by shifting the 0-db axis. The gain margin for each of the amplitude ratios is included in the figure.

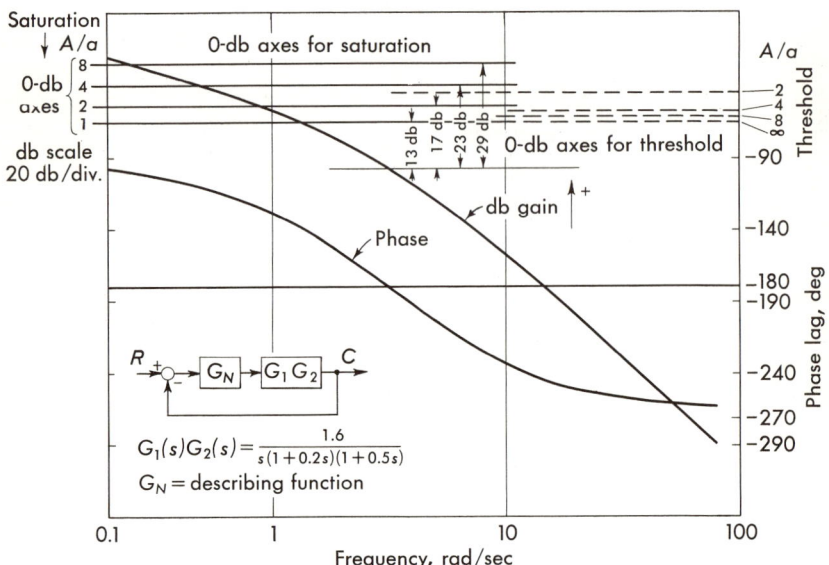

Fig. 8-16 Frequency analysis of a simple position servo using the describing-function approximation for a nonlinear amplifier.

The root-locus diagram, shown in Fig. 8-17, is found in the standard manner (cf. Chap. 4). The gain changes as a function of the signal level A/a, as shown by the describing function of Fig. 8-11, and the response of the system changes with signal level. The loop gain at various amplitude ratios A/a is included in Fig. 8-17.

The describing function for threshold is similar to G_n for saturation, since both are simply gain changes. The frequency response, shown in Fig. 8-16, is made to reflect the effect of threshold by changing the 0-db axes. The root-locus method is applied to this system, as shown in Fig. 8-18. As the signal level A/a changes, the system operates at new values of loop gain with corresponding changes in system response. As the system approaches a null, $A/a \to 1$, the loop gain decreases, and

Nonlinearities in Control-system Design

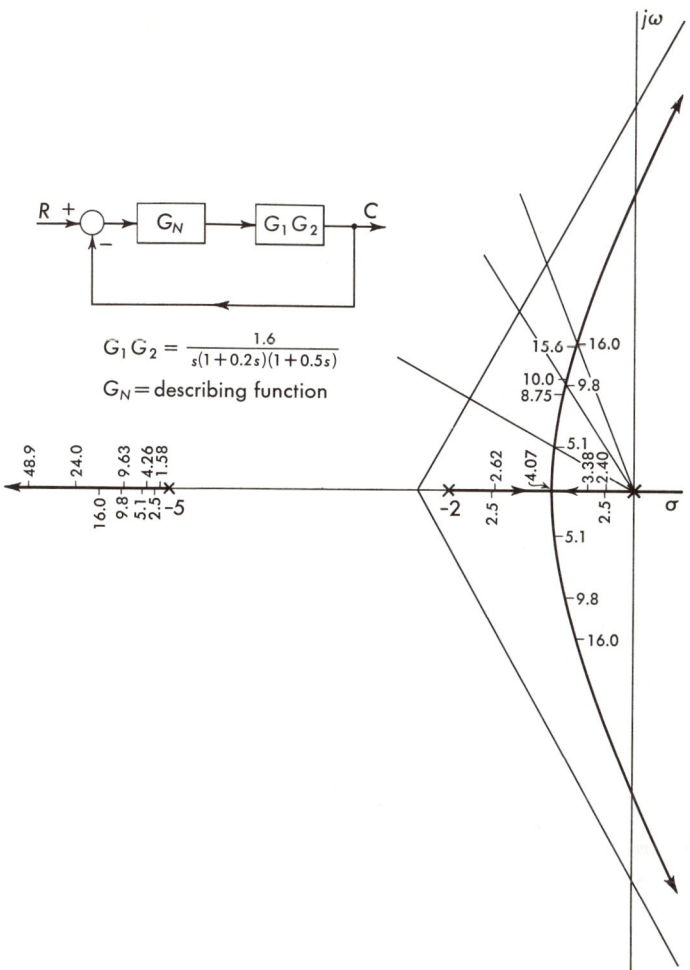

Fig. 8-17 Root-locus–describing-function solution of a simple position servo containing a saturating amplifier.

when A/a becomes less than 1, the loop gain reduces to zero. The case of amplifier saturation is different in that as the system approaches a null, the gain increases and the system operates linearly.

8-10 Limitations of Describing Functions. The basic assumption underlying the describing-function approach to the analysis of nonlinear systems is that, if the input to the nonlinear element is a sinusoidal signal, the fundamental component of the output is sufficient to describe the nonlinearity. Use of the fundamental component is justified on the grounds that:

1. The control system usually acts like a low-pass filter that attenuates the higher harmonics.
2. The harmonics generated from the nonlinearities are usually of smaller amplitude than the fundamental.

Since the input is sinusoidal, the describing function is best suited to the frequency-analysis stability method. However, the Nyquist criterion,

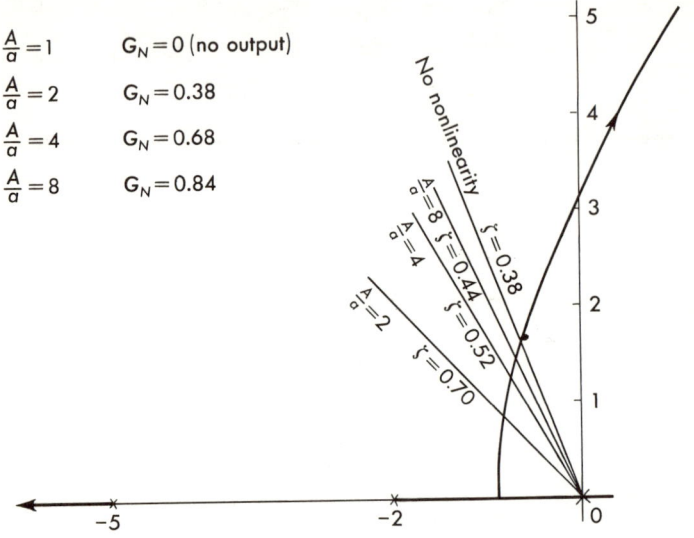

Fig. 8-18 Root-locus diagram for threshold.

i.e., the number of encirclements of the point $-1 + j0$ corresponds to the number of roots in the right half plane, is derived for linear systems. The extension to nonlinear systems is not well defined.

For cases where G_n is simply a change in gain, i.e., saturation and threshold, frequency methods give reasonable information. The root locus is less valuable, since the describing function is not valid away from the $j\omega$ axis. For the case of describing functions which are complex the reader is referred to the literature.[15,21,38]

The describing-function analysis requires a sinusoidal input. This condition exists only if the system is oscillating or is being driven with a sinusoid. The describing function is not easily extended to a system with another input. As the system approaches null, the amplitude, and hence G_n, changes. As the loop gain varies, the response changes with signal level. Use of the describing function must certainly be considered as an approximate method of analysis.

8-11 Topological Solution of Feedback Control Systems.[2,24,36] In subsequent sections emphasis is placed upon transient analysis of nonlinear systems. The topological, or "phase plane," methods presented in these next few sections are subject to the following limitations:

1. They are easily applicable to second-order systems only.
2. Only driving functions that can be represented by initial conditions can be used.

In practice, of course, few feedback control systems can be represented by a second-order differential equation. Often, however, a complex system can be approximated by a second-order system comprising a pair of least damped roots. In any case, investigation of the effects of nonlinearities on the response of second-order systems sheds light on effects of nonlinearities in higher-order systems.

Limitations on the driving function are less serious. Since step, ramp, and impulse functions can usually be represented by initial conditions, these inputs can be studied.

Under the above limitations, topological methods, and especially the all-graphical methods of Sec. 8-14, provide a practical means of investigating some nonlinear equations.

8-12 The Phase Plane. The response of a second-order system, after application of a disturbance, can be described by the variable* x and its first derivative $\dot{x}$. The differential equation solution can be reduced to a form where time is a parameter. The system response is conveniently plotted on a "phase plane" (the vertical axis is the velocity $y = \dot{x}$ and the horizontal axis the displacement x). The path which represents the motion x and $\dot{x}$ is termed the "phase trajectory."

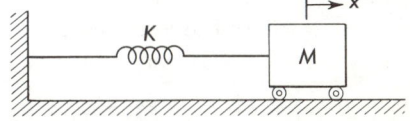

Fig. 8-19 Undamped mechanical system.

To demonstrate the phase-plane solution, consider the undamped mechanical system of Fig. 8-19. The second-order differential equation

---

* The dependent variable is labeled x, and its derivative $dx/dt = \dot{x} = y$. Any particular variable c or ϵ can be later substituted for x.

for the nonforced motion is

$$M \frac{d^2x}{dt^2} + Kx = 0 \tag{8-28}$$

Time is eliminated from this equation by multiplying through by $\dot{x}$ and integrating with respect to time:

$$M \int_0^t \dot{x}\ddot{x}\, dt + K \int_0^t \dot{x}x\, dt = 0$$

which reduces to

$$\frac{M\dot{x}^2}{2} + \frac{Kx^2}{2} = h \tag{8-29}$$

where h is the constant of integration and depends upon the initial conditions. Equation (8-29) can be rewritten

$$\frac{y^2}{a^2} + \frac{x^2}{b^2} = 1 \tag{8-30}$$

where $y = \dot{x}$, $a^2 = 2h/M$, and $b^2 = 2h/K$. Equation (8-29) is the sum of the kinetic ($\tfrac{1}{2}M\dot{x}^2$) and potential ($\tfrac{1}{2}Kx^2$) energies. For a conservative system, i.e., one which has no energy loss (i.e., no damping), the sum of the kinetic and potential energies is constant. This constant h_i depends upon the amount of energy supplied to the system by the initial conditions.

The phase trajectories for various initial conditions, h_i, are plotted from Eq. (8-30) in Fig. 8-20. This equation represents a family of ellipses which are plotted with time indicated by time markers (small marks shown on the h_4 trajectory). These time marks are placed on the curve in such a way that the time solutions

$$x = A \sin\left(\sqrt{\frac{K}{M}}\, t + \phi\right)$$

$$y = \dot{x} = A\sqrt{\frac{K}{M}} \cos\left(\sqrt{\frac{K}{M}}\, t + \phi\right) \tag{8-31}$$

are satisfied. Motion in the phase plane is always in a clockwise direction if the positive directions of x and $\dot{x}$ are assumed as shown in Fig. 8-20.

For the example of Fig. 8-20, the phase plane contours are closed. The time solution, that is, x versus t, or $\dot{x}$ versus t, is found by projecting the motion of the point on the phase trajectory (the time markers) on the x axis to yield x versus t and projecting the motion on the y axis to yield $\dot{x}$ versus t. The initial conditions determine the point at which the motion begins. For example, if the system has, as initial conditions, $\dot{x} = 0$ and $x = +d$ at $t = 0$, the starting point is as shown in Fig. 8-20.

The method of determining the phase plane and hence the solution for feedback control systems becomes useful only when it is plotted by

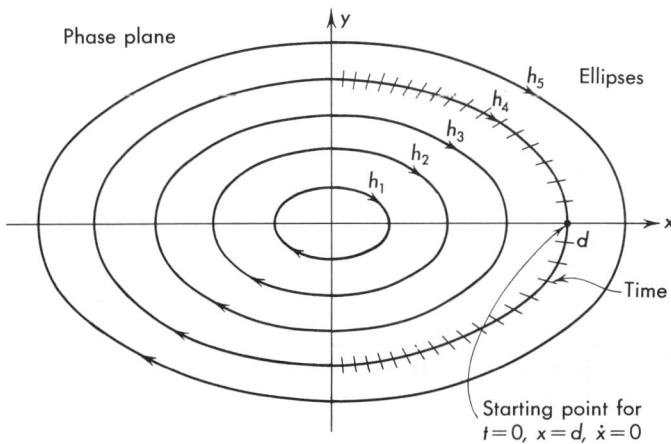

Fig. 8-20 Phase trajectories for the undamped second-order system.

graphical techniques. The following sections present two graphical methods for constructing phase-plane solutions.

8-13 The Method of Isoclines. Suppose that the entire phase plane is covered with line segments and that each has a slope corresponding to a slope at some point of the phase trajectory. With these "slope lines" plotted, the phase trajectories are constructed by connecting the slope lines.

The slope at a point (x, y) in the phase plane is

$$\lambda = \frac{dy}{dx} = \frac{dy}{dt}\frac{1}{dx/dt} = \frac{\dot{y}}{\dot{x}} \tag{8-32}$$

The lines along which λ is a constant are termed "isoclines." In the general case both $\dot{y}$ and $\dot{x}$ can be represented as functions of x and y:

$$\dot{x} = P(x, y) \qquad \text{and} \qquad \dot{y} = Q(x, y) \tag{8-33}$$

where P and Q are functions of x and y, not necessarily linear. The time variable is eliminated by forming the slope of the phase trajectory:

$$\frac{dy}{dx} = \lambda = \frac{\dot{y}}{\dot{x}} = \frac{Q(x, y)}{P(x, y)} \tag{8-34}$$

For various constant values of λ the isoclines (lines along which the slope is a constant $\lambda = \lambda_i$) are plotted from

$$\lambda_i P(x, y) = Q(x, y) \tag{8-35}$$

By connecting the slope lines, the desired phase trajectories are constructed.

Consider as an example a pendulum, shown in Fig. 8-21, under the influence of gravity. The equation for the angular position of the pendulum is

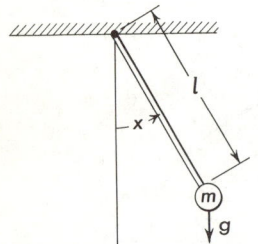

Fig. 8-21 Simple pendulum under the influence of gravity.

$$\ddot{x} + \frac{g}{l} \sin x = 0 \tag{8-36}$$

Let $\dot{x} = y$, then $\dot{y} = \ddot{x} = -(g/l) \sin x$ and $\lambda_1 = \dot{y}/\dot{x} = -(g/ly) \sin x$. This equation is rearranged:

$$y = -\frac{g}{l\lambda_i} \sin x \tag{8-37}$$

This is the equation for the isoclines. When various values of λ_i are substituted, $\lambda_i = 0, \pm\frac{1}{2}, \pm 1, \pm 3, \infty$, the curves along which λ_i is a constant are found and plotted as shown in Fig. 8-22. As the sine curves are drawn, the small slope lines are drawn at the slope λ_i that corresponds to the particular isocline. Phase trajectories are constructed by connecting the slope lines, as shown by the one trajectory in Fig. 8-22. With $g/l = 1$, the trajectory is a circle for small oscillations.

The complete family of phase trajectories is plotted in Fig. 8-23. The closed curves correspond to finite oscillatory motion, while the upper curves represent the motion of the pendulum rotating in a complete circle about the pivot. The set of trajectories that separates these two types of motion follows the application of initial conditions

$$x = \pi \text{ rad} \qquad \dot{x} = 0 \tag{8-38}$$

The motion in the phase plane is clockwise, as can be verified by considering the relation between x and $\dot{x}$. When x is at x_0 and $\dot{x}$ is zero,

Nonlinearities in Control-system Design 371

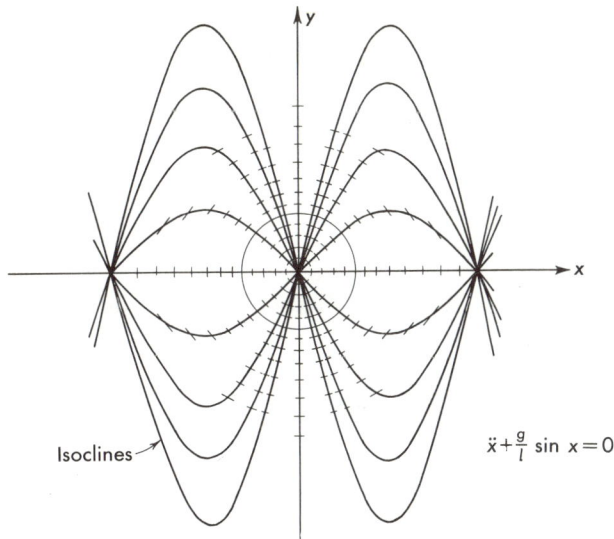

Fig. 8-22 Solution of pendulum problem by method of isoclines.

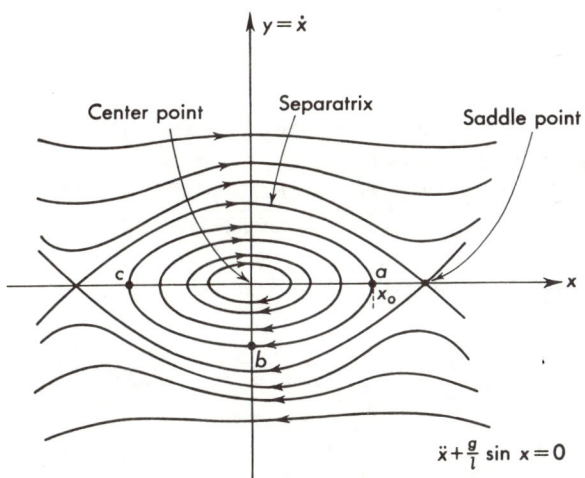

Fig. 8-23 Phase trajectories for the simple pendulum.

the point on the trajectory corresponds to the point a in Fig. 8-23. At this position, the pendulum is displaced to the right (cf. Fig. 8-21) an angular distance x_0. When the pendulum moves, x decreases and $\dot{x}$ begins to increase in the negative direction. When the pendulum reaches $x = 0$, the velocity is maximum negative. This point corresponds to the point b in Fig. 8-23. Continuing on, the velocity starts to decrease and x becomes negative. When the pendulum has reached its maximum angular displacement to the left, in the negative direction, the velocity is zero and the point on the trajectory is c. The motion continues in this fashion.

8-14 The Lienard Construction. This construction is a more direct method of plotting phase trajectories. The accuracy is comparable with that obtained by the method of isoclines but the Lienard method is faster and easier to use. The construction described in this section is applicable to the following types of nonlinear second-order differential equations:

$$\ddot{x} + F(\dot{x}) + x = 0 \tag{8-39}$$

$$\ddot{x} + \dot{x} + g(x) = 0 \tag{8-40}$$

$$\ddot{x} + F(\dot{x}) + g(x) = 0 \tag{8-41}$$

$$\ddot{x} + f(x)\dot{x} + g(x) = 0 \tag{8-42}$$

The method consists of constructing the slope of the phase trajectory graphically. For example, the slope for Eq. (8-39) is found by first eliminating time:

$$\dot{x} = y$$

$$\dot{y} = \ddot{x} = -[F(y) + x] \tag{8-43}$$

and the slope is

$$\lambda = \frac{\dot{y}}{\dot{x}} = \frac{-[F(y) + x]}{y} \tag{8-44}$$

The normal slope λ_n is related to λ as follows:

$$\lambda_n = -\frac{1}{\lambda} = \frac{y}{x + F(y)} = \frac{y}{x - [-F(y)]} \tag{8-45}$$

The last form of Eq. (8-45) is used for the graphical construction. The auxiliary curve $x = -F(y)$ is plotted, as shown in Fig. 8-24, with $+F(y)$ horizontal and y vertical. Scales of both axes are the same, and $F(y)$ has the same scale as x. Slope of the trajectory at the point (x_0, y_0) is constructed in Fig. 8-24. The normal slope λ_{n0} at the point (x_0, y_0) is found from the ratio

$$\lambda_{n0} = \frac{y_0}{x_0 - [-F(y_0)]} \tag{8-46}$$

The denominator consists of two terms. $-F(y_0)$ is found by projecting y_0 back along a line $y = y_0$ until the curve of $-F(y)$ is intersected as shown in Fig. 8-24. The distance from the point $(x_0, 0)$ to $-[F(y_0), 0]$, as determined graphically, is the denominator of Eq. (8-46). The numerator y_0 is the vertical distance from the point (y_0, x_0) to the x axis. Hence, the normal slope is along the line from the point $-[F(y_0), 0]$ to $P(x_0, y_0)$, as shown in Fig. 8-24. The desired slope of the phase trajectory is found by constructing a small line perpendicular to λ_{n0} at $P(x_0, y_0)$. This is easily done with a compass.

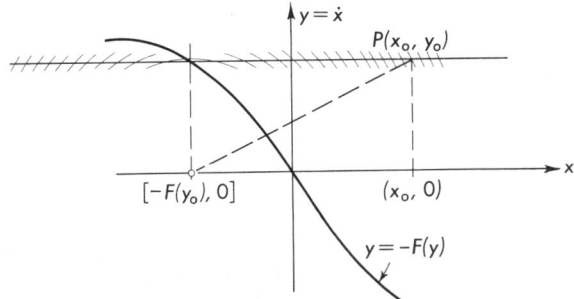

Fig. 8-24 Lienard construction for $\ddot{x} + F(\dot{x}) + x = 0$.

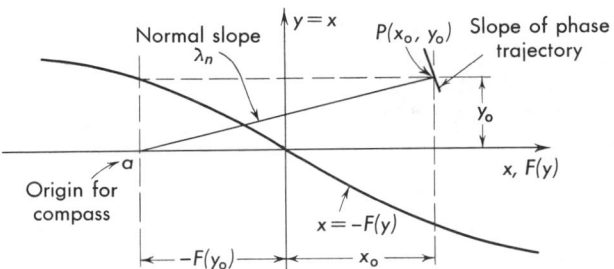

Fig. 8-25 Lienard construction for all points along the line $y = y_0$.

As an example, suppose the slope lines for all points along the line $y = y_0$ are constructed. This is accomplished, in Fig. 8-24, by putting the point of the compass at $[-F(y_0), 0]$ and drawing all intersections with $y = y_0$. The solution for particular initial conditions is found by following one trajectory around. In this case, start at the initial condition $P(x_0, y_0)$ and go step by step until the trajectory is traced out. Each slope line is constructed as shown in Fig. 8-25. It is not necessary to draw the line that represents λ_{n0}, since only the origin for the compass, point a, is needed to strike the small slope line λ_0 at the point $P(x_0, y_0)$.

The method for constructing the solution of Eq. (8-40) is similar to that for Eq. (8-39). Again let

$$\dot{x} = y$$
$$\dot{y} = \ddot{x} = -y - g(x) \qquad (8\text{-}47)$$

and the normal slope is

$$\lambda_n = -\frac{1}{\lambda} = -\frac{\dot{x}}{\dot{y}} = \frac{y}{y - [-g(x)]} \qquad (8\text{-}48)$$

The auxiliary curve $y = -g(x)$ is plotted on the phase plane to the same scale as y. Both x and y axes must have the same scale so that angles

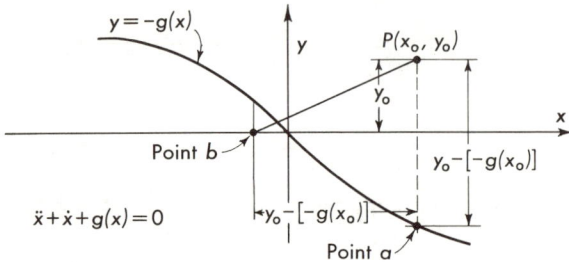

Fig. 8-26 Lienard construction for $\ddot{x} + \dot{x} + g(x) = 0$.

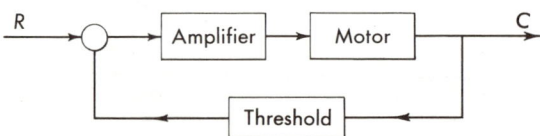

Fig. 8-27 Position servo with threshold in the feedback path.

are preserved. For a point $P(x_0, y_0)$, the numerator of Eq. (8-48) represents the distance from the point $P(x_0, y_0)$ to the x axis. The denominator of Eq. (8-48) for $x = x_0$ and $y = y_0$ is composed of two parts: y_0, which is the distance from the point $P(x_0, y_0)$ to the x axis, and $-g(x_0)$, which is the distance from the x axis to the point where the line $x = x_0$ intersects the $-g(x_0)$ curve (point a on Fig. 8-26). The algebraic sum $y_0 - [-g(x_0)]$ is measured with a pair of dividers or a compass and is transferred to the x axis, as shown in Fig. 8-26. This locates the center point of the compass, point b in Fig. 8-26. A small line drawn through $P(x_0, y_0)$ with the center at point b is the slope of the phase trajectory at that point.

As an example of this construction, consider the position servo of Fig. 8-27. The motor transfer function is

$$\frac{4.1}{s(0.016s + 1)} \qquad (8\text{-}49)$$

and the amplifier transfer function A is set for a damping ratio of 0.20 with no threshold. The open-loop transfer function with this gain setting is

$$G = \frac{24{,}400}{s(s + 62.5)} \qquad (8\text{-}50)$$

The threshold transfer function, plotted in Fig. 8-28, is represented by $f(c)$. The differential equation is

$$\ddot{c} + 62.5\dot{c} + 24{,}400 H(c) = 24{,}400 r \qquad (8\text{-}51)$$

The system response to a step input is desired. Since the phase plane can be used only for zero driving functions, a set of initial conditions that will duplicate the step input must be found. In this case, set $r = 0$ and take for initial conditions the following:

At $t = 0 \qquad c = c_0$

$\qquad\qquad\qquad \dot{c} = 0 \qquad (8\text{-}52)$

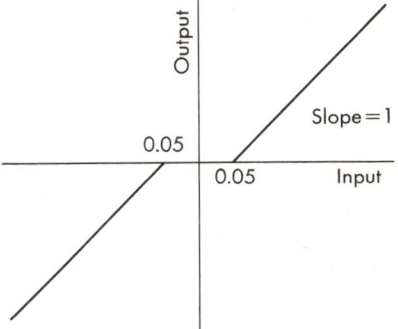

Fig. 8-28 Threshold in feedback loop of position servo.

These initial conditions are such that as c returns to the null position the solution passes back and forth across the nonlinearity. This duplicates the response to an input step function. Making the usual substitutions,

$$y = \dot{c}$$
$$\dot{y} = \ddot{c} = -62.5y - 24{,}400 H(c) \qquad (8\text{-}53)$$

and the normal slope is

$$\lambda_n = -\frac{\dot{c}}{\dot{y}} = \frac{y}{62.5y + 24{,}400 H(c)} \qquad (8\text{-}54)$$

Equation (8-54) is put in a more suitable form by dividing out the constant 62.5:

$$\lambda_n = \frac{1}{62.5} \frac{y}{y + 391H(c)} \quad (8\text{-}55)$$

Dividing numerator and denominator by 62.5 and rearranging,

$$62.5\lambda_n = \frac{y/62.5}{y/62.5 + 6.25H(c)} \quad (8\text{-}56)$$

The normalized phase plane is plotted with the variables $y' = y/62.5$ and

$$\lambda'_n = \frac{c}{y'} = \frac{c}{y/62.5} = \frac{62.5c}{y} = 62.5\lambda_n$$

and with the equation

$$\lambda'_n = \frac{y'}{y' + 6.25H(c)}$$

$$= \frac{y'}{y' - [-6.25H(c)]} \quad (8\text{-}57)$$

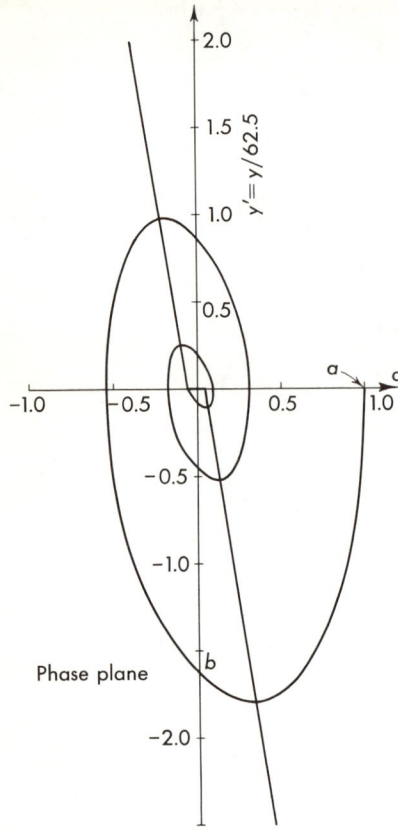

Fig. 8-29 Phase trajectory for system with threshold.

Both axes y' and c on the phase plane now have the same dimension, radians (since 62.5 has the dimension seconds^{-1}). The resulting construction for a step change of 1 rad is shown in Fig. 8-29. The threshold function $-6.25H(c)$ is plotted first, and the construction follows the scheme outlined in Fig. 8-26. The system response to large input amplitude is similar to a linear system because of the relatively small threshold. As the input signal becomes smaller, the nonlinear effect becomes more pronounced. The system does not return to a complete null but sticks at a point where the phase trajectory passes through the threshold with zero velocity (cf. Fig. 8-29).

The solution for Eq. (8-41) is constructed in a manner similar to that of the preceding two cases. The normal slope is found from the equation

$$\lambda_n = -\frac{\dot{x}}{\dot{y}} = \frac{y}{x - [x - g(x) \quad F(y)]} \qquad (8\text{-}58)$$

where $\dot{x} = y$ and $-\dot{y} = F(y) + g(x) = x - [x - g(x) - F(y)]$. The construction is similar to the two previous cases except that the procedure is more complicated, since the horizontal position of the center of the compass is located at $[x - g(x)] - F(y)$. Both the curves $x' = x - g(x)^*$ and $x = F(y)$ are plotted on the phase trajectory, and the distances are measured and summed with a compass.

The solution of Eq. (8-42) is constructed in the phase plane by defining the variable y not as $y = \dot{x}$, but through the equation

$$\dot{x} = y - \int_0^x f(x)\, dx = y - F(x) \qquad (8\text{-}59)$$

where $F(x) = \int_0^x f(x)\, dx$. When Eq. (8-59) is differentiated with respect to time,

$$\ddot{x} = \frac{d}{dt}\dot{x} = \dot{y} - \frac{d}{dt}F(x) = \dot{y} - \frac{d}{dx}F(x)\frac{dx}{dt} = \dot{y} - \dot{x}f(x) \qquad (8\text{-}60)$$

Substituting Eq. (8-60) into Eq. (8-39) gives

$$\dot{y} = -g(x) \qquad (8\text{-}61)$$

The normal slope is found from Eqs. (8-59) and (8-60). The normal slope at a point (x_0, y_0) is given by the ratio

$$\lambda_{n0} = \frac{y_0 - F(x_0)}{x_0 - [x_0 - g(x_0)]} = \frac{y_0 - F(x_0)}{x_0 - x'(x_0)} \qquad (8\text{-}62)$$

The denominator construction is similar to previous cases [cf. the construction of Eq. (8-40)]. The numerator is constructed by subtracting the lengths y_0 and $F(x_0)$. The desired center is found from the ratio of Eq. (8-62) with the graphical construction shown on Fig. 8-30. Both $F(x) = \int_0^x f(x)\, dx$ and $g(x)$ are plotted on the phase plane, and the

---

* This is plotted graphically by adding, point by point, the curve of $-g(x)$ to the linear function x.

378 Control System Design

construction proceeds as shown on Fig. 8-30. Because of the complexity of the slope equation, it is desirable to plot the slope lines along vertical lines ($x = $ const). For these lines the radius is centered at the same point for all y.

8-15 Determination of the Time Markers. The phase-plane solution of a differential equation is a plot of velocity* $\dot{x} = y$ versus displacement x with time t as a parameter. The time solutions x versus t and $\dot{x}$ versus t can be plotted from the phase trajectory when time markers are placed upon the phase plane. Two methods of determining time

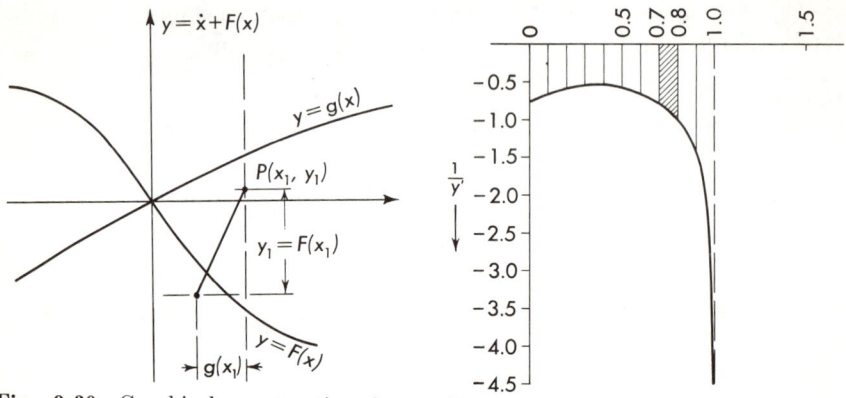

Fig. 8-30 Graphical construction for $\ddot{x} + f(x)\dot{x} + g(x) = 0$.

Fig. 8-31 Time found graphically.

explicitly are presented. The first method involves plotting $1/y$ as a function of x. Since $y = \dot{x} = dx/dt$, the time is found as follows:

$$dt = \frac{1}{y} dx \qquad (8\text{-}63)$$

When Eq. (8-63) is integrated,

$$t_2 - t_1 = \int_{x_1}^{x_2} \frac{1}{y} dx \qquad (8\text{-}64)$$

When $1/y$ is plotted as a function of x, the area under the curve between two values of x is equal to the elapsed time $t_2 - t_1$ between the two points. The time is found for the position servo shown in the phase plane of Fig. 8-29. In Fig. 8-31, $1/y'$ is plotted against x for the interval from

---

* For the last construction of Sec. 8-14, $\dot{x}$ does not equal y, but is equal to $y - F(x)$. The determination of time markers is still valid, but a plot of y versus t is not readily obtainable from the trajectory.

a to b in Fig. 8-29. The time between any two points, say, 0.7 to 0.8, is proportional to the area shaded in the figure. The entire area between $x = 1$ and $x = 0$ is approximately 0.75, in dimensionless form. To convert this into seconds, the area must be multiplied by $1/62.5$, or

$$t = (0.75)\left(\frac{1}{62.5}\right) = 12 \text{ msec} \tag{8-65}$$

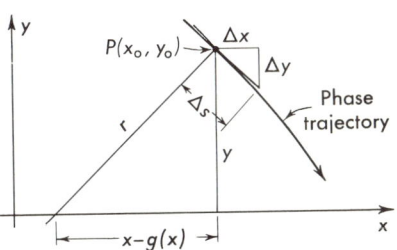

Fig. 8-32 Construction of isochrones.

The second method of determining the time markers, termed "isochrones," is based upon a completely graphical construction. The basis for this construction is shown in the phase plane of Fig. 8-32, where, from the two similar triangles,

$$\frac{r}{\Delta s} = \frac{y}{\Delta x} \tag{8-66}$$

But $y = dx/dt$, which is approximately equal to $\Delta x/\Delta t$, or

$$\frac{r}{\Delta s} = \frac{\Delta x}{\Delta t}\frac{1}{\Delta x} = \frac{1}{\Delta t} \tag{8-67}$$

and the distance Δs along the trajectory that corresponds to a time interval Δt is found from Eq. (8-67):

$$\Delta s = r\,\Delta t \tag{8-68}$$

If a constant value for Δt, for example, 0.1, is chosen, Δs is proportional to r, the distance from the trajectory to the construction point along the x axis. Notice that r is not the radius of curvature of the phase trajectory but is the radius used to define the normal slope in the Lienard construction. In some cases these are the same.

It is usually more convenient to find the time marks while constructing the phase trajectory. This method of construction saves time, since the radius r must be determined before the particular slope line can be drawn. Also, if a constant Δt is chosen, setting the isochrones on the trajectory as it is being constructed helps to determine the spacing of the slope lines. The time solution for the position servo pictured on the phase plane of Fig. 8-29 is presented in Fig. 8-33. For purposes of comparison, the time solution of a system without threshold is also plotted. Equation (8-42) requires a different construction to determine the time markers. In that case the radius r is found from $y - F(x)$ in the numerator rather than just y.

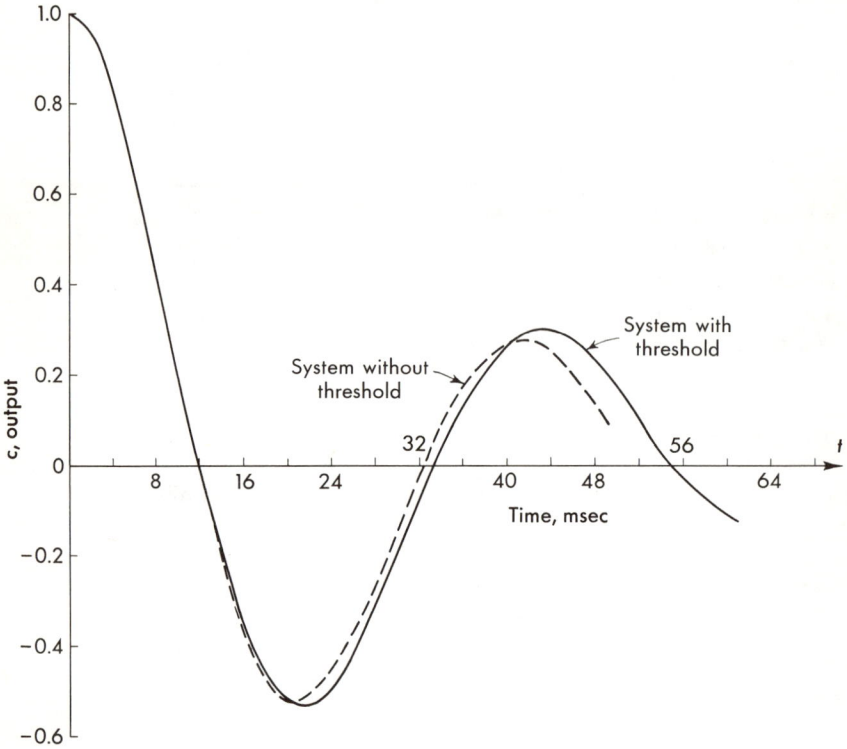

Fig. 8-33 Time response, second-order system with and without threshold.

8-16 Comparison of Several Methods. Some of the more commonly used methods available for the analysis of systems defined by nonlinear differential equations have been presented in this chapter. Various analytic methods, notably iteration methods,[36] perturbation methods,[36] and step-by-step solutions,[24] have been neglected. Although both the describing-function approach and the topological methods are limited in their scope, they are used more than the analytic methods for nonlinear control-system design.

Because of the difficulty of analyzing complex nonlinear systems, general-purpose analogue and digital computation is often used to effect detailed and accurate investigations of system performance. However, the analytical and topological procedures advanced in this chapter can often be used to gain insight into the gross behavior and major aspects of the problem.

PROBLEMS

8-1. Calculate and plot the describing function for spring preload which has the input-output relation shown by Fig. 8P-1.

8-2. Calculate and plot the second and third harmonics for Prob. 8-1.

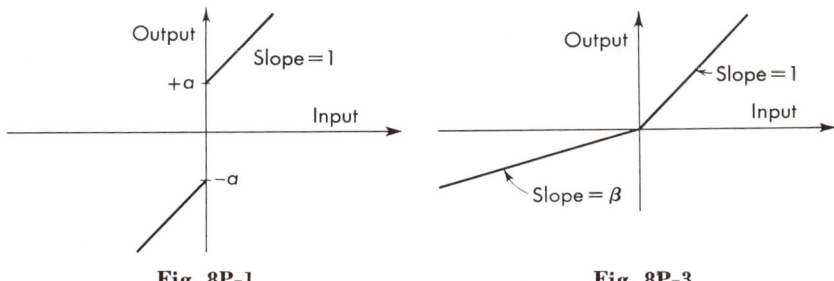

Fig. 8P-1   Fig. 8P-3

8-3. Calculate and plot the describing function for the nonbilateral element shown in Fig. 8P-3.

8-4. Calculate and plot the second and third harmonics for Prob. 8-3.

8-5. Subject the simple position servo given by $G_1 = 1.0/s(1 + 0.1s)(1 + s)^2$ to amplifier saturation. The block diagram is given in Fig. 8P-5. Determine the effect on the system when the nonlinearity is included in the system. Solve the problem by both frequency-analysis and root-locus methods. For amplifier saturation take the ratio of input amplitude A to saturation level equal to 1, 2, 4, and 8.

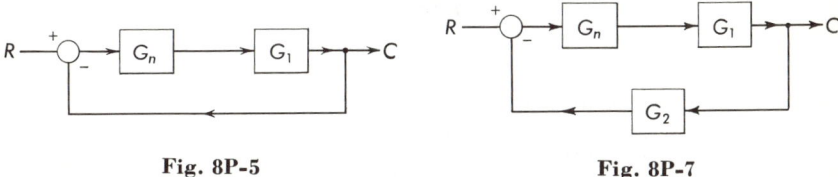

Fig. 8P-5   Fig. 8P-7

8-6. Repeat Prob. 8-5 for motor threshold. Take the ratio of input amplitude A to dead bandwidth a equal to 1, 2, 4, and 8.

8-7. Consider the effect of saturation upon the simplified gyro-stabilized platform shown in Fig. 8P-7, where

$$G_1 = \frac{K}{s[(s+5)^2 + 30^2]} \qquad G_2 = s + 1$$

G_n is the describing function. Take the ratio of $A/a = 1, 2, 4,$ and 8, where A is the amplitude of the input sinusoid. Solve the problem by frequency-response techniques.

8-8. Plot accurately the phase trajectory for a pendulum of mass 1 slug and length 1 ft. At $t = 0$ the pendulum is started from rest at an angle of 170° from the downward vertical. Neglect friction.

8-9. The motion of a ship proceeding on a straight course with rudder amidships can be described by the equation

$$J\ddot{\alpha} + C(\dot{\alpha}) - M(\alpha) = 0$$

where J = moment of inertia of ship about vertical axis through center of ship
$C(\dot{\alpha})$ = resistance to turning
$M(\alpha)$ = moment of leeway force (restoring torque)
α = angular deviation from course

Take $C(\dot{\alpha}) = 0$ and assume $M(\alpha) = M_0\alpha - M_1\alpha^3$. Sketch the trajectories and state the nature of the stability of the ship on its course.

8-10. Using the method of isoclines, plot the family of phase trajectories for the Van der Pol equation

$$\ddot{x} - \mu(1 - x^2)\dot{x} + x = 0$$

Take $\mu = 1$.

8-11. Use the Lienard construction to find the trajectories for the following differential equation:

$$\ddot{x} + F_1(\dot{x}) + x = 0 \qquad F_1(\dot{x}) = 2 + \tfrac{1}{2}\dot{x} \qquad \text{for } \dot{x} > 0$$
$$F_1(\dot{x}) = -2 + \tfrac{1}{2}\dot{x} \qquad \text{for } \dot{x} < 0$$

8-12. Repeat Prob. 8-11 with the following differential equation:

$$\ddot{x} + F_2(\dot{x}) + x = 0 \qquad F_2(\dot{x}) = 2 - \tfrac{1}{2}\dot{x} \qquad \text{for } \dot{x} > 0$$
$$F_2(\dot{x}) = -2 - \tfrac{1}{2}\dot{x} \qquad \text{for } \dot{x} < 0$$

Choose two sets of initial conditions so that at least two trajectories are obtained. Discuss and compare the resulting solutions of Prob. 8-11 and 8-12.

8-13. Construct isochrones on the phase trajectories of Probs. 8-11 and 8-12. Choose $\Delta t = 0.1$.

8-14. (a) By use of the theory of isoclines, plot the trajectory of the $L = C = 1$ series circuit having an equation $\ddot{q} + q = 0$ and passing through $q = 5$, $\dot{q} = 0$ at $t = 0$.

(b) Derive for this case $t = \int_5^q dq/\dot{q}$.

(c) Plot the curve of $1/\dot{q}$ versus q by taking coordinates from the isocline plot. The area under this curve from $q = 5$ to q is the time elapsed since $t = 0$. Evaluate this area graphically from the plotted curve.

(d) Plot on one sheet the q versus t curves obtained as follows:

(1) By the graphical integration of $t = \int_5^q \frac{1}{\dot{q}} dq$

(2) By using isochrones

(3) Analytic solution of equation $\ddot{q} + q = 0$ using stated initial conditions

Compare the various methods as to accuracy, ease, etc.

Nonlinearities in Control-system Design

8-15. Consider the servo of Fig. 8P-15, where

$$\text{Motor transfer function} = \frac{4.1}{s(1 + 0.016s)}$$

$$\text{Amplifier transfer function} = A \text{ (a constant)}$$

$$\text{Feedback function } b = \alpha c^n$$

Take $\alpha = 1$, and compare the time response to a unit step function for

$$n = 1 \qquad n = 3 \qquad n = \tfrac{1}{3}$$

Choose A for a $\zeta = 0.2$ for $n = 1$ (the linear case). Use the same gain for $n = 3$ and $n = \tfrac{1}{3}$. Would any other gain yield a better system?

On the basis of this problem draw conclusions concerning the use of nonlinear feedback for control-system stability.

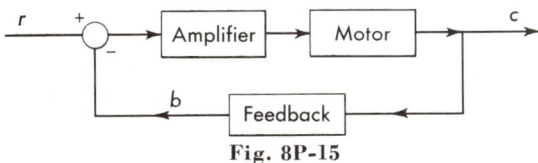

Fig. 8P-15

8-16. Plot accurately the family of phase trajectories for the system

$$G = \frac{1}{s(5s + 1)}$$

Choose $|a| = 1$ and take initial conditions $x_0 = 2, 5, 10$. Plot isochrones and the x versus t curves for these initial conditions. The block diagram is shown in Fig. 8P-16.

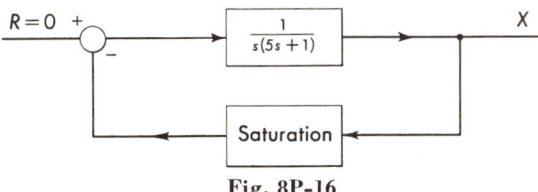

Fig. 8P-16

8-17. Consider the monorail suspension system in Fig. 8P-17. Contact is maintained by a loose-fitting shoe with an appreciable dead space d. How does the presence of the dead space affect the operation of the suspension? Assume the mass of the shoe is negligible. Compare the response of the system (on a phase plane plot) with and without dead space d. The system can be characterized by two parameters

$$L = \frac{\text{static spring deflection}}{\text{dead space}} = 1$$

$\zeta = \%$ of critical damping of the linear system $= 0.2$

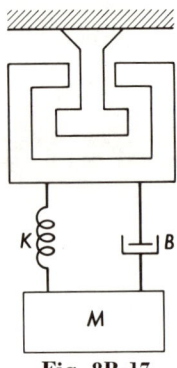

Fig. 8P-17

8-18. Sketch $x(t)$ in the steady state and compute the period of oscillation for the system in Fig. 8P-18a. The friction force is combined coulomb and stiction, as shown in Fig. 8P-18b.

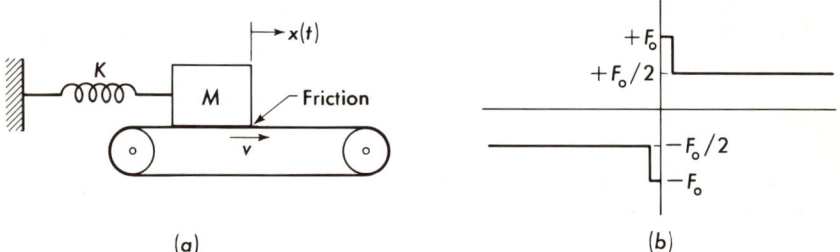

Fig. 8P-18

appendix I

LAPLACE-TRANSFORM METHOD

I-1 Introduction. The Laplace transformation is a mathematical operation useful in solving ordinary differential equations. The transformation is defined by the following integral:

$$\mathcal{L}[f(t)] = \int_0^\infty f(t)e^{-st}\,dt = F(s) \tag{I-1}$$

where $f(t)$ is a function of time which is zero for $t < 0$, s is a complex variable, $F(s)$ is the function of s which results when $f(t)$ is Laplace-transformed, $\mathcal{L}$ is an operational symbol indicating that the function following is to be Laplace-transformed. Since many excellent texts[8,13,27] have been written on this subject, only the mechanics of the Laplace-transform theory are treated in this appendix.

A linear differential equation of the form

$$\frac{d^n y}{dt^n} + a_{n-1}\frac{d^{n-1}y}{dt^{n-1}} + \cdots + a_1\frac{dy}{dt} + a_0 y = f(t) \tag{I-2}$$

is often encountered in control-system design. The coefficients a_i are real constants and independent of t or y. Both the dependent variable y and the driving function $f(t)$ are functions of time. When Eq. (I-2) is Laplace-transformed, a new, or "transformed," equation results. The differential equation is reduced to an algebraic equation with dependent s. This transformed equation can be manipulated algebraically to solve for the desired quantity in transformed form.

If the time solution is required, the function of s must be inverse-transformed into a function of time. This process, termed "finding the

386   Control System Design

inverse Laplace transform," is most easily accomplished with the aid of Laplace-transform tables.*

I-2 The Laplace Transform of Functions. Application of Eq. (I-1) to various $f(t)$ serves as the basis for a table of "transform pairs."

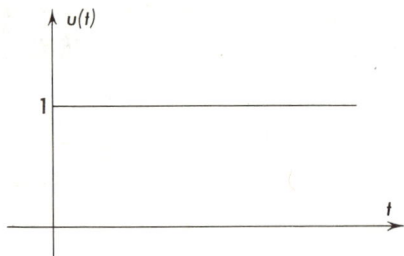

Fig. I-1   Plot of a unit step function $u(t)$.

When the function is known, for example, consider the unit step function $u(t)$ shown in Fig. I-1. The mathematical representation is

$$u(t) = \begin{cases} 0 & t < 0 \\ 1 & t > 0 \end{cases} \tag{I-3}$$

The Laplace transform is

$$\mathcal{L}[u(t)] = \int_0^\infty e^{-st}\, dt = \frac{e^{-st}}{s}\bigg|_0^\infty = \frac{1}{s} \tag{I-4}$$

Table I-1 is a short table of Laplace transforms. This table is satisfactory for most driving functions encountered in control-system design.

I-3 The Laplace Transform of Operations. Besides finding the Laplace transforms of known functions of time $f(t)$, it is also necessary to transform dependent variables $y(t)$ and derivatives of $y(t)$:

$$\mathcal{L}\left[\frac{dy}{dt}\right] = \int_0^\infty \frac{dy}{dt} e^{-st}\, dt \tag{I-5}$$

This integral is evaluated by parts as follows:

$$\int_a^b u\, dv = uv\bigg|_a^b - \int_a^b v\, du \tag{I-6}$$

* An excellent table of Laplace transforms is to be found in Ref. 27.

Table I-1 Laplace-transform Table

Ref. no.	$f(t)$	$F(s)$	Ref. no.	$f(t)$	$F(s)$
1	$u(t)$	$\dfrac{1}{s}$	10	$\sin \omega t$	$\dfrac{\omega}{s^2 + \omega^2}$
2	t	$\dfrac{1}{s^2}$	11	$\cos \omega t$	$\dfrac{s}{s^2 + \omega^2}$
3	t^n	$\dfrac{n!}{s^{n+1}}$	12	$\dfrac{K}{\omega} \sin(\omega t + \psi)$ $\psi = \tan^{-1} \dfrac{\omega}{a_0}$ $K = (a_0^2 + \omega^2)^{\frac{1}{2}}$	$\dfrac{s + a_0}{s^2 + \omega^2}$
4	$\delta(t)$ unit impulse	1	13	$\sinh \omega t$	$\dfrac{\omega}{s^2 - \omega^2}$
5	e^{-at}	$\dfrac{1}{s + a}$	14	$\cosh \omega t$	$\dfrac{s}{s^2 - \omega^2}$
6	$\dfrac{e^{-at} - e^{-bt}}{b - a}$	$\dfrac{1}{(s + a)(s + b)}$	15	$e^{-\alpha t} \sin \omega t$	$\dfrac{\omega}{(s + \alpha)^2 + \omega^2}$
7	te^{-at}	$\dfrac{1}{(s + s)^2}$	16	$\dfrac{K}{\omega} e^{-\alpha t} \sin(\omega t + \psi)$ $\psi = \tan^{-1} \dfrac{\omega}{a_0 - \alpha}$ $K = [(a_0 - \alpha)^2 + \omega^2]^{\frac{1}{2}}$	$\dfrac{s + a_0}{(s + \alpha)^2 + \omega^2}$
8	$t^n e^{-at}$	$\dfrac{n!}{(s + a)^{n+1}}$			
9	$(Kt + 1)e^{-\alpha t}$ $K = a_0 - \alpha$	$\dfrac{s + a_0}{(s + \alpha)^2}$	17	$t \cos \omega t$	$\dfrac{s^2 - \omega^2}{(s^2 + \omega^2)^2}$
			18	$t \sin \omega t$	$\dfrac{2\omega s}{(s^2 + \omega^2)^2}$

where the following substitutions are made:

$$u = e^{-st} \qquad du = -se^{-st}\, dt$$
$$dv = \frac{dy}{dt}\, dt \qquad v = y(t) \tag{I-7}$$

The integral is now evaluated:

$$\mathcal{L}\left[\frac{dy}{dt}\right] = y(t)e^{-st}\Big|_0^\infty + s \int_0^\infty y(t)e^{-st}\, dt \tag{I-8}$$

Since the last integral in Eq. (I-8) is the Laplace transform of $y(t)$,

$$\mathcal{L}[y(t)] = \int_0^\infty y(t)e^{-st}\,dt = Y(s) \tag{I-9}$$

and since $y(t)e^{-st}$ is zero at the upper limit, provided the real part of $s > 0$,

$$\mathcal{L}\left[\frac{dy}{dt}\right] = -y(0) + sY(s) \tag{I-10}$$

In a similar fashion, the Laplace transform of higher derivatives can be found:

$$\mathcal{L}\left[\frac{d^2y}{dt^2}\right] = s^2 Y(s) - sy(0) - \left.\frac{dy}{dt}\right|_{t=0} \tag{I-11}$$

$$\mathcal{L}\left[\frac{d^n y}{dt^n}\right] = s^n Y(s) - s^{n-1} y(0) - s^{n-2}\left.\frac{dy}{dt}\right|_{t=0} - \cdots - \left.\frac{d^{n-1}y}{dt^{n-1}}\right|_{t=0} \tag{I-12}$$

For the case of zero initial conditions, which is the case in control-system analysis, this last equation reduces to

$$\mathcal{L}\left[\frac{d^n y}{dt^n}\right] = s^n Y(s) \tag{I-13}$$

Hence, for this case, taking the Laplace transform is equivalent to replacing the derivative with respect to time d/dt by the Laplace-transform operator s.

Another important Laplace-transform pair, which is required for the solution of linear differential equations, is the transform of the sum of two time functions as follows:

$$\mathcal{L}[a_1 y_1(t) + a_2 y_2(t)] = a_1 Y_1(s) + a_2 Y_2(s) \tag{I-14}$$

This relation is easily proved by substitution into the defining equation (I-1).

The Laplace transform of a definite integral of a function of time is also found by integrating by parts with the result

$$\mathcal{L}\left[\int_0^t y(\tau)\,d\tau\right] = -\frac{1}{s} e^{-st} \int_0^t y(\tau)\,d\tau \Big|_0^\infty + \frac{1}{s}\int_0^\infty e^{-st} y(t)\,dt \tag{I-15}$$

The first term on the right is zero at the upper limit, since $e^{-\infty} = 0$. If $\int_0^0 y(t)\,dt = 0$, that is, if the integral has no initial value, the first term vanishes at the lower limit and the result simplifies to

$$\mathcal{L}\left[\int_0^t y(\tau)\,d\tau\right] = \frac{Y(s)}{s} \qquad (\text{I-16})$$

These transform pairs are summarized in Table I-2. The initial conditions are set equal to zero for the pairs in this table, since in most control-system applications zero initial conditions are suitable to solve the problem.

Table I-2 Operation Transform Pairs*

Function of t	Transformed function of s
$\dfrac{dy}{dt}$	$sY(s)$
$\dfrac{d^2y}{dt^2}$	$s^2Y(s)$
$a_1y_1(t) + a_2y_2(t)$	$a_1Y_1(s) + a_2Y_2(s)$
$\int_0^t y(t)\,dt$	$\dfrac{1}{s}Y(s)$

* All initial conditions are taken equal to zero in this table.

The Laplace-transform operator s, which has the dimension seconds^{-1}, is often termed the "complex frequency," since it has a real and imaginary parts as follows:

$$s = \sigma + j\omega \qquad (\text{I-17})$$

I-4 Solution of Ordinary Linear Differential Equations Utilizing Laplace Transforms. As an example of the transform method of solving differential equations, consider the following simple equation:

$$\frac{d^2y}{dt^2} + 9y = u(t) \qquad t > 0 \qquad (\text{I-18})$$

with all initial conditions taken equal to zero. $u(t)$ is a unit step function. The Laplace transform of both sides is taken:

$$s^2Y + 9Y = \frac{1}{s} \qquad (\text{I-19})$$

and the result is solved for the transformed variable:

$$Y = \frac{1}{s(s^2 + 9)} \qquad (\text{I-20})$$

The solution $y(t)$ is found by breaking Eq. (I-17) into a sum of two parts by partial fractions:*

$$\frac{1}{s(s^2+9)} = \frac{A}{s} + \frac{B}{s+j3} + \frac{C}{s-j3} \qquad \text{(I-21)}$$

The constants A, B, and C are evaluated (as in the next section), and $Y(s)$ is written

$$Y(s) = \frac{1}{9}\left(\frac{1}{s} - \frac{s}{s^2+9}\right) \qquad \text{(I-22)}$$

The reader can verify that Eq. (I-22) is equivalent to Eq. (I-21) by putting Eq. (I-22) over a common denominator:

$$\frac{1}{9}\left(\frac{1}{s} - \frac{s}{s^2+9}\right) = \frac{1}{9}\frac{s^2+9-s^2}{s(s^2+9)} = \frac{1}{s(s^2+9)} \qquad \text{(I-23)}$$

The inverse transforms of the terms in Eq. (I-22) are found in Table I-1. The inverse transform yields

$$y(t) = \tfrac{1}{9}[u(t) - u(t)\cos 3t] = \tfrac{1}{9}u(t)(1 - \cos 3t) \qquad \text{(I-24)}$$

As a second example consider the differential equation

$$\frac{d^2y}{dt^2} + \omega^2 y = \cos \omega t \quad t > 0 \qquad \text{(I-25)}$$

with $y(0) = 0 = dy/dt\big|_{t=0}$, that is, zero initial conditions. Taking the Laplace transform of both sides and solving for $Y(s)$,

$$Y(s) = \frac{s}{(s^2+\omega^2)^2} \qquad \text{(I-26)}$$

The inverse Laplace transform is found from Table I-1:

$$y(t) = \mathcal{L}^{-1}\frac{s}{(s^2+\omega^2)^2} = \frac{t}{2\omega}\sin \omega t \qquad \text{(I-27)}$$

where the symbol $\mathcal{L}^{-1}$ means "inverse Laplace transform of."

* A discussion of partial fractions is included in Sec. I-5.

As a third example consider the second-order servo system discussed in Chap. 1. The differential equation with a step input and with zero initial condition is written

$$\frac{d^2c}{dt^2} + 2\zeta\omega_n \frac{dc}{dt} + \omega_n^2 c = \omega_n^2 u(t) \tag{I-28}$$

The Laplace transform of both sides of the equation is taken:

$$(s^2 + 2\zeta\omega_n s + \omega_n^2) C(s) = \frac{\omega_n^2}{s} \tag{I-29}$$

This is solved for the variable C:

$$C(s) = \frac{\omega_n^2}{s(s^2 + 2\zeta\omega_n s + \omega_n^2)} = \frac{\omega_n^2}{s[(s + \zeta\omega_n)^2 + \omega_n^2(1 - \zeta^2)]} \tag{I-30}$$

This equation is separated by partial fractions:

$$C(s) = \frac{1}{s} - \frac{s + 2\zeta\omega_n}{(s + \zeta\omega_n)^2 + \omega_n^2(1 - \zeta^2)} \tag{I-31}$$

as the reader can verify by combining Eq. (I-31) over a common denominator. The solution is found by referring to two transform pairs (1 and 16) in Table I-1:

$$c(t) = u(t) - \frac{1}{\sqrt{1 - \zeta^2}} e^{-\zeta\omega_n t} \sin(\omega_n \sqrt{1 - \zeta^2}\, t + \phi) \tag{I-32}$$

where

$$\phi = \tan^{-1} \frac{\sqrt{1 - \zeta^2}}{\zeta} \tag{I-33}$$

I-5 Partial-fraction Expansion. The inverse Laplace transformation of rational fractions is required to find the time solution from the transformed equation. The general rational fraction is written

$$Y(s) = \frac{A(s)}{B(s)} \tag{I-34}$$

where $A(s)$ and $B(s)$ are polynomials in s. When the roots of $B(s) = 0$ are found, Eq. (I-34) can be written

$$Y(s) = \frac{A(s)}{B(s)} = \frac{A(s)}{(s + s_1)(s + s_2)(s + s_3) \cdots (s + s_q)} \tag{I-35}$$

The inverse transformation is carried out by expanding Eq. (I-35) in partial fractions as follows:

$$\frac{A(s)}{B(s)} = \frac{A(s)}{(s+s_1)(s+s_2)(s+s_3)\cdots(s+s_q)}$$

$$= \frac{K_1}{s+s_1} + \frac{K_2}{s+s_2} + \frac{K_3}{s+s_3} + \cdots + \frac{K_q}{s+s_q} \qquad (\text{I-36})$$

Each term in Eq. (I-36) can be found by reference to Table I-1. The total time solution is found by summing the time solutions found from the separate terms in Eq. (I-36).

The partial-fraction expansion is an important step in the solution. Depending upon the form of $B(s)$, this expansion is carried out as follows:

a. *If $B(s)$ Contains Only Simple Roots.* In this case the K_i are evaluated by multiplying each side of Eq. (I-36) by $s + s_i$:

$$\frac{(s+s_i)A(s)}{B(s)} = K_1 \frac{s+s_i}{s+s_1} + K_2 \frac{s+s_i}{s+s_2} + \cdots + K_i \frac{s+s_i}{s+s_i}$$

$$+ \cdots + K_q \frac{s+s_i}{s+s_q} \qquad (\text{I-37})$$

Since $s + s_i$ is a factor in $B(s)$, it is divided out. When s is set equal to $-s_i$, the term on the left becomes a constant. All terms on the right reduce to zero except K_i. Hence each constant can be evaluated from the equation

$$K_i = \frac{A(s)(s+s_i)}{B(s)} \bigg|_{s=-s_i} \qquad (\text{I-38})$$

The procedure is unaltered if one of the roots is located at the origin. The constant K_0 is evaluated similarly:

$$K_0 = \frac{sA(s)}{B(s)} \qquad (\text{I-39})$$

When complex conjugate roots $(s+\alpha)^2 + \beta^2$ exist, the procedure is similar:

$$K_{j1} = \frac{(s+\alpha+j\beta)A(s)}{B(s)} \bigg|_{s=-\alpha-j\beta} \qquad (\text{I-40})$$

and

$$K_{j2} = \frac{(s+\alpha-j\beta)A(s)}{B(s)} \bigg|_{s=-\alpha+j\beta} \qquad (\text{I-41})$$

Since K_{j1} and K_{j2} are complex conjugate, the sum of the imaginary parts is zero and the sum of the real parts is twice the real part of either K. The two terms

$$\frac{K_{j1}}{s+\alpha+j\beta} \quad \text{and} \quad \frac{K_{j2}}{s+\alpha-j\beta} \tag{I-42}$$

combine into a single term which inverse-transforms into an exponentially bounded sinusoid.

b. If $B(s)$ Contains Multiple-order Roots. If the denominator of $Y(s)$ has multiple-order zeros, the procedure of the preceding section cannot be used. For example, if

$$\frac{A(s)}{B(s)} = \frac{1}{(s+s_1)(s+s_2)^2} \tag{I-43}$$

the partial expansion must include a second-order term:

$$\frac{1}{(s+s_1)(s+s_2)^2} = \frac{K_1}{s+s_1} + \frac{K_{12}}{s+s_2} + \frac{K_{22}}{(s+s_2)^2} \tag{I-44}$$

In general an nth-order root is expanded:

$$\frac{1}{(s+s_1)\cdots(s+s_i)^n} = \frac{K_1}{s+s_1} + \cdots + \frac{K_{ni}}{(s+s_i)^n}$$
$$+ \frac{K_{(n-1)i}}{(s+s_i)^{n-1}} + \cdots + \frac{K_{1i}}{s+s_i} \tag{I-45}$$

The constants K_i associated with first-order roots are evaluated as above. The constant associated with the highest power K_{ni} is evaluated in the same manner as a simple pole. That is, multiply both sides of Eq. (I-45) by $(s+s_i)^n$ and let $s = -s_i$. All terms on the right side are zero except the K_{ni} term. The left side reduces to a number. K_{ni} is evaluated as follows:

$$K_{ni} = \frac{(s+s_i)^n A(s)}{B(s)}\bigg|_{s=-s_i} \tag{I-46}$$

The procedure used for simple roots and used to evaluate the constant associated with the highest power K_{ni} is insufficient to evaluate any of the other coefficients. These constants are evaluated by differentiation. Both sides of Eq. (I-45) are multiplied by $(s+s_i)^n$. The resulting equation is differentiated once with respect to s:

$$\frac{d}{ds}\frac{(s+s_i)^n A(s)}{B(s)} = \frac{d}{ds}\frac{(s+s_i)^n}{s+s_1}K_1 + \cdots + K_{(n-1)i}$$
$$+ 2(s+s_i)K_{(n-2)i} + \cdots + (n-1)(s+s_i)^{n-2}K_{1i} \quad \text{(I-47)}$$

By letting $s = s_i$, all terms on the right except $K_{(n-1)i}$ vanish:

$$K_{(n-1)i} = \frac{1}{(n-1)!}\frac{d}{ds}\left[\frac{(s+s_i)^n A(s)}{B(s)}\right]_{s=-s_i} \quad \text{(I-48)}$$

The process of differentiating and then setting $s = -s_i$ can be repeated until all the unknown constants are determined.

As an example, consider the partial-fraction expansion of the following transfer function:

$$\frac{A(s)}{B(s)} = \frac{4s^3 + s^2 - 22s + 16}{s(s+2)(s-2)^2} \quad \text{(I-49)}$$

The fraction is broken up as follows:

$$\frac{4s^3 + s^2 - 22s + 16}{s(s+2)(s-2)^2} = \frac{K_1}{s} + \frac{K_2}{s+2} + \frac{K_{13}}{s-2} + \frac{K_{23}}{(s-2)^2} \quad \text{(I-50)}$$

The constants corresponding to the simple poles are easily found:

$$K_1 = \frac{s(4s^3 + s^2 + 22s + 16)}{s(s+2)(s-2)^2}\bigg|_{s=0} = 2$$

$$K_2 = \frac{(s+2)(4s^3 + s^2 - 22s + 16)}{s(s+2)(s-2)^2}\bigg|_{s=-2} = -1 \quad \text{(I-51)}$$

K_{23} is found in a similar fashion:

$$K_{23} = \frac{(s-2)^2(4s^3 + s^2 - 22s + 16)}{s(s+2)(s-2)^2}\bigg|_{s=2} = 1 \quad \text{(I-52)}$$

K_{13} is found by multiplying Eq. (I-50) by $(s-2)^2$ and then differentiating with respect to s:

$$\frac{d}{ds}\left[\frac{4s^3 + s^2 - 22s + 16}{s(s+2)}\right] = \frac{d}{ds}\left[\left(\frac{K_1}{s} + \frac{K_2}{s+2}\right)(s-2)^2\right]_{s=2}$$
$$+ \frac{d}{ds}(s-2)K_{13} + \frac{d}{ds}K_{23} \quad \text{(I-53)}$$

which reduces to

$$K_{13} = \frac{d}{ds}\left(\frac{4s^3 + s^2 - 22s + 16}{s^2 + 2s}\right)$$

$$= \left[\frac{(s^2 + 2s)(12s^2 + 2s - 22) - (4s^3 + s^2 - 22s + 16)(2s + 2)}{(s^2 + 2s)^2}\right]_{s-2}$$

$$= 3 \quad\quad\quad\quad\quad\quad\quad\quad\quad\quad\quad\quad\quad\quad\quad\quad\quad\quad\quad\quad\quad\quad\quad\quad\quad\quad\quad\quad\text{(I-54)}$$

Hence the partial-fraction expansion of Eq. (I-49) is written

$$\frac{A(s)}{B(s)} = \frac{2}{s} - \frac{1}{s+2} + \frac{1}{(s-2)^2} + \frac{3}{s-2} \quad\quad\text{(I-55)}$$

I-6 Additional Properties of the Laplace Transform. Some of the relationships involving the Laplace transformation are included in this section. These are often termed "theorems."* Detailed proofs are not given in this text for most of these theorems.

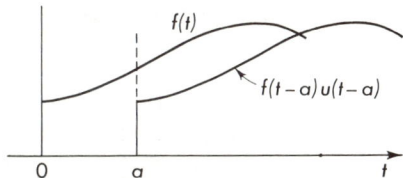

Fig. I-2 Real translation in time.

a. Real Translation. If $F(s)$ is the Laplace transform of $f(t)$, then

$$\mathcal{L}[f(t-a)u(t-a)] = e^{-as}F(s) \quad\quad\text{(I-56)}$$

Multiplication by e^{-as} in the complex frequency plane (s plane) results in translation in the time domain. The function $f(t-a)u(t-a)$ is shown shifted in Fig. I-2. As an example of the use of this theorem, suppose it is necessary to find the Laplace transform of one period of a sine wave, as shown in Fig. I-3. The time function can be formed by subtracting from a sine wave another sinusoid which has been shifted in time $2\pi/\omega$. Mathematically, this is written

$$f(t) = A \sin \omega t - A \sin \omega \left(t - \frac{2\pi}{\omega}\right) u\left(t - \frac{2\pi}{\omega}\right) \quad\quad\text{(I-57)}$$

* See Ref. 13, chap 8.

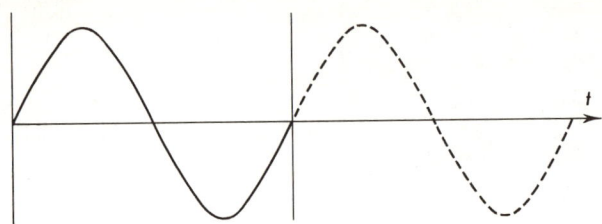

Fig. I-3   One period of a sine wave.

These are transformed as follows:

$$\mathcal{L}[f(t)] = \frac{A\omega}{s^2 + \omega^2} - \frac{A\omega}{s^2 + \omega^2} e^{-(2\pi/\omega)s} = \frac{a\omega}{s^2 + \omega^2}(1 - e^{-(2\pi/\omega)s}) \quad \text{(I-58)}$$

b. Second Independent Variable.   If $F(s, a)$ is the transform of $f(t, a)$, the following relation holds:

$$\mathcal{L}_t[\lim_{a \to a_0} f(t, a)] = \lim_{a \to a_0} F(s, a) \quad \text{(I-59)}$$

where $\mathcal{L}_t$ means the Laplace transform with respect to time.   As an example of the use of this equation consider the transform pair

$$\mathcal{L}_t e^{-\alpha t} \sin \omega t = \frac{\omega}{(s + \alpha)^2 + \omega^2} \quad \text{(I-60)}$$

given in Table I-1.   By taking the limit as α approaches zero, a second pair results:

$$\mathcal{L} \sin \omega t = \frac{\omega}{s^2 + \omega^2} \quad \text{(I-61)}$$

In a similar fashion, differentiation with respect to the quantity a is permissible:

$$\mathcal{L}_t \left[\frac{df(t, a)}{da} \right] = \frac{dF}{da}(s, a) \quad \text{(I-62)}$$

As an example differentiate the transform pair

$$\mathcal{L}_t e^{-at} = \frac{1}{s + a} \quad \text{(I-63)}$$

and obtain another pair

$$\mathcal{L}_t t e^{-at} = \frac{1}{(s+a)^2} \qquad (I\text{-}64)$$

In a similar fashion, the integral, with respect to the quantity a, can be found:

$$\mathcal{L}_t \left[\int_{a_1}^{a} f(t, \alpha) \, d\alpha \right] = \int_{a_1}^{a} F(x, \alpha) \, d\alpha \qquad (I\text{-}65)$$

c. Final-value and Initial-value Theorems. These theorems are valuable for control-system design. They are unique, since they permit finding a time function at either $t = 0$ or $t = \infty$ directly from the transform without inverting the transformed equation.

Final-value theorem:

$$\lim_{t \to \infty} y(t) = \lim_{s \to 0} sY(s) \qquad (I\text{-}66)$$

provided $y(t)$ is stable, i.e., all poles of $sY(s)$ are in the left half plane.

Initial-value theorem:

$$\lim_{t \to 0} y(t) = \lim_{s \to \infty} sY(s) \qquad (I\text{-}67)$$

provided the limit exists. Application of these theorems is shown in Chap. 3.

d. The Convolution Integral. The convolution integral is expressed as follows:

If $F_1(s)$ is the Laplace transform of $f_1(t)$ and $F_2(s)$ is the Laplace transform of $f_2(t)$, then

$$\mathcal{L} \int_0^t f_1(\tau) f_2(t - \tau) \, d\tau = F_1(s) F_2(s) \qquad (I\text{-}68)$$

The integral on the left is termed the "convolution integral," and the proof is included in many texts on Laplace transforms.[27]

The importance of this integral can be understood from a physical argument. In Fig. I-4 is shown a driving function $f(t)$ which is the input to a linear system and which is approximated by a series of pulses. Since the system which is shown in Fig. I-5 is linear, the total output can be approximated by a sum of the outputs due to each of these pulses. Suppose $h(t)$ is the response of the system to a unit pulse at the origin. The response or output of a system at time t to a unit pulse applied at time $n \, \Delta t$ is

$$h(t - n \, \Delta t) \qquad (I\text{-}69)$$

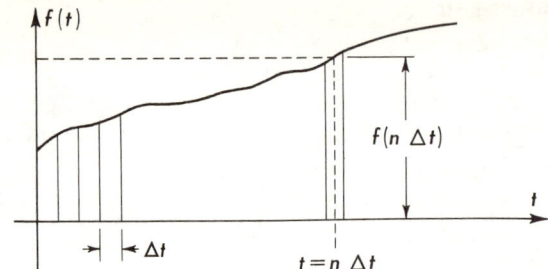

Fig. I-4 Driving function represented by a series of pulses.

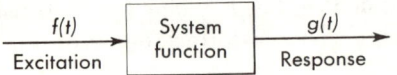

Fig. I-5 Output from a linear system.

The response or output of the system due to a pulse of area $\Delta t\, f(n\, \Delta t)$ is

$$h(t - n\, \Delta t) f(n\, \Delta t)\, \Delta t \tag{I-70}$$

The output due to all pulses up to time t is

$$g(t) = \sum_{n=0}^{N} h(t - n\, \Delta t) f(n\, \Delta t)\, \Delta t \tag{I-71}$$

If the variable τ equals $n\, \Delta t$ and if the limit is approached so that $\Delta t \to 0$, as $n \to \infty$,

$$n\, \Delta t \to \tau \quad \text{and} \quad N\, \Delta t \to t \tag{I-72}$$

The sum becomes an integral, and

$$g(t) = \int_0^t h(t - \tau) f(\tau)\, d\tau \tag{I-73}$$

where $h(t)$ is the response of the system to a unit impulse $\delta(t)$.

Hence the output of a linear system is found by "convolving" the input function $f(t)$ with the impulse response $h(t)$. The Laplace transform of Eq. (I-73) yields the important result

$$G(s) = H(s) F(s) \tag{I-74}$$

Hence the Laplace-transformed output $G(s)$ of a linear system is the product of the Laplace-transformed input $F(s)$ and the transformed impulse response (transfer function) $H(s)$.

appendix II

ROOTS OF EQUATIONS

II-1 Introduction. Basic to the solution of ordinary differential equations is the problem of finding roots of high-degree polynomial expressions. The Laplace-transform method is based upon knowledge of the roots of the characteristic equation. This appendix considers several techniques for finding roots.

II-2 The Quadratic. The second-degree polynomial, the quadratic

$$as^2 + bs + c = 0 \tag{II-1}$$

is solved easily by completing the square. The roots are given by the familiar equation

$$s_{1,2} = \frac{-b \pm \sqrt{b^2 - 4ac}}{2a} \tag{II-2}$$

II-3 The Cubic. For polynomials with real coefficients, at least one root of a cubic is real. If this one real root is found, the remaining two roots are easily found from the quadratic equation (II-2). To find the real root for the equation

$$f(s) = s^3 + as^2 + bs + c = 0 \tag{II-3}$$

follow this procedure:

1. Make a rough plot of $f(s)$ as a function of s to determine a value of s where $f(s)$ changes sign. This may be done with a table instead of a plot.
2. Estimate a value s_1 that may make $f(s) = 0$.

400 Control System Design

3. To compute a second, more accurate, estimate utilize Newton's method as follows:

$$s_2 = s_1 - \frac{f(s_1)}{f'(s_1)} \tag{II-4}$$

where

$$f'(s_1) = \frac{df}{ds}\bigg|_{s=s_1} \tag{II-5}$$

and continue the process until the desired accuracy is achieved.

As an example, consider the cubic

$$s^3 - 3s^2 - 2s + 5 = 0 \tag{II-6}$$

Corresponding values of s and $f(s)$ are

s	$f(s)$
0	5
1	1
2	-3

(II-7)

Since $f(s)$ changes sign between 1 and 2, a root must lie there. As a first estimate take $s = 1.5$. To find the second approximation evaluate

$$f(1.5) = (1.5)^3 - 3(1.5)^2 - 2(1.5) + 5 = -1.37$$
$$f'(1.5) = 3(1.5)^2 - 6(1.5) - 2 = -4.25 \tag{II-8}$$

and the second estimate is

$$s_2 = s_1 - \frac{f(s_1)}{f'(s_1)} = 1.5 - \frac{-1.37}{-4.25} = 1.18 \tag{II-9}$$

To find the next approximation,

$$f(1.18) = (1.18)^3 - 3(1.18)^2 - 2(1.18) + 5 = +0.10$$
$$f'(1.18) = 3(1.18)^2 - 6(1.18) - 2 = -4.90 \tag{II-10}$$

The third approximation is

$$s_3 = s_2 - \frac{f(s_2)}{f'(s_2)} = 1.18 - \frac{+0.10}{-4.90} = 1.20 \tag{II-11}$$

If more accuracy is desired, the process can be repeated as necessary. The remaining roots are found from the quadratic that remains when $s - 1.20$ is factored from the original cubic:

$$s^3 - 3s^2 - 2s + 5 = (s - 1.20)(s^2 - 1.80s - 4.16) \quad \text{(II-12)}$$

The quadratic factors into

$$s_1 = 3.125 \qquad s_2 = -1.325 \qquad s_3 = 1.20 \quad \text{(II-13)}$$

Proof of the approximation expression follows from a consideration of Fig. II-1. If s_1 is the first estimate, the second estimate s_2 is

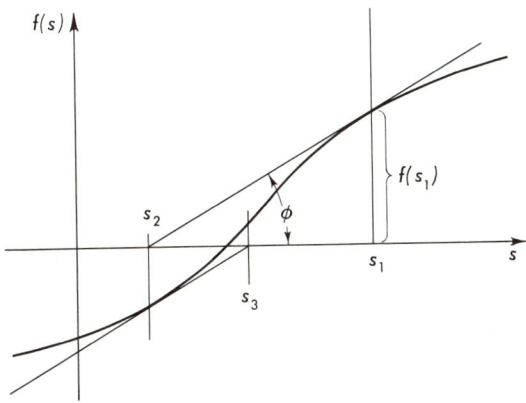

Fig. II-1 A sketch which geometrically exhibits Newton's method.

found from the equation

$$\tan \phi = \frac{f(s_1)}{s_2 - s_1} \quad \text{(II-14)}$$

But $\tan \phi$ is the slope of the $f(s)$ curve at the point s_1, or

$$\tan \phi = f'(s_1) = \frac{f(s_1)}{s_1 - s_2} \quad \text{(II-15)}$$

This is rearranged into the more usable form

$$s_2 = s_1 - \frac{f(s_1)}{f'(s_1)} \quad \text{(II-16)}$$

II-4 Descartes' Rules. Often the problem of finding roots of high-order polynomials is aided by means of the following rules due to Descartes:

1. The number of positive real roots of an equation $f(s) = 0$ is no more than the number of variations in signs among the coefficients of $f(s)$.
2. The number of negative real roots of the equation is no more than the number of variations in the signs of the equation $f(-s) = 0$.

For example, consider the polynomial

$$f(s) = s^4 + 7s^2 - 15s - 2 = 0 \tag{II-17}$$

for which
$$f(-s) = s^4 + 7s^2 + 15s - 2 \tag{II-18}$$

The sign changes of $f(s)$ are $++--$ or one. Hence $f(s) = 0$ has at most one positive real root. The sign changes of $f(-s)$ are $+++-$ or one, and $f(s) = 0$ has at most one negative real root.

II-5 Higher-degree Algebraic Equations. The difficulty of finding the roots of higher-degree equations increases considerably as the degree increases. Of the methods available, the root-locus method, described in Chap. 4, is suggested. Of the analytical methods, the Graeffe, or root-squaring, method is available. Because good accounts of the latter method appear in many places in the mathematical and engineering literature,[40] it will not be detailed here.

appendix III

USE OF DETERMINANTS

III-1 Definition of a Determinant. Use of determinants provides a mathematical means for the systematic solution of simultaneous equations, as required in Chap. 2. A determinant is an ordered array of quantities y_{ij} symbolized by straight-line brackets as follows:

$$\begin{vmatrix} y_{11} & y_{12} & \cdots & y_{1n} \\ y_{21} & y_{22} & \cdots & y_{2n} \\ y_{31} & y_{32} & \cdots & y_{3n} \\ \cdots & \cdots & \cdots & \cdots \\ y_{n1} & y_{n2} & \cdots & y_{nn} \end{vmatrix} \qquad \text{(III-1)}$$

The determinant indicated in Eq. (III-1) is of nth order, since there are n (horizontal) rows and n (vertical) columns. The determinant, which is a square array having the same number of columns as rows, contains n^2 elements. The position of an element in the determinant is identified by the subscripts:

$$\text{Row} \longrightarrow \underset{\uparrow}{y_i} \; \underset{\uparrow}{j} \longleftarrow \text{Column} \qquad \text{(III-2)}$$

The major diagonal of a determinant is the line through the elements y_{ii}. These elements form a sloping line starting with y_{11} and extending to y_{nn}.

III-2 Expansion of a Determinant. The value of a determinant is obtained by expansion. For example, a second-order determinant is expanded:

$$\begin{vmatrix} y_{11} & y_{12} \\ y_{21} & y_{22} \end{vmatrix} = y_{12}y_{22} - y_{12}y_{21} \qquad \text{(III-3)}$$

and a third-order determinant is expanded:

$$\Delta = \begin{vmatrix} y_{11} & y_{12} & y_{13} \\ y_{21} & y_{22} & y_{23} \\ y_{31} & y_{32} & y_{33} \end{vmatrix} = y_{11}\begin{vmatrix} y_{22} & y_{23} \\ y_{32} & y_{33} \end{vmatrix} - y_{21}\begin{vmatrix} y_{12} & y_{13} \\ y_{32} & y_{33} \end{vmatrix} + y_{31}\begin{vmatrix} y_{12} & y_{13} \\ y_{22} & y_{23} \end{vmatrix}$$

(III-4)

Each of the second-order determinants is easily expanded, yielding

$$y_{11}y_{22}y_{33} - y_{11}y_{32}y_{23} - y_{21}y_{12}y_{33} + y_{21}y_{32}y_{13} + y_{31}y_{12}y_{23} - y_{31}y_{22}y_{13}$$

(III-5)

In general a determinant is expanded in terms of its principal minors. The minor Δ_{ij} of an element of a determinant Δ is the determinant found by canceling the ith row and the jth column. For the fourth-order determinant

$$\begin{vmatrix} y_{11} & y_{12} & y_{13} & y_{14} \\ y_{21} & y_{22} & y_{23} & y_{24} \\ y_{31} & y_{32} & y_{33} & y_{34} \\ y_{41} & y_{42} & y_{43} & y_{44} \end{vmatrix}$$

(III-6)

the minor of Δ_{21} as found by canceling the second row and first column is

$$\Delta_{21} = \begin{vmatrix} y_{12} & y_{13} & y_{14} \\ y_{32} & y_{33} & y_{34} \\ y_{42} & y_{43} & y_{44} \end{vmatrix}$$

(III-7)

The principal minor is one less order than the original determinant. When a minor Δ_{ij} is multiplied by $(-1)^{i+j}$, the result

$$(-1)^{i+j}\Delta_{ij}$$

(III-8)

is termed the "cofactor." Hence the cofactor is a signed minor:

$$\text{Cofactor} = (-1)^{i+j}(\text{minor})$$

(III-9)

The $(-1)^{i+j}$ quantity can be determined by counting—plus, minus, plus, minus—starting plus on y_{11} and continuing in a horizontal or vertical path until the y_{ij} element is reached. For example, Δ_{21} [Eq. (II-7)] is negative.

The expansion of a determinant by means of minors consists of a series of consecutive reductions along either a column or a row, such as

$$\Delta = y_{11}\Delta_{11} - y_{21}\Delta_{21} + y_{31}\Delta_{31}$$

(III-10)

for a third-order determinant. Here the determinant Δ is expanded along a column. An expansion along a row is as follows:

$$\Delta = y_{11}\Delta_{11} - y_{12}\Delta_{12} + y_{13}\Delta_{13} \tag{III-11}$$

III-3 Theorems Concerning Determinants. The following theorems are stated for the purpose of simplifying the evaluation of a determinant. These theorems can easily be verified by expanding a determinant.

1. The value of a determinant is not changed if its rows are changed to columns and columns to rows:

$$\begin{vmatrix} a_1 & b_1 & c_1 \\ a_2 & b_2 & c_2 \\ a_3 & b_3 & c_3 \end{vmatrix} = \begin{vmatrix} a_1 & a_2 & a_3 \\ b_1 & b_2 & b_3 \\ c_1 & c_2 & c_3 \end{vmatrix} \tag{III-12}$$

2. The sign of a determinant is changed if any two rows or columns are interchanged:

$$\begin{vmatrix} a_1 & b_1 & c_1 \\ a_2 & b_2 & c_2 \\ a_3 & b_3 & c_3 \end{vmatrix} = - \begin{vmatrix} a_1 & c_1 & b_1 \\ a_2 & c_2 & b_2 \\ a_3 & c_3 & b_3 \end{vmatrix} \tag{III-13}$$

3. If each element of one column or of one row is multiplied by a function f, the value of the determinant is multiplied by f:

$$\begin{vmatrix} a_1 & fb_1 & c_1 \\ a_2 & fb_2 & c_2 \\ a_3 & fb_3 & c_3 \end{vmatrix} = f \begin{vmatrix} a_1 & b_1 & c_1 \\ a_2 & b_2 & c_2 \\ a_3 & b_3 & c_3 \end{vmatrix} \tag{III-14}$$

4. The value of a determinant is not changed if all the elements of one column or row are multiplied by a quantity K and are added to or subtracted from the corresponding elements of another column or row:

$$\begin{vmatrix} a_1 & b_1 & c_1 \\ a_2 & b_2 & c_2 \\ a_3 & b_3 & c_3 \end{vmatrix} = \begin{vmatrix} a_1 + Kc_1 & b_1 & c_1 \\ a_2 + Kc_2 & b_2 & c_2 \\ a_3 + Kc_3 & b_3 & c_3 \end{vmatrix} \tag{III-15}$$

appendix IV

ROUTH-HURWITZ STABILITY CRITERION

IV-1 Introduction. The general form of the closed-loop transfer function of a feedback system is expressed as a quotient of two polynomials in the operator s. If R is the input to the system and C is the output, then C and R are related in the following manner:

$$\frac{C(s)}{R(s)} = \frac{N(s)}{D(s)} = \frac{b_l s^l + b_{l-1} s^{l-1} + \cdots + b_1 s + b_0}{a_n s^n + a_{n-1} s^{n-1} + \cdots + a_1 s + a_0} \qquad \text{(IV-1)}$$

The system stability depends on the roots of the characteristic polynomial $D(s)$,

$$a_n s^n + a_{n-1} s^{n-1} + \cdots + a_1 s + a_0 = 0 \qquad \text{(IV-2)}$$

In a physical system, the coefficients $a_0, a_1 \ldots, a_n$ are usually real numbers. Since a polynomial equation with real coefficients has either real or complex conjugate roots, certain restrictions are placed upon the possible roots of $D(s) = 0$. Factors of $D(s)$ corresponding to the form $s + \gamma_1$ give rise to time solutions of the form $A_1 \epsilon^{-\gamma_1 t}$. If $D(s)$ contains pairs of factors of the form $(s + \gamma_q + j\beta_q)(s + \gamma_q - j\beta_q)$, the time solutions have the form $B_q e^{-\gamma_q t} \sin(\beta_q t + \phi_q)$. In either case, for roots with negative real parts, the transient component decays to zero as time approaches $+\infty$ and the system is stable. Therefore, any root with a negative real part is termed a "stable root."

In contrast, roots that have positive real parts can exist. The result of such a condition is that the system is unstable. Therefore, roots having positive real parts are termed "unstable roots." The boundary between stable and unstable roots is termed "marginal stability" (or neutral stability). If roots with zero real parts exist, they are termed "marginally stable roots." It is possible to detect the presence of roots with positive real parts without actually finding the roots. To this end the Routh-Hurwitz stability criterion is employed.

IV-2 Routh-Hurwitz Stability Criterion. This method provides a simple and direct means for determining the number of roots with positive real parts (i.e., roots which lie in the right half plane). Although we cannot actually locate the roots, we can determine, without factoring the characteristic equation, if any of the roots lie in the right half plane and hence give rise to an unstable system.

The Laplace transformed response of a linear system is a ratio of polynomials in s, and it can be written

$$X(s) = \frac{N(s)}{D(s)} = \frac{N(s)}{s^n + a_{n-1}s^{n-1} + a_{n-2}s^{n-2} + \cdots + a_1 s + a_0} \quad \text{(IV-3)}$$

where all the a's are real positive constants. All powers of s in the denominator polynomial (the characteristic equation) must be present (from s^n to s^0). If any power of s is missing (i.e., any coefficient is zero) or if any coefficient is negative, we know immediately that $D(s)$ has roots in the right half plane or on the j axis, and the system is unstable or marginally stable. In this case, it is not necessary to continue unless we wish to determine the actual number of roots in the right half plane.

The Routh-Hurwitz method centers about an array which is formed as follows:

s^n	a_n	a_{n-2}	a_{n-4}	a_{n-6}	$\cdots$
s^{n-1}	a_{n-1}	a_{n-3}	a_{n-5}	a_{n-7}	$\cdots$
s^{n-2}	b_1	b_2	b_3	b_4	$\cdots$
s^{n-3}	c_1	c_2	c_3	$\cdots$	
$\cdots$					
s^1	i_1				
s^0	j_1				

(IV-4)

where the constants in the third row are found by cross-multiplying as follows:

$$b_1 = \frac{a_{n-1}a_{n-2} - a_n a_{n-3}}{a_{n-1}}$$

$$b_2 = \frac{a_{n-1}a_{n-4} - a_n a_{n-5}}{a_{n-1}} \qquad \text{(IV-5)}$$

$$b_3 = \frac{a_{n-1}a_{n-6} - a_n a_{n-7}}{a_{n-1}}$$

$$\ldots \ldots \ldots \ldots$$

We continue the pattern until all remaining b's are equal to zero. The next row, the c row, is formed by cross-multiplying, using the s^{n-1} and the s^{n-2} rows. These constants are evaluated as follows:

$$c_1 = \frac{b_1 a_{n-3} - b_2 a_{n-1}}{b_1}$$

$$c_2 = \frac{b_1 a_{n-5} - b_3 a_{n-1}}{b_1} \qquad \text{(IV-6)}$$

$$c_3 = \frac{b_1 a_{n-7} - b_4 a_{n-1}}{b_1}$$

$$\ldots \ldots \ldots \ldots$$

This is continued until all remaining c's are equal to zero. The remainder of the rows down to s^0 is formed in similar fashion. Each of the last two rows (s^1 and s^0) contains only one nonzero term.

Having formed the coefficient array, we can determine the number of roots in the right half plane from the Routh-Hurwitz criterion as follows: *The number of roots of the characteristic equation with positive real parts is equal to the number of changes of sign of the coefficients in the first column.* Hence, if all terms in the first column have the same sign, the system is stable.

As an example, consider the following polynomial:

$$D_1(s) = s^5 + s^4 + 3s^3 + 9s^2 + 16s + 10 \qquad \text{(IV-7)}$$

We form the Routh-Hurwitz array by using Eqs. (IV-4) to (IV-6) with the result

s^5	1	3	16
s^4	1	9	10
s^3	$\dfrac{(1)(3) - (1)(9)}{1} = -6$	$\dfrac{(1)(16) - (1)(10)}{1} = +6$	$\ldots$
s^2	$+10$	10	$\ldots$
s^1	$+12$	$\ldots$	$\ldots$
s^0	$+10$	$\ldots$	$\ldots$

There are two changes of sign in the first column: from $+1$ to -6 and from -6 to $+10$. From the theorem, we conclude that there are two roots of $D_1(s)$ in the right half plane (positive real part). This is verified by factoring Eq. (IV-7). The roots are

$$s = -1$$
$$s = -1 \pm j1$$
$$s = +1 \pm j2$$

The labor involved in calculating the coefficients in the array may be reduced with the aid of the following rule: *The coefficients of any row may be multiplied or divided by a positive number without changing the signs of the first column.* If appropriate numbers are chosen, the coefficients are reduced in size and the computational job is simplified. For example, in computing the array for $D_1(s)$ of Eq. (IV-7), we can use this rule to good advantage as follows:

s^5	1	3	16
s^4	1	9	10
s^3	$-\cancel{6}$	$+\cancel{6}$	
	-1	$+1$	after dividing by $+6$
s^2	$+\cancel{10}$	$\cancel{10}$	
	$+1$	$+1$	after dividing by $+10$
s^1	$+2$		
s^0	$+1$		

As before, the number of sign changes is two, and we reach the same conclusion; i.e., there are two roots with positive real parts.

The following two special cases must be considered:

1. A zero coefficient in the first column
2. A row in which all coefficients are zero

Special techniques are used when either of these conditions prevails.

Special Case 1. When the first term in a row is zero but other terms in that row are not zero, one of the following two methods can be used to obviate the difficulty:

1. Replace the zero with a small positive number ϵ and proceed to compute the remaining terms in the array.

410   Control System Design

2. In the original polynomial, substitute $1/x$ for s and find the number of roots of x which have positive real parts. This is also the number of roots of s with positive real parts.

We illustrate both methods on the polynomial

$$D_2(s) = s^5 + 3s^4 + 4s^3 + 12s^2 + 35s + 25 \qquad \text{(IV-8)}$$

METHOD 1. In this method, the coefficient array is formed in the normal fashion as follows:

$$
\begin{array}{c|ccc}
s^5 & 1 & 4 & 35 \\
s^4 & 3 & 12 & 25 \\
s^3 & 0 & \frac{80}{3} & \cdots \\
\text{Replace 0 by } \epsilon & \epsilon & \frac{80}{3} & \cdots \\
s^2 & 12 - \dfrac{80}{\epsilon} & 25 & \cdots \\
s^1 & \dfrac{(12 - 80/\epsilon)\frac{80}{3} - 25\epsilon}{12 - 80/\epsilon} & & \\
s^0 & 25 & &
\end{array}
$$

The first term in the s^2 row is large but negative, and, as $\epsilon \to 0$, the first term in the s^1 row becomes approximately $+80/3$. The signs of the first column are

$$
\begin{array}{c|c}
s^5 & + \\
s^4 & + \\
s^3 & + \\
s^2 & - \\
s^1 & + \\
s^0 & +
\end{array}
$$

There are two changes of sign; hence there are two roots in the right half plane.

METHOD 2. We replace s by $1/x$ in the original polynomial:

$$D_2\left(\frac{1}{x}\right) = \left(\frac{1}{x}\right)^5 + 3\left(\frac{1}{x}\right)^4 + 4\left(\frac{1}{x}\right)^3 + 12\left(\frac{1}{x}\right)^2 + 35\left(\frac{1}{x}\right) + 25$$

We multiply by x^5 with the result

$$25x^5 + 35x^4 + 12x^3 + 4x^2 + 3x + 1 \qquad \text{(IV-9)}$$

This equation is now considered as any other, and the coefficient array is formed in the usual manner:

$$\begin{array}{c|ccc}
x^5 & 25 & 12 & 3 \\
x^4 & 35 & 4 & 1 \\
x^3 & \frac{320}{35} & \frac{80}{35} & \cdots \\
 & 4 & 1 & \cdots \\
x^2 & -\frac{19}{4} & 1 & \cdots \\
x^1 & \frac{35}{19} & \cdots & \cdots \\
x^0 & 1 & \cdots & \cdots \\
\end{array}$$ multiply by $\frac{35}{80}$

Since there are two changes of sign, there are two roots of x with positive real parts. Hence, there are two roots of s with positive real parts.

Special Case 2. When all coefficients in any one row are zero, use is made of the following procedure:

1. The coefficient array is formed in the usual manner until the all-zero-coefficient row appears.
2. The all-zero row is replaced by the coefficients obtained by differentiating an auxiliary equation which is formed from the preceding row. The roots of the auxiliary equation, which are roots of the original equation, occur in pairs and are the negatives of each other.

The occurrence of an all-zero row usually means that two roots lie on the j axis ($\pm j\omega$). This condition occurs, however, any time the polynomial has two equal roots with opposite sign or two pairs of complex conjugate roots. As an example of the special technique needed here, consider the following characteristic equation:

$$D_3(s) = s^6 + 4s^5 + 11s^4 + 22s^3 + 30s^2 + 24s + 8 \qquad \text{(IV-10)}$$

The coefficient array is written

$$\begin{array}{c|cccc}
s^6 & 1 & 11 & 30 & 8 \\
s^5 & 4 & 22 & 24 & \\
s^4 & \frac{11}{2} & 24 & 8 & \\
s^3 & \frac{50}{11} & \frac{200}{11} & & \\
 & 1 & 4 & & \\
s^2 & 2 & 8 & & \\
 & 1 & 4 & & \\
s^1 & 0 & 0 & & \\
\end{array}$$

multiply by $\frac{11}{50}$

divide by 2

412 Control System Design

The existence of the all-zero s^1 row indicates the presence of equal roots of opposite sign. We must follow the special procedure. An auxiliary equation is formed from the preceding row, as follows:

$$s^2 + 4 = 0 \tag{IV-11}$$

The auxiliary equation is formed by taking as the highest power in s the power of the s which appears in that row. In this case, the power is 2 and the highest term is s^2. Only alternate powers, either all odd or all even, are taken for the rest of the terms. Here, the next term following s^2 is $4s^0 = 4$. The roots of this equation, which are also the roots of the original polynomial, are

$$s = \pm j2$$

Equation (IV-11) is differentiated, yielding $2s + 0$, and the coefficients of this equation are inserted in the all-zero row, in this case the s^1 row, and the table is completed in the usual manner:

$$
\begin{array}{c|cc}
\vdots & & \\
\vdots & & \\
\vdots & & \\
s^2 & 1 & 4 \\
s^1 & 2 & \\
s^0 & 4 & \\
\end{array}
$$

Since there are no changes of sign in the first column, there are no roots in the right half plane. We have found, however, that in this case a pair

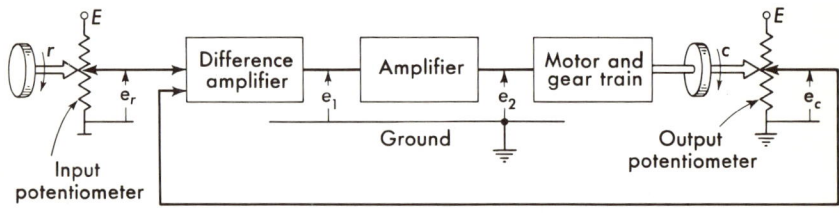

Fig. IV-1 A position servo.

of roots lies on the j axis at $\pm j2$. These roots give rise to an undamped sinusoidal response.

As a practical example of the use of the Routh-Hurwitz method, consider the servomechanism of Fig. IV-1. This system is used to position a large inertia such as a radar antenna. The load is driven with a motor and gear train. The output position is measured with a potentiometer that converts the position signal to a voltage. This voltage is subtracted from the input voltage in a difference amplifier, and the difference (or error) is amplified and used to drive the motor. The differential equation for this system is obtained from the block diagram of Fig. IV-2. The transfer functions are obtained as shown in Chaps. 3

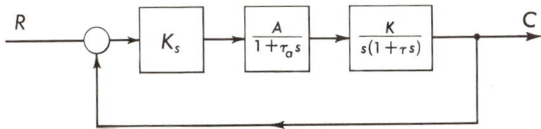

Fig. IV-2 Block diagram of position servo.

and 5. The overall differential equation which relates the input R to the output shaft position C is

$$\frac{C(s)}{R(s)} = \frac{AK_sK_m}{\tau_a\tau_m s^3 + (\tau_a + \tau_m)s^2 + s + AK_sK_m} \qquad \text{(IV-12)}$$

where A = amplifier gain
K_s = potentiometer sensitivity, volts/rad
K_m = motor constant, rad/(volt/sec)
τ_a = amplifier time constant, sec
τ_m = motor time constant, sec
C = Laplace-transformed output position
R = Laplace-transformed input position

Suppose we wish to determine the effect of the amplifier time constant τ_a upon the system stability. To do so, we shall find the relationship between the variables for marginal stability (i.e., two equal imaginary roots). The Routh-Hurwitz array is established:

$$\begin{array}{c|cc} s^3 & \tau_a\tau_m & 1 \\ s^2 & \tau_a\tau_m & AK_sK_m \\ s^1 & \dfrac{\tau_a + \tau_m - \tau_a\tau_m AK_sK_m}{\tau_a + \tau_m} & \cdots \\ s^0 & \cdots \end{array} \qquad \text{(IV-13)}$$

We know that if all coefficients in a row are zero, we have two equal roots of opposite sign. We can obtain all zeros in a single row by setting the first term in the s^1 row equal to zero with the result

$$\frac{1}{A} = \frac{\tau_a\tau_m}{\tau_a + \tau_m} K_s K_m \qquad \text{(IV-14)}$$

All other coefficients in this row are zero. We need not continue, since Eq. (IV-14) gives us the relationship between A and τ_a. Equation

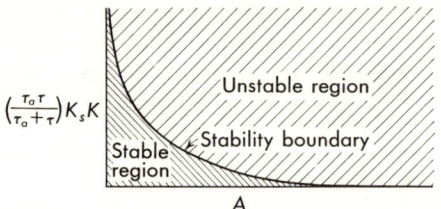

Fig. IV-3 Stability boundary for position servo.

(IV-14) is plotted in Fig. IV-3, which shows the effect of τ_a upon system stability: the larger the amplifier time constant, the smaller the amplifier gain for a given degree of stability.

appendix V

FREQUENCY-RESPONSE DERIVATIONS

V-1 Derivation of the Nyquist Criterion. The formal derivation of the Nyquist criterion is based upon complex-variable theory. The residue theorem* states: If C is a closed curve within and on which $G(s)$ is analytic except for a finite number of singular points $s_1, s_2, \ldots, s_n$ inside C, and if $b_1, b_2 \ldots, b_n$ denote the residues of $G(s)$ at these points, then

$$\int_c G(s)\, ds = 2\pi j(b_1 + v_2 + \cdots + b_n) \tag{V-1}$$

The residue for simple poles of $G(s)$ is the coefficient of the $(s - s_1)^{-1}$ term in the Laurent series

$$G(s) = \frac{b_1}{s - s_1} + \frac{b_2}{(s - s_1)^2} + \cdots + \frac{b_m}{(s - s_1)^m} \tag{V-2}$$

where b_1 is the residue.

In the derivation of the Nyquist stability criterion, let:

1. $G(s)$ have a number k of distinct poles of multiplicity m_i at a point s_i.
2. $G(s)$ have a number l of distinct zeros of multiplicity n_j at a point s_j.

Then the ratio $G'(s)/G(s)$ has a pole of order 1 at $s = s_i$ and also a pole of order 1 at $s = s_j$. This can be shown by letting

$$G(s) = \frac{(s - s_j)^{n_j}}{(s - s_i)^{m_i}} \tag{V-3}$$

* See Ref. 7, p. 118.

The derivative of Eq. (V-3) is written

$$\frac{dG(s)}{ds} = G'(s) = \frac{(s - s_i)^{m_i} n_j (s - s_j)^{n_j - 1} - (s - s_j)^{n_j} m_i (s - s_i)^{m_i - 1}}{(s - s_i)^{2m_i}} \quad \text{(V-4)}$$

and the ratio $G'(s)/G(s)$ is found from Eq. (V-4):

$$\frac{G'(s)}{G(s)} = \frac{(s - s_i)^{m_i} n_j (s - s_j)^{n_j - 1} - (s - s_j)^{n_j} m_i (s - s_i)^{-1}}{(s - s_j)^{n_j} (s - s_i)^{m_i}} \quad \text{(V-5)}$$

Equation (V-5) is simplified as follows:

$$\frac{G'(s)}{G(s)} = \frac{n_j}{s - s_j} - \frac{m_i}{s - s_i} \quad \text{(V-6)}$$

Equation (V-6) is easily extended to any number of poles and zeros. Hence $G'(s)/G(s)$ has a pole of order 1 at $s = s_j$ and a pole of order 1 at $s = s_i$.

Hence from Eq. (V-6) and the residue theorem

$$\frac{1}{2\pi j} \oint \frac{G'(s)}{G(s)} ds = \sum_{i=1}^{k} (-m_i) + \sum_{j=1}^{l} (n_j) \quad \text{(V-7)}$$

But $\sum_{i=1}^{k} (m_i)$ is the number of poles and $\sum_{j=1}^{l} (n_j)$ is the number of zeros on or within the contour of integration. Hence the contour integral of Eq. (V-7) can be written

$$\frac{1}{2\pi j} \oint \frac{G'(s)}{f(s)} ds = -n_P + n_Z \quad \text{(V-8)}$$

where n_P is the number of poles of $G(s)$ and n_Z is the number of zeros of $G(s)$ within or on the contour.

To establish physical significance for Eq. (V-8), consider the logarithm of $G(s)$:

$$\ln G(s) = \ln |G(s)| + j\phi(s) \quad \text{(V-9)}$$

The logarithm of a complex function is equal to the log of the magnitude plus the phase angle of the function. Equation (V-9) is related to Eq.

(V-8) by differentiating as follows:

$$d\,[\ln G(s)] = \frac{G'(s)}{G(s)}\,ds \tag{V-10}$$

By forming the integral of Eq. (V-10) around the closed contour,

$$\frac{1}{2\pi j}\oint \frac{G'(s)}{G(s)}\,ds = \frac{1}{2\pi j}\oint d[\ln G(s)]$$

$$\frac{1}{2\pi j}\oint \frac{G'(s)}{G(s)}\,ds = \frac{1}{2\pi j}[\ln G(s)]_{s_{\text{start}}}^{s_{\text{finish}}} = \frac{1}{2\pi j}[\ln |G(s)| + j\phi(s)]_{s_{\text{start}}}^{s_{\text{finish}}} \tag{V-11}$$

Since the contour of integration is closed, the only contribution comes from the imaginary part; the contribution from the real part is zero. Hence

$$\frac{1}{2\pi j}\oint \frac{G'(s)}{G(s)}\,ds = \frac{1}{2\pi j}j\,\Delta\phi = \frac{\Delta\phi}{2\pi} \tag{V-12}$$

where $\Delta\phi$ is the net change in angle as the function $d[\ln G(s)]$ is integrated around the closed contour and $\Delta\phi/2\pi$ is the number of encirclements N' of the origin. Combination of Eqs. (V-8) and (V-12) yields the following expression:

$$N' = \frac{\Delta\phi}{2\pi} = -n_P + n_Z \tag{V-13}$$

In control-system applications, the location of the roots of the function $1 + G(s)H(s)$ is of importance. If N is defined as the number of clockwise encirclements of the $-1 + j0$ point rather than the origin, then the above derivation is extended to control-system problems with the equation

$$N = n_R - n_P \tag{V-14}$$

When the contour encircles the entire right half plane, n_R is the number of roots of $1 + G(s)H(s)$ in the right half plane and n_P is the number of poles of $G(s)H(s)$ in the right half plane.

V-2 Derivation of M and N Circles. The equations of the M circles are found from the definition

$$M = \left|\frac{C(j\omega)}{R(j\omega)}\right| = \frac{G(j\omega)}{1 + G(j\omega)} \tag{V-15}$$

418   Control System Design

Since $G(j\omega)$ is a complex function, it can be represented by a real part x and imaginary part y:

$$G(j\omega) = x + jy \qquad \text{(V-16)}$$

When Eq. (V-16) is substituted into the defining equation, Eq. (V-15),

$$M = \left| \frac{x + jy}{1 + x + jy} \right| \qquad \text{(V-17)}$$

The magnitude is found by taking the square of both sides of Eq. (V-17):

$$M^2 = \frac{x^2 + y^2}{(1 + x)^2 + y^2} \qquad \text{(V-18)}$$

Equation (V-18) is multiplied and simplified with the result

$$y^2 + x^2 + 2x \frac{M^2}{M^2 - 1} + \frac{M^2}{M^2 - 1} = 0 \qquad \text{(V-19)}$$

This last expression is the equation of a circle and is put in normal form as follows:

$$y^2 + \left(x + \frac{M^2}{M^2 - 1}\right)^2 = \frac{M^2}{(M^2 - 1)^2} \qquad \text{(V-20)}$$

as can be verified by multiplying out the term in x. Hence the radii of the circles are given by

$$R = \left| \frac{M}{(M^2 - 1)} \right| \qquad \text{(V-21)}$$

and the centers are located at

$$x_0 = \frac{-M^2}{M^2 - 1} \qquad y = 0 \qquad \text{(V-22)}$$

These expressions are included in Chap. 5.

The equations of the N circles are found in a similar fashion. The definition of the quantity N is given by the expression

$$N = \tan \phi = \frac{\text{Im}\, [C(j\omega)/R(j\omega)]}{\text{Re}\, [C(j\omega)/R(j\omega)]} \qquad \text{(V-23)}$$

But, from Eq. (V-16),

$$\frac{C(j\omega)}{R(j\omega)} = \frac{R(j\omega)}{C(j\omega)} = \frac{x+jy}{(1+x)+jy} = \frac{(x+jy)[(1+x)-jy]}{(1+x)^2+y^2} \quad \text{(V-24)}$$

which simplifies to

$$\frac{C(j\omega)}{R(j\omega)} = \frac{x^2+x+y^2+jy}{(1+x)^2+y^2} \quad \text{(V-25)}$$

Substituting the imaginary and real parts of Eq. (V-25) into Eq. (V-23),

$$N = \frac{y}{x^2+x+y^2} \quad \text{(V-26)}$$

When N is taken as a constant, Eq. (V-26) is multiplied and simplified as

$$(x^2+x)+\left(y^2-\frac{y}{N}\right)=0 \quad \text{(V-27)}$$

Adding $\frac{1}{4}+\frac{1}{4}N^2$ to both sides of Eq. (V-27) completes the square in x and y:

$$\left(x^2+x+\frac{1}{4}\right)+\left(y^2-\frac{y}{N}+\frac{1}{4N^2}\right)=\frac{1}{4}+\frac{1}{4N^2} \quad \text{(V-28)}$$

which reduces to

$$\left(x+\frac{1}{2}\right)^2+\left(y-\frac{1}{2N}\right)^2=\frac{1}{4}\left(1+\frac{1}{N^2}\right) \quad \text{(V-29)}$$

Equation (V-29) is the equation of a circle whose center is located at

$$x = -\frac{1}{2} \qquad y = +\frac{1}{2N} \quad \text{(V-30)}$$

The radius of the circle is given by

$$R = \pm\frac{1}{2}\sqrt{1+\frac{1}{N^2}} \quad \text{(V-31)}$$

V-3 Relations between ζ and Other Stability Quantities. The phase margin ϕ_m and the M criterion can be used to estimate the value of the damping ratio. These two relations are derived for a second-order system.

a. *Damping Ratio ζ versus Phase Margin.* The curve of damping ratio vs. phase margin is based upon the second-order system. The open-loop transfer function for a second-order system is

$$G(s) = \frac{\omega_n^2}{s(s + 2\zeta\omega_n)} \tag{V-32}$$

which for a sinusoidal input becomes

$$G(j\omega) = \frac{\omega_n^2}{j\omega(j\omega + 2\zeta\omega_n)} = \frac{\omega_n^2}{-\omega^2 + j2\zeta\omega_n\omega} \tag{V-33}$$

The value of ω when the magnitude of $G(j\omega) = 1$ is found as follows:

$$\left|\frac{\omega_n^2}{\omega_1(-\omega_1 + j2\zeta\omega_n)}\right| = 1 \tag{V-34}$$

which when squared and multiplied out results in the following quadratic in the variable ω_1^2:

$$\omega_1^4 + 4\zeta^2\omega_n^2\omega_1^2 - \omega_n^4 = 0 \tag{V-35}$$

The solution of this quadratic yields

$$\frac{\omega_1^2}{\omega_n^2} = \sqrt{4\zeta^4 + 1} - 2\zeta^2 \tag{V-36}$$

The phase margin ϕ_m, which is 180° minus the phase lag when the magnitude of G is unity, is given by

$$\phi = \tan^{-1}\frac{\text{Im }[G(j\omega)]}{\text{Re }[G(j\omega)]} = \tan^{-1} 2\zeta\frac{\omega_n}{\omega} \tag{V-37}$$

When the frequency ω_1 is substituted, the phase margin is

$$\phi_m = \tan^{-1} 2\zeta \left[\frac{1}{(4\zeta^4 + 1)^{\frac{1}{2}} - 2\zeta^2}\right]^{\frac{1}{2}} \tag{V-38}$$

This expression is plotted in Chap. 5.

For small damping the phase margin is related to ζ by the following expressions:

$$\zeta \approx \tfrac{1}{2}\phi_m$$

when ϕ_m is in radians, or

$$\zeta = \frac{\phi_m 180}{2\pi} \qquad (V\text{-}39)$$

when ϕ_m is in degrees.

b. *Damping Ratio ζ versus M_p.* The damping ratio versus M_p curve is derived exactly for the second-order system:

$$G = \frac{\omega_n{}^2}{s(s + 2\zeta\omega_n)} \qquad (V\text{-}40)$$

For a sinusoidal input the magnitude of M^2 is given by

$$M^2 = \left|\frac{G}{1+G}\right|^2 = \frac{\omega_n{}^4}{(\omega_n{}^2 - \omega^2)^2 + 4\zeta^2\omega_n{}^2\omega^2} \qquad (V\text{-}41)$$

The value of ω yielding the peak value of M is found by differentiating M^2, setting it equal to zero, and solving for ω:

$$\frac{dM^2}{d\omega} = \frac{\omega_n{}^4[2(\omega_n{}^2 - \omega^2)(-2\omega) + 8\zeta^2\omega_n{}^2\omega]}{[(\omega_n{}^2 - \omega^2)^2 + 4\zeta^2\omega_n{}^2\omega^2]^2} = 0 \qquad (V\text{-}42)$$

Simplification of this expression yields the frequency at which the peak M occurs:

$$\omega_p = \pm\omega_n\sqrt{1 - 2\zeta^2} \qquad (V\text{-}43)$$

Only the plus sign can be used, since a negative frequency is not possible. When this value of frequency is substituted into the expression for M, the peak value of M^2 results:

$$M_p{}^2 = \frac{\omega_n{}^4}{[\omega_n{}^2 - \omega_n{}^2(1 - 2\zeta^2)]^2 + 4\zeta^2\omega_n{}^2\omega_n{}^2(1 - 2\zeta^2)} \qquad (V\text{-}44)$$

Reduction yields

$$M_p = \frac{1}{2\zeta\sqrt{1 - \zeta^2}} \qquad (V\text{-}45)$$

This expression is plotted in Chap. 5.

appendix VI

DESIGN OF BRIDGED-T AND PARALLEL-T NETWORKS[28,29]

VI-1 Introduction. The resistor-shunt and capacitor-shunt bridged-T and parallel-T networks are shown in Fig. VI-1, and the transfer functions are included in Chap. 6. Design of these networks is based upon the following criteria:

1. If infinite attenuation is desired, a symmetrical parallel-T (or twin-T) network is suggested.
2. If only a partial notch is desired, the bridged-T network is used, since it requires fewer components.

The basic assumption used in the first design attempt is that the load impedance on the network is infinite and the source impedance of the driving voltage is zero. An approximation of low source and large load impedance, with respect to network impedance, is adequate. Section VI-4 discusses the design of loaded networks.

In order to design either of the bridged-T networks, four design parameters must be specified:

1. f_0 is the frequency where the notch is to occur.
2. r is the notch ratio, i.e., the ratio of the amplitude at f_0 to the amplitude at zero frequency. Figure 6-53 shows the effect of varying this quantity.
3. n is the relative width of the notch. Figures VI-2 and VI-3 show the effect on the notch width of varying this parameter. The choice of

this parameter is not completely independent but depends on the notch ratio. The choice of n is discussed under Design Procedure for the Bridged-T Networks (Sec. VI-2).

1. *Direct-current impedance level.* Since the input and output impedances vary with frequency, it is felt that the d-c impedance (the series impedance of the network at 0 frequency) is the one of greatest use to the designer in choosing one parameter.

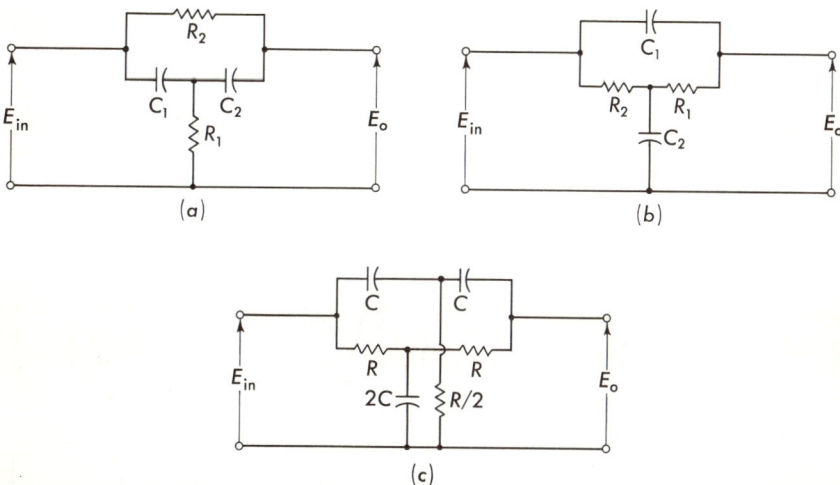

Fig. VI-1   (a) Resistor-shunt bridged-T network; (b) capacitor-shunt bridged-T network; (c) parallel-T network.

In the design of an infinite-attenuation parallel-T network with a minimum width, only two parameters need be specified: the notch frequency f_0 and the impedance level. The notch ratio is zero, and n is chosen as a minimum so that the sharpest possible notch is obtained.

VI-2   Design Procedure for the Bridged-T Networks

a. Capacitor-shunt Bridged-T Network

1. With the known notch ratio r choose the ratio C_1/C_2, from Fig. VI-4, corresponding to the value of n desired or obtainable. Although the minimum width corresponds to low values of n, a limit is reached beyond which the ratio tends to zero. Thus it is desirable to choose values of n

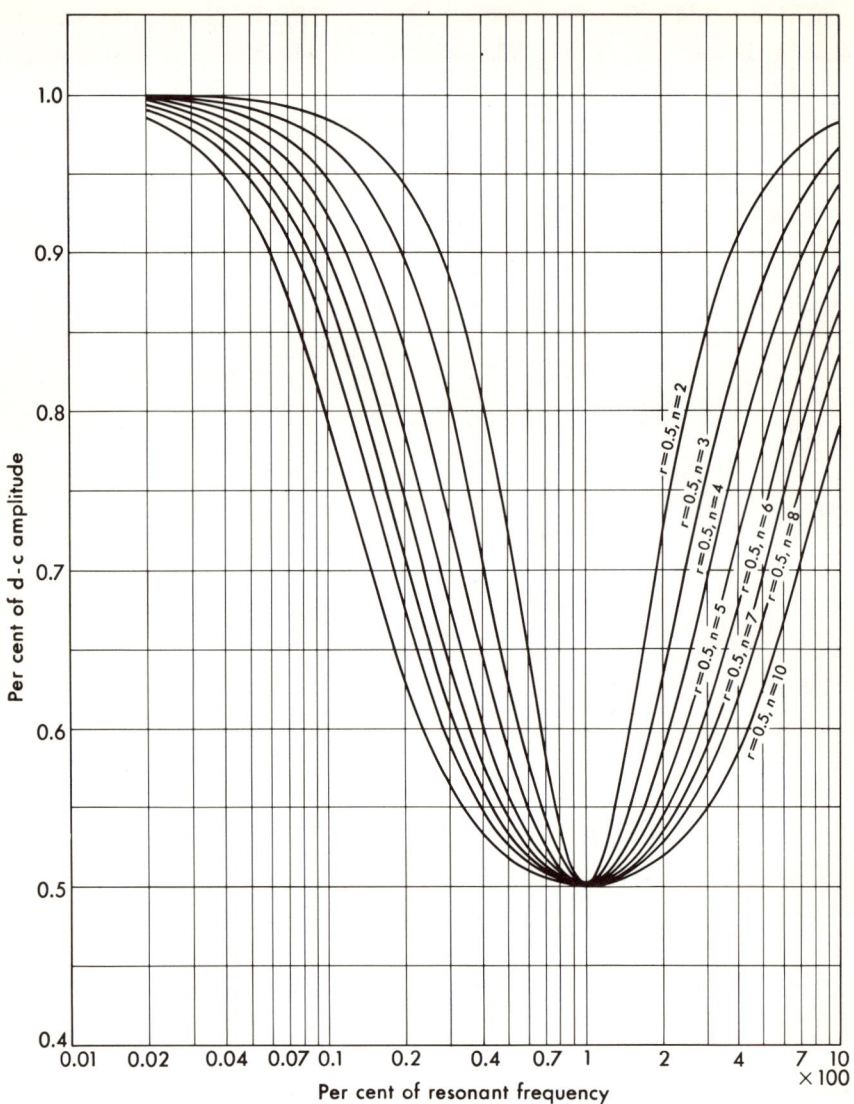

Fig. VI-2 Amplitude response for the bridged-T networks.

such that the ratio is greater than 0.04 (shown by a dashed line in Fig. VI-4).

2. Calculate the quantity $n(1 - r)$ and obtain both CR products from Fig. VI-5. Note that $C_1 R_1$ is obtained when $\gamma = 1/n(1 - r)$ and that $C_2 R_2$ is obtained when $\gamma = n(1 - r)$.

3. When $R_1 + R_2 = R_{\text{d-c}}$ is chosen according to the desired d-c imped-

Appendix VI   425

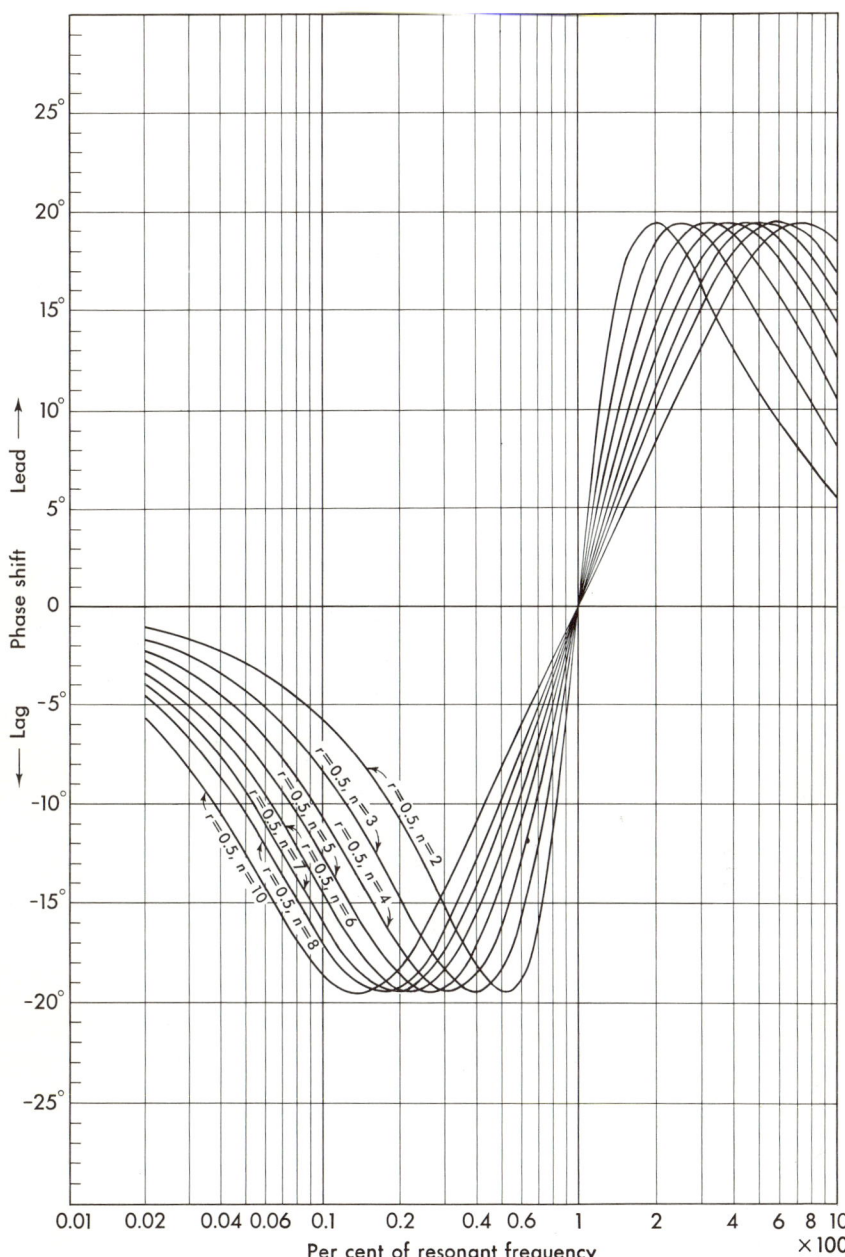

Fig. VI-3 Phase-angle response for the bridged-T networks.

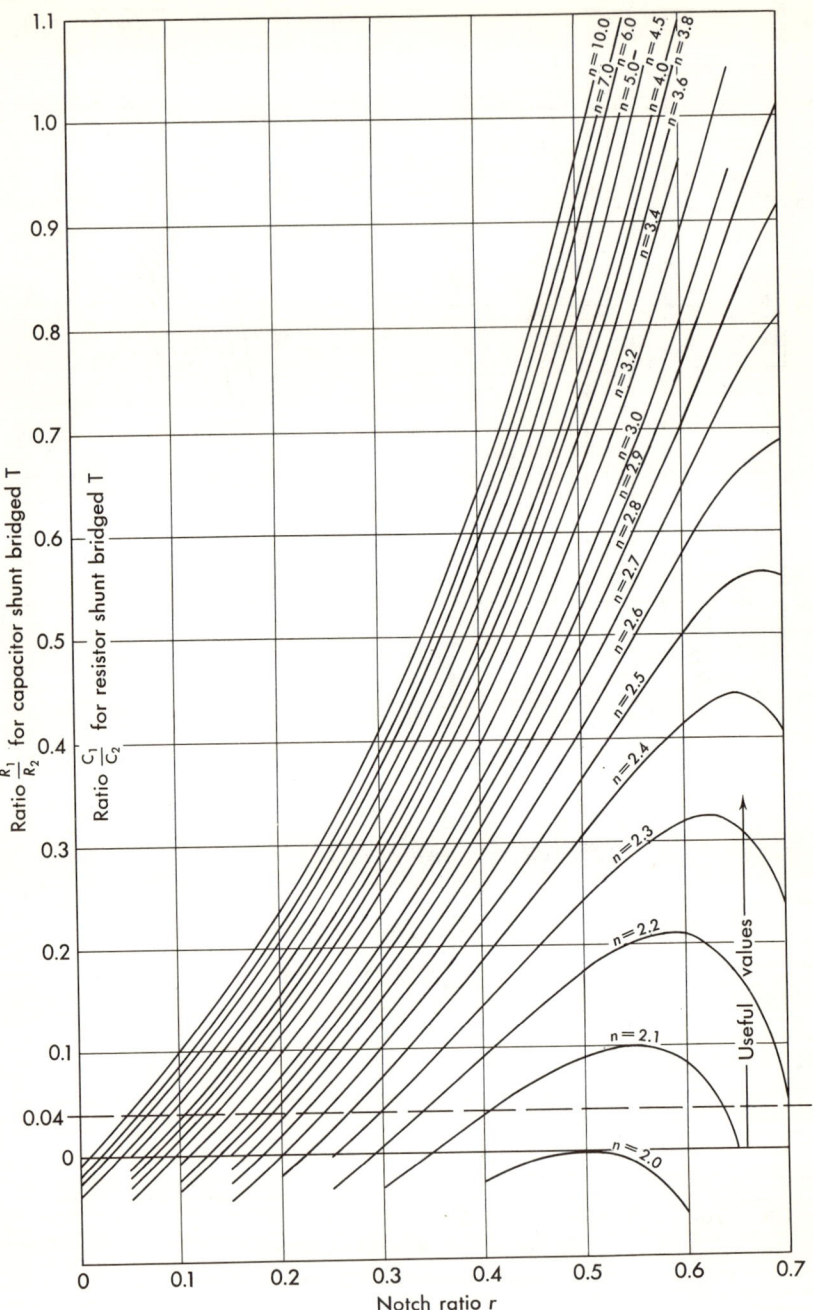

Fig. VI-4 Bridged-T design chart; notch ratio.

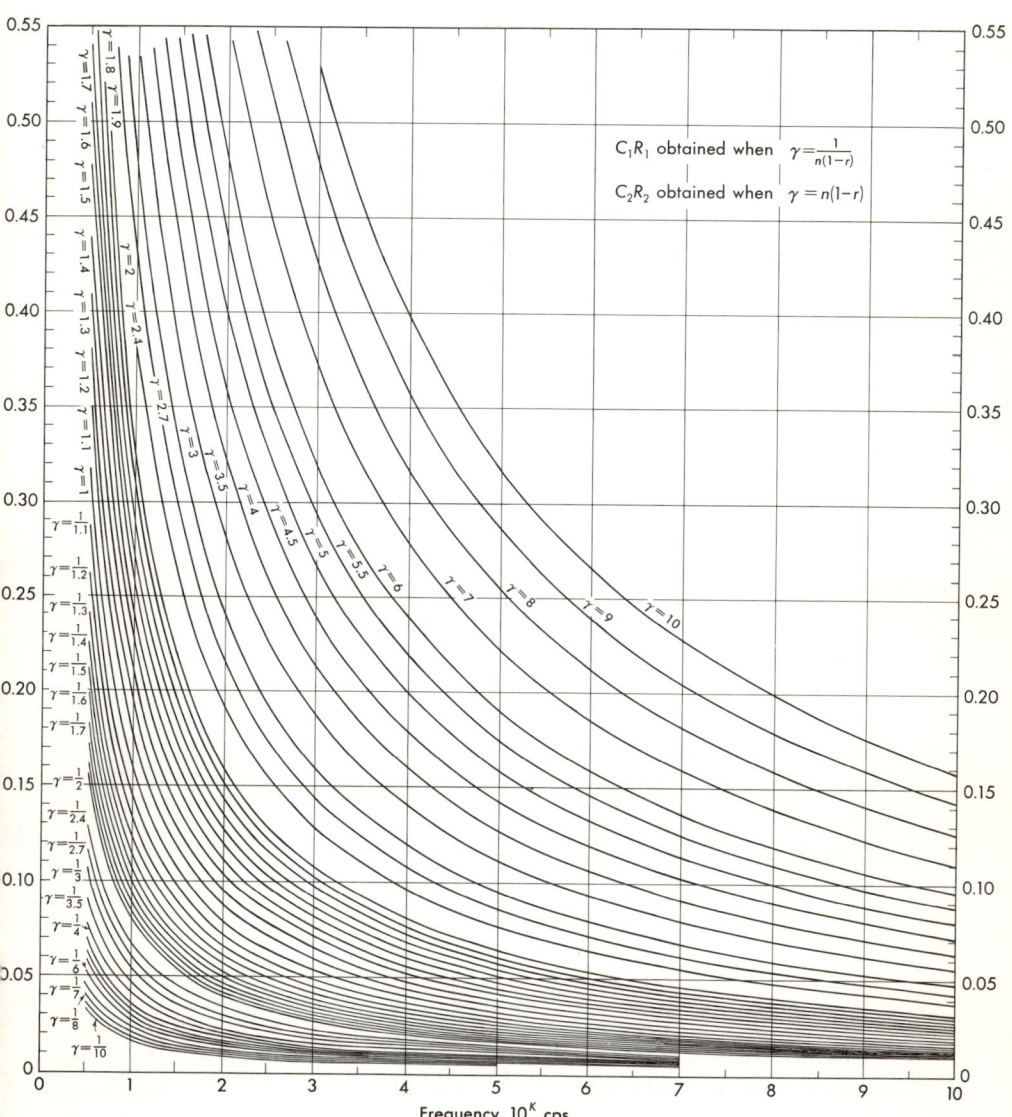

Fig. VI-5 Bridged-T design chart; CR product.

ance level, the necessary parameters can be found from the following equations:

$$R_1 + R_2 = R_{\text{d-c}} \qquad C_1 = (C_1R_1)\frac{1}{R_1}$$

$$\frac{R_1}{R_2} = \frac{(C_1R_1)}{(C_2R_2)}\frac{C_2}{C_1} \qquad C_2 = (C_2R_2)\frac{1}{R_2}$$

(VI-1)

If it is necessary to interpolate more closely between curves, the network components can be found from the following equations:

$$R_{\text{d-c}} = R_1 + R_2$$

$$C_1 R_1 = \frac{1}{2\pi f_0} \frac{1}{n(1-r)}$$

$$C_2 R_2 = \frac{1}{2\pi f_0} n(1-r) \qquad \text{(VI-2)}$$

$$\frac{C_1}{C_2} = \frac{r}{1-r} - \frac{1}{n^2(1-r)^2}$$

when n is chosen from Fig. VI-4. As an example of the use of this graphical construction, consider the following requirements:

$$R_{\text{d-c}} = 200 \text{ kilohms} \qquad r = 0.2 \qquad f_0 = 20 \text{ cps}$$

From Fig. VI-4, $C_1/C_2 = 0.05$ and $n = 2.8$. Calculating,

$$n(1-r) = (2.8)(0.8) = 2.24$$

From Fig. VI-5,

$$R_1 C_1 = 0.0036 \qquad R_2 C_2 = 0.0175$$

Hence

$$\frac{R_1}{R_2} = \frac{0.0036}{0.0175} \, 20 = 4.11$$

$$R_2 = \frac{R_{\text{d-c}}}{1 + 4.11} = \frac{200 \text{ kilohms}}{5.11} = 39.2 \text{ kilohms}$$

$$R_1 = R_{\text{d-c}} - R_2 = 200 \text{ kilohms} - 39.2 \text{ kilohms} = 160.8 \text{ kilohms} \qquad \text{(VI-3)}$$

$$C_1 = \frac{0.0036 \times 10^{-3}}{160.8} = 0.0224 \, \mu\text{f}$$

$$C_2 = \frac{0.0175}{39.2\text{K}} = 0.445 \, \mu\text{f}$$

Before passing to the next network, it must be pointed out that if any one value proves to be higher or lower than that which is practical, it may be necessary to use a larger value of n.

b. Resistor-shunt Bridged-T Network

1. With the known notch ratio r choose the ratio R_1/R_2, from Fig. VI-4, corresponding to the permissible value of n. As described in the preceding section, the minimum width corresponds to the smallest value of n consistent with a nonzero value of R_1/R_2.

2. Calculate the quantity $n(1-r)$ and obtain both CR products from Fig. VI-5. Note that C_1R_1 is obtained when $\gamma = 1/n(1-r)$ and that C_2R_2 is obtained when $\gamma = n(1-r)$.

3. Choosing $R_2 = R_{\text{d-c}}$ according to the desired impedance level, the necessary parameters may be found from the following equations:

$$R_2 = R_{\text{d-c}} \qquad C_2 = (C_2R_2)\frac{1}{R_2}$$
$$R_1 = \left(\frac{R_1}{R_2}\right)R_2 \qquad C_1 = (C_1R_1)\frac{1}{R_1} \qquad \text{(VI-4)}$$

If it is necessary to interpolate more closely between curves, the network components may be found from the following equations:

$$C_1R_1 = \frac{1}{(2\pi f_0)n(1-r)} \qquad C_2R_2 = \frac{n}{2\pi f_0}(1-r)$$
$$\frac{R_1}{R_2} = \frac{r}{1-r} - \frac{1}{n^2(1-r)^2} \qquad \text{(VI-5)}$$

where n must be chosen from Fig. VI-4.

As pointed out before, if any one value calculates out too high or too low, then it may be necessary to use a larger value of n.

As an example of the design of a resistor-shunt bridged-T consider the following problem:

$$R_{\text{d-c}} = 575 \text{ kilohms} \qquad r = 0.25 \qquad f_0 = 26 \text{ cps}$$

From Fig. VI-4

$$\frac{R_1}{R_2} = 0.048 \qquad n = 2.5$$
$$n(1-r) = (2.5)(0.75) = 1.875$$

From Fig. VI-5

$$C_1R_1 = 0.0032 \qquad C_2R_2 = 0.0115$$

The parameters are calculated:

$$R_2 = R_{\text{d-c}} = 575 \text{ kilohms}$$
$$R_1 = \left(\frac{R_1}{R_2}\right)R_2 = (0.048)(575) = 27.6 \text{ kilohms}$$
$$C_1 = (C_1R_1)\frac{1}{R_1} = \frac{0.0032 \times 10^{-3}}{27.6} = 0.116 \text{ µf} \qquad \text{(VI-6)}$$
$$C_2 = (C_2R_2)\frac{1}{R_2} = \frac{0.0115 \times 10^{-3}}{575} = 0.02 \text{ µf}$$

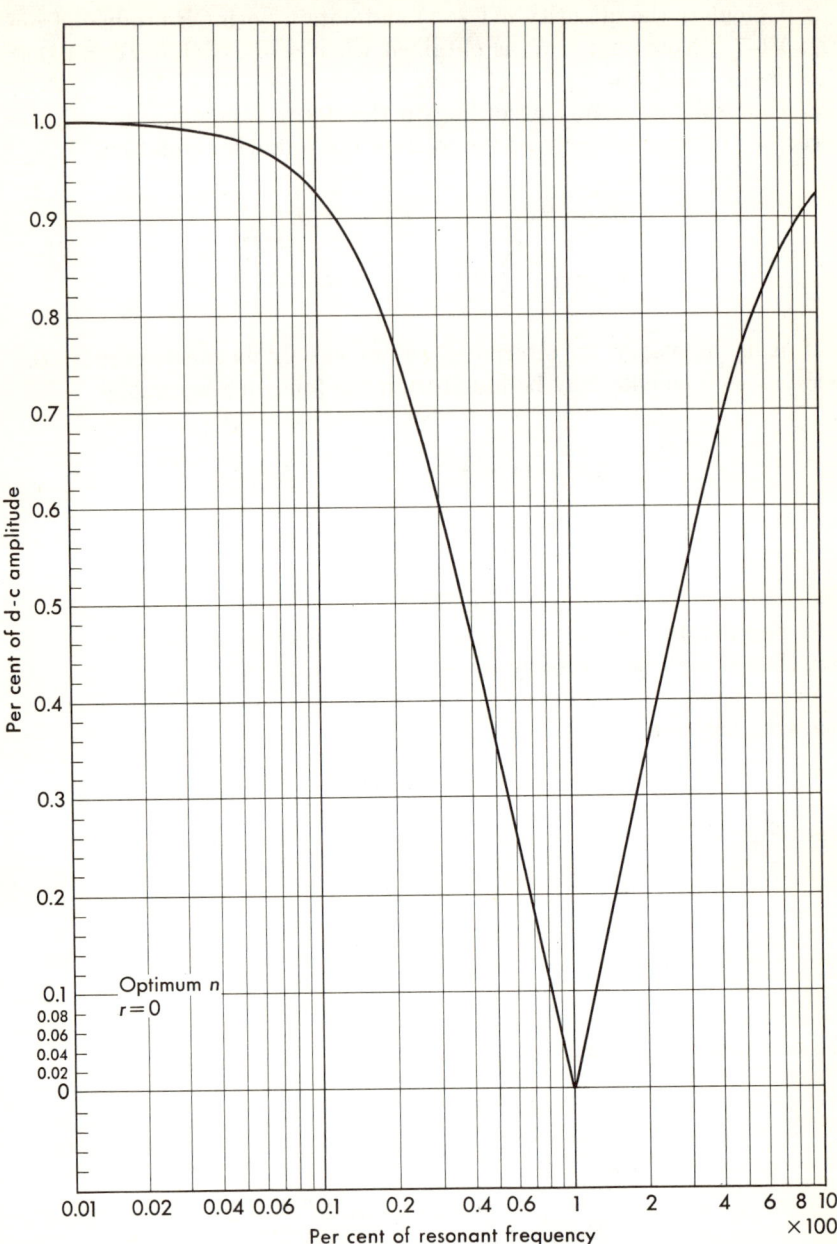

Fig. VI-6a Amplitude response for the parallel-T networks.

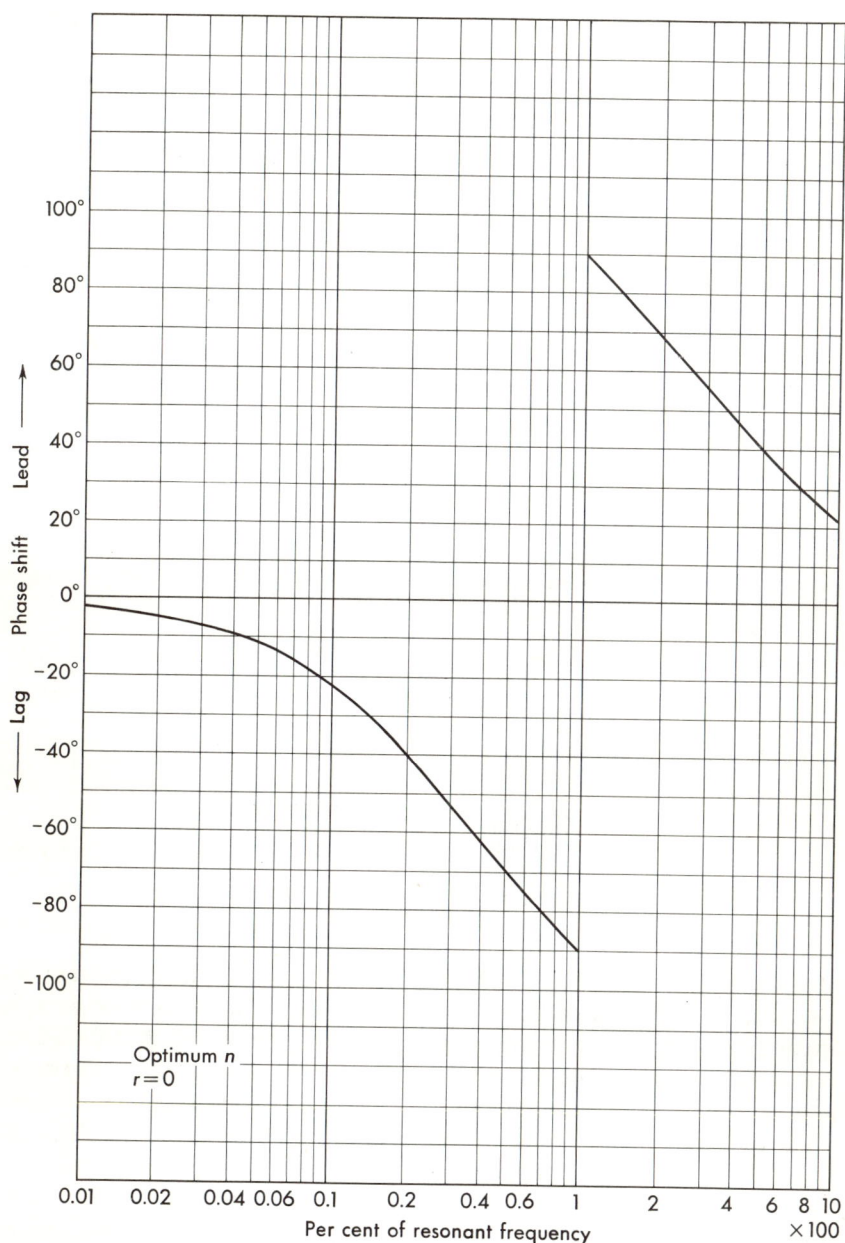

Fig. VI-6b Phase response for the parallel-T networks.

432 Control System Design

VI-3 Design Procedure for the Infinite-attenuation Parallel-T Network. It was stated at the outset that only the infinite-attenuation minimum-width parallel-T network is considered. Because of this, only two parameters are necessary to determine the constants: the resonant frequency and the d-c impedance level. The amplitude and phase response of this network are shown in Fig. VI-6. The design procedure is simply the following:

1. Enter Fig. VI-5 and read from the $\gamma = 1$ curve the value of RC product corresponding to the desired resonant frequency.
2. When $R_{\text{d-c}}$ is chosen according to the desired impedance level, the necessary parameters are found from the following equations:

$$R = \frac{R_{\text{d-c}}}{2} \qquad R_3 = \frac{R}{2} \qquad C = CR\frac{1}{R} \qquad C_3 = 2C \qquad \text{(VI-7)}$$

As an example of this calculation consider the following example:

$$R_{\text{d-c}} = 26.6 \text{ kilohms} \qquad f_0 = 400 \text{ cps}$$

From Fig. VI-5

$$RC = 0.0004$$

The parameters are calculated:

$$R = \frac{R_{\text{d-c}}}{2} = 13.3 \text{ kilohms} \qquad C = \frac{0.0004 \cdot 10^{-3}}{13.3} = 0.03 \text{ μf}$$
$$R_3 = \frac{R}{2} = 6.65 \text{ kilohms} \qquad C_3 = 2C = 0.06 \text{ μf} \qquad \text{(VI-8)}$$

VI-4 Design of Loaded Bridged-T Networks. The performance, and hence the design, of these networks is altered when the network is driven from a high source impedance and loaded with a low impedance. The design procedure of this section is based upon operating a network between equal impedances, i.e., source impedance equal to load impedance.

The capacitor-shunt bridged-T network is shown in Fig. VI-7, where all component values are defined, along with the source impedance A and the load impedance B. In this design procedure, resistances A and B are taken equal. This advantageous choice permits the cascading of a number of networks all of the same impedance level. Also, when a load has been matched to its source, the insertion of this equalizer network does not disturb the match.

Appendix VI 433

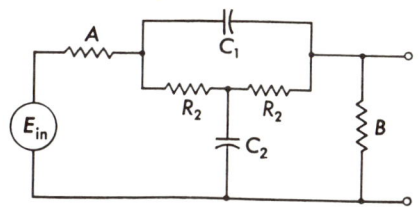

A = source impedance   B = load impedance
Fig. VI-7 Loaded bridged-T network.

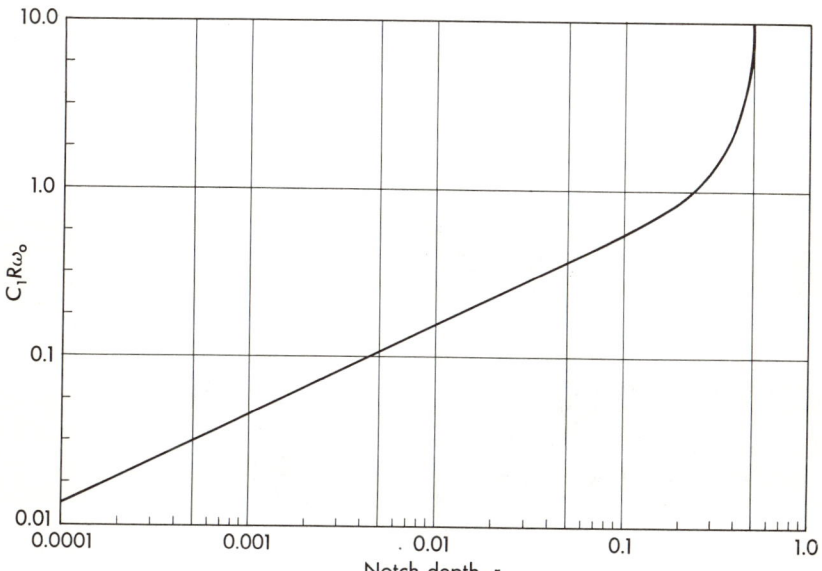

Fig. VI-8 $C_1 R \omega_0$ versus notch depth r.

To design the capacitor-shunt bridged T, which has equal input and load impedance, proceed as follows:

1. Enter Figs. VI-8 and VI-9 with the notch depth r and read the values of $C_1 R \omega_0$ and $C_2 R \omega_0$.
2. Set $R_1 = R_2 = A = B = R$.
3. Calculate C_1 and C_2 from the values read from the curve as follows:

$$C_1 = \frac{C_1 R \omega_0}{A \omega_0} \qquad C_2 = \frac{C_2 R \omega_0}{A \omega_0} \qquad \text{(VI-9)}$$

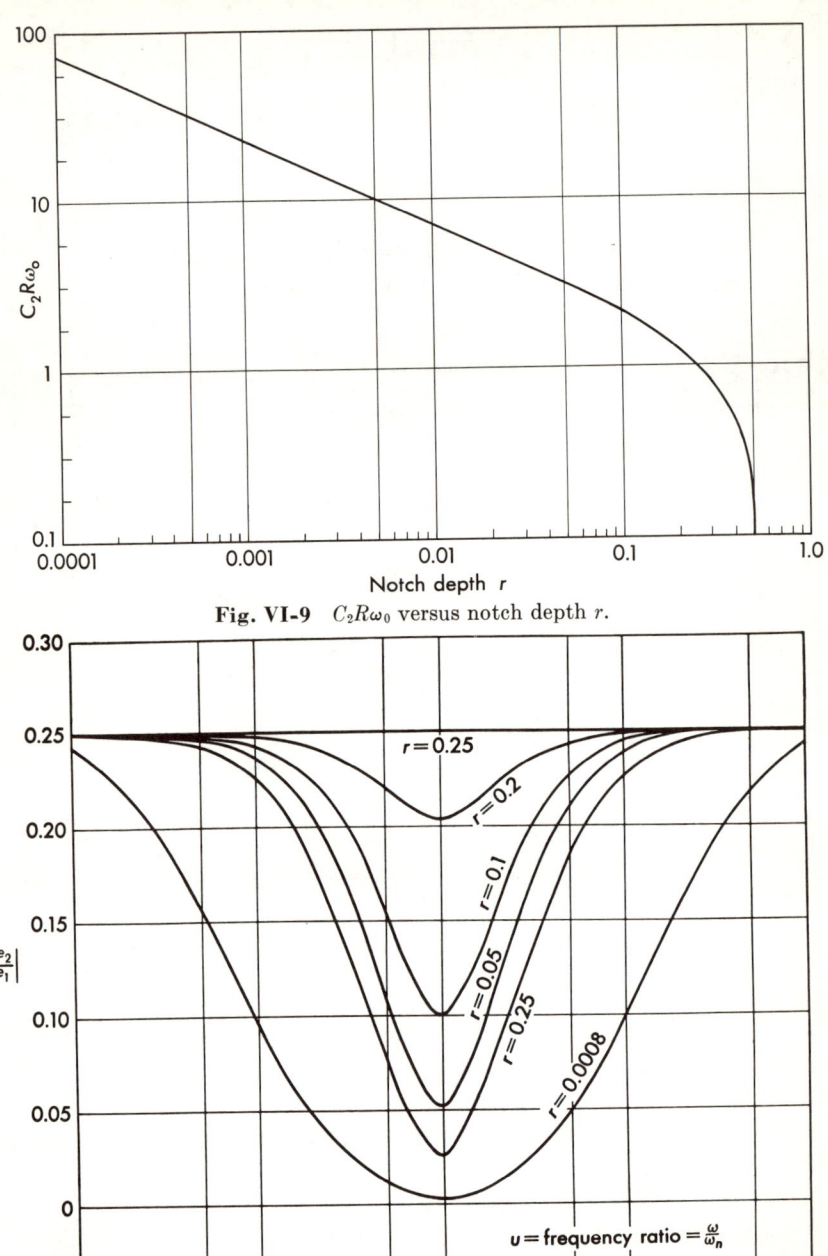

Fig. VI-9 $C_2R\omega_0$ versus notch depth r.

Fig. VI-10 Amplitude-response loaded bridged-T network.

Appendix VI 435

For closer interpolation between curves, the network components can be calculated from the following:

$$C_1 = \frac{1}{R\omega}\left(\frac{1}{2r} - 1\right)^{\frac{1}{2}} \quad C_2 = \frac{1}{R\omega_0}\left(\frac{1}{2r} - 1\right)^{\frac{1}{2}} \quad R_1 = R_2 = A = B$$
(VI-10)

If necessary, the source impedance A can be made equal to the load impedance B by padding.

EXAMPLE. $A = B = 300$ kilohms, $r = 0.25$, $f_0 = 26$ cps. From Figs. VI-8 and VI-9

$$C_1 R\omega_0 = 1.0 \qquad A = B = R_1 = R_2 = 300 \text{ kilohms}$$
$$C_2 R\omega_0 = 1.0 \qquad \omega_0 = (6.28)(26) = 166 \text{ rad/sec}$$

The attenuation and phase response for the bridged-T network of Fig. VI-7 can be seen in Figs. VI-10 and VI-11, plotted as a function of

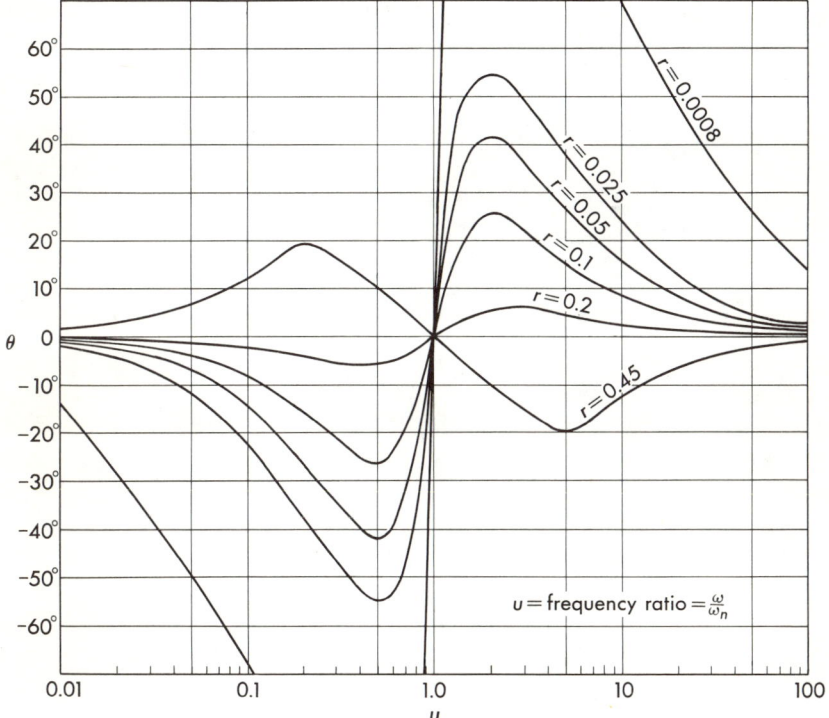

Fig. VI-11 Phase-response loaded bridged-T network.

the normalized frequency $u = \omega/\omega_0$. Notice that the amplitude curves have the same symmetrical appearance and form as in the unloaded case with two exceptions:

1. The d-c and infinite frequency gain is $\frac{1}{4}$. This means that the network acts as a 4:1 voltage divider at any frequency.

2. The minimum notch width becomes excessively wide for values of $r < 0.025$.

appendix VII[*]

A RULE FOR DETERMINING STABILITY OF A LINEAR SYSTEM

This stability rule is applicable to a general linear feedback system of the type shown in Fig. VII-1. The proof of the method depends upon polynomial theory, conformal mapping, and root locus.

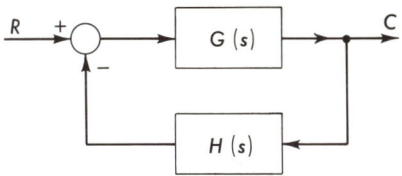

Fig. VII-1 Block diagram of a linear feedback control system.

For Fig. VII-1, the output C has the solution

$$\frac{C}{R}(s) = \frac{G(s)}{1 + GH(s)} \qquad \text{(VII-1)}$$

Let

$$GH(s) = \frac{KN(s)}{D(s)} \qquad \text{(VII-2)}$$

[*] The method is the discovery of Peter W. Stephan of North American Aviation, Inc., of Downey, Calif. The author is indebted to Mr. Stephan for writing this appendix.

$N(s)$ and $D(s)$ are polynomials in s; so $GH(s)$ is a rational function. The solutions to

$$1 + GH(s) = 0 \qquad \text{(VII-3)}$$

which are the same as the solutions to

$$D(s) + KN(s) = 0 \qquad \text{(VII-4)}$$

determine system stability. The quantity K is the lumped-system gain. For a fixed K, if the n solutions to Eq. (VII-4) are

$$s_k = \sigma_k + j\omega_k \qquad k = 1, 2, \ldots, n \qquad \text{(VII-5)}$$

the system is stable only if all σ_k's are negative. The solutions to Eq. (VII-3) or (VII-4) for a given value of K are termed roots of the characteristic equation. The rule is used to find the intervals of K values for

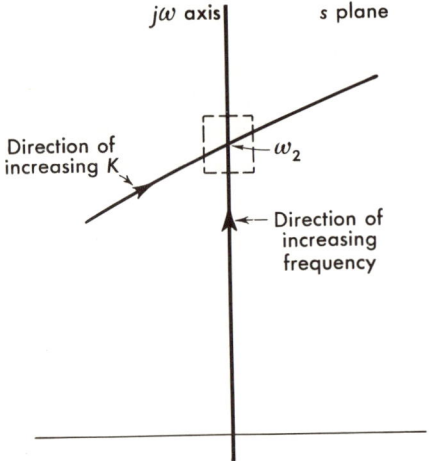

Fig. VII-2 Part of a root-locus diagram of a linear $GH(s)$.

which all solutions to Eq. (VII-4) are stable, for K varying from zero to infinity.

Although the rule can be applied in the Nyquist, Nichols, or Bode plots, the rule is developed only for the Nichols chart, which is the most logical diagram. A digital computer not only computes the log gain (in decibels) and phase of a given $GH(j\omega)$, but also plots the result in a Nichols chart form.

The rule depends strongly on how the root-locus crossing of the $j\omega$ axis in Fig. VII-2 is transformed to the Nichols chart. To this end, Fig. VII-2 is enlarged in Fig. VII-3 to show detail.

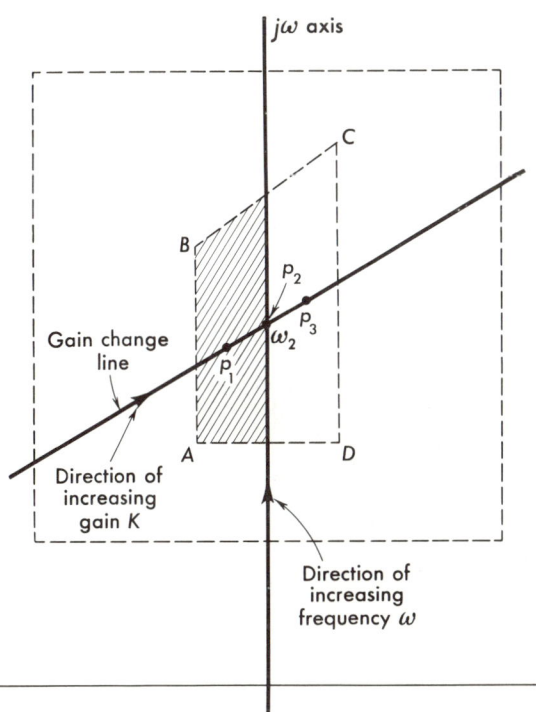

Fig. VII-3 Enlargement of part of Fig. VII-2.

In Fig. VII-3, a "small" trapezoid has been superimposed on the root locus, centered at ω_2, the frequency at which the root locus crosses the $j\omega$ axis in the s plane. The part of the trapezoid in the left half plane has been shaded. Three points have been drawn in Fig. VII-3: p_1, p_2, and p_3. These are for three values of gain K along the root locus, where

$$K_{p_1} < K_{p_2} < K_{p_3} \tag{VII-6}$$

For gain K_{p_1}, the roots at p_1, $\bar{p}_1$ (conjugate of p_1) are stable. At p_2, the roots are marginally stable. At p_3, for gain K_{p_3}, the roots are unstable.

440 Control System Design

By complex-variable theory, the mapping of log $GH(j\omega)$ is conformal except for the neighborhoods* of the open-loop zeros and poles of $GH(s)$. In mapping log $GH(j\omega)$, the neighborhoods of all zeros and poles are avoided. "Conformal" mapping has the property that the image of a "small" figure has the same shape as the original. Hence for the transformation to the Nichols chart, the area inside the square in Fig. VII-3 transforms to the square in Fig. VII-4.

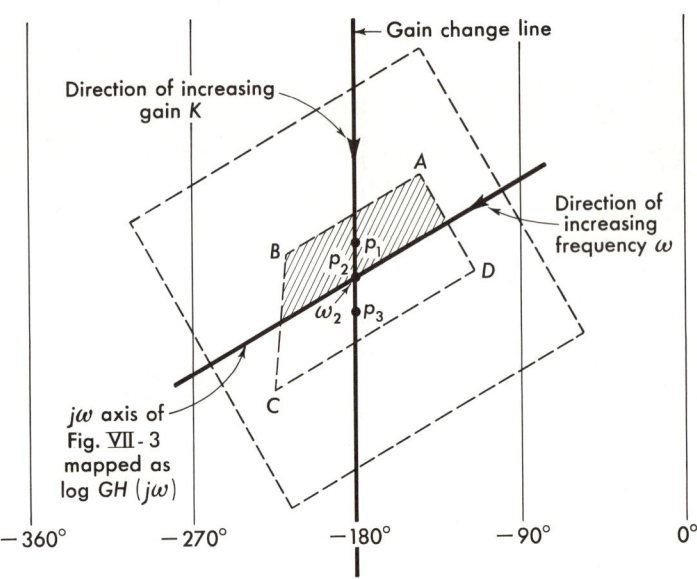

Fig. VII-4 Transformation of square area of Fig. VII-3 to the log magnitude–phase plane.

For the coordinate system used for the Nichols chart, the transformation actually is the mirror image (about the zero phase-margin line) of the conformal mapping. The angles and areas are still preserved, but the sense of direction around the periphery of the trapezoid has been reversed. p_1 is to the left of the $j\omega$ axis in Fig. VII-3, whereas in Fig. VII-4 it is to the right.

From Figs. VII-3 and VII-4, the deduction is made that in the Nichols chart, as K increases from K_{p_1} to K_{p_3}, two roots migrate from the left-half s plane to the right-half s plane. The above explanation shows the key to the rule.

* "Neighborhoods" is discussed in several complex-variable texts, e.g., Refs. 7 and 26.

The Rule

Part 1. In the regular Nichols chart (axes as in Fig. VII-5) a crossing of the zero phase-margin line by $GH(j\omega)$ going toward the *left* for *increasing* frequency causes the number of unstable roots to *increase* by 2 (except if $\omega_2 = 0$, whereupon only *one* root becomes unstable).

Part 2. When $GH(j\omega)$ crosses the zero phase-margin line going toward the *right* for *increasing* frequency, the numbers of unstable roots are

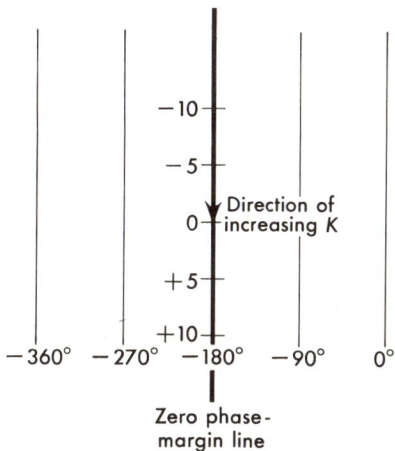

Fig. VII-5 The log magnitude–phase plane, showing region where $K \to 0$.

decreased by 2 (except if $\omega_2 = 0$, whereupon only *one* root becomes stable).

The philosophy of applying the rule is not complete until the following observation is made. For $K \to 0$, the roots are always arbitrarily near the poles. The region in the Nichols chart where K is near zero is the upper part, shown in Fig. VII-5. For a given $GH(s)$, the number of poles in the right half plane is known. Thus, for $K \to 0$, the number of roots in the right half plane is also known. It is usually the same as the number of open-loop poles in the right half plane. For special $GH(s)$ such as

$$GH(s) = \frac{s^2 + s + 2}{s^3(s + 4)} \qquad \text{(VII-7)}$$

some knowledge of root-locus theory is helpful in finding the number of roots in the right half plane for small K. For $GH(s)$ of Eq.

(VII-7), it is known that for $K_1 = \epsilon$ (ϵ being an infinitesimal >0) two roots exist in the right half plane. This is because of the triple pole at the origin. The root locus for this $GH(s)$ is shown in Fig. VII-6. The gain where the root locus crosses the $j\omega$ axis in Fig. VII-6 is termed K_2. For gains greater than K_2, two closed-loop poles move from the right half plane to the left half plane; i.e., the number of unstable roots has decreased by 2.

From the root locus, the result follows that for $K \to 0$ (the top region of Fig. VII-7) there are two roots in the right half plane. At gain K_2 in Fig. VII-7, for frequency ω_2, the transformed point p_2 of Fig. VII-6 is

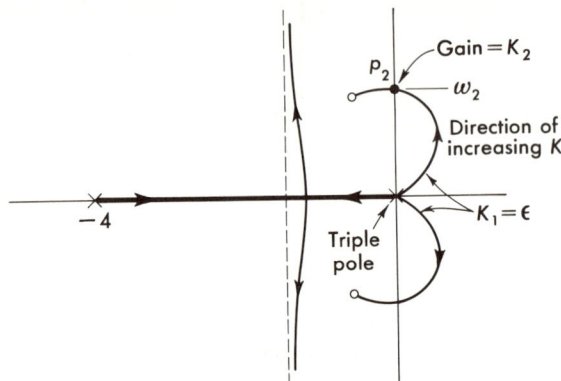

Fig. VII-6 Root locus for $GH(s)$ of Eq. (VII-7).

mapped. From part 2 of the rule, at p_2 in Fig. VII-7 the number of roots in the right half plane has been decreased by 2. Note that the *identical* information obtained at point p_2 of Fig. VII-6 is obtained from the transformed point p_2 of Fig. VII-7. Hence, for all gains higher than K_2, all the roots are stable. Both stable and unstable values of K are marked in Fig. VII-7.

Although the gains in Figs. VII-6 and VII-7 are both termed K_2, the one in Fig. VII-6 is the root-locus gain, whereas the one in Fig. VII-7 is the Bode gain [excluding poles and zero of $GH(s)$ at the origin]. Root-locus gain and Bode gain are related as follows:

$$K_B = K_{RL} \left| \frac{\prod_{i=1}^{m} Z_i}{\prod_{i=1}^{n} P_i} \right| \quad \text{excluding poles and zeros of } GH(s) \text{ at origin} \quad \text{(VII-8)}$$

where Z_i are the m solutions to $N(s) = 0$, considering $N(s)$ to be of degree m (but excluding any solutions at the origin), and P_i are the n solutions to $D(s) = 0$ (but excluding any solutions at the origin). The absolute-

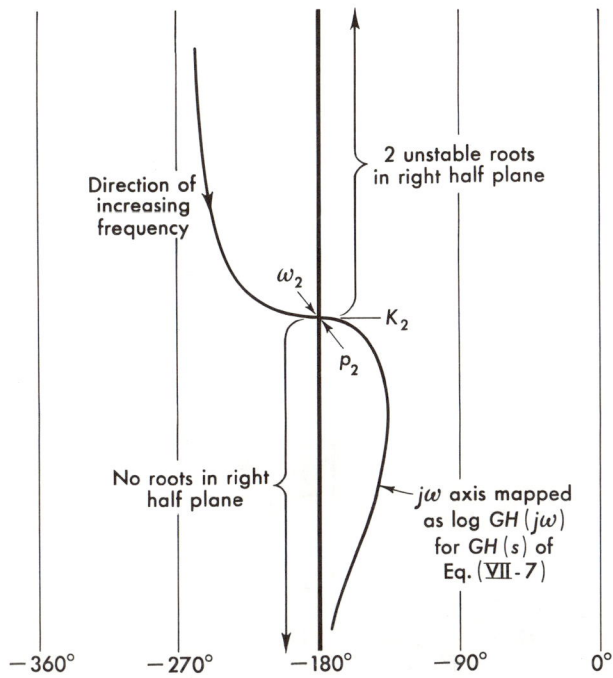

Fig. VII-7 The log magnitude–phase diagram for $GH(s)$ of Eq. (VII-7).

value signs are used to ensure that the Bode gain always has the same sign as the root-locus gain.

Consider another example to illustrate the method,

$$GH(s) = \frac{1}{(s+5)(s+4)(s-1)} \qquad \text{(VII-9)}$$

whose root-locus diagram is Fig. VII-8. Gain $K_1 < K_2$, where gain K_1 occurs for $s = 0$. As seen from Fig. VII-8, at a gain slightly higher than K_1, a root has moved to the left half plane. Figure VII-9 is the Nichols plot for $GH(s)$ of Eq. (VII-9). Since the first crossing of the zero phase-margin line occurs for $K = K_1$, for smaller K value (same idea as considering $K \to 0$) there is one root in the right half plane since $GH(s)$ has a pole in the right half plane. For $\omega = 0$, with gain

K_1, only *one* root goes from the right to the left half plane. This reduces the roots in the right half plane to zero for $K = K_1+$. This is an application of the special form of part 2 of the rule (for $\omega_2 = 0$). Thus, for any gain between K_1 and K_2 the closed-loop system is stable. At

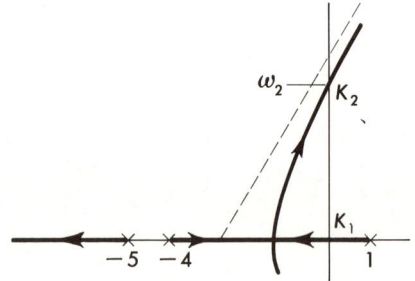

Fig. VII-8 Root locus for $GH(s)$ of Eq. (VII-9).

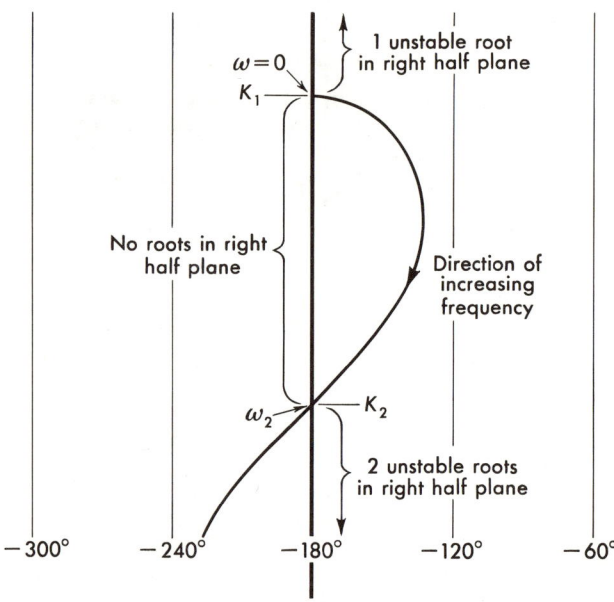

Fig. VII-9 The log magnitude–phase diagram of $GH(s)$ of Eq. (VII-9).

K_2 the roots in the right half plane are increased by 2 and the closed-loop system continues to be unstable for any $K > K_2$.

As another example, which shows that the use of the rule of thumb "Always keep the -1 point to the right in the Nichols chart" cannot

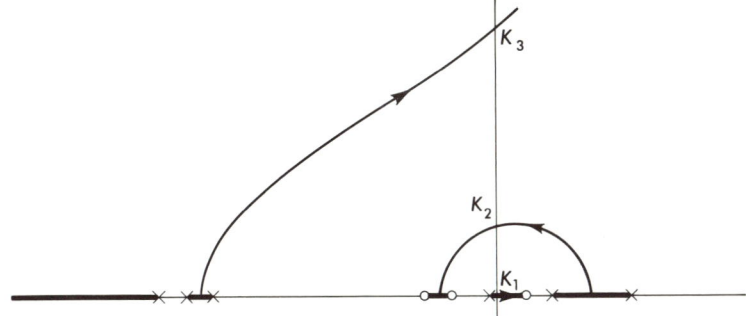

Fig. VII-10 The root-locus diagram of a given $GH(s)$.

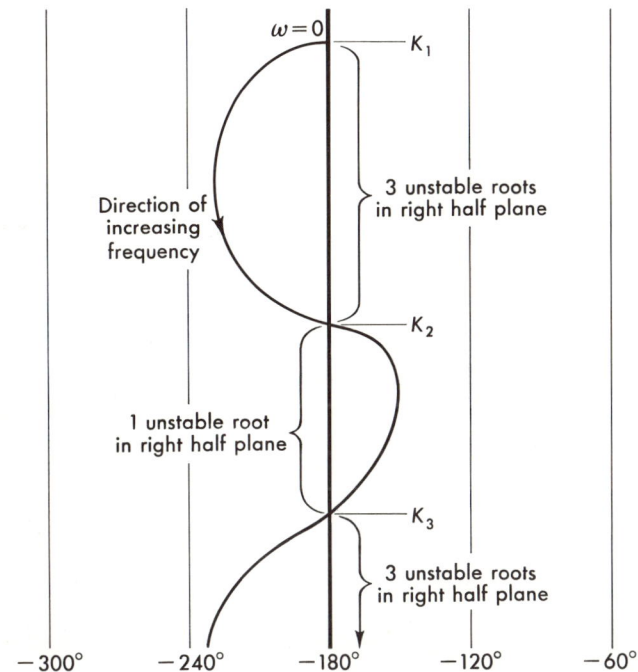

Fig. VII-11 The log magnitude–phase diagram for $GH(s)$ of Fig. VII-10.

always be relied on, consider the root-locus diagram which is shown in Fig. VII-10. The log $GH(j\omega)$ diagram is shown in Fig. VII-11. Superficially, one might expect that for gains between K_2 and K_3 the system is stable. But by use of the rule it is determined that one root in the right half plane exists for gains from K_2 to K_3. Actually, the given system is unstable for any value of K.

446   Control System Design

The rule is easily extended to handle open-loop poles and zeros on the $j\omega$ axis and double poles at the origin. However, the explanation becomes tedious, and since these cases are rare the explanation is not attempted here.

Application of the new criterion, especially in Nichols charts, results in both stable gain information and relative stability information being rapidly ascertained from one convenient diagram.

bibliography

1. Ahrendt, W. R., and C. J. Savant, Jr.: "Servomechanism Practice," 2d ed., McGraw-Hill Book Company, Inc., New York, 1960.
2. Andronow, A. A., and C. E. Chaikin: "Theory of Oscillations," English translation by S. Lefschetz, Princeton University Press, Princeton, N.J., 1949.
3. Bode, H. W.: "Network Analysis and Feedback Amplifier Design," D. Van Nostrand Company, Inc., Princeton, N.J., 1945.
4. Bronwell, A.: "Advanced Mathematics in Physics and Engineering," McGraw-Hill Book Company, Inc., New York, 1953.
5. Brown, G. S., and D. P. Campbell: "Principles of Servomechanisms," John Wiley & Sons, Inc., New York, 1948.
6. Chesnut, H., and R. W. Mayer: "Servomechanisms and Regulating Systems Design," John Wiley & Sons, Inc., New York, 1951.
7. Churchill, R. V.: "Complex Variables and Applications," 2d ed., McGraw-Hill Book Company, Inc., New York, 1960.
8. Churchill, R. V.: "Operational Mathematics," 2d ed., McGraw-Hill Book Company, Inc., New York, 1958.
9. Coblenz, A., and H. L. Owens: "Transistors: Theory and Applications," McGraw-Hill Book Company, Inc., New York, 1955.
10. Davis, S. A.: Rotating Components for Automatic Control, *Prod. Eng.*, November, 1953.
11. Evans, W. R.: "Control-system Dynamics," McGraw-Hill Book Company, Inc., New York, 1954.
12. Evans, W. R.: Control System Synthesis by Root Locus Method, *Trans. AIEE*, vol. 69, pp. 66–69, 1950.
13. Gardner, M. F., and J. L. Barnes: "Transients in Linear Systems," John Wiley & Sons, Inc., New York, 1948.
14. Goldman, S.: "Frequency Analysis, Modulation, and Noise," McGraw-Hill Book Company, Inc., New York, 1948.
15. Greif, H. D.: Describing Function Method of Servo-mechanism Analysis Applied to Most Commonly Encountered Nonlinearities, *Trans. AIEE*, pt. 2, vol. 72, pp. 243–248, 1953.
16. Guillemin, E. A.: "Introductory Circuit Theory," John Wiley & Sons, Inc., New York, 1953.
17. Guillemin, E. A.: "A Summary of Modern Methods of Network Synthesis," vol. 3 in "Advances in Electronics," Academic Press Inc., New York, 1951.

18. Hurwitz, A.: Über die Bedingungen, unter Welchen eine Gleichung mit negativen reelen Teilen Besitzt (The Conditions under Which an Equation Has Only Roots with Negative Real Parts), *Math. Ann.*, vol. 46, pp. 273–284, 1895.
19. IRE Standards on Terminology for Feedback Control Systems, *Proc. IRE*, vol. 44, no. 1, January, 1956.
20. James, H. M., N. B. Nichols, and R. S. Phillips: "Theory of Servomechanisms," McGraw-Hill Book Company, Inc., New York, 1947.
21. Kochenburger, R. J.: A Frequency Response Method for Analyzing and Synthesizing Contactor Servomechanisms, *Trans. AIEE*, vol. 69, pp. 270–284, 1950.
22. Liénard, A.: Étude des oscillations entretenues (Study of Self-excited Oscillations), *Rev. Gén. Élec.*, vol. 23, 1928.
23. McLachlan, N. W.: "Nonlinear Differential Equations," Oxford University Press, New York, 1950.
24. Minorsky, N.: "Introduction to Nonlinear Mechanics," J. W. Edwards, Publisher, Incorporated, Ann Arbor, Mich., 1947.
25. Nixon, F. E.: "Principles of Automatic Controls," Prentice-Hall, Inc., Englewood Cliffs, N.J., 1953.
26. Reddick, H. W., and F. H. Miller: "Advanced Mathematics for Engineers," John Wiley & Sons, Inc., New York, 1947.
27. Savant, C. J., Jr.: "Fundamentals of the Laplace Transformation," McGraw-Hill Book Company, Inc., New York, 1962.
28. Savant, C. J.: How to Design Notch Networks, in "Electronics Engineering Manual," vol. 7, pp. 242–245, McGraw-Hill Book Company, Inc., New York, 1953.
29. Savant, C. J., and C. A. Savant: Notch Network Design, *Electronics*, vol. 28, no. 9, p. 172, 1955.
30. Savant, C. J., Jr., R. C. Howard, C. B. Solloway, and C. A. Savant: "Principles of Inertial Navigation," McGraw-Hill Book Company, Inc., New York, 1961.
31. Schmidt, H. A.: The Precision Potentiometer as a Voltage Divider, *Prod. Eng.*, Annual Handbook of Product Design for 1954, McGraw-Hill Book Company, Inc., New York.
32. Seely, S.: "Electron-tube Circuits," 2d ed., McGraw-Hill Book Company, Inc., New York, 1958.
33. Shea, R. F.: "Transistor Circuits," John Wiley & Sons, Inc., New York, 1953.
34. Slater, J. M., and D. B. Duncan: Inertial Navigation, *Aeron. Eng. Rev.*, January, 1956.
35. Karplus, W. J., and W. W. Soroka: "Analog Methods: Computation and Simulation," 2d ed., McGraw-Hill Book Company, Inc., New York, 1959.
36. Stoker, J. M.: "Nonlinear Vibrations in Mechanical and Electrical Systems," Interscience Publishers, Inc., New York, 1950.
37. Thaler, G. J., and R. G. Brown: "Analysis and Design of Feedback Control Systems," 2d ed., McGraw-Hill Book Company, Inc., New York, 1960.
38. Truxal, J. C.: "Automatic Feedback Control System Synthesis," McGraw-Hill Book Company, Inc., New York, 1955.
39. Van Valkenburg, M. E.: "Network Analysis," Prentice-Hall, Inc., Englewood Cliffs, N.J., 1955.
40. Kármán, T. V., and M. A. Biot: "Mathematical Methods in Engineering," McGraw-Hill Book Company, Inc., New York, 1940.

index

Absolutely stable system, 7
A-c control motors, 325, 326
A-c servo systems, 252–276
   design of, 271
   equalization of, 256
A-c tachometer, 306–309
Acceleration, measurement of, 310–316
Acceleration damping, 267–271
Acceleration-error constant, 82, 85
Accelerometer, force-balance, 314–316
   mechanical, 311
   seismic, 312
   transfer function for, 315
Active networks, 51–56
   synthesis of, 238
Actuating signal, 27, 79
Adjustable networks, 225, 226, 275
Adjustment, gain, 210, 211
Admittance, 47, 48
   common, 49
   mechanical, 63
      common, 59
      self-, 59
   self-, 49, 63
Algebra, block diagram, 20–22
All-pass network, 226
Alnico, 310
Amplifier, difference, 344
   one-stage, 53
Amplifier saturation, 356
Amplitude and phase response test, 180
Amplitude ratio, 94
Analogies, 69

Analogue, loop, 70
   nodal, 69
Angle of arrival at complex zeros, 117
Angle of departure, of branch, 129
   from complex poles, 116
Angular momentum, 331
Angular motion, 61–66
Antenna drive, 283
Asymptotic angles for root locus, 109
Asymptotic approximation, 158–167
   for frequency-invariant factors, 159
   for quadratic zeros and poles, 165
   for simple poles, 164
   for simple zeros, 161
Autosyn, 302–305

Back emf, 242
Backlash, 357
Bandwidth, 94
Base resistance, 53
Bilateral elements, 38
Bilateral networks, 39
Block diagram, 19–22, 24, 45, 48, 59
Block-diagram algebra, 20
Bode diagrams, 155–158
Bourdon tube, 316
Break point, 164
Breakaway point, 111–114
   from real axis, 112
Bridged-T network, 223–225, 259, 276, 422–428
   capacitor-shunt, 423–428

449

450   Control System Design

Bridged-T network, design chart, 426–427
  loaded, 432
  resistor-shunt, 428

Calibration, 2
Capacitive reactance, 149
Capacitor-shunt bridged-T network, 423–428
Capacitors, 38, 46
Carpenter metal, 310
Carrier networks, 258
Carrier systems (see A-c servo systems)
Center of gravity, 110, 126
  location of, 111
  of roots, 126
Centrifuge, 284
Characteristic equation, roots of, 399
Chart, Nichols, 186–194, 201, 438–446
Closed-loop system, 1
  frequency response of, 201, 202
Cofactor, 404
Collector resistance, 53
Common admittance, 49
Compensation, series-parallel, 252
Complex pole, angle of departure from, 116
Complex zero, angle of arrival at, 117
Component, steady-state, 78
Conditionally stable system, 30
Conformity, 293
Constant, acceleration-error, 82, 85
  gain, 107
  position-error, 82
  velocity-error, 82–83
Constant coefficients, 9
Contours, M and N, 186–188, 417–419
Control motor, 317–327
  a-c type, 325, 326
  d-c type, 326, 327
Control phase, 319
Control synchro, 304
Control system, 1–6
  a-c, 252–276
    design, 271–276

Control system, conventions, 27
  equalization, 210–225
  feedback, 5, 210–225
  light-seeker, 281
  multiple-loop, 143, 247–250
  specifications, 93
  stability, 7
Control voltage, 319
Conventions, control system, 27
Convolution, 24–27
Convolution integral, 397
Corner frequency, 94, 162–164
Coulomb friction, 355
Coulomb-friction damper, 266
Coupling, mechanical, 63
  transformer, 276
Cramer's rule, 39
Critical damping, 16
Cubic equation, 399–401
Curie shunt, 316
Current, loop, 35, 36

Damper, coulomb-friction, 266
  dry-friction, 355
  eddy-current, 266
  Foucault, 266
  vibration, 60
  viscous-friction, 266
Damping, critical, 16
  equivalent, 354
  proportional to acceleration, 267–271
Damping constant, 16
Damping ratio ζ, 12, 17, 95, 134
  versus phase margin, 420
Dashpot, 8
Data, experimental, 202
Data transmission, 301
D-c control motors, 326, 327
D-c motor, 326, 327
D-c rate generators, 309, 310
D-c tachometers, 309, 310
Dead space, 361
Decibels, 94, 155
Degree of instability, 187
Degree of stability, 186, 194

Demodulation, phase-sensitive, 346
Demodulation-modulation technique, 258
Demodulator, 345
 phase-sensitive, 256–258, 345
 ring, 347
Demodulator circuits, 346
Derivation, of M and N circles, 417–419
 of Nyquist criterion, 415–417
Descartes' rules, 402
Describing function, 357–367
 limitations of, 365
 for saturation, 358
 for threshold, 360
Design, of a-c control system, 271–276
 of bridged-T network, 427
 feedback control system, 210–225
Design specifications, 78
Determinant, definition of, 403
 expansion of, 403
 theorems concerning, 405
Diagram, block, 19–22, 24, 45, 48, 59
 free-body, 57, 62
Difference amplifier, 344
Differential equations, linear, 9–11
 nonlinear, 353–381
Differential gearbox, 342
Differential synchro, 303
Differentiator, 218, 239
 suppressed-carrier, 264
Direct current through motor windings, 267
Disturbance, load, 90–93
Drag cup, 325
Drag-cup tachometer, 307
Dry-friction damper, 355
Duals, 70, 71
Dynamometer, 322

E pickoff, 305
Eddy-current damper, 266
Electrical analogue, 69
Electromechanical integrator, 263
Electromechanical network equalizer, 264

Electromechanical networks, 66–69, 262–265
Electrostatic support, 277
Emitter resistance, 53
Encirclement of origin, 173
Equalization, a-c system, 256
 control system, 210–217
 definition of, 210
 lead network, 226
 parallel, 240
 rate feedback, 243–247
 series, 240
Equations, linear, 9–11
 nonlinear differential, 353–381
 roots of, 140–142, 399–402
Equivalent damping, 354
Equivalent network for d-c equalizer, 262
Equivalent spring constant, 356
Equivalent viscous damping, 355
Erection, 341
Error, 3
 loading, 290
 steady-state, 78–96
Error detector, 20
Error signal, 27
Evans, W. R., 130n.
Expansion of determinant, 403
Experimental data, 202
Experimental speed-torque curve, 10, 323

Factoring of polynomials, 140–142, 399–402
Feedback, definition of, 4
Feedback control system, 5
 classification by types, 81–82
 design of, 210–225
Feedback loop transfer function, 102
Feedback signal, 27
Final-value theorem, 80, 397
Flotation fluid, 339
Flow meter, 282
Force-balance accelerometer, 314–316
Forcer, 327

Forward-loop transfer function, 102
Foucault damping, 266
Four-winding resolver, 298
Four-wire resolver system, 301
Fraction, partial, 390
Free-body diagram, 57, 62
Free gyro, 335
Frequency, closed-loop, 201
   corner, 162–164
   cutoff, 94
   natural resonant, 95
   nutation, 339
   undamped natural, 12–16
Frequency analysis method, 149–204
Frequency-domain specifications, 93
Frequency gradient method, 197, 198
Frequency-invariant factors, 159
Frequency response, 150
Frequency spectrum, 255
Frequency transformation, 261, 262
Friction, coulomb, 266
Function, weighting, 25

Gain, measurement of, 119–121
Gain-adjustment, 210, 211
Gain constant, 107
Gain margin, 194, 195
Gain-variation method, 114
Gear ratio, optimum, 328
Gear train, 63, 327
Generalized steady-state errors, 88
Generator, rate, 307–309
Gimbal lock, 332
Gimbals, 330
Graph, signal-flow, 22–24
Gravity, center of, 110, 111, 126
Grounded emitter, 55
Gyro gimbal system, 333
Gyroscope, 330
   free, 335
   integrating, 338
   rate, 309, 336
   restrained, 338
   single-degree-of-freedom, 332–338, 341

Gyroscope, two-degree-of-freedom, 335
   vertical, 341
Gyroscopic law, fundamental, 331

High-degree polynomials, factoring of, 140–142

Impedance, concept of, 149–151
   input, 42
   operational, 39
   self-, 41
Impulse function, 25
Independent linearity, 289
Inductance, mutual, 41–44
   self-, 44
Induction components, 295–313
Induction generator, two-phase, 308
Induction motor, two-phase, 317
Induction tachometers, 306
Induction-type transducers, 295–308
Inductive reactance, 149
Inductors, 38
Inertia, products of, 332
Initial-value theorem, 397
Input disturbances, 78
Input signal, 27
Instability, 7
   degree of, 187
Instrument servo, 265
Integral, convolution, 24–27, 397
Integrating gyro, 338
Integrator, 220, 239
   electromechanical, 263
Inverse Laplace transformation, 391–395
Isochrones, 379
Isoclines, 369

Kirchhoff's laws, first, 35–39
   second, 35, 44–49
Kochenburger, R. J., 356n.

Ladder network, 234
Laplace transform, 385–398
  of functions, 386, 387
  of operations, 386–389
  solution of differential equation, 13–17
  tables, 387, 389
Lattice network, 229
Law, Kirchhoff's, 35–39
    second, 35, 44–49
    Newton's, 56–61
Lead network equalization, 226
Lienard construction, 372–378
Light-seeker control system, 281
Limitedly stable system, 7
Linear differential equation, 9–11
Linear potentiometer, 287–293
Linear spring, 56
Linearity, 9, 38
  independent, 289
  zero-based, 290
Load disturbances, 90
Loaded bridged-T network, 432
Loaded ladder network, 234
Loading error, 290
Locus of roots (see Root-locus method)
Loop analogue, 70
Loop analysis of electrical networks, 35–43
Loop currents, 35, 36
Loop method, 35
Loudspeaker, 66, 67
Low-Q networks, 260

M criterion, 196–197
M and N contours, 186–188
  derivation of, 417–419
Mach number, 348
Magnet, stabilization of, 327
Magnetic amplifiers, 347
Margin, gain, 194, 195
Marginal stability, 30
Mechanical admittance, 63
  common, 59
Mechanical couplings, 63

Mechanical networks, 56–66
Mechanical self-admittances, 63
Mechanical suspension, 20
Mechanical system, 7, 56
Minor, 404
Minor loop, 247
Modulation, suppressed-carrier, 253–256
Modulation-demodulation, 258
Modulators, 345
Momentum, angular, 331
Motion, angular, 61–66
Motor, control, 317–327
  a-c, 325, 326
  d-c, 326, 327
  direct current through, 267
  induction, two-phase, 317
  speed-torque curve for, 330
Motor constant, 322
Motor-tachs, 326
Multiple-loop system, 143, 144, 251
Mutual inductance, 41–44
Mutual resistance, 53

N circles, 190
N contours (see M and N contours)
Natural resonant frequency ω_n, 95
Negative K, 108
Networks, active, 51–56, 238
  adjustable, 225, 226, 275
  all-pass, 226
  bilateral, 39
  bridged-T, 223–225, 259, 276, 422–428
  carrier, 258
  electromechanical, 66–69, 262–265
  equalizer, 223
  ladder, 234
  lattice, 229–234
  loaded bridged-T, 432
  loaded ladder, 234
  loop analysis of, 35–43
  low-Q, 260
  mechanical, 56
  n-loop, 40
  nonreciprocal, 54

Networks, nonseparable, 71
  parallel-T, 422
  passive lag, 220
  passive lead, 218
  phase-shift, 273
  planar, 36
  resistor-shunt bridged-T, 428
Neutral stable roots, 30, 407
Newton's law, 56–61
Newton's method, 400, 401
Newton's second law of motion, 8
Nichols chart, 186–194, 201, 438–446
Nodal analogue, 69
Nodal analysis of electrical networks, 44–51
  example, 45
Nonlinear differential equations, 353–381
Nonlinear potentiometer, 293
Nonlinear systems, 353–381
Nonlinearities, 352, 353
  small, 354–356
Nonreciprocal network, 54
Nonseparable networks, 71
Normal slope, 372
Nutation frequency, 339
Nyquist stability criterion, 174–176
  derivation of, 415–417

Octave, 94, 160
One-stage amplifier, 53
Open-loop systems, 1, 2
  disadvantages of, 3
Open-loop transfer function, 81, 104
Operational impedances, 39
Optimum gear ratio, 328
Oscillating system, 7
Oscillator, voltage, 7
Output signal, 27
Overshoot, 95
  per cent, 15

Parallel equalization, 240–243
Parallel-T network, 422

Partial fraction, 390–395
Partial-fraction expansion, 391
Passive lag network, 220
Passive lead network, 218
Per cent overshoot, 15
Permanent-magnet tachometers, 310
Phase margin, 195
Phase plane, 367–369
Phase-plane methods, 367–381
Phase-sensitive demodulation, 346
Phase-sensitive demodulator, 256–258, 345
Phase-shift network, 273
Phase trajectory, 367, 369–372
  Lienard construction of, 372–378
Pitot pressure, 348
Planar networks, 36
Plane, root, 29
Plots, Bode, 156–158
Point, break, 164
  breakaway, 111–114
Polarity markings, 43
Poles, 82, 105
  complex, 116
  quadratic, 165–168
  simple, 164
Polynomial factor, 140
Polynomials, factoring of, 140–142, 399–402
  roots of, 399
Position-error constant, 82
Position servo, 9, 20
  with acceleration damper, 268
Potentiometer, 287–295
  accuracy of, 289
  conformity of, 293
  linear, 287–293
  linear motion, 294
  multiturn, 288
  resolution of, 289, 293
  subminiature, 293
Potentiometer loading error, 291
Precession velocity, 333
Pressure, pitot, 348
  stagnation, 348
Pressure transducers, 316–317

Index

Quadratic, 399
Quadratic zeros and poles, 165–168

Rate feedback equalization, 243–247
Rate generator, 306–310
Rate gyro, 336
Ratio, amplitude, 94
 damping, 12–17, 95, 134, 420
Reactance, capacitive, 149
 inductive, 149
Real translation, 395
Recorder, strip-chart, 278
Reluctance, 305
Resistance, base, 53
 collector, 53
 emitter, 53
 mutual, 53
Resistance subtraction, 344
Resistor-shunt bridged-T network, 428
Resistors, 38, 46
Resolution, 288
Resolver, 296
 four-winding, 298
 two-winding, 297
Reponse, amplitude, 180
 frequency, 150
 phase, 180
 to unit step function, 14, 95
Restrained gyro, 338
Ring demodulator, 347
Rise time, 95
rms defined, 149n.
Root, 105
 neutral stable, 407
 unstable, 407
Root-locus method, 99–144
 for factoring polynomials, 140–142
 for multiple-loop system, 143
 for parameter variations, 135, 140
 proof of rules, 121–130
 rules, 108–121
Root-mean-square, 149
Root plane, 29
Roots, of characteristic equation, 399
 of equations, 140–142, 399–402

Roots, of high-degree polynomials, 399–402
Routh-Hurwitz stability criterion, 406–408
Rule, Cramer's, 39
 Descartes', 402

Saturation, amplifier, 356
Scale factor, 133
Second independent variable, 396
Second-order system, 9–18, 99–102
Seismic accelerometer, 312
Self-admittance, 49, 59, 63
Self-impedance, 41
Self-inductance, 44
Selsyn, 302–305
Separation of networks, 71, 72
Series equalization, 240
Series-parallel compensation, 252
Servo, instrument-type, 265
 position, 9, 20
Servo conventions, 27
Servo systems (*see* A-c servo systems)
Settling time, 96
Shock mount problem, 137
Signal, actuating, 27, 79
 error, 27
 feedback, 27
 output, 27
 suppressed-carrier, 253
Signal-flow diagram, 24
Signal-flow graphs, 22–24
Simple, meaning of, 158
Simple poles, 164
Simple zeros, 161
Single-degree-of-freedom gyroscope, 332–338, 341
Singularities, 105
Specifications, control system, 93
 design, 78
 frequency-domain, 93
 time-domain, 95, 96
Spectrum, frequency, 255
Speed regulator, 279
Speed-torque curve, 10, 318–324

Speed-torque curve, experimental, 323
Spirule, 130–135
  used sum angles, 131
Spring, linear, 56
Spring constant, equivalent, 356
Squirrel cage, 325
Stability, degree of, 7, 30, 186, 194
    absolute, 7
    conditional, 30
    limited, 7
    marginal, 30, 407
  relative, 197
  Routh-Hurwitz, 406–408
Stability boundary, 414
Stability criterion, 27, 182
Stabilization (see Compensation; Equalization)
Stable root, 406
Stagnation pressure, 348
Stall torque versus voltage, 11
Steady-state component, 78
Steady-state error, 78–96
  generalized, 88
Strip-chart recorder, 278
Subminiature potentiometer, 293
Subtractor, 20, 342
Support, electrostatic, 277
Suppressed-carrier differentiator, 264
Suppressed-carrier modulation, 253–256
Suppressed-carrier signal, 253
Synchro, 302–305
  control, 304
  differential, 303
  as position indicator, 304
Synchro torque gradient, 302
Synthesis, active network, 238
  ladder, 234
  lattice, 229–234
System, closed-loop, 1
  control (see Control system)
  mechanical, 7, 56
    n-mass, 59
  open-loop, 1–3
  oscillating, 7
  rotational, 62

System, second-order, 9–18, 99–102
System optimization, 216
System stability, 28–30

Tachometers, 306–310
  a-c, 306–309
  d-c, 309
  drag-cup, 307
  induction, 306
  permanent-magnet, 310
Telesyn, 302
Theorem, 395–398
  final-value, 80, 397
  initial-value, 397
Theorems concerning determinants, 405
Three-winding resolver, 298
Threshold, 361
Time constant, 322
Time-domain specifications, 95
Time-phase shifter, 300
Topological solution, 367–380
Toroids, 343
Torque gradient, 304
  synchro, 302
Torquer, 327
Transducer, 286
  induction-type, 295
  pressure, 316
Transfer function, 18, 19, 24
  accelerometer, 315
  feedback-loop, 102
  forward-loop, 102
  open-loop, 81, 104
Transform pairs, 387, 389
Transformation, of coordinates, 299
  Laplace (see Laplace transform)
Transformation ratio of resolver, 297
Transformer, 38, 50, 343
  polarity markings, 43
Transformer coupling, 276
Transistor, 53, 54
Triode, 52
Two-degree-of-freedom gyro, 335
Two-phase induction generator, 308

Two-phase motor, 317–325, 326
Two-winding resolver, 298
Type 0 system, 82, 83
Type 1 system, 82–84

Undamped natural frequency, 12, 16
Unit impulse, 25
Unstable roots, 407

Vacuum tube, 52, 54
Variable-reluctance pickoffs, 305
Velocity, precision, 333
Velocity coefficient, 212
Velocity constant, 84, 87
Velocity-error constant, 82, 83
Velocity pickoff, 306

Vertical gyro, 341
Vibration damper, 60
Viscous-friction damper, 266
Voltage, control, 319
Voltage gradient, 307, 310
Voltage oscillator, 7

Weighting function, 25

Zero-based linearity, 290
Zeros, 82, 105
  complex, 117
  at origin, 159
  quadratic, 165–168
  simple, 161